Transformations of Myth Through Time

An Anthology of Readings

Transformations of Myth Through Time

An Anthology of Readings

Diane U. Eisenberg — *Eisenberg Associates*
George deForest Lord — *Yale University*
Peter Markman — *Fullerton College*
Roberta H. Markman — *California State University, Long Beach*
Robert Merrill — *The Catholic University of America*
Megan Scribner — *Eisenberg Associates*
Charles S. J. White — *American University*

HBJ

Harcourt Brace Jovanovich, Publishers

San Diego New York Chicago Austin Washington, D.C.
London Sydney Tokyo Toronto

Cover art: Rich Richardson, Harry Shiotani, Susan Wilder

ISBN: 0-15-592335-8

Library of Congress Catalog Card Number: 89-84200

Printed in the United States of America

Preface

Transformations of Myth Through Time: An Anthology of Readings offers a wide-ranging selection of readings on the world's mythologies. This self-contained anthology also serves as part of a television course based on a series of lectures by Joseph Campbell, world-renowned scholar and mythologist. Joseph Campbell—author, scholar, teacher, and storyteller, inspiration to people of all walks of life the world over—firmly believed in the importance and relevance of myths to our lives. He wrote: "Religions, philosophies, arts, the social forms of primitive and historic man, prime discoveries in science and technology, the very dreams that blister sleep boil up from the basic magic ring of myth." His life's work was devoted to studying the various mythologies of the world and learning how they reconcile human beings to the mysteries of life.

A National Academic Advisory Committee of leading scholars has guided the selection of the readings for this anthology:

Dr. Charles Briggs
Department of Anthropology
Vassar University

Dr. Joseph Epes Brown
Department of Religious Studies
University of Montana

Dr. Grace Burford
Department of Theology
Georgetown University

Dr. Marcus Cunliffe
University Professor
George Washington University

Dr. Wendy Doniger
Mircea Eliade Professor of the History of Religion
University of Chicago

Dr. Marija Gimbutas
Department of European Archaeology
University of California, Los Angeles

Dr. James Jackson
Humanities Department
Johnson County Community College

The committee's goal was to create a collection of readings by Campbell and other scholars and writers with varying perspectives on mythology's role in human history, a collection to serve as an introduction to the study of the world's mythologies.

Transformations of Myth Through Time, the title of this anthology of readings, refers to the fundamental idea that all myths of humanity are expressions, in one way or another, of certain basic, perennial themes. These are themes which Joseph Campbell believed were deeply embedded in the human psyche and which naturally found expression in the myths of each of the cultures created throughout our long history. Each culture transformed the basic patterns by embodying them in their own images, but the patterns remained, and still remain, the same.

The selections in the *Anthology of Readings* explore the cultural manifestations of these basic patterns. They also provide a general background to these mythologies, including primal mythologies, the grounding principles and early expressions of myth; American Indian myths; goddesses and gods; Egyptian mythology; Hinduism and Buddhism; Kundalini Yoga, an Eastern system of meditation; the *Tibetan Book of the Dead;* ancient Greece's mystery cult; the historical and mythological significance of the Quest for the Grail; and the famous Arthurian Romances on love, marriage, and the individual.

We hope that these readings will captivate you and serve as the beginning point for your study of myth and for your better understanding of yourself and your culture.

THE EDITORS

Editors' Acknowledgments

The editors take full responsibility for selecting the excerpts that have been included in the *Anthology of Readings*, but we wish to acknowledge the many people who helped us accomplish this task. For their dedicated commitment to preserving the work of Joseph Campbell through their production of the television programs for this course, we would like to thank Stuart Brown and William Free. To Laurance S. Rockefeller we express our appreciation for providing the funding that made this course possible. We thank Dee Brock and Jinny Goldstein, of the PBS Adult Learning Service, for their advice and support during the development of *Transformations of Myth Through Time: An Anthology of Readings*. Additionally, we gained very valuable suggestions and insights from the members of the National Academic Advisory Committee for this television course. We also would like to thank the staff at Harcourt Brace Jovanovich, especially Bill Barnett, Joan Harlan, Eleanor Garner, Don Fujimoto, and Diane Southworth. Finally, we thank our families and friends for their support and understanding.

Contents

Introduction

The Hero's Journey: The World of Joseph Campbell

The Inner Reaches of Outer Space: Myth as Metaphor and as Religion

JOSEPH CAMPBELL

In this selection, Joseph Campbell discusses the various influences on the development of his own thought such as the ideas of Frobenius, Spengler, and Bastian. He distinguishes between the universal and local aspects of world mythologies, pointing out the universal compulsions from which myths have been generated. Campbell presents his concern that the old myths are dying, and that new mythologies are needed with new symbols and metaphors that will inject the mythology with renewed vitality. The development of such a new mythology would illustrate how myth is transformed through time.

Reviewing with unprejudiced eye the religious traditions of mankind, one becomes very soon aware of certain mythic motifs that are common to all, though differently understood and developed in the differing traditions: ideas, for example, of a life beyond death, or of malevolent and protective spirits. Adolf Bastian (1826–1905), a medical man, world traveler, and leading ethnologist of the last century, for whom the chair in anthropology at the University of Berlin was established, termed these recurrent themes and features "elementary ideas," *Elementargedanken*, designating as "ethnic" or "folk ideas," *Völkergedanken*, the differing manners of their representation, interpretation, and application in the arts and customs, mythologies and theologies, of the peoples of this single planet.

Such a recognition of two aspects, a universal and a local, in the constitution of religions everywhere clarifies at one stroke those controversies touching eternal and temporal values, truth and falsehood, which forever engage theologians; besides setting apart, as of two distinct yet related sciences, studies on the one hand of the differing "ethnic" or "folk ideas," which are the concern properly of historians and ethnologists, and on the other hand, of the *Elementargedanken*, which pertain to psychology. A number of leading psychologists of the past century addressed themselves to the analysis of these universals, of whom Carl G. Jung (1875–1961), it seems to me, was the most insightful and illuminating. The same mythic motifs that Bastian had termed "elementary ideas," Jung called "archetypes of the collective unconscious," transferring emphasis, thereby, from the mental sphere of rational ideation to the obscure subliminal abysm out of which dreams arise.

For myths and dreams, in this view, are motivated from a single psychophysiological source—namely, the human imagination moved by the conflicting urgencies of the organs (including the brain) of the human body, of which the anatomy has remained pretty much the same since *ca.* 40,000 B.C. Accordingly,

Imagery of a dream is metaphorical of ψ of dreamer

as the imagery of a dream is metaphorical of the psychology of its dreamer, that of a mythology is metaphorical of the psychological posture of the people to whom it pertains. The sociological structure coordinate to such a posture was termed by the Africanist Leo Frobenius (1873–1938) a cultural "monad." Every feature of such a social organism is, in his sense, expressive and therefore symbolic of the informing psychological posture. In *The Decline of the West*, Oswald Spengler (1880–1936) identified eight colossal monads of great majesty, with a ninth now in formation, as having shaped and dominated world history since the rise, in the fourth millennium B.C., of the first literate high cultures—(1) the Sumero-Babylonian, (2) the Egyptian, (3) the Greco-Roman (Apollonian), (4) the Vedic-Aryan, of India, (5) the Chinese, (6) the Maya-Aztec/Incan, (7) the Magian (Persian-Arabian, Judeo–Christian-Islamic), (8) the Faustian (Gothic Christian to modern European-American), and now, beneath the imposed alien crust of a Marxian cultural pseudomorphosis, (9) the germinating Russian-Christian.

Long antecedent, however, to the world-historical appearances, flowerings, and inevitable declines of these monumental monads, an all but timeless period is recognized of nonliterate, aboriginal societies—some, nomadic hunters, others, settled horticulturalists; some of no more than a half dozen related families, others of tens of thousands. And each had its mythology—some, pitifully fragmentary, but others, marvelously rich and magnificently composed. These mythologies were all conditioned, of course, by local geography and social necessities. Their images have been derived from the local landscapes, flora and fauna, from recollections of personages and events, shared visionary experiences, and so forth. Narrative themes and other mythic features, furthermore, have passed from one domain to another. However, the definition of the "monad" is not a function of the number and character of such influences and details, but of the psychological stance in relation to their universe of the people, whether great or small, of whom the monad is the cohering life. The study of any mythology from the point of view of an ethnologist or historian, therefore, is of the relevance of its metaphors to a disclosure of the structure and force of the nucleating monad by which every feature of the culture is invested with its spiritual sense. Out of this emerge the forms of its art, its tools, and its weapons, ritual forms, musical instruments, social regulations, and ways of relating in war and in peace to its neighbors.

monad = psych. stance of people in relation to their universe.

In terms of Bastian's vocabulary, these monads are local organizations of the number of "ethnic" or "folk ideas" of the represented cultures, constellating variously in relation to current needs and interests the primal energies and urges of the common human species: bioenergies that are of the essence of life itself, and which, when unbridled, become terrific, horrifying, and destructive.

The first, most elementary and horrifying of all, is the innocent voraciousness of life which feeds on lives and provides the first interest of the infant feeding on its mother. The peace of sleep shatters in nightmare into apparitions of the cannibal ogress, cannibal giant, or approaching crocodile, which are features, also, of the fairy tale. In Dionysiac orgies the culminating frenzies issue still, in some parts of the world, in the merciless group-cannibalizing of living bulls. The most telling mythological image of this grim first premise of life is to be seen in the Hindu figure of the world-mother herself as Kālī, "Black Time," licking up with her extended, long, red tongue the lives of all the living of this world of her creation. For, as noticed in a paper on "ritual killing" by the late director, Adolf E. Jensen, of the Frobenius Institute in Frankfurt-am-Main, "it is the common mark of all animal life that it can maintain itself only by destroying life;" citing to this point an Abyssinian song in celebration of the joys of life: "He who has not yet killed, shall kill. She who has not yet given birth, shall bear."

The second primal compulsion, linked almost in identity with the first (as recognized in this Abyssinian paean), is the sexual, generative urge, which during the years of passage out of infancy comes to knowledge with such urgency that in its seasons it overleaps the claims even of the first. For here the species talks. The individual is surpassed. In the quiver of the Hindu god Kāma, whose name means "desire" and "longing," and who is a counterpart of Cupid—no child, however, but a splendid youth, emitting a fragrance of blossoms, dark and magnificent as an elephant stung with vehement desire—there are five flowered arrows to be sent flying from his flowery bow, and their names are "Open Up!" "Exciter of the Paroxysm of Desire," "The Inflamer," "The Parcher," and "The Carrier of Death." Orgies of whole companies overtaken by the released zeal of the arrows of this god are reported from every quarter of the globe.

A third motivation, which on the world stage of world history has been the unique generator of the action—since the period, at least, of Sargon I of Akkad, in southern Mesopotamia, *ca.* 2300 B.C.—is the apparently irresistible impulse to plunder. Psychologically, this might perhaps be read as an extension of the bioenergetic command to feed upon and consume; however, the motivation here is not of any such primal biological urgency, but of an impulse launched from the eyes, not to consume, but to possess. An ample anthology of exemplary texts to this purpose, readily at hand, will be found in the Bible; for example:

> When the Lord your God brings you into the land which you are entering to take possession of it, and clears away many nations before you, the Hittites, the Girgashites, the Amorites, the Canaanites, the Perizzites, the Hivites, and the Jebusites, seven nations greater and mightier than yourselves, and when the Lord your God gives them over to you, and you defeat them; then you must utterly destroy them; you shall make no covenant with them and show them no mercy. You shall not make marriages with them, giving your daughters to their sons or taking their daughters for your sons. For they would turn away your sons from following me, to serve other gods; then the anger of the Lord would be kindled against you, and he would destroy you utterly. But thus shall you deal with them; you shall break down their altars, and dash in pieces their pillars, and hew down their Asherim, and burn their graven images with fire. For you are a people holy to the Lord your God; the Lord your God has chosen you to be a people for his own possession, out of all the peoples that are on the face of the earth. (Deuteronomy 7:1–6)
>
> When you draw near to a city to fight against it, offer terms of peace to it. And if its answer to you is peace and it opens to you, then all the people who are found in it shall do forced labor for you and shall serve you. But if it makes no peace with you, but makes war against you, then you shall besiege

> it; and when the Lord your God gives it into your hand you shall put all its males to the sword, but the women and the little ones, the cattle, and everything else in the city, all its spoil, you shall take as booty for yourselves; and you shall enjoy the spoil of your enemies, which the Lord your God has given you. Thus you shall do to all the cities which are very far from you, which are not cities of the nations here. But in the cities of these people that the Lord your God gives you for an inheritance, you shall save alive nothing that breathes, but you shall utterly destroy them, the Hittites and the Amorites, the Canaanites and the Perizzites, the Hivites and the Jebusites, as the Lord your God has commanded. (Deuteronomy 20:10–18)
>
> And when the Lord your God brings you into the land which he swore to your fathers, to Abraham, to Isaac, and to Jacob, to give you, with great and goodly cities, which you did not build, and houses full of all good things, which you did not fill, and cisterns hewn out, which you did not hew, and vineyards and olive trees, which you did not plant, and when you eat and are full, then take heed lest you forget the Lord, who brought you out of the land of Egypt, out of the house of bondage. (Deuteronomy 6:10–12)

War gods of this kind, always tribal in their ranges both of mercy and of power, have abounded over the earth as the fomenting agents of world history. Indra of the Vedic Aryans, Zeus and Ares of the Homeric Greeks, were deities of this class, contemporary with Yahweh; and in the period (sixteenth to twentieth centuries A.D.) of the Spanish, Portuguese, French, and Anglo-Saxon struggles for hegemony over the peoples of the planet, even Christ, his saints, and the Virgin Mary were converted into the tutelaries of pillaging armies.

In the Arthaśāstra, "Textbook on the Art of Winning," which is a classic Indian treatise on polity believed to have been compiled by the counselor, Kauṭilya, of the founder of the Maurya dynasty, King Chandragupta I (reigned *c.* 321–297 B.C.), the moral order by which all life is governed, and according to which kings and princes are therefore to be advised, is recognized and expounded as the "Law of the Fish" (*matsya-nyāya*), which is, simply: "The big ones eat the little ones and the little ones have to be numerous and fast."

For, whether in the depths of the forgotten sea out of which life originated, or in the jungle of its evolution on land, or now in these great cities that are being built to be demolished in our recurrent wars, the same dread triad of god-given urgencies, of feeding, procreating, and overcoming, are the motivating powers. And for the proper functioning of at least the first and third of these motivations in the fish pond of world history, the first requirement in the order of nature—as already recognized in the passage just quoted from Deuteronomy 7:1–6 (seventh century B.C.)–is suppression of the natural impulse to mercy.

For the quality of mercy, empathy, or compassion is also a gift of nature, late to appear in the evolution of species, yet evident already in the play and care of their young of the higher mammals. In contrast to the bioenergetic urge to procreate, however, which is an immediate urgency of the organs, compassion, like the will to plunder, is an impulse launched from the eyes. Moreover, it is not tribal- or species-oriented, but open to the appeal of the whole range of living beings. So that one of the first concerns of the elders, prophets, and established priesthoods of tribal or institutionally oriented mythological systems has always been to limit and define the permitted field of expression of this expansive faculty of the heart, holding it to a fixed focus within the field exclusively of the ethnic monad, while deliberately directing outward every impulse to violence. Within the monadic horizon deeds of violence are forbidden: "Thou shalt not kill . . . Thou shalt not covet thy neighbor's wife" (Exodus 20:13, 17; also, Deuteronomy 5:17, 21), whereas abroad, such acts are required: "you shall put all its males to the sword, but the women . . . you shall take as booty to yourselves" (Deuteronomy 20:13–14). In Islamic thought the nations of the earth are distinguished as of two realms: *dar al'islām,* "the realm of submission [to Allah]," and *dar al'harb* "the realm of war," which is to say, the rest of the world. And in Christian thought, the words reported of the resurrected Christ to his eleven remaining apostles—"Go ye therefore, and make disciples of all the nations" (Matthew 28:19)—have been interpreted as a divine mandate for a conquest of the planet.

In our present day, when this same planet, Earth, rocking slowly on its axis in its course around the sun, is about to pass out of astrological range of the zodiacal sign of the Fish (Pisces) into that of the Water Bearer (Aquarius), it does indeed seem that a fundamental transformation of the historical conditions of its inhabiting humanity is in prospect, and that the age of the conquering armies of the contending monster monads—which in the time of Sargon I of Akkad, some 4,320 years ago, was inaugurated in southern Iraq—may be about to close.

For there are no more intact monadic horizons: all are dissolving. And along with them, the psychological hold is weakening of the mythological images and related social rituals by which they were supported. As already recognized half a century ago by the Irish poet Yeats in his foreboding vision *"The Second Coming:"*

> Turning and turning in the widening gyre
> The falcon cannot hear the falconer;
> Things fall apart; the center cannot hold;
> Mere anarchy is loosed upon the world,
> The blood-dimmed tide is loosed, and everywhere
> The ceremony of innocence is drowned;
> The best lack all conviction, while the worst
> Are full of passionate intensity,
>
> Surely some revelation is at hand . . .

The old gods are dead or dying and people everywhere are searching, asking: What is the new mythology to be, the mythology of this unified earth as of one harmonious being?

One cannot predict the next mythology any more than one can predict tonight's dream; for a mythology is not an ideology. It is not something projected from the brain, but something experienced from the heart, from recognitions of identities behind or within the appearances of nature, perceiving with love a "thou" where there would have been otherwise only an "it." As stated already centuries ago in the Indian Kena Upanishad: "That which in the lightning flashes forth, makes one blink, and say 'Ah!'—that 'Ah!' refers to divinity." And centuries before that, in the Chhāndogya Upanishad (*c.* ninth century B.C.):

> When [in the world] one sees nothing else, hears nothing else, recognizes nothing else: that is [participation in] the Infinite. But when one sees, hears, and recognizes only otherness: that is smallness. The Infinite is the immortal. That which is small is mortal.
>
> But sir, that Infinite: upon what is it established?
>
> Upon its own greatness—or rather, not upon greatness. For by greatness people here understand cows and horses, elephants and gold, slaves, wives, mansions and estates. That is not what I mean; not that! For in that context everything is established on something else.
>
> This Infinite of which I speak is below. It is above. It is to the west, to the east, to the south, to the north. It is, in fact, this whole world. And accordingly, with respect to the notion of ego (*ahaṁkārādeśa*): I also am below, above, to the east, to the south, and to the north. I, also, am this whole world.
>
> Or again, with respect to the Self (*ātman*): The Self (the Spirit) is below, above, to the west, to the east, to the south, and to the north. The Self (the Spirit), indeed, is the whole world.
>
> Verily, the one who sees this way, thinks and understands this way, takes pleasure in the Self, delights in the Self, dwells with the Self and knows bliss in the Self; such a one is autonomous (*svarāj*), moving through all the world at pleasure (*kāmacāra*). Whereas those who think otherwise are ruled by others (*anya-rājan*), know but perishable pleasures, and are moved about the world against their will (*akāmacāra*).

The life of a mythology derives from the vitality of its symbols as metaphors delivering, not simply the idea, but a sense of actual participation in such a realization of transcendence, infinity, and abundance, as this of which the upanishadic authors tell. Indeed, the first and most essential service of a mythology is this one, of opening the mind and heart to the utter wonder of all being. And the second service, then, is cosmological: of representing the universe and whole spectacle of nature, both as known to the mind and as beheld by the eye, as an epiphany of such kind that when lightning flashes, or a setting sun ignites the sky, or a deer is seen standing alerted, the exclamation "Ah!" may be uttered as a recognition of divinity.

This suggests that in the new mythology, which is to be of the whole human race, the old Near Eastern desacralization of nature by way of a doctrine of the Fall will have been rejected; so that any such limiting sentiment as that expressed in II Kings 5:15, "there is no God in all the earth but in Israel," will be (to use a biblical term) an abomination. The image of the universe will no longer be the old Sumero-Babylonian, locally centered, three-layered affair, of a heaven above and abyss below, with an ocean-encircled bit of earth between; nor the later, ptolemaic one, of a mysteriously suspended globe enclosed in an orderly complex of revolving crystalline spheres; nor even the recent heliocentric image of a single planetary system at large within a galaxy of exploding stars; but (as of today, at least) an inconceivable immensity of galaxies, clusters of galaxies, and clusters of clusters (superclusters) of galaxies, speeding apart into expanding distance, with humanity as a kind of recently developed scurf on the epidermis of one of the lesser satellites of a minor star in the outer arm of an average galaxy, amidst one of the lesser clusters among the thousands, catapulting apart, which took form some fifteen billion years ago as a consequence of an inconceivable preternatural event.

Introduction to *The Power of Myth*

BILL MOYERS

Acclaimed television journalist Bill Moyers is widely respected for his efforts to bring outstanding thinkers to television. In this selection, Moyers explains how Campbell's work has influenced him personally and demonstrates in great detail how it can teach us "the experience *of being alive." Moyers synthesizes the essential assumptions underlining Campbell's thought, showing us the important relevance of myths to our lives today.*

For weeks after Joseph Campbell died, I was reminded of him just about everywhere I turned.

Coming up from the subway at Times Square and feeling the energy of the pressing crowd, I smiled to myself upon remembering the image that once had appeared to Campbell there: "The latest incarnation of Oedipus, the continued romance of Beauty and the Beast, stands this afternoon on the corner of Forty-second Street and Fifth Avenue, waiting for the traffic light to change."

At a preview of John Huston's last film, *The Dead,* based on a story by James Joyce, I thought again of Campbell. One of his first important works was a key to *Finnegans Wake.* What Joyce called "the grave and constant" in human sufferings Campbell knew to be a principal theme of classic mythology. "The secret cause of all suffering," he said, "is mortality itself, which is the prime condition of life. It cannot be denied if life is to be affirmed."

Once, as we were discussing the subject of suffering, he mentioned in tandem Joyce and Igjugarjuk. "Who is Igjugarjuk?" I said, barely able to imitate the pronunciation. "Oh," replied Campbell, "he was the shaman of a Caribou Eskimo tribe in northern Canada, the one who told European visitors that the only true wisdom 'lives far from mankind, out in the great loneliness, and can be reached only through suffering. Privation and suffering alone open the mind to all that is hidden to others.'"

"Of course," I said, "Igjugarjuk."

Joe let pass my cultural ignorance. We had stopped walking. His eyes were alight as he said, "Can you imagine a long evening around the fire with Joyce and Igjugarjuk? Boy, I'd like to sit in on *that.*"

Campbell died just before the twenty-fourth anniversary of John F. Kennedy's assassination, a tragedy he had discussed in mythological terms during our first meeting years earlier. Now, as that melancholy remembrance came around again, I sat talking with my grown children about Campbell's reflections. The solemn state funeral he had described as "an illustration of the high service of ritual to a society," evoking mythological themes rooted in human need. "This was a ritualized occasion of the greatest social necessity," Campbell had written. The public murder of a president, "representing our whole society, the living social organism of which ourselves were the members, taken away at a moment of exuberant life, required a compensatory rite to reestablish the sense of solidarity. Here was an enormous nation, made those four days into a unanimous community, all of us participating in the same way, simultaneously, in a single symbolic event." He said it was "the first and only thing of its kind in peacetime that has ever given me the sense of being a member of this whole national community, engaged as a unit in the observance of a deeply significant rite."

That description I recalled also when one of my colleagues had been asked by a friend about our collaboration with Campbell: "Why do you need the mythology?" She held the familiar, modern opinion that "all these Greek gods and stuff" are irrelevant to the human condition today. What she did not know—what most do not know—is that the remnants of all that "stuff" line the walls of our interior system of belief, like shards of broken pottery in an archaeological site. But as we are organic beings, there is energy in all that "stuff." Rituals evoke it. Consider the position of judges in our society, which Campbell saw in mythological, not sociological, terms. If this position were just a role, the judge could wear a gray suit to court instead of the magisterial black robe. For the law to hold authority beyond mere coercion, the power of the judge must be ritualized, mythologized. So must much of life today, Campbell said, from religion and war to love and death.

Walking to work one morning after Campbell's death, I stopped before a neighborhood video store that was showing scenes from George Lucas' *Star Wars* on a monitor in the window. I stood there thinking of the time Campbell and I had watched the movie together at Lucas' Skywalker Ranch in California. Lucas and Campbell had become good friends after the filmmaker, acknowledging a debt to Campbell's work, invited the scholar to view the *Star Wars* trilogy. Campbell reveled in the ancient themes and motifs of mythology unfolding on the wide screen in powerful contemporary images. On this particular visit, having again exulted over the perils and heroics of Luke Skywalker, Joe grew animated as he talked about how Lucas "has put the newest and most powerful spin" to the classic story of the hero.

"And what is that?" I asked.

"It's what Goethe said in Faust but which Lucas has dressed in modern idiom—the message that technology is not going to save us. Our computers, our tools, our machines are not enough. We have to rely on our intuition, our true being."

"Isn't that an affront to reason?" I said. "And aren't we already beating a hasty retreat from reason, as it is?"

"That's not what the hero's journey is about. It's not to deny reason. To the contrary, by overcoming the dark passions, the hero symbolizes our ability to control the irrational savage within us." Campbell had lamented on other occasions our failure "to admit within ourselves the carnivorous, lecherous fever" that is endemic to human nature. Now he was describing the hero's journey not as a courageous act but as a life lived in self-discovery, "and Luke Skywalker was never more rational than when he found within himself the resources of character to meet his destiny."

Ironically, to Campbell the end of the hero's journey is not the aggrandizement of the hero. "It is," he said in one of his lectures, "not to identify oneself with any of the figures or powers experienced. The Indian yogi, striving for release, identifies himself with the Light and never returns. But no one with a will to the service of others would permit himself such an escape. The ultimate aim of the quest must be neither release nor ecstasy for oneself, but the wisdom and the power to serve others." One of the many distinctions between the celebrity and the hero, he said, is that one lives only for self while the other acts to redeem society.

Joseph Campbell affirmed life as adventure. "To hell with it," he said, after his university adviser tried to hold him to a narrow academic curriculum. He gave up on the pursuit of a doctorate and went instead into the woods to read. He continued all his life to read books about the world: anthropology, biology, philosophy, art, history, religion. And he continued to remind others that one sure path into the world runs along the printed page. A few days after his death, I received a letter from one of his former students who now helps to edit a major magazine. Hearing of the series on which I had been working with Campbell, she wrote to share how this man's "cyclone of energy blew across all the intellectual possibilities" of the students who sat "breathless in his classroom" at Sarah Lawrence College. "While all of us listened spellbound," she wrote, "we did stagger under the weight of his weekly reading assignments. Finally, one of our number stood up and confronted him (Sarah Lawrence style), saying, 'I *am* taking three other courses, you know. All of them assigned reading, you know. How do you expect me to complete all this in a week?' Campbell just laughed and said, 'I'm astonished you tried. You have the rest of your life to do the reading.'"

She concluded, "And I still haven't finished—the never ending example of his life and work."

One could get a sense of that impact at the memorial service held for him at the Museum of Natural History in New York. Brought there as a boy, he had been transfixed by the totem poles and masks. Who made them? he wondered. What did they mean? He began to read everything he could about Indians, their myths and legends. By ten he was into the pursuit that made him one of the world's leading scholars of mythology and one of the most exciting teachers of our time; it was said that "he could make the bones of folklore and anthropology live." Now, at the memorial service in the museum where three quarters of a century earlier his imagination had first been excited, people gathered to pay honor to his memory. There was a performance by Mickey Hart, the drummer for the Grateful Dead, the rock group with whom Campbell shared a fascination with percussion. Robert Bly played a dulcimer and read poetry dedicated to Campbell. Former students spoke, as did friends whom he had made after he retired and moved with his wife, the dancer Jean Erdman, to Hawaii. The great publishing houses of New York were represented. So were writers and scholars, young and old, who had found their pathbreaker in Joseph Campbell.

And journalists. I had been drawn to him eight years earlier when, self-appointed, I was attempting to bring to television the lively minds of our time. We had taped two programs at the museum, and so compellingly had his presence permeated the screen that more than fourteen thousand people wrote asking for transcripts of the conversations. I vowed then that I would come after him again, this time for a more systematic and thorough exploration of his ideas. He wrote or edited some twenty books, but it was as a teacher that I had experienced him, one rich in the lore of the world and the imagery of language, and I wanted others to experience him as teacher, too. So the desire to share the treasure of the man inspired my PBS series and this book.

A journalist, it is said, enjoys a license to be educated in public; we are the lucky ones, allowed to spend our days in a continuing course of adult education. No one has taught me more of late than Campbell, and when I told him he would have to bear the responsibility for whatever comes of having me as a pupil, he laughed and quoted an old Roman: "The fates lead him who will; him who won't they drag."

He taught, as great teachers teach, by example. It was not his manner to try to talk anyone into anything (except once, when he persuaded Jean to marry him). Preachers err, he told me, by trying "to talk people into belief; better they reveal the radiance of their own discovery." How he did reveal a joy for learning and living! Matthew Arnold believed the highest criticism is "to know the best that is known and thought in the world, and by in its turn making this known, to create a current of true and fresh ideas." This is what Campbell did. It was impossible to listen to him—truly to hear him—without realizing in one's own consciousness a stirring of fresh life, the rising of one's own imagination.

He agreed that the "guiding idea" of his work was to find "the commonality of themes in world myths, pointing to a constant requirement in the human psyche for a centering in terms of deep principles."

"You're talking about a search for the meaning of life?" I asked.

"No, no, no," he said. "For the *experience* of being alive."

I have said that mythology is an interior road map of experience, drawn by people who have traveled it. He would, I suspect, not settle for the journalist's prosaic definition. To him mythology was "the song of the universe," "the music of the spheres"—music we dance to even when we cannot name the tune. We are hearing its refrains "whether we listen with aloof amusement to the mumbo jumbo of some witch doctor of the Congo, or read with cultivated rapture translations from sonnets of Lao-tsu, or now and again crack the hard nutshell of an argument of Aquinas, or catch suddenly the shining meaning of a bizarre Eskimoan fairy tale."

He imagined that this grand and cacophonous chorus began when our primal ancestors told stories to themselves about the animals that they killed for food and about the supernatural world to which the animals seemed to go when they died. "Out there somewhere," beyond the visible plain of existence, was the "animal master," who held over human beings the power of life and death: if he failed to send the beasts back to be sacrificed again, the hunters and their kin would starve. Thus early societies learned that "the essence of life is that it lives by killing and eating; that's the great mystery that the myths have to deal with." The hunt became a ritual of sacrifice, and the hunters in turn performed acts of atonement to the departed spirits of the animals, hoping to coax them into returning to be sacrificed again. The beasts were seen as envoys from that other world, and Campbell surmised "a magical, wonderful accord" growing between the hunter and the hunted, as if they were locked in a "mystical, timeless" cycle of death, burial, and resurrection. Their art—the paintings on cave walls—and oral literature gave form to the impulse we now call religion.

As these primal folk turned from hunting to planting, the stories they told to interpret the mysteries of life changed, too. Now the seed became the magic symbol of the endless cycle. The plant died, and was buried, and its seed was born again. Campbell was fascinated by how this symbol was seized upon by the world's great religions as the revelation of eternal truth—that from death comes life, or as he put it: "From sacrifice, bliss."

"Jesus had the eye," he said. "What a magnificent reality he saw in the mustard seed." He would quote the words of Jesus from the gospel of John—"Truly, truly, I say unto you, unless a grain of wheat falls into the earth and dies, it remains alone; but if it dies, it bears much fruit"—and in the next breath, the Koran: "Do you think that you shall enter the Garden of Bliss without such trials as came to those who passed away before you?" He roamed this vast literature of the spirit, even translating the Hindu scriptures from Sanskrit, and continued to collect more recent stories which he added to the wisdom of the ancients. One story he especially liked told of the troubled woman who came to the Indian saint and sage Ramakrishna, saying, "O Master, I do not find that I love God." And he asked, "Is there nothing, then, that you love?" To this she answered, "My little nephew." And he said to her, "There is your love and service to God, in your love and service to that child."

"And there," said Campbell, "is the high message of religion: 'Inasmuch as ye have done it unto one of the least of these . . .'"

A spiritual man, he found in the literature of faith those principles common to the human spirit. But they had to be liberated from tribal lien, or the religions of the world would remain—as in the Middle East and Northern Ireland today—the source of disdain and aggression. The images of God are many, he said, calling them "the masks of eternity" that both cover and reveal "the Face of Glory." He wanted to know what it means that God assumes such different masks in different cultures, yet how it is that comparable stories can be found in these divergent traditions—stories of creation, of virgin births, incarnations, death and resurrection, second comings, and judgment days. He liked the insight of the Hindu scripture: "Truth is one; the sages call it by many names." All our names and images for God are masks, he said, signifying the ultimate reality that by definition transcends language and art. A myth is a mask of God, too—a metaphor for what lies behind the visible world. However the mystic traditions differ, he said, they are in accord in calling us to a deeper awareness of the very act of living itself. The unpardonable sin, in Campbell's book, was the sin of inadvertence, of not being alert, not quite awake.

I never met anyone who could better tell a story. Listening to him talk of primal societies, I was transported to the wide plains under the great dome of the open sky, or to the forest dense, beneath a canopy of trees, and I began to understand how the voices of the gods spoke from the wind and thunder, and the spirit of God flowed in every mountain stream, and the whole earth bloomed as a sacred place—the realm of mythic imagination. And I asked: Now that we moderns have stripped the earth of its mystery—have made, in Saul Bellow's description, "a housecleaning of belief"—how are our imaginations to be nourished? By Hollywood and made-for-TV movies?

Campbell was no pessimist. He believed there is a "point of wisdom beyond the conflicts of illusion and truth by which lives can be put back together again." Finding it is the "prime question of the time." In his final years he was striving for a new synthesis of science and spirit. "The shift from a geocentric to a heliocentric world view," he wrote after the astronauts touched the moon, "seemed to have removed man from the center—and the center seemed so important. Spiritually, however, the center is where sight is. Stand on a height and view the horizon. Stand on the moon and view the whole earth rising—even, by way of television, in your parlor." The result is an unprecedented expansion of horizon, one that could well serve in our age, as the ancient mythologies did in theirs, to cleanse the doors of perception "to the wonder, at once terrible and fascinating, of ourselves and of the universe." He argued that it is not science that has diminished human beings or divorced us from divinity. On the contrary, the new discoveries of science "rejoin us to the ancients" by enabling us to recognize in this whole universe "a reflection magnified of our own most inward nature; so that we are indeed its ears, its eyes, its thinking, and its speech—or, in theological terms, God's ears, God's eyes, God's thinking, and God's Word." The last time I saw him I asked him if he still believed—as he once had written—"that we are at this moment participating in one of the very greatest leaps of the human spirit to a knowledge not only of outside nature but also of our own deep inward mystery."

He thought a minute and answered, "The greatest ever."

When I heard the news of his death, I tarried awhile in the copy he had given me of *The Hero with a Thousand Faces.* And I thought of the time I first discovered the world of the mythic hero. I had wandered into the little public library of the town where I grew up and, casually exploring the stacks, pulled down a book that opened wonders to me: Prometheus, stealing fire from the gods for the sake of the human race; Jason, braving the dragon to seize the Golden Fleece; the Knights of the Round Table, pursuing the Holy Grail. But not until I met Joseph Campbell did I understand that the Westerns I saw at the Saturday matinees had borrowed freely from those ancient tales. And that the stories we learned in Sunday school corresponded with those of other cultures that recognized the soul's high adventure, the quest of mortals to grasp the reality of God. He helped me to see the connections, to understand how the pieces fit, and not merely to fear less but to welcome what he described as "a mighty multicultural future."

He was, of course, criticized for dwelling on the psychological interpretation of myth, for seeming to confine the contemporary role of myth to either an ideological or a therapeutic function. I am not competent to enter that debate, and leave it for others to wage. He never seemed bothered by the controversy. He just kept on teaching, opening others to a new way of seeing.

It was, above all, the authentic life he lived that instructs us. When he said that myths are clues to our deepest spiritual potential, able to lead us to delight, illumination, and even rapture, he spoke as one who had been to the places he was inviting others to visit.

What did draw me to him?

Wisdom, yes; he was very wise.

And learning; he did indeed "know the vast sweep of our panoramic past as few men have ever known it."

But there was more.

A story's the way to tell it. He was a man with a thousand stories. This was one of his favorites. In Japan for an international conference on religion, Campbell overheard another American delegate, a social philosopher from New York, say to a Shinto priest, "We've been now to a good many ceremonies and have seen quite a few of your shrines. But I don't get your ideology. I don't get your theology." The Japanese paused as though in deep thought and then slowly shook his head. "I think we don't have ideology," he said. "We don't have theology. We dance."

And so did Joseph Campbell—to the music of the spheres.

Patterns in Comparative Religion

MIRCEA ELIADE

The work of Mircea Eliade, past professor of history of religion at the University of Chicago, focuses on the elements of religious experience in both myth and ritual as symbolic realities. His study of Shamans has been acknowledged as the definitive work on the subject. In this selection, Eliade discusses the fundamental unity of seemingly opposite concepts such as life and death and masculine and feminine principles as they are manifested in the great mythological structures. As Eliade explores the relationship between myth and history, he enhances the reader's understanding of the concept that myth surpasses history by being transformed through time.

WHAT MYTHS REVEAL

The myth, whatever its nature, is always a precedent and an example, not only for man's actions (sacred or profane), but also as regards the condition in which his nature places him; a precedent, we may say, for the expressions of reality as a whole. "We must do what the gods did in the beginning"; "Thus the gods acted, thus men act." Statements of this kind give a perfect indication of primitive man's conduct, but they do not necessarily exhaust the content and function of myths; indeed one whole series of myths, recording what gods or mythical beings did *in illo tempore*, discloses a level of reality quite beyond any empirical or rational comprehension. There are, among others, the myths we may lump together as the myths of polarity (or bi-unity) and reintegration, which I have studied by themselves in another book. There is a major group of mythological traditions about the "brotherhood" of gods and demons (for instance, the *devas*[1] and *asuras*), the "friendship" or consanguinity between heroes and their opponents (as with Indra and Namuki), between legendary saints and she-devils (of the type of folklore's Saint Sisinius and his sister the she-devil Uerzelia), and so on. The myths giving a common "father" to figures embodying diametrically opposed principles still survive in the religious traditions which lay the stress on dualism, like the Iranian theology. Zervanism calls Ormuzd and Ahriman brothers, both sons of Zervan, and even the Avesta bears traces of the same idea. This myth has in some cases also passed into popular traditions: there are a number of Rumanian beliefs and proverbs calling God and Satan brothers.

There is another category of myths and legends illustrating not merely a brotherhood between opposing figures, but their paradoxical convertibility. The sun, prototype of the gods, is sometimes called "Serpent" and Agni, the god of fire, is at the same time a "priest Asura"—essentially a demon; he is sometimes described as "without feet or head, hiding both his heads" just like a coiled snake. The *Aitareya Brāhmana* declares that Ahi-Budhnya is invisibly (*parokṣena*) what Agni is visibly (*pratyakṣa*); in other words, the serpent is simply a virtuality of fire, while darkness is light in its latent state. In the *Vājasaneyī Saṁhitā*, Ahi-Budhnya is identified with the sun. *Soma*, the drink which bestows immortality, is supremely "divine," "solar," and yet we read in the *Ṛg Veda* that *soma*, "like Agni, slips out of its old skin," which seems to give it a serpentine quality. Varuna, sky god and archetype of the "Universal Sovereign," is also the god of the ocean, where serpents dwell, as the *Mahābhārata* explains; he is the "king of serpents" (*nāgarājā*) and the *Atharva Veda* even goes so far as to call him "viper."

In any logical perspective, all these reptilian attributes *ought not* to fit a sky divinity like Varuna. But myth reveals a region of ontology inaccessible to superficial logical experience. The myth of Varuṇa discloses the divine bi-unity, the apposition of contraries, all attributes whatever brought to their totality within the divine nature. Myth expresses in action and drama what metaphysics and theology define dialectically. Heraclitus saw that "God is day and night, winter and summer, war and peace, satiety and hunger: all opposites are in him." We find a similar formulation of this idea in an Indian text which tells us that the goddess "is *śrī* [splendour] in the house of those who do good, but *alaksmi* [the opposite of Laksmī, goddess of good luck and prosperity] in the house of the wicked." But this text is simply making clear in it own way the fact that the Indian Great Godesses (Kālī and the rest) like all other Great Goddesses, possess at once both the attributes of gentleness and of dread. They are at once divinities of fertility and destruction, of birth and of death (and often also of war). Kālī, for instance, is called "the gentle and benevolent," although the mythology and iconography connected with her are terrifying (Kālī is covered in blood, wears a necklace of human skulls, holds a cup made out of a skull, and so on), and her cult is the bloodiest anywhere in Asia. In India, every divinity has a "gentle form" and equally a "terrible form" (*krodha-mūrti*). In this, Śiva may be looked on as the archetype of a tremendous series of gods and goddesses for he rhythmically creates and destroys the entire universe.

COINCIDENTIA OPPOSITORUM—THE MYTHICAL PATTERN

All these myths present us with a twofold revelation: they express on the one hand the diametrical opposition of two divine figures sprung from one and the same principle and destined, in many versions, to be reconciled at some *illud tempus* of eschatology, and on the other, the *coincidentia oppositorum* in the very nature of the divinity, which shows itself, by turns or even simultaneously, benevolent and terrible, creative and destructive, solar and serpentine, and so on (in other words, actual and potential). In this sense it is true to say that myth reveals more profoundly than any rational experience ever could, the actual structure of the divinity, which transcends all attributes and reconciles all contraries. That this mythical experience is no mere deviation is proved by the fact that it enters into almost all the religious experience of mankind, even within as strict a tradition as theJudæo-Christian. Yahweh is both kind and wrathful; the God of the Christian mystics and theologians is terrible and gentle at once and it is this *coincidentia oppositorum* which is the starting point for the boldest speculations of such men as the pseudo-Dionysius, Meister Eckhardt, and Nicholas of Cusa.

The *coincidentia oppositorum* is one of the most primitive ways of expressing the paradox of divine reality. We shall be returning to this formula when we come to look at divine "forms," to the peculiar structure revealed by every divine "personality," given of course that the divine personality is not to be simply looked upon as a mere projection of human personality. However, although this conception, in which all contraries are reconciled (or rather, transcended), constitutes what is, in fact, the most basic definition of divinity, and shows how utterly different it is from humanity, the *coincidentia oppositorum* becomes nevertheless an archetypal model for certain types of religious men, or for certain of the forms religious experience takes. The *coincidentia oppositorum* or transcending of all attributes can be achieved by man in all sorts of ways. At the most elementary level of religious life there is the orgy: for it symbolizes a return to the amorphous and the indistinct, to a state in which all attributes disappear and contraries are merged. But exactly the same doctrine can also be discerned in the highest ideas of the eastern sage and ascetic, whose contemplative methods and techniques are aimed at transcending all attributes of every kind. The ascetic, the sage, the Indian or Chinese "mystic" tries to wipe out of his experience and consciousness every sort of "extreme," to attain to a state of perfect indifference and neutrality, to become insensible to pleasure and pain, to become completely self-sufficient. This transcending of extremes through asceticism and contemplation also results in the "coinciding of opposites"; the consciousness of such a man knows no more conflict, and such pairs of opposites as pleasure and pain, desire and repulsion, cold and heat, the agreeable and the disagreeable are expunged from his awareness, while something is taking place within him which parallels the total realization of contraries within the divinity. As we saw earlier, the oriental mind cannot conceive perfection unless all opposites are present in their fulness. The neophyte begins by identifying all his experience with the rhythms governing the universe (sun and moon), but once this "cosmisation" has been achieved, he turns all his efforts toward *unifying* the sun and moon, towards taking into himself the *cosmos as a whole*; he remakes in himself and for himself the primeval unity which was before the world was made; a unity which signifies not the chaos that existed before any forms were created but the undifferentiated *being* in which all forms are merged.

THE MYTH OF DIVINE ANDROGYNY

Another example will illustrate more clearly still the efforts made by religious man to imitate the divine archetype revealed in myth. Since all attributes exist together in the divinity, then one must expect to see both sexes more or less clearly expressed together. Divine androgyny is simply a primitive formula for the divine bi-unity; mythological and religious thought, before expressing this concept of divine two-in-oneness in metaphysical terms (*esse* and *non esse*), or theological terms (the revealed and unrevealed), expressed it first in the biological terms of bisexuality. We have already noted on more than one occasion how archaic ontology was expressed in biological terms. But we must not make the mistake of taking the terminology superficially in the concrete, profane ("modern") sense of the words. The word "woman," in myth or ritual, is never just woman: it includes the cosmological principle woman embodies. And the divine androgyny which we find in so many myths and beliefs has its own theoretical, metaphysical significance. The real point of the formula is to express—in biological terms—the coexistence of contraries, of cosmological principles (male and female) within the heart of the divinity.

This is not the place to consider the problem which I discussed in my *Mitul Reintegrarii*. We must simply note that the divinities of cosmic fertility are, for the most part, either hermaphrodites or male one year and female the next (cf. for instance the Estonians' "spirit of the forest"). Most of the vegetation divinities (such as Attis, Adonis, Dionysos) are bisexual, and so are the Great Mothers (like Cybele). The primal god is androgynous in as primitive a religion as the Australian as well as in the most highly developed religions in India and elsewhere (sometimes even Dyaus; Puruṣa, the cosmic giant of the *Rg Veda,* etc.). The most important couple in the Indian pantheon, Siva-Kālī, are sometimes represented as a single being (*ardhanarīśvara*). And Tantric iconography swarms with pictures of the God Śiva closely entwined with Śakti, his own "power," depicted as a feminine divinity (Kālī). And then, too, all of Indian erotic mysticism is expressly aimed at perfecting man by identifying him with a "divine pair," that is, by way of androgyny.

Divine bisexuality is an element found in a great many religions and—a point worth noting—even the most supremely masculine or feminine divinities are androgynous. Under whatever form the divinity manifests itself, he or she is ultimate reality, absolute power, and this reality, this power, will not let itself be limited by any attributes whatsoever (good, evil, male, female, or anything else). Several of the most ancient Egyptian gods were bisexual. Among the Greeks, androgyny was acknowledged even down to the last centuries of Antiquity. Almost all the major gods in Scandinavian mythology always preserved traces of androgyny: Odhin, Loki, Tuisco, Nerthus, and so on. The Iranian god of limitless time, Zervan, whom the Greek historians rightly saw as Chronos, was also androgynous and Zervan, as we noted earlier, gave birth to twin sons, Ormuzd and Ahriman, the god of "good" and the god of "evil," the god of "light" and the god of "darkness." Even the Chinese had a hermaphrodite Supreme Divinity, who was the god of darkness and of light; the symbol is a consistent one, for light and darkness are simply successive aspects of one and the same reality; seen apart the two might seem separate and opposed, but in the sight of the wise man they are not merely "twins" (like Ormuzd and Ahriman), but form a single essence, now manifest, now unmanifest.

Divine couples (like Bel and Belit, and so on) are most usually later fabrications or imperfect formulations of the primeval androgyny that characterizes all divinities. Thus, with the Semites, the goddess Tanit was nicknamed "daughter of Ba'al" and Astarte the "name of Ba'al." There are innumerable cases of the divinity's being given the title of "father and mother"; worlds, beings, men, all were born of his own substance with no other agency involved. Divine androgyny would include as a logical consequence monogeny or autogeny, and very many myths tell how the divinity drew his existence from himself—a simple and dramatic way of explaining that he is totally self-sufficient. The same myth was to appear again, though based this time on a complex metaphysic, in the neo-Platonic and gnostic speculations of late Antiquity.

THE MYTH OF HUMAN ANDROGYNY

Corresponding to this myth of divine androgyny—which reveals the paradox of divine existence more clearly than any of the other formulæ for the *coincidentia oppositorum*—there is a whole series of myths and rituals relating to human androgyny. The divine myth forms the paradigm for man's religious experience. A great many traditions hold that the "primeval man," the ancestor, was a hermaphrodite (Tuisco is the type) and later mythical variants speak of "primeval pairs" (Yama—that is, "twin"—and his sister, Yami, or the Iranian pair Yima and Yimagh, or Mashyagh and Mashyanagh). Several rabbinical commentaries give us to understand that even Adam was sometimes thought of as androgynous. In this case, the "birth" of Eve would have been simply the division of the primeval hermaphrodite into two beings, male and female. "Adam and Eve were made back to back, attached at their shoulders; then God separated them with an axe, or cut them in two. Others have a different picture: the first man, Adam, was a man on his right side, a woman on his left; but God split him into two halves." The bisexuality of the first man is an even more living tradition in the societies we call primitive (for instance in Australia, and Oceania) and was even preserved, and improved upon, in anthropology as advanced as that of Plato and the gnostics.

We have a further proof that the androgyny of the first man must be seen as one of the expressions of perfection and totalization in the fact that the first hermaphrodite was very often thought of as spherical (Australia; Plato): and it is well known that the sphere symbolized perfection and totality from the time of the most ancient cultures (as in China). The myth of a primordial hermaphrodite spherical in form thus links up with the myth of the cosmogonic egg. For instance, in Taoist tradition, "breathing"—which embodied, among other things, the two sexes—merged together and formed an egg, the Great One; from this heaven and earth were later detached. This cosmological schema was clearly the model for the Taoist techniques of mystical physiology.

The myth of the hermaphrodite god and bisexual ancestor (or first man) is the paradigm for a whole series of ceremonies which are directed towards a *periodic returning* to this original condition which is thought to be the perfect expression of humanity. In addition to the circumcision and subincision which are performed on young aboriginals, male and female respectively, with the aim of transforming them into hermaphrodites, I would also mention all the ceremonies of "exchanging costume" which are lesser versions of the same thing. In India, Persia, and other parts of Asia, the ritual of "exchanging clothes" played a major part in agricultural feasts. In some regions of India, the men even wore false bosoms during the feast of the goddess of vegetation who was, herself, also, of course, androgynous.

In short, from time to time man feels the need to return—if only for an instant—to the state of perfect humanity in which the sexes exist side by side as they coexist with all other qualities, and all other attributes, in the Divinity. The man dressed in woman's clothes is not trying to make himself a woman, as a first glance might suggest, but for a moment he is effecting the unity of the sexes, and thus facilitating his total understanding of the cosmos. The need man feels to cancel periodically his differentiated and determined condition so as to return to primeval "totalization" is the same need which spurs him to periodic orgies in which all forms dissolve, to end by recovering that "oneness" that was before the creation. Here again we come upon the need to destroy the past, to expunge "history" and to start a new life in a new Creation. The ritual of "exchanging costumes" is similar in essence to the ceremonial orgy; and, indeed, these disguises were very often the occasion for actual orgies to break loose. However, even the wildest variants on these rituals never succeeded in abolishing their essential significance—of making their participants once more share in the paradisal condition of "primeval man." And all these rituals have as their exemplar model the myth of divine androgyny.

If I wished to give more examples of the paradigmatic function of myths I should only have to go again through the larger part of the material given in the preceding chapters. As we have seen, it is not simply a question of a paradigm for ritual, but for other religious and metaphysical experience as well, for "wisdom," the techniques of mystical physiology and so on. The most fundamental of the myths reveal archetypes which man labours to reenact, often quite outside religious life properly so called. As a single example: androgyny may be attained not only by the surgical operations that accompany Australian initiation ceremonies, by ritual orgy, by "exchanging costumes" and the rest, but also by means of alchemy (cf. *rebis,* formula of the Philosophers' Stone, also called the "hermetic hermaphrodite"), through marriage (e.g., in the Kabbala), and even, in romantic German ideology, by sexual intercourse. Indeed, we may even talk of the "androgynization" of man through love, for in love each sex attains, conquers the "characteristics" of the opposite sex (as with the grace, submission, and devotion achieved by a man when he is in love, and so on).

MYTHS OF RENEWAL, CONSTRUCTION, INITIATION, AND SO ON

In no case can a myth be taken as merely the fantastic projection of a "natural" event. On the plane of magico-religious experience, as I have already pointed out, nature is never "natural." What looks like a natural situation or process to the empirical and rational mind, is a kratophany or hierophany in magico-religious experience. And it is by these kratophanies or hierophanies alone that "nature" becomes something magico-religious, and, as such, of interest to religious phenomenology and the history of religions. The myths of the "gods of vegetation" constitute, in this regard, an excellent example of the transformation and significance of a "natural" cosmic event. It was not the periodic disappearing and reappearing of vegetation which produced the figures and myths of the vegatation gods (Tammuz, Attis, Osiris, and the rest); at least, it was not the mere empirical, rational, observation of the "natural" phenomenon. The appearing and disappearing of vegetation were always felt, in the perspective of magico-religious experience, to be a *sign* of the periodic creation of the Universe. The sufferings, death and resurrection of Tammuz, as they appear in myth and in the things they reveal, are as far removed from the "natural phenomenon" of winter and spring as *Madame Bovary* or *Anna Karenina* from an adultery. Myth is an autonomous act of creation by the mind: it is through that act of creation that revelation is brought about—not through the things or events it makes use of. In short, the drama of the death and resurrection of vegetation is revealed by the myth of Tammuz, rather than the other way about.

Indeed, the myth of Tammuz, and the myths of gods like him, disclose aspects of the nature of the cosmos which extend far beyond the sphere of plant life; it discloses on the one hand, the fundamental *unity* of life and death, and on the other, the hopes man draws, with good reason, from that fundamental unity, for his own life after death. From this point of view, we may look upon the myths of the sufferings, deaths and resurrections of the vegetation gods as paradigms of the state of mankind: they reveal "nature" better and more intimately than any empirical or rational experience and observation could, and it is to maintain and renew that revelation that the myth must be constantly celebrated and repeated; the appearing and disappearing of vegetation, in themselves, as "cosmic phenomena," signify no more than they actually are: a periodic appearance and disappearance of plant life. Only myth can transform this *event* into a *mode of being*: on one hand, of course, because the death and resurrection of the vegetation gods become archetypes of all deaths and all resurrections, whatever form they take, and on whatever plane they occur, but also because they are better than any empirical or rational means of revealing human destiny.

In the same way, some of the cosmogonic myths, telling how the universe was made out of the body of a primeval giant, if not of the body and blood of the creator himself, became the model not only for the "rites for building" (involving, as we know, the sacrifice of a living being to accompany the setting up of a house, bridge or sanctuary) but also for all forms of "creation" in the broadest sense of the word. The myth revealed the nature of all "creations"—that they cannot be accomplished without an "animation," without a direct giving of life by a creature already possessing that life; at the same time it revealed man's powerlessness to create apart from reproducing his own species—and even that, in many societies, is held to be the work of religious forces outside man (children are thought to come from trees, stones, water, "ancestors," and so on).

A mass of myths and legends describe the "difficulties" demi-gods and heroes meet with in entering a "forbidden domain," a transcendent place—heaven or hell. There is a bridge that cuts like a knife to cross, a quivering creeping plant to get through, two almost touching rocks to pass between, a door to be entered that is open only for an instant, a place surrounded by mountains, by water, by a circle of fire, guarded by monsters, or a door standing at the spot "where Sky and Earth" meet, or where "the ends of the Year" come together. Some versions of this myth of trials, like the labours and adventures of Hercules, the expedition of the Argonauts and others, even had a tremendous literary career in antiquity, being constantly used and remodelled by poets and mythographers; they, in their turn, were imitated in the cycles of semi-historic legend, like the cycle of Alexander the Great, who also wandered through the land of darkness, sought the herb of life, fought with monsters, and so on. Many of these myths were, without doubt, the archetypes of initiation rites (as for instance, the duel with a three-headed monster, that classic "trial" in military initiations). But these myths of the "search for the transcendent land" explain something else in addition to initiation dramas; they show the paradox of getting beyond opposition, which is a necessary part of any world (of any "condition"). Going through the "narrow door," through the "eye of the needle," between "rocks that touch," and all the rest, always involves a pair of opposites (like good and evil, night and day, high and low, etc.). In this sense, it is true to say that the myths of "quest" and of "initiation trials" reveal, in artistic or dramatic form, the actual act by which the mind gets beyond a conditioned, piece-meal universe, swinging between opposites, to return to the fundamental oneness that existed before creation.

THE MAKE-UP OF A MYTH: VARUNA AND VṚTRA

Myth, like symbol, has its own particular "logic," its own intrinsic consistency which enables it to be "true" on a variety of planes, however far removed they may be from the plane upon which the myth originally appeared. I remarked earlier in how many ways and from how many differing standpoints the creation myth is "true"—and therefore effective, "usable." For another example, let us turn once again to the myth and structure of Varuṇa, sovereign sky god, all-powerful, and, upon occasions, one who "binds" by his "spiritual power," by "magic." But his cosmic aspect is more complex still: as we saw, he is not only a sky god, but also a moon and water god. There was a certain "nocturnal" keynote in Varuṇa—possibly from a very early date indeed—upon which Bergaigne, and, more recently, Coomaraswamy, have laid great stress. Bergaigne pointed out that the commentator of *Taittiriya Saṁhitā* speaks of Varuṇa as "he who envelops like darkness." This "nocturnal" side of Varuṇa must not be interpreted only in the atmospheric sense of the "night sky," but also in a wider sense more truly cosmological and even metaphysical: night, too, *is* potentiality, seed, non-manifestation, and it is because Varuṇa has this "nocturnal" element that he can become a god of water, and can be assimilated with the demon Vṛtra.

This is not the place to go into the "Vṛtra-Varuna" problem, and we must be content with noting the fact that the two beings have more than one trait in common. Even leaving aside the probable etymological relationship between their two names, we should note that both are connected with water, and primarily with "water held back" ("Great Varuṇa has hidden the sea") and that Vṛtra, like Varuṇa, is sometimes called *māyin*, or magician. From one point of view, these various identifications with Vṛtra, like all Varuṇa's other attributes and functions, fit together and help to explain each other. Night (the non-manifest), water (the potential, seeds), transcendence and impassivity (both marks of supreme gods and sky gods) are linked both mythologically and metaphysically with, on the one hand, every kind of being that "binds" and, on the other, the Vṛtra who "holds back," "stops" or "imprisons" the waters. At the cosmic level, Vṛtra, too, is a "binder." Like all the great myths, the myth of Vṛtra has thus got many meanings, and no single interpretation exhausts it. We might even say that one of the main functions of myth is to determine, to authenticate the levels of reality which both a first impression and further thought indicate to be manifold and heterogeneous. Thus, in the myth of Vṛtra, besides other significations, we note that of returning to the non-manifest, of a "stopping," of a "bond" preventing "forms"—the life of the Cosmos, in fact—from manifestation. Obviously we must not push the parallel between Vṛtra and Varuṇa too far. But there can be no denying the structural connection between the "nocturnal," "impassive," "magician," Varuṇa, "binding" the guilty from afar, and Vṛtra "imprisoning" the waters. Both are acting so as to stop life, and bring death, the one on an individual, the other, on a cosmic scale.

MYTH AS "EXEMPLAR HISTORY"

Every myth, whatever its nature, recounts an event that took place *in illo tempore*, and constitutes as a result, a precedent and pattern for all the actions and "situations" later to repeat that event. Every ritual, and every meaningful act that man performs, repeats a mythical archetype; and, as we saw, this repetition involves the abolition of profane time and placing of man in a magico-religious time which has no connection with succession in the true sense, but forms the "eternal now" of mythical time. In other words, along with other magico-religious experiences, myth makes man once more exist in a timeless period, which is in effect an *illud tempus*, a time of dawn and of "paradise," outside history. Anyone who performs any rite transcends profane time and space; similarly, anyone who "imitates" a mythological model or even ritually assists at the retelling of a myth (taking part in it), is taken out of profane "becoming," and returns to the Great Time.

We moderns would say that myth (and with it all other religious experiences) abolishes "history." But note that the majority of myths, simply because they record what took place "in illo tempore," themselves constitute an *exemplar history* for the human society in which they have been preserved, and for the world that society lives in. Even the cosmogonic myth is history, for it recounts all that took place *ab origine*; but we must, I need hardly say, remember that it is not "history" in our sense of the word—things that took place once and will never take place again—but exemplar history which can be repeated (regularly or otherwise), and whose meaning and value lie in that very repetition. The history that took place at the beginning must be repeated because the primeval epiphanies were prolific—they could not be fully expressed in a single manifestation. And myths too are prolific in their content, for it is paradigmatic, and therefore presents a *meaning*, creates something, tells us something.

The function of myths as exemplar history is further apparent from the need primitive man feels to show "proofs" of the event recorded in the myth. Suppose that it is a well known mythological theme: such and such a thing happened, men became mortal, seals lost their toes, a mark appeared on the moon, or something similar. To the primitive mind this theme can be clearly "proved" by the fact that man *is* mortal, seals *have* no toes, the moon has *got* marks on it. The myth which tells how the island of Tonga was fished up from the bottom of the sea is proved by the fact that you can still see the line used to pull it up, and the rock where the hook caught. This need to prove the truth of myth also helps us to grasp what history and "historical evidence" mean to the primitive mind. It shows what an importance primitive man attaches to things that have *really happened*, to the events which actually took place in his surroundings; it shows how his mind hungers for what is "real," for what *is* in the fullest sense. But, at the same time, the archetypal function given to these events of *illud tempus* give us a glimpse of the interest primitive people take in realities that are significant, creative, paradigmatic. This interest survived even in the first historians of the ancient world, for to them the "past" had meaning only in so far as it was an example to be imitated, and consequently formed the *summa* of the learning of all mankind. This work of "exemplar history" which devolved upon myth must, if we are to understand it properly, be seen in relation to primitive man's tendency to effect a concrete realization of an ideal archetype, to live eternity "experientially" here and now—an aspiration which we studied in our analysis of sacred time.

THE CORRUPTION OF MYTHS

A myth may degenerate into an epic legend, a ballad or a romance, or survive only in the attenuated form of "superstitions," customs, nostalgias, and so on; for all this, it loses neither its essence nor its significance. Remember how the myth of the Cosmic Tree was preserved in legend and in the rites for gathering simples. The "trials," sufferings, and journeyings of the candidate for initiation survived in the tales of the sufferings and obstacles undergone by heroes of epic or drama before they gained their end (Ulysses, Aeneas, Parsifal, certain of Shakespeare's characters, Faust, and so on). All these "trials" and "sufferings" which make up the stories of epic, drama or romance can be clearly connected with the ritual sufferings and obstacles on the "way to the centre." No doubt the "way" is not on the same initiatory plane, but, typologically, the wanderings of Ulysses, or the search for the Holy Grail, are echoed in the great novels of the nineteenth century, to say nothing of paperback novels, the archaic origins of whose plots are not hard to trace. If to-day, detective stories recount the contest between a criminal and a detective (the good genie and wicked genie, the dragon and fairy prince of the old stories), whereas a few generations back, they preferred to show an orphan prince or innocent maiden at grips with a "villain," while the fashion of a hundred and fifty years ago was for "black" and turgid romances with "black monks," "Italians," "villains," "abducted maidens," "masked protectors," and so on, such variations of detail are due to the different colouring and turn of popular sentiment; the theme does not change.

Obviously, every further step down brings with it a blurring of the conflict and characters of the drama as well as a greater number of additions supplied by "local colour." But the patterns that have come down from the distant past never disappear; they do not lose the possibility of being brought back to life. They retain their point even for the "modern" consciousness. To take one of a thousand examples: Achilles and Søren Kierkegaard. Achilles, like many other heroes, did not marry, though a happy and fruitful life had been predicted for him had he done so; but in that case he would have given up becoming a *hero*, he would not have realized his unique success, would not have gained immortality. Kierkegaard passed through exactly the same existential drama with regard to Regina Olsen; he gave up marriage to remain himself, unique, that he might hope for the eternal by refusing the path of a happy life with the general run of men. He makes this clear in a fragment of his private Journal: "I should be happier, in a finite sense, could I drive out this thorn I feel in my flesh; but in the infinite sense, I should be lost." In this way a mythical pattern can still be realized, and is in fact realized, on the plane of existential experience, and, certainly in this case, with no thought of or influence from the myth.

The archetype is still creative even though sunk to lower and lower levels. So, for instance, with the myth of the Fortunate Islands or that of the Earthly Paradise, which obsessed not only the imagination of the secular mind, but nautical science too right up to the great age of seafaring discoveries. Almost all navigators, even those bent on a definite economic purpose (like the Indian route), *also* hoped to discover the Islands of the Blessed or the Earthly Paradise. And we all know that there were many who thought they had actually found the Island of Paradise. From the Phoenicians to the Portuguese, all the memorable geographical discoveries were the result of this myth of the land of Eden. And these voyages, searches, and discoveries were the only ones to acquire a spiritual meaning, to create culture. If the memory of Alexander's journey to India never faded it was because, being classed with the great myths, it satisfied the longing for "mythical geography"—the only sort of geography man could never do without. The Genoese commercial ventures in Crimea and the Caspian Sea, and the Venetian in Syria and in Egypt, must have meant a very advanced degree of nautical skill, and yet the mercantile routes in question "have left no memory in the history of geographical discovery." On the other hand, expeditions to discover the mythical countries did not only create legends: they also brought an increase of geographical knowledge.

These islands and these new lands preserved their mythical character long after geography had become scientific. The "Isles of the Blessed" survived till Camoens, passed through the age of enlightenment and the romantic age, and have their place even in our own day. But the mythical island no longer means the garden of Eden: it is Camoens' Isle of Love, Daniel Defoe's island of the "good savage," Eminescu's Island of Euthanasius, the "exotic" isle, a land of dreams with hidden beauties, or the island of liberty, or perfect rest—or ideal holidays, or cruises on luxury steamers, to which modern man aspires in the mirages offered to him by books, films or his own imagination. The *function* of the paradisal land of perfect freedom remains unchanged; it is just that man's view of it has undergone a great many displacements—from Paradise in the biblical sense to the exotic paradise of our contemporaries' dreams. A decline, no doubt, but a very prolific one. At all levels of human experience, however ordinary, archetypes still continue to give meaning to life and to create "cultural values": the paradise of modern novels and the isle of Camoens are as significant culturally as any of the isles of medieval literature.

In other words man, whatever else he may be free of, is forever the prisoner of his own archetypal intuitions, formed at the moment when he first perceived his position in the cosmos. The longing for Paradise can be traced even in the most banal actions of the modern man. Man's concept of the *absolute* can never be completely uprooted: it can only be debased. And primitive spirituality lives on in its own way not in action, not as a thing man can effectively accomplish, but as a *nostalgia* which creates things that become values in themselves: art, the sciences, social theory, and all the other things to which men will give the whole of themselves.

NOTES

Guide to Pronunciation

[1]Sanskrit vowels are pronounced very much like Italian vowels, with the exception of the short *a* which is pronounced like the *u* in the English word 'but'; long *ā* is pronounced like the *a* in 'father'. As for the consonants, a reasonable approximation will be obtained by pronouncing *c* as in 'church', *j* as in 'jungle', *ṣ* as in 'shun', *s* as in 'sun', and *ś* as something half-way between the other two *s*'s. The aspirated consonants should be pronounced distinctly: *th* as in goat-herd, *ph* as in top-hat, *gh* as in dog-house, *dh* as in mad-house, *bh* as in cab-horse. ṛ is a vowel, pronounced mid-way between 'ri' as in 'rivet' and 'er' as in 'father'.

Unit I

In the Beginning: Origins of Man and Myth

The Moon and Its Mystique

MIRCEA ELIADE

In this selection Mircea Eliade identifies the mythic patterns associated with lunar symbolism, rain, fertility, the feminine, the serpent, death, and regeneration. He shows that these symbols have meaning only as parts of the pattern in which they exist, thus demonstrating to the reader that both the symbols and the myths are transformed through time. Eliade's distinction between masculine and feminine principles is basic to understanding myths through metaphor.

THE MOON AND TIME

The sun is always the same, always itself, never in any sense "becoming." The moon, on the other hand, is a body which waxes, wanes and disappears, a body whose existence is subject to the universal law of becoming, of birth and death. The moon, like man, has a career involving tragedy, for its failing, like man's, ends in death. For three nights the starry sky is without a moon. But this "death" is followed by a rebirth: the "new moon." The moon's going out, in "death," is never final. One Babylonian hymn to Sin sees the moon as "a fruit growing from itself." It is reborn of its own substance, in pursuance of its own destined career.

This perpetual return to its beginnings, and this ever-recurring cycle make the moon *the* heavenly body above all others concerned with the rhythms of life. It is not surprising, then, that it governs all those spheres of nature that fall under the law of recurring cycles: waters, rain, plant life, fertility. The phases of the moon showed man time in the concrete sense—as distinct from astronomical time which certainly only came to be realized later. Even in the Ice Age the meaning of the moon's phases and their magic powers were clearly known. We find the symbolism of spirals, snakes and lightning—all of them growing out of the notion of the moon as the measure of rhythmic change and fertility—in the Siberian cultures of the Ice Age. Time was quite certainly measured everywhere by the phases of the moon. Even to-day there are nomad tribes living off what they can hunt and grow who use only the lunar calendar. The oldest Indo-Aryan root connected with the heavenly bodies is the one that means "moon": it is the root *me*, which in Sanskrit becomes *māmi*, "I measure." The moon becomes the universal measuring gauge. All the words relating to the moon in the Indo-European languages come from that root: *mās* (Sanskrit), *mah* (Avestic), *mah* (Old Prussian), *menu* (Lithuanian), *mena* (Gothic), *mene* (Greek), *mensis* (Latin). The Germans used to measure time by nights. Traces of this ancient way of reckoning are also preserved in popular European traditions; certain feasts are celebrated at night as, for instance, Christmas night, Easter, Pentecost, Saint John's Day and so on.

Time as governed and measured by the phases of the moon might be called "living" time. It is bound up with the reality of life and nature, rain and the tides, the time of sowing, the menstrual cycle. A whole series of phenomena belonging to totally different "cosmic levels" are ordered according to the rhythms of the moon or are under their influence. The "primitive mind," once having grasped the "powers" of the moon, then establishes connections of response and even interchange between the moon and those phenomena. Thus, for instance, from the earliest times, certainly since the Neolithic Age, with the discovery of agriculture, the same symbolism has linked together the moon, the sea waters, rain, the fertility of women and of animals, plant life, man's destiny after death and the ceremonies of initiation. The mental syntheses made possible by the realization of the moon's rhythms connect and unify very varied realities; their structural symmetries and the analogies in their workings could never have been seen had not "primitive" man intuitively perceived the moon's law of periodic change, as he did very early on.

The moon measures, but it also unifies. Its "forces" or rhythms are what one may call the "lowest common denominator" of an endless number of phenomena and symbols. The whole universe is seen as a pattern, subject to certain laws. The world is no longer an infinite space filled with the activity of a lot of disconnected autonomous creatures: within that space itself things can be seen to correspond and fit together. All this, of course, is not the result of a reasoned analysis of reality, but of an ever clearer intuition of it in its totality. Though there may be a series of ritual or mythical side-commentaries on the moon which are separate from the rest, with their own somewhat specialized function (as, for instance, certain mythical lunar beings with only one foot or one hand, by whose magic power one can cause rain to fall), there can be no symbol, ritual or myth of the moon that does not imply all the lunar values known at a given time. There can be no part without the whole. The spiral, for instance, which was taken to be a symbol of the moon as early as the Ice Age, relates to the phases of the moon, but also includes erotic elements springing from the vulva-shell analogy, water elements (the moon = shell), and some to do with fertility (the double volute, horns and so on). By wearing a pearl as an amulet a woman is united to the powers of water (shell), the moon (the shell a symbol of the moon; created by the rays of the moon, etc.), eroticism, birth and embryology. A medicinal plant contains in itself the threefold effectiveness of the moon, the waters and vegetation, even when only one of these powers is explicitly present in the mind of the user. Each of these powers or "effectivenesses" in its turn works on a number of different levels. Vegetation, for instance, implies notions of death and rebirth, of light and darkness (as zones of the universe), of fecundity and abundance, and so on. There is no such thing as a symbol, emblem or power with only one kind of meaning. Everything hangs together, everything is connected, and makes up a cosmic whole.

THE COHERENCE OF ALL LUNAR EPIPHANIES

Such a whole could certainly never be grasped by any mind accustomed to proceeding analytically. And even by intuition modern man cannot get hold of all the wealth of meaning and harmony that such a cosmic *reality* (or, in fact, sacred reality) involves in the primitive mind. To the primitive, a lunar symbol, (an amulet or iconographic sign) does not merely contain in itself all the lunar forces at work on every level of the cosmos—but actually, by the power of the ritual involved, places the wearer himself at the centre of those forces, increasing his vitality, making him more *real,* and guaranteeing him a happier state after death. It is important to keep stressing this fact that every religious act (that is, every act with a meaning) performed by primitive man has a character of *totality,* for there is always a danger of our looking upon the functions, powers and attributes of the moon as we discuss them in this chapter in an *analytic* and *cumulative* manner. We tend to divide what is and must remain a whole. Where we use the words "because," and "therefore," the mind of the primitive man would phrase it perhaps as "in the same way" (for instance, I say: because the moon governs the waters, plants are subject to it, but it would be more correct to say: plants and the waters are subject to it *in the same way* . . .).

The "powers" of the moon are to be discovered not by means of a succession of analytical exercises, but by intuition; *it reveals itself* more and more fully. The analogies formed in the primitive mind are as it were orchestrated there by means of symbols; for instance, the moon appears and disappears; the snail shows and withdraws its horns; the bear appears and disappears with the seasons; thus, the snail becomes the scene of a lunar theophany, as in the ancient religion of Mexico in which the moon god, Tecciztecatl, is shown enclosed in a snail's shell; it also becomes an amulet, and so on. The bear becomes the ancestor of the human race, for man, whose life is similar to that of the moon, must have been created out of the very substance or by the magic power of that orb of living reality.

The symbols which get their meaning from the moon *are* at the same time the moon. The spiral is both a lunar hierophany—expressing the light-darkness cycle—and a sign by which man can absorb the moon's powers into himself. Lightning, too, is a kratophany of the moon, for its brightness recalls that of the moon, and it heralds rain, which is governed by the moon. All these symbols, hierophanies, myths, rituals, amulets and the rest, which I call lunar to give them one convenient name, form a whole in the mind of the primitive; they are bound together by harmonies, analogies, and elements held in common, like one great cosmic "net," a vast web in which every piece fits and nothing is isolated from the rest. If you want to express the multiplicity of lunar hierophanies in a single formula, you may say that they reveal life repeating itself rhythmically. All the values of the moon, whether cosmological, magic or religious, are explained by its modality of *being*: by the fact that it is "living," and inexhaustible in its own regeneration. In the primitive mind, the intuition of the cosmic destiny of the moon was equivalent to the first step, the foundation of an anthropology. Man saw himself reflected in the "life" of the moon; not simply because his own life came to an end, like that of all organisms, but because his own thirst for regeneration, his hopes of a "rebirth," gained confirmation from the fact of there being always a new moon.

It does not matter to us a great deal whether, in the innumerable beliefs centring upon the moon, we are dealing with adoration of the moon itself, of a divinity inhabiting it, or of a mythical personification of it. Nowhere in the history of religions do we find an adoration of any natural object in itself. A sacred thing, whatever its form and substance, is sacred because it reveals or shares in ultimate *reality*. Every religious object is always an "incarnation" of something: of the *sacred*. It incarnates it by the quality of its being (as for instance, the sky, the sun, the moon or the earth), or by its form (that is, symbolically: as with the spiral-snail), or by a hierophany (a *certain* place, a *certain* stone, etc. becomes sacred; a certain object is "sanctified" or "consecrated" by a ritual, or by contact with another sacred object or person, and so on).

Consequently, the moon is no more adored in *itself* than any other object, but in what it reveals of the sacred, that is, in the power centred in it, in the inexhaustible life and reality that it manifests. The sacred reality of the moon was recognized either immediately, in the lunar hierophany itself, or in the forms created by that hierophany over the course of thousands of years—that is in the representations to which it had given birth: personifications, symbols or myths. The differences between these various forms are not the concern of this chapter. After all, what we are seeking here is mainly to explain the hierophany of the moon and all that it involves. There is no need even to confine ourselves to evidence that is obviously "sacred," lunar gods, for example, and the rituals and myths consecrated to them. To the primitive mind, I repeat, everything that had a meaning, everything connected with *absolute reality*, had sacred value. We can observe the religious character of the moon with as much precision in the symbolism of the pearl or of the lightning as we can by studying a lunar divinity like the Babylonian Sin or the goddess Hecate.

THE MOON AND THE WATERS

Both because they are subject to rhythms (rain and tides), and because they sponsor the growth of living things, waters are subject to the moon. "The moon is in the waters" and "rain comes from the moon"; those are two *leitmotiven* in Indian thought. *Apām napāt*, "the son of water," was in primitive times the name of a spirit of vegetation, but was later applied also to the moon and to the nectar of the moon, *soma*. Ardvisura Anahita, the Iranian goddess of water, was a lunar being; Sin, the Babylonian moon god, also governed the waters. One hymn brings out how fruitful his theophany is: "When thou floatest like a boat on the waters . . . the pure river Euphrates is filled with water to the full. . . ." One text of the "Langdon Epic" speaks of the place "whence the waters flow from their source, from the moon's reservoir."

All the moon divinities preserve more or less obvious water attributes or functions. To certain American Indian tribes, the moon, or the moon god, is at the same time the god of water. (This is true in Mexico, and among the Iroquois, to name two instances.) One tribe in central Brazil call the moon-god's daughter "Mother of the Waters." Hieronymo de Chaves said (in 1576), speaking of ancient Mexican beliefs concerning the moon, "the moon makes all things grow and multiply . . ." and "all moisture is governed by it." The link between the moon and the tides which both the Greeks and the Celts observed, was also known to the Maoris of New Zealand and the Eskimos (whose moon divinities govern the tides).

From the earliest times it was recognized that rainfall followed the phases of the moon. A whole series of mythical characters, belonging to cultures as varied as the Bushman, Mexican, Australian, Samoyed and Chinese, were marked by their power to cause rain and by having only one foot or only one hand. Hentze establishes quite fully that they are lunar in essence. And, too, there are numerous moon symbols in all images of them, and all their various rites and myths have a definite lunar character. While the waters and the rain are governed by the moon, and normally follow a fixed order—that is, they follow the phases of the moon—all disasters connected with them, on the other hand, display the moon's other aspect, as the periodic destroyer of outworn "forms" and, we may say, of effecting regeneration on the cosmic scale.

Flood corresponds to the three days of darkness, or "death," of the moon. It is a cataclysm, but never a final one, for it takes place under the seal of the moon and the waters, which are pre-eminently the sign of growth and regeneration. A flood destroys simply because the "forms" are old and worn out, but it is always followed by a new humanity and a new history. The vast majority of deluge myths tell how a single individual survived, and how the new race was descended from him. This survivor—man or woman—occasionally marries a lunar animal, which thus becomes ancestor to the race. So, for instance, one Dyak legend tells how a woman was the only survivor of a flood which followed upon the slaying of an immense boa constrictor, a "lunar animal," and gave birth to a new humanity by mating with a dog (or, in some variants, with a stick for firing, found next to a dog).

Of the numerous variants on the Deluge myth we will look at one—an Australian version (that of the Kurnai tribe). One day all the waters were swallowed by an immense frog, Dak. In vain the parched animals tried to make her laugh. Not until the eel (or serpent) began to roll about and twist itself round did Dak burst out laughing, and the waters thus rushed out and produced the flood. The frog is a lunar animal, for a great many legends speak of a frog to be seen in the moon and it is always present in the innumerable rites for inducing rain. Father Schmidt explains the Australian myth by the fact that the new moon halts the flow of the waters (Dak swallowing them). And Winthuis, who disagrees with Schmidt's interpretation, discerns an erotic meaning in this frog myth; but that would not, of course, disprove its lunar nature, nor the anthropogonic function of the deluge (which "creates" a new, regenerated, humanity).

Again in Australia we find another variant on the watery disaster produced by the moon. The moon asked man one day for some oppossum skins to wear at night as it was cold, and man refused; to avenge itself, the moon caused torrents of rain to fall and flood the whole area. The Mexicans also believed that the moon caused the disaster, but under the guise of a young and beautiful woman. However, there is one thing to note with all these catastrophes induced by the moon (most of them provoked by some insult paid to it, or by ignorance of some ritual prohibition—that is, by a "sin" indicating that man is backsliding spiritually, abandoning law and order, putting the rhythms of nature out of joint): and that is the myth of regeneration, and the appearance of a "new man." This myth fits in perfectly, as we shall see, with the redemptive functions of the moon and the waters.

THE MOON AND VEGETATION

That there was a connection between the moon, rain and plant life was realized before the discovery of agriculture. The plant world comes from the same source of universal fertility, and is subject to the same recurring cycles governed by the moon's movements. One Iranian text says that plants grow by its warmth. Some tribes in Brazil call it "mother of grasses" and in a great many places (Polynesia, Moluccas, Melanesia, China, Sweden, etc.) it is thought that grass grows on the moon. French peasants, even today, sow at the time of the new moon; but they prune their trees and pick their vegetables when the moon is on the wane, presumably in order not to go against the rhythm of nature by damaging a living organism when nature's forces are on the upward swing.

The organic connection between the moon and vegetation is so strong that a very large number of fertility gods are also divinities of the moon; for instance the Egyptian Hathor, Ishtar the Iranian Anaitis, and so on. In almost all the gods of vegetation and fecundity there persist lunar attributes or powers—even when their divine "form" has become completely autonomous. Sin is also the creator of the grasses; Dionysos is both moon-god and god of vegetation; Osiris possesses all these attributes—moon, water, plant life and agriculture. We can discern the moon-water-vegetation pattern particularly clearly in the religious nature of certain beverages of divine origin, such as the Indian *soma,* and the Iranian *haoma*; these were even personified into divinities—autonomous, though less important than the major gods of the Indo-Iranian pantheon. And in this divine liquor which confers immortality on all who drink it, we can recognize the sacredness that centres round the moon, water and vegetation. It is supremely the "divine substance," for it transmutes life into absolute reality—or immortality. *Amṛta,* ambrosia, *soma, haoma* and the rest all have a celestial prototype drunk only by gods and heroes, but there is a similar power in earthly drinks too—in the *soma* the Indians drank in Vedic times, in the wine of the Dionysiac orgies. Furthermore, these earthly drinks owe their potency to their corresponding celestial prototype. Sacred inebriation makes it possible to share—though fleetingly and imperfectly—in the divine mode of being; it achieves, in fact, the paradox of at once possessing the fulness of existence, and becoming; of being at once dynamic and static. The metaphysical role of the moon is to *live* and yet remain *immortal,* to undergo death, but as a rest and regeneration, never as a conclusion. This is the destiny which man is trying to conquer for himself in all the rites, symbols and myths—rites, symbols and myths in which, as we have seen, the sacred values of the moon exist together with those of water and of vegetation, whether the latter derive their sacredness from the moon, or constitute autonomous hierophanies. In either case we are faced with an *ultimate reality,* a source of power and of life from which all living forms spring, either of its substance, or as a result of its blessing.

The connections seen between the different cosmic levels that the moon governs—rain, plant life, animal and human fecundity, the souls of the dead—enter into even as primitive a religion as that of the Pygmies of Africa. Their feast of the new moon takes place a little before the rainy season. The moon, which they call Pe, is held to be the "principle of generation, and the mother of fecundity." The feast of the new moon is reserved exclusively for the women, just as the feast of the sun is celebrated exclusively by the men. Because the moon is both "mother and the refuge of ghosts" the women honour it by smearing themselves with clay and vegetable juices, thus becoming white like ghosts and moonlight. The ritual consists in preparing an alcoholic liquid from fermented bananas, which the women drink when they are wearied by their dancing, and of dances and prayers addressed to the moon. The men do not dance, nor even accompany the ritual on their tom-toms. The moon, "mother of living things," is asked to keep away the souls of the dead and to bring fertility, to give the tribe a lot of children, and fish, game and fruit.

THE MOON AND FERTILITY

The fertility of animals, as well as that of plants, is subject to the moon. The relationship between the moon and fecundity occasionally becomes somewhat complicated owing to the appearance of new religious "forms"—like the Earth-Mother, and the various agricultural divinities. However, there is one aspect of the moon that remains permanently evident, however many religious syntheses have gone towards making up these new "forms;" and that is the prerogative of fertility, of recurring creation, of inexhaustible life. The horns of oxen, for instance, which are used to characterize the great divinities of fecundity, are an emblem of the divine *Magna Mater.* Wherever they are to be found in Neolithic cultures, either in iconography, or as part of idols in the form of oxen, they denote the presence of the Great Goddess of fertility. And a horn is always the image of the new moon: "Clearly the ox's horn became a symbol of the moon because it brings to mind a crescent; therefore both horns together represent two crescents, or the complete career of the moon." And also in the iconography of the prehistoric Chinese cultures of Kansu and Yang-kao you will often find symbols of the moon and symbols of fertility together—stylized horns are framed by a pattern of lightning-flashes (signifying the rain and the moon) and lozenges (which are a symbol of femininity).

Certain animals become symbols of even "presences" of the moon because their shape or their behaviour is reminiscent of the moon's. So with the snail which goes in and out of its shell; the bear, which disappears in midwinter and reappears in the spring; the frog because it swells up, submerges itself, and later returns to the surface of the water; the dog, because it can be seen in the moon, or because it is supposed in some myths to be the ancestor of the race; the snake, because it appears and disappears, and because it has as many coils as the moon has days (this legend is also preserved in Greek tradition); or because it is "the husband of all women," or because it sloughs its skin (that is to say, is periodically reborn, is "immortal"), and so on. The symbolism of the snake is somewhat confusing, but all the symbols are directed to the same central idea: it is immortal because it is continually reborn, and therefore it is a moon "force," and as such can bestow fecundity, knowledge (that is, prophecy) and even immortality. There are innumerable myths telling the disastrous story of how the serpent stole the immortality given to man by his god. But they are all later variants on a primitive myth in which the serpent (or a sea monster) guarded the sacred spring and the spring of immortality (the Tree of Life, the Fountain of Youth, the Golden Apples).

I can only mention here a few of the myths and symbols connected with the serpent, and only those which indicate its character of a lunar animal. In the first place, its connection with women and with fecundity: the moon is the source of all fertility, and also governs the menstrual cycle. It is personified as "the master of women." A great many peoples used to think—and some think it to this day—that the moon, in the form of a man, or a serpent, copulates with their women. That is why, among the Eskimos for instance, unmarried girls will not look at the moon for fear of becoming pregnant. The Australians believe that the moon comes down to earth in the form of a sort of Don Juan, makes women pregnant and then deserts them. This myth is still current in India.

Since the serpent is an epiphany of the moon, it fulfills the same function. Even today it is said in the Abruzzi that the serpent copulates with all women. The Greeks and Romans also believed it. Alexander the Great's mother, Olympia, played with snakes. The famous Aratus of Sicyon was said to be a son of Aesculapius because, according to Pausanias, his mother had conceived him of a serpent. Suetonius and Dio Cassius tell how the mother of Augustus conceived from the embrace of a serpent in Apollo's temple. A similar legend was current about the elder Scipio. In Germany, France, Portugal, and elsewhere, women used to be afraid that a snake would slip into their mouths when they were asleep, and they would become pregnant, particularly during menstruation. In India, when women wanted to have children, they adored a cobra. All over the East it was believed that woman's first sexual contact was with a snake, at puberty or during menstruation. The Komati tribe in the Mysore province of India use snakes made of stones in a rite to bring about the fertility of the women. Claudius Aelianus declares that the Hebrews believed that snakes mated with unmarried girls; and we also find this belief in Japan. A Persian tradition says that after the first woman had been seduced by the serpent, she immediately began to menstruate. And it was said by the rabbis that menstruation was the result of Eve's relations with the serpent in the Garden of Eden. In Abyssinia it was thought that girls were in danger of being raped by snakes until they were married. One Algerian story tells how a snake escaped when no one was looking and raped all the unmarried girls in a house. Similar traditions are to be found among the Mandi Hottentots of East Africa, in Sierra Leone and elsewhere.

Certainly the menstrual cycle helps to explain the spread of the belief that the moon is the first mate of all women. The Papoos thought menstruation was a proof that women and girls were connected with the moon, but in their iconography (sculptures on wood) they pictured reptiles emerging from their genital organs, which confirms that snakes and the moon are identified. Among the Chiriguanoes, various fumigations and purifications are performed after a woman's first menstrual period, and after that the women of the house drive away every snake they come upon, as responsible for this evil. A great many tribes look upon the snake as the cause of the menstrual cycle. Its phallic character, which Crawley was one of the first ethnologists to demonstrate, far from excluding its connection with the moon, only confirms it. A great deal of the iconographical documentation which remains—both of the Neolithic civilizations of Asia (such as the idol of the Panchan culture, at Kansu, and the sculptured gold of Ngan-Yang) and of the Amerindian civilizations (such as the bronze discs of Calchaqui)—show the double imagery of the snake decorated with lozenges (symbolizing the vulva). The two together undoubtedly have an erotic meaning, but the coexistence of the snake (phallus) and lozenges also expresses an idea of dualism and reintegration which is a supremely lunar notion, for we find that same motif in the lunar imagery of "rain," of "light and darkness," and the rest.

THE MOON, WOMAN, AND SNAKES

The moon then can also be personified as reptile and masculine, but such personifications (which often break away from the original pattern and follow a path of their own in myth and legend), are still fundamentally based on the notion of the moon as source of living reality, and basis of all fertility and periodic regeneration. Snakes are thought of as producing children; in Guatemala, for instance, in the Urabunna tribe of central Australia (who believe themselves to be descended from two snakes which travelled about the world and left *maiaurli*, or "the souls of children" wherever they stopped), among the Togos in Africa (a giant snake dwells in a pool near the town of Klewe, and receiving children from the hands of the supreme god Namu, brings them into the town before their birth). In India, from Buddhist times (cf. the Jātakas), snakes were held to be the givers of all fertility (water, treasures). Some of the Nagpur paintings depict the mating of women with cobras. A mass of beliefs in present day India evince the beneficent and fertilizing power of snakes: they prevent women from being sterile and ensure that they will have a large number of children.

There are a great many different woman-snake relationships, but none of them can be fully explained by any purely erotic symbolism. The snake has a variety of meanings, and I think we must hold its "regeneration" to be one of the most important. The snake is an animal that "changes." Gressman tried to see in Eve a primitive Phoenician goddess of the underworld, personified by a snake. The Mediterranean deities are represented with snakes in their hands (the Arcadian Artemis, Hecate, Persephone, and so on), or with snakes for hair (the Gorgon, Erinyes and others). And there are some central European superstitions to the effect that if, when a woman is under the moon's influence (that is, when she is menstruating), you pull out some of her hair and bury it, the hairs will turn into snakes.

One Breton legend says the hair of a witch turns into snakes. This cannot therefore happen to ordinary women, except when under the influence of the moon, when sharing in the magic power of "change." There is a great deal of ethnological evidence to show that witchcraft is a thing bestowed by the moon (either directly, or through the intermediary of snakes). To the Chinese, for instance, snakes are at the bottom of all magic power, while the Hebrew and Arabic words for magic come from words that mean "snakes." Because snakes are "lunar"—that is, eternal—and live underground, embodying (among many other things) the souls of the dead, they know all secrets, are the source of all wisdom, and can foresee the future. Anyone, therefore, who eats a snake becomes conversant with the language of animals, and particularly of birds (a symbol which can also have a metaphysical meaning: access to the transcendent reality); this is a belief held by a tremendous number of races, and it was accepted even by the learned of antiquity.

The same central symbolism of fertility and regeneration governed by the moon, and bestowed by the moon itself or by forms the same in substance (*magna mater, terra mater*) explains the presence of snakes in the imagery and rites of the Great Goddesses of universal fertility. As an attribute of the Great Goddess, the snake keeps its lunar character (of periodic regeneration) in addition to a telluric one. At one stage the moon was identified with the earth and itself considered the origin of all living forms. Some races even believe that the earth and the moon are formed of the same substance. The Great Goddesses share as much in the sacred nature of the moon as in that of the earth. And because these goddesses are also funeral goddesses (the dead disappear into the ground or into the moon to be reborn and reappear under new forms), the snake becomes very specially the animal of death and burial, embodying the souls of the dead, the ancestor of the tribe, etc. And this symbolism of regeneration also explains the presence of snakes in initiation ceremonies.

LUNAR SYMBOLISM

What emerges fairly clearly from all this varied symbolism of snakes is their lunar character—that is, their powers of fertility, of regeneration, of immortality through metamorphosis. We could, of course, look at a series of their attributes or functions and conclude that all these various relationships and significations have developed one from another by some method of logical analysis. You can reduce any religious system to nothing by methodically breaking it down to its component parts and studying them. In reality, all the meanings in a symbol are present together, even when it may look as if only some of them are effective. The intuition of the moon as the measure of rhythms, as the source of energy, of life, and of rebirth, has woven a sort of web between the various levels of the universe, producing parallels, similarities and unities among vastly differing kinds of phenomena. It is not always easy to find the centre of such a "web;" secondary centres will sometimes stand out, looking like the most important, or perhaps the oldest starting point. Thus the erotic symbolism of snakes has in its turn "woven" a system of meanings and associations which in some cases at least push its lunar connections into the background. What in fact we are faced with is a series of threads running parallel to or across each other, all fitting together, some connected directly with the "centre" on which they all depend, others developing within their own systems.

Thus the whole pattern is moon-rain-fertility-woman-serpent-death-periodic-regeneration, but we may be dealing with one of the patterns within a pattern such as Serpent-Woman-Fertility, or Serpent-Rain-Fertility, or perhaps Woman-Serpent-Magic, and so on. A lot of mythology has grown up around these secondary "centres," and if one does not realize this, it may overshadow the original pattern, though that pattern is, in fact, fully implicated in even the tiniest fragments. So, for instance, in the snake-water (or rain) binomial, the fact that both these are subject to the moon is not always obvious. Innumerable legends and myths show snakes or dragons governing the clouds, dwelling in pools and keeping the world supplied with water. The link between snakes and springs and streams has been kept to this day in the popular beliefs of Europe. In American Indian iconography, the serpent-water connection is very often found; for instance, the Mexican rain-god, Tlaloc, is represented by an emblem of two snakes twisted together; in the same Borgia Codex a snake wounded by an arrow means rainfall; the Dresden Codex shows water in a vessel shaped like a reptile; the Codex-Tro-Cortesianus shows water flowing from a vase in the shape of a snake, and there are many more examples.

Hentze's researches have quite conclusively proved that this symbolism is based on the fact that the moon supplies the rains. Sometimes the Moon-Snake-Rain pattern has even been kept in the ritual: in India, for instance, the annual ceremony of venerating the snake (*sarpabali*), as it is given in the *Gṛhyasūtras*, lasts for four months; it starts at the full moon in Śravaṇa (the first month of the rainy season) and finishes at the full moon in Mārgaśīrṣa (the first month of winter). The *sarpabali* thus includes all three elements of the original pattern. Yet it is not quite right to think of them as three separate elements: it is a *triple repetition*, a "concentration" of the moon, for the rain and the snakes are not merely things that follow the rhythms of the moon, but are in fact of the same substance. Like every sacred thing, and every symbol, these waters and snakes achieve that paradox of being at once *themselves* and *something other*—in this case the moon.

THE MOON AND DEATH

As E. Seler, the student of Americana, wrote long ago, the moon is the first of the dead. For three nights the sky is dark; but as the moon is reborn on the fourth night, so shall the dead achieve a new sort of existence. Death, as we shall see later, is not an extinction, but a change—and generally a provisional one—of one's level of existence. Death belongs to another kind of "life." And because what happens to the moon, and to the earth (for as people discovered the agricultural cycle they came to see the Earth as related to the Moon) proves that there is a "life in death" and gives the idea meaning, the dead either go to the moon or return to the underworld to be regenerated and to absorb the forces needed to start a new existence. That is why so many lunar divinities are in addition chthonian and funereal divinities (Min, Persephone, probably Hermes, and so on). And why, too, so many beliefs see the moon as the land of the dead. Sometimes the privilege of repose on the moon after death is reserved to political or religious leaders; that is what the Guaycurus believe, and the Polynesians of Tokelau and others. This is one of those aristocratic, or heroic systems, which concede immortality only to the privileged rulers or the initiated ("magicians"), which we also find elsewhere.

This journey to the moon after death was also preserved in highly developed cultures (India, Greece, Iran), but something else was added. To the Indians, it is the "path of the *manes*" (*pitṛyāna*), and souls reposed in the moon while awaiting reincarnation, whereas the sun road or "path of the gods" (*devayāna*) was taken by the initiated, or those set free from the illusions of ignorance. In Iranian tradition, the souls of the dead, having passed the Cinvat bridge, went towards the stars, and if they had been good they went to the moon and then into the sun, and the most virtuous of all entered the *garotman,* the infinite light of Ahura Mazda. The same belief was kept in Manicheeism and existed in the East as well. Pythagorism gave astral theology a further impulse by popularizing the idea of the empyrean: the Elysian Fields, where heroes and Caesars went after death, were in the moon. The "Isles of the Blessed," and all the mythical geography of death, were set in the sky utilizing the moon, the sun, the Milky Way. Here, of course, we have clearly got formulæ and cults impregnated with astronomical speculation and eschatological gnosis. But even in such late developments as that it is not hard to identify the traditional key ideas: the moon as land of the dead, the moon as receiver and regenerator of souls.

The lunar sphere was only one stage in an ascension including several others (the sun, the Milky Way, the "outer sphere"). The soul rested in the moon, but as in the Upaniṣad tradition, only to await reincarnation, and a return to the round of life. That is why the moon has the chief place in forming organisms, but also in breaking them apart: "Omnia animantium corpora et concepta procreat et generata dissolvit." Its task is to "reabsorb" forms and recreate them. Only what is beyond the moon is beyond becoming: "Supra lunam sunt aeterna omnia." To Plutarch, who believed man to be made up of three parts, body (*soma*), soul (*psyche*), and mind (*nous*), this meant that the souls of the just were purified in the moon while their bodies were given back to the earth and their minds to the sun.

To the duality of soul and mind correspond the two different itineraries after death to the moon and sun, rather like the Upaniṣad tradition of the "path of souls" and the "path of the gods." The path of souls is a lunar one because the "soul" has not the light of reason, or in other words, because man has not come to know the ultimate metaphysical reality: Brahman. Plutarch wrote that man had two deaths: the first took place on earth, in the domain of Demeter, when the body became cut off from the psyche and the *nous* and returned to dust (which is why the Athenians called the dead *demetreioi*); the second takes place in the moon, in the domain of Persephone, when the psyche separates from the *nous* and returns into the moon's substance. The soul, or psyche, remains in the moon, though holding on to dreams and memories of life for some time. The righteous are soon reabsorbed; the souls which have been ambitious, self-willed, or too fond of their own bodies are constantly drawn towards earth and it is a long time before they can be reabsorbed. The *nous* is drawn towards the sun, which receives it, to whose substance it corresponds. The process of birth is the exact reverse: the moon receives sun from the *nous,* and, coming to fruition there, it gives birth to a new soul. The earth furnishes the body. Note the symbolism of the moon rendered fertile by the sun, and its relation to the regeneration of the *nous* and psyche, the first integration of the human personality.

Cumont thinks that the dividing of the mind into *nous* and psyche comes from the East and is Semitic in nature, and he reminds us that the Jews believed in a "vegetative soul" (*nephesh*), which continued to dwell on earth for some time and a "spiritual soul" (*ruah*) which departed from the body immediately after death. He finds a confirmation of his theory in eastern theology as it became popularized under the Roman Empire, which describes the influences exercised by the layers of atmosphere, the sun and the moon on a soul coming down from the empyrean to the earth. It may be objected to this hypothesis that this duality in the soul and its destiny after death is to be found, in embryo at least, in the oldest Hellenic traditions. Plato held both the duality of the soul (*Phaedo*) and its separation later into three. With regard to the astral eschatology, the successive journeys of *nous* and psyche and its elements from moon to sun and back cannot be found in the *Timaeus,* and probably comes from some Semitic influence. But what we are specially concerned with at the moment is the conception of the moon as the dwelling-place of the souls of the dead, which we find expressed iconographically in the carvings of the Assyrians and Babylonians, the Phoenicians, the Hittites and the Anatolians, and which was later used in funeral monuments all over the Roman Empire. Everywhere in Europe the half-moon is to be found as a funeral symbol. This does not mean that it came in with the Roman and Eastern religions fashionable under the Empire; for in Gaul for instance, the moon was a local symbol in use long before any contact with the Romans. The "fashion" merely brought primitive notions up to date by formulating in new language a tradition older than history.

THE MOON AND INITIATION

Death, however, is not final—for the moon's death is not. "As the moon dieth and cometh to life again, so we also, having to die, will again rise," declare the Juan Capistrano Indians of California in ceremonies performed when the moon is new. A mass of myths describe a "message" given to men by the moon through the intermediary of an animal (hare, dog, lizard or another) in which it promises that "as I die and rise to life again, so you shall also die and rise to life again." From either ignorance or ill-will, the "messenger" conveys the exact opposite, and declares that man, unlike the moon, will never live again once he is dead. This myth is extremely common in Africa, but it is also to be found in Fiji, Australia, among the Ainus and elsewhere. It justifies the concrete fact that man dies, as well as the existence of initiation ceremonies. Even within the framework of Christian apologetics, the phases of the moon provide a good exemplar for our belief in resurrection. "Luna per omnes menses nascitur, crescit, perficitur, minuitur, consumitur, innovatur," wrote Saint Augustine. "Quod in luna per menses, hoc in resurrectione semel in toto tempore." It is therefore quite easy to understand the role of the moon in initiations, which consist precisely in undergoing a ritual death followed by a "rebirth," by which the initiate takes on his true personality as a "new man."

In Australian initiations, the "dead man" (that is, the neophyte), rises from a tomb as the moon rises from darkness. Among the Koryaks of north-eastern Siberia, the Gilyaks, Tlingits, Tongas and Haidas, a bear—a "lunar animal" because it appears and disappears with the seasons—is present in the initiation ceremonies, just as it played an essential part in the ceremonies of Paleolithic times. The Pomo Indians of Northern California have their candidates initiated by the Grizzly Bear, which "kills" them and "makes a hole" in their backs with its claws. They are undressed, then dressed in new clothes, and they then spend four days in the forest while ritual secrets are revealed to them. Even when no lunar animals appear in the rites and no direct reference is made to the disappearance and reappearance of the moon, we are driven to connect all the various initiation ceremonies with the lunar myth throughout the area of southern Asia and the Pacific, as Gahs has shown in a yet unpublished monograph.

In certain of the shaman initiation ceremonies, the candidate is "broken in pieces" just as the moon is divided into parts (innumerable myths represent the story of the moon being broken or pulverized by God, by the sun and so on). We find the same archetypal model in the osirian initiations. According to the tradition recorded by Plutarch, Osiris ruled for twenty-eight years and was killed on the seventeenth of the month, when the moon was on the wane. The coffin in which Isis had hidden him was discovered by Set when he was hunting by moonlight; Set divided Osiris' body into fourteen and scattered the pieces throughout Egypt. The ritual emblem of the dead god is in the shape of the new moon. There is clearly an analogy between death and initiation. "That is why," Plutarch tells us, "there is such a close analogy between the Greek words for dying and initiating. If mystical initiation is achieved through a ritual death, then death can be looked upon as an initiation. Plutarch calls the souls that attain to the upper part of the moon "victorious," and they wear the same crown on their heads as the initiate and the triumphant.

THE SYMBOLISM OF LUNAR "BECOMING"

"Becoming" is the lunar order of things. Whether it is taken as the playing-out of a drama (the birth, fulness and disappearance of the moon), or given the sense of a "division" or "enumeration," or intuitively seen as the "hempen rope" of which the threads of fate are woven, depends, of course, on the myth-making and theorizing powers of individual tribes, and their level of culture. But the formulæ used to express that "becoming" are heterogeneous on the surface only. The moon "divides," "spins," and "measures;" or feeds, makes fruitful, and blesses; or receives the souls of the dead, initiates and purifies—because it is living, and therefore in a perpetual state of rhythmic becoming. This rhythm always enters into lunar rituals. Sometimes the ceremonial will re-enact the phases of the moon as a whole, as does for instance the Indian *pūjā* introduced by

Tantrism. The goddess Tripurasundari must, says one Tantric text, be considered as actually being *in* the moon. One Tantric writer, Bhaskara Rājā, states definitely that the goddess' *pūja* must begin on the first day of the new moon, and last for the whole fifteen days of moonlight; this calls for sixteen Brahmans, each representing one aspect of the goddess (that is, one phase of the moon, one *tithi*). Tučci notes quite rightly that the presence of the Brahmans can only be a recent innovation, and that in the primitive *pūjā*, other figures represented the "becoming" of the moon goddess. And, indeed, in *Rudrayamālā*, a treatise of undoubted authority, we find the description of the traditional ceremony, *kumarī-pūjā*, or "the adoration of a maiden." And that *pūjā* always started at the new moon and lasted fifteen nights. But instead of the sixteen Brahmans, there must be sixteen *kumarī* to represent the sixteen *tithi* of the moon. The adoration is *vṛddhibhedana*, that is, in order of age, and the sixteen maidens must be aged from one to sixteen. Each evening the *pūjā* represents the corresponding *tithi* of the moon. Tantric ceremonial in general gives tremendous importance to woman and to female divinities; in this case the parallel between the lunar form and the feminine is complete.

That the moon "measures" and "divides" is shown by primitive classifications as well as etymologies. In India, again, the *Bṛhadārańyaka Upaniṣad* says that "Prajāpati is the year. It has sixteen parts; the nights are fifteen of them, the sixteenth is fixed. It is by night that it grows and decreases" and so on. The *Chāndogya Upaniṣad* tells us that man is made up of sixteen parts and grows at the same time as food does. Traces of the octaval system abound in India: eight *mālā*, eight *mūrti*, etc.; sixteen *kāla*, sixteen *śakti*, sixteen *mātrikā*, etc.; thirty-two sorts of *dikṣa*, etc.; sixty-four *yoginī*, sixty-four *upacāra*, etc. In Vedic and Brahman literature the number four prevails. *Vāc* (the "logos") is made up of four parts, *puruṣa* ("man," the "macanthrope") also.

The phases of the moon give rise to the most complex relationships in later speculative thought. Stuchen devoted a whole book to a study of the relations between the letters of the alphabet and the phases of the moon as conceived by the Arabs. Hommel has shown that ten or eleven Hebrew characters indicate phases of the moon (for instance, *aleph*, which means "bull," is the symbol of the moon in its first week and also the name of the sign of the zodiac where the moon's mansions begin, and so on). Among the Babylonians, too, there is a relationship between graphic signs and the phases of the moon, and among the Greeks and Scandinavians (the twenty-four runes are divided into three sorts or *aettir*, each containing eight runes). One of the clearest and most complete assimilations of the alphabet (as a collection of sounds, that is, not as written) with the phases of the moon is to be found in a scholium of Dionysius Thrax, in which the vowels correspond to the full moon, the hard consonants to the half moon (the quarters), and the soft consonants to the new moon.

COSMO-BIOLOGY AND MYSTICAL PHYSIOLOGY

These assimilations do not simply serve a function of classification. They are obtained by an attempt to integrate man and the universe fully into the same divine rhythm. Their meaning is primarily magic and redemptive; by taking to himself the powers that lie hidden in "letters" and "sounds," man places himself in various central points of cosmic energy and thus effects complete harmony between himself and all that is. "Letters" and "sounds" do the work of images, making it possible, by contemplation or by magic, to pass from one cosmic level to another. To give only one example: in India, when a man is going to make a divine image, he must first meditate, and his meditation will include, among others, the following exercise (in which the moon, mystical physiology, the written symbol and the sound value together form a pattern of consummate subtlety): "Conceiving in his own heart the moon's orb as developed from the primal sound [*prathamasvara-pariṇatam*, i.e., evolved from the letter 'A'], let him visualize therein a beautiful blue lotus, within its filaments the moon's unspotted orb, and thereon the yellow seed-syllable *Tam*. . . ."

Clearly, man's integration into the cosmos can only take place if he can bring himself into harmony with the two astral rhythms, "unifying" the sun and moon in his living body. The "unification" of the two centres of sacred and natural energy aims—in this technique of mystical physiology—at reintegrating them in the primal undifferentiated unity, as it was when not yet broken up by the act that created the universe; and this "unification" realizes a transcendence of the cosmos. In one Tantric text, an exercise in mystical physiology seeks to change "vowels and consonants into bracelets, the sun and moon into rings." The Tantric and Haṭhayoga schools developed to a very high degree these complex analogies between the sun, the moon and various "mystical" centres or arteries, divinities, blood and *semen virile*, etc. The point of these analogies is first of all to unite man with the rhythms and energies of the cosmos, and then to unify the rhythms, fuse the centres and finally effect that leap into the transcendent which is made possible when all "forms" disappear and the primal unity is re-established. A technique like this is of course the polished product of a long mystical tradition, but we find the rudimentary groundwork of it as often among primitive peoples as in the syncretist periods of the Mediterranean religions (the moon influences the left eye and the sun the right: the moon and the sun in funeral monuments as a symbol of eternity; and so on).

By its mode of being, the moon "binds" together a whole mass of realities and destinies. The rhythms of the moon weave together harmonies, symmetries, analogies and participations which make up an endless "fabric," a "net" of invisible threads, which "binds" together at once mankind, rain, vegetation, fertility, health, animals, death, regeneration, after-life, and more. That is why the moon is seen in so many traditions personified by a divinity, or acting through a lunar animal, "weaving" the cosmic veil, or the destinies of men. It was lunar goddesses who either invented the profession of weaving (like the Egyptian divinity Neith), or were famous for their ability to weave (Athene punished Arachne, for daring to rival her, by turning her into a spider), or wove a garment of cosmic proportions (like Proserpine and Harmonia), and so on. It was believed in medieval Europe that Holda was patroness of weavers, and we see beyond this figure to the chthonian and lunar nature of the divinities of fertility and death.

We are obviously dealing here with extremely complex forms in which myths, ceremonials and symbols from different religious structures are crystallized, and they have not always come directly from the intuition of the moon as the measure of cosmic rhythms and the support of life and death. On the other hand, we find in them the syntheses of the moon and Mother Earth with all that they imply (the ambivalence of good and evil, death, fertility, destiny). Similarly, you cannot always limit every mythological intuition of a cosmic "net" to the moon. In Indian thought, for instance, the universe was "woven" by the air just as breath (*prāṇa*) "wove" human life. Corresponding to the five winds that divide the Cosmos and yet preserve its unity, there are five breaths (*prāṇas*) "weaving" human life into a whole (the identity of breath and wind can be found as early as in Vedic writings). What we have got in these traditions is the primitive conception of the living whole—whether cosmic or microcosmic—in which the different parts are held together by a breathing force (wind or breath) that "weaves" them together.

THE MOON AND FATE

The moon, however, simply because she is mistress of all living things and sure guide of the dead, has "woven" all destinies. Not for nothing is she envisaged in myth as an immense spider—an image you will find used by a great many peoples. For to weave is not merely to predestine (anthropologically), and to join together differing realities (cosmologically) but also to *create*, to make something of one's own substance as the spider does in spinning its web. And the moon is the inexhaustible creator of all living forms. But, like everything woven, the lives thus created are fixed in a pattern: they have a destiny. The Moirai, who spin fates, are lunar divinities. Homer calls them "the spinners," and one of them is even called Clotho, which means "spinner." They probably began by being goddesses at birth, but the later development of thought raised them to the position of personifications of fate. Yet their lunar nature was never totally lost to view. Porphyry said the Moirai were dependent on the forces of the moon, and an Orphic text looks on them as forming part (*ta mere*) of the moon. In the old Germanic languages, one of the words for fate (Old High German *wurt*, Old Norse, *urdhr*, Anglo-Saxon *wyrd*) comes from an Indo-European verb *uert*, "turn," whence we get the Old High German words *wirt, wirtel*, "spindle," "distaff," and the Dutch *worwelen*, "turn."

Needless to say, in those cultures in which Great Goddesses have absorbed the powers of the moon, the earth and vegetation, the spindle and distaff with which they spin the fates of men become two more of their many attributes. Such is the case of the goddess with the spindle found at Troy, dating from the period between 2000 and 1500 B.C. This iconographic figure is common in the East: we find a distaff in the hand of Ishtar, of the Hittite Great Goddess, of the Syrian goddess Atargatis, of a primitive Cypriot divinity, of the goddess of Ephesus. Destiny, the thread of life, is a long or short period of *time*. The Great Goddesses consequently become mistresses of time, of the destinies they create according to their will. In Sanskrit Time is *kāla*, and the word is very close to the name of the Great Goddess, *Kālī*. (In fact, a connection has been suggested between the two words.) *Kāla* also means black, darkened, stained. Time is black because it is irrational, hard, merciless. Those who live under the dominion of time are subject to every kind of suffering, and to be set free consists primarily in the abolition of time, in an escape from the law of change. Indian tradition has it that mankind is at present in the *Kālī-yuga*, that is, "the dark era," the period of total confusion and utter spiritual decadence, the final stage in the completion of a cosmic cycle.

LUNAR METAPHYSICS

We must try to get a general picture of all these lunar hierophanies. What do they reveal? How far do they fit together and complement each other, how far do they make up a "theory"—that is, express a succession of "truths" which, taken together, could constitute a system? The hierophanies of the moon that we have noted may be grouped round the following themes: (*a*) fertility (waters, vegetation, women; mythological "ancestor"); (*b*) periodic regeneration (the symbolism of the serpent and all the lunar animals; "the new man" who has survived a watery catastrophe caused by the moon; the death and resurrection of initiations; etc.); (*c*) time and destiny (the moon "measures," or "weaves" destinies, "binds" together diverse cosmic levels and heterogeneous realities); (*d*) change, marked by the opposition of light and darkness (full moon—new moon; the "world above" and the "underworld;" brothers who are enemies, good and evil), or by the balance between being and non-being, the virtual and the actual (the symbolism of hidden things: dusky night, darkness, death, seeds and larvæ). In all these themes the dominant idea is one of *rhythm* carried out by a succession of contraries, of "becoming" through the succession of opposing modalities (being and non-being; forms and hidden essences; life and death; etc.). It is a becoming, I need hardly add, that cannot take place without drama or *pathos*; the sub-lunar world is not only the world of change but also the world of suffering and of "history." Nothing that happens in this world under the moon can be "eternal," for its law is the law of becoming, and no change is final; every change is merely part of a cyclic pattern.

The phases of the moon give us, if not the historical origin, at least the mythological and symbolic illustration of all dualisms. "The underworld, the world of darkness, is typified by the waning moon (horns = crescents, the sign of the double volute = two crescents facing the opposite way, placed one on top of the other and fastened together = lunar change, a decrepit and bony old man). The higher world, the world of life and of growing light, is typified by a tiger (the monster of darkness and of the new moon) letting humanity, represented by a child, escape its jaws (the child being the ancestor of the tribe, likened to the new moon, the 'Light that returns')." These images come from the cultural area of primitive China, but light and darkness symbols were complementary there; the owl, a symbol of darkness, is to be found beside the pheasant, symbol of light. The cicada, too, is at once related to the demon of darkness and to the demon of light. At every cosmic level a "dark" period is followed by a "light," pure regenerate period. The symbolism of emerging from the "darkness" can be found in initiation rituals as well as in the mythology of death, and the life of plants (buried seed, the "darkness" from which the "new plant" (*neophyte*) arises), and in the whole conception of "historical" cycles. The "dark age," *Kālī-yuga*, is to be followed, after a complete break-up of the cosmos, (*mahāpralaya*), by a new, regenerate era. The same idea is to be found in all the traditions that tell of cosmic historic cycles, and though it does not seem to have first entered the human mind with the discovery of the moon's phases, it is undoubtedly illustrated perfectly by their rhythm.

It is in this sense that we can talk of the positive value of periods of shadow, times of large-scale decadence and disintegration; they gain a suprahistorical significance, though in fact it is just at such times that "history" is most fully accomplished, for then the balance of things is precarious, human conditions infinitely varied, new developments are encouraged by the disintegration of the laws and of all the old framework. Such dark periods are a sort of darkness, of universal night. And as such, just as death represents a positive value in itself, so do they; it is the same symbolism as that of larvæ in the dark, of hibernation, of seeds bursting apart in the earth so that a new form can appear.

It might be said that the moon shows man his true human condition; that in a sense man looks at himself, and finds himself anew in the life of the moon. That is why the symbolism and mythology of the moon have an element of *pathos* and at the same time of consolation, for the moon governs both death and fertility, both drama and initiation. Though the modality of the moon is supremely one of change, of rhythm, it is equally one of periodic returning; and this pattern of existence is disturbing and consoling at the same time—for though the manifestations of life are so frail that they can suddenly disappear altogether, they are restored in the "eternal returning" regulated by the moon. Such is the law of the whole sublunary universe. But that law, which is at once harsh and merciful, can be abolished, and in some cases one may "transcend" this periodic becoming and achieve a mode of existence that is absolute. We saw how, in certain Tantric techniques, an attempt is made to "unify" the moon and the sun, to get beyond the opposition between things, to be reintegrated in the primeval unity. This myth of reintegration is to be found almost everywhere in the history of religion in an infinity of variations—and fundamentally it is an expression of the thirst to abolish dualisms, endless returnings and fragmentary existences. It existed at the most primitive stages; which indicates that man, from the time when he first realized his position in the universe, desired passionately and tried to achieve concretely (i.e., by religion and by magic together) a passing beyond his human status ("reflected" so exactly by the moon's). We shall be dealing with myths of this nature elsewhere, but I note them here, for they mark man's first attempt to get beyond his "lunar mode of being."

The Emergence of Mankind

JOSEPH CAMPBELL

Joseph Campbell identifies the four functions of myth in this selection. He establishes the historical and geographical factors that are involved with the rise and diffusions of specific myths and mythological systems. An understanding of the four functions of mythology enables the reader to understand how and what myths mean in a given culture. This selection explains the basic need of myths to be transformed through time since the function that myths serve, although constant in their universal implications, must change with the changing specifics in various cultures.

What, then, are the earliest evidences of the mythological thinking of mankind?

As already remarked, among the earliest evidences we can cite today of emergent manlike creatures on this earth are the relics recently unearthed in the Olduvai Gorge of East Africa by Dr. L. S. B. Leakey: distinctly humanoid jaws and skulls discovered in earth strata of about 1,800,000 years ago. That is a long, long drop into the past. And from that period on, until the rise in the Near East of the arts of grain agriculture and domestication of cattle, man was dependent absolutely for his food supply on foraging for roots and fruit and on hunting and fishing. In those earliest millenniums, furthermore, men dwelt and moved about in little groups as a minority on this earth. Today we are the great majority, and the enemies that we face are of our own species. Then, on the other hand, the great majority were the beasts, who, furthermore, were the "old-timers" on earth, fixed and certain in their ways, at home here, and many of them extremely dangerous. Only relatively rarely would one community of humans have to face and deal with another. Normally, it would be with animals that their encounters—desperate and otherwise—would occur. And as we today confront our human neighbors variously with fear, respect, revulsion, affection, or indifference, so also then—for all those millenniums of centuries—it was normally animal neighbors that were thus experienced. Moreover, as we today have understandings with our neighbors—or at least imagine that we have—so also those earliest ape-men seem to have imagined that there were certain mutual understandings which they shared with the animal world.

Our first tangible evidences of mythological thinking are from the period of Neanderthal Man, which endured from *c.* 250,000 to *c.* 50,000 B.C.; and these comprise, first, burials with food supplies, grave gear, tools, sacrificed animals, and the like; and second, a number of chapels in high-mountain caves, where cave-bear skulls, ceremonially disposed in symbolic settings, have been preserved. The burials suggest the idea, if not exactly of immortality, then at least of some kind of life to come; and the almost inaccessible high-mountain bear-skull sanctuaries surely represent a cult in honor of that great, upright, manlike, hairy personage, the bear. The bear is still revered by the hunting and fishing peoples of the far North, both in Europe and Siberia and among our North American Indian tribes; and we have reports of a number among whom the heads and skulls of the honored beasts are preserved very much as in those early Neanderthal caves.

Particularly instructive and well reported is the instance of the bear cult of the Ainu of Japan, a Caucasoid race that entered and settled Japan centuries earlier than the Mongoloid Japanese, and are confined today to the northern islands, Kokkaido and Sakhalin—the latter now, of course, in Russian hands. These curious people have the sensible idea that *this* world is more attractive than the next, and that godly beings residing in that other, consequently, are inclined to come pay us visits. They arrive in the shapes of animals, but, once they have donned their animal uniforms, are unable to remove them. They therefore cannot return home without human help. And so the Ainu do help—by killing them, removing and eating the uniforms, and ceremonially bidding the released visitors *bon voyage*.

We have a number of detailed accounts of the ceremonials, and even now one may have the good fortune to witness such an occasion. The bears are taken when still cubs and are raised as pets of the captor's family, affectionately nursed by the womenfolk and allowed to tumble about with the youngsters. When they have become older and a little too rough, however, they are kept confined in a cage, and when the little guest is about four years old, the time arrives for him to be sent home. The head of the household in which he has been living will prepare him for the occasion by advising him that although he may find the festivities a bit harsh, they are unavoidably so and kindly intended. "Little divinity," the caged little fellow will be told in a public speech, "we are about to send you home, and in case you have never experienced one of these ceremonies before, you must know that it has to be this way. We want you to go home and tell your parents how well you have been treated here on earth. And if you have enjoyed your life among us and would like to do us the honor of coming to visit again, we in turn shall do you the honor of arranging for another bear ceremony of this kind." The little fellow is quickly and skillfully dispatched. His hide is removed with head and paws attached and arranged upon a rack to look alive. A banquet is then prepared, of which the main dish is a chunky stew of his own meat, a lavish bowl of which is placed beneath his snout for his own last supper on earth; after which, with a number of farewell presents to take along, he is supposed to go happily home.

Now a leading theme, to which I would call attention here, is that of the invitation to the bear to return to earth. This implies that in the Ainu view there is no such thing as death. And we find the same thought expressed in the final instructions delivered to the departed in the Ainu rites of burial. The dead are not to come back as haunts or possessing spirits, but only by the proper natural course, as babies. Moreover, since death alone would be no punishment for an Ainu, their extreme sentence for serious crimes is death by torture.

A second essential idea is that of the bear as a divine visitor whose animal body has to be "broken" (as they say) to release him for return to his other-world home. Many edible plants, as well as hunted beasts, are believed to be visitors of this kind; so that the Ainu, killing and eating them, are doing them no harm, but actually a favor. There is here an obvious psychological defense against the guilt feelings and fears of revenge of a primitive hunting and fishing folk whose whole existence hangs upon acts of continual merciless killing. The murdered beasts and consumed plants are thought of as willing victims; so that gratitude, not malice, must be the response of their liberated spirits to the "breaking and eating" of their merely provisional material bodies.

There is a legend of the Ainu of Kushiro (on the southeastern coast of Hokkaido) which purports to explain the high reverence in which the bear is held. It tells of a young wife who used to go every day with her baby to the mountains to search for lily roots and other edibles; and when she had gathered her fill, she would go to a stream to wash her roots, removing the baby from her back and leaving it wrapped in her clothes on the bank, while she went naked into the water. One day thus in the stream she began to sing a beautiful song, and when she had waded to shore, still singing, commenced dancing to its tune, altogether enchanted by her own dance and song and unaware of her surroundings, until, suddenly, she heard a frightening sound, and when she looked, there was the bear-god coming. Terrified, she ran off, just as she was. And when the bear-god saw the abandoned child by the stream, he thought: I came, attracted by that beautiful song, stepping quietly, not to be heard. But alas! Her music was so beautiful it moved me to rapture and inadvertently I made a noise.

The infant having begun to cry, the bear-god put his tongue into its mouth to nourish and to quiet it, and for a number of days, tenderly nursing it this way, never leaving its side, contrived to keep it alive. When, however, a band of hunters from the village approached, the bear took off, and the villagers, coming upon the abandoned child alive, understood that the bear had cared for it, and, marveling, said to one another, "He took care of this lost baby. The bear is good. He is a worthy deity, and surely deserving of our worship." So they pursued and shot him, brought him back to their village, held a bear festival, and, offering good food and wine to his soul, as well as loading it with fetishes, sent him homeward on his way in wealth and joy.

Since the bear, the leading figure of the Ainu pantheon, is regarded as a mountain god, a number of scholars have suggested that a like belief may account for the selection of lofty mountain caves to be the chapels of the old Neanderthal bear cult. The Ainu too preserve the skulls of the bears they sacrifice. Moreover, signs of fire hearths have been noted in the high Neanderthal chapels; and in the ocurse of the Ainu rite the fire-goddess Fuji is invited to share with the sacrificed bear the banquet of his meat. The two, the fire-goddess and the mountain god, are supposed to be chatting together while their Ainu hosts and hostesses entertain them with song the night long, and with food and drink. We cannot be certain, of course, that the old Neanderthalers of some two hundred thousand years ago had any such ideas. A number of authoritative scholars seriously question the propriety of interpreting prehistoric remains by reference to the customs of modern primitive peoples. And yet, in the present instance the analogies are truly striking. It has even been remarked that in both contexts the number of neck vertebrae remaining attached to the severed skulls is generally two. But in any case, we can surely say without serious doubt that the bear is in both contexts a venerated beast, that his powers survive death and are effective in the preserved skull, that rituals serve to link those powers to the aims of the human community, and that the power of fire is in some manner associated with the rites.

The earliest known evidences of the cultivation of fire go back to a period as remote from that of Neanderthal Man as is his dim day from our own, namely, that of Pithecanthropus, some five hundred thousand years ago, in the dens of the ravenous low-browed cannibal known as Peking Man, who was particularly fond, apparently, of brains *à la nature*, gobbled raw from freshly opened skulls. His fires were not used for cooking. Neither were those of the Neanderthalers. For what, then? To furnish heat? Possibly! But possibly, also, as a fascinating fetish, kept alive in its hearth as on an altar. And this conjecture is the more likely in the light of the later appearance of domesticated fire, not only in the high Neanderthal bear sanctuaries but also in the context of the Ainu bear festivals, where it is identified explicitly with the manifestation of a goddess. Fire, then, may well have been the first enshrined divinity of prehistoric man. Fire has the property of not being diminished when halved, but increased. Fire is luminous, like the sun and lightning, the only such thing on earth. Also, it is alive: in the warmth of the human body it is life itself, which departs when the body gets cold. It is prodigious in volcanoes, and, as we know from the lore of many primitive traditions, it has been frequently identified with a demoness of volcanoes, who presides over an afterworld where the dead enjoy an everlasting dance in marvelously dancing volcanic flames.

The rugged race and life style of Neanderthal Man passed away and even out of memory with the termination of the Ice Ages, some forty thousand years ago; and there appeared then, rather abruptly, a distinctly superior race of man, Homo sapiens proper, which is directly ancestral to ourselves. It is with these men—significantly—that the beautiful cave paintings are associated of the French Pyrenees, French Dordogne, and Spanish Cantabrian hills; also, those little female figurines of stone, or of mammoth bone or ivory, that have been dubbed—amusingly—paleolithic Venuses and are, apparently, the earliest works ever produced of human art. A worshiped cave-bear skull is not an art object, nor is a burial, or a flaked tool, in the sense that I am here using the term. The figurines were fashioned without feet, because they were intended to be pressed into the earth, set up in little household shrines.

And it seems to me important to remark that, whereas when masculine figures appear in the wall paintings of the same period they are always clothed in some sort of costume, these female figurines are absolutely naked, simply standing, unadorned. This says something about the psychological and consequently mythical values of, respectively, the male and the female presences. The woman is immediately mythic in herself and is experienced as such, not only as the source and giver of life, but also in the magic of her touch and presence. The accord of her seasons with the cycles of the moon is a matter of mystery too.

Whereas the male, costumed, is one who has *gained* his powers and represents some specific, limited, social role or function. In infancy—as both Freud and Jung have pointed out—the mother is experienced as a power of nature and the father as the authority of society. The mother has brought forth the child, provides it with nourishment, and in the infant's imagination may appear also (like the witch of Hansel and Gretel) as a consuming mother, threatening to swallow her product back. The father is, then, the initiator, not only inducting the boy into his social role, but also, as representing to his daughter her first and foremost experience of the character of the male, awakening her to her social role as female to male. The paleolithic Venuses have been found in the precincts always of domestic hearths, while the figures of the costumed males, on the other hand, appear in the deep, dark interiors of the painted temple-caves, among the wonderfully pictured animal herds. They resemble in their dress and attitudes, furthermore, the shamans of our later primitive tribes, and were undoubtedly associated with rituals of the hunt and of initiation.

Let me here review a legend of the North American Blackfoot tribe that I have already recounted in *The Masks of God,* Volume 1, *Primitive Mythology*; for it suggests better than any other legend I know the manner in which the artist-hunters of the paleolithic age must have interpreted the rituals of their mysteriously painted temple-caves. This Blackfoot legend is of a season when the Indians found themselves, on the approach of winter, unable to lay up a supply of buffalo meat, since the animals were refusing to be stampeded over the buffalo fall. When driven toward the precipice, they would swerve at the edge to right or left and gallop away.

And so it was that, early one morning, when a young woman of the hungering village encamped at the foot of the great cliff went to fetch water for her family's tent and, looking up, spied a herd grazing on the plain above, at the edge of the precipice, she cried out that if they would only jump into the corral, she would marry one of them. Whereupon, lo! the animals began coming over, tumbling and falling to their deaths. She was, of course, amazed and thrilled, but then, when a big bull with a single bound cleared the walls of the corral and came trotting in her direction, she was terrified. "Come along!" he said. "Oh no!" she answered, drawing back. But insisting on her promise, he led her up the cliff, onto the prairie, and away.

That bull had been the moving spirit of the herd, a figure rather of mythic than of material dimension. And we find his counterparts everywhere in the legends of primitive hunters: semi-human, semi-animal, shamanistic characters (like the serpent of Eden), difficult to picture either as animal or as man; yet in the narratives we accept their parts with ease.

When the happy people of the village had finished slaughtering their windfall, they realized that the young woman had disappeared. Her father, discovering her tracks and noticing beside them those of the buffalo, turned back for his bow and quiver, and then followed the trail on up the cliff and out onto the plain. It was a considerable way that he had walked before he came to a buffalo wallow and, a little way off, spied a herd. Being tired, he sat down and, while considering what to do, saw a magpie flying, which descended to the wallow close by and began picking about.

"Ha!" cried the man. "You handsome bird! As you fly around, should you see my daughter, would you tell her, please, that her father is here, waiting for her at the wallow?"

The beautiful black and white bird with long graceful tail winged away directly to the herd and, seeing a young woman there, fluttered to earth nearby and resumed his picking, turning his head this way and that, until, coming very close to her, he whispered, "Your father is waiting for you at the wallow."

She was frightened and glanced about. The bull, her husband, close by, was asleep. "Sh-h-h! Go back," she whispered, "and tell my father to wait."

The bird returned with her message to the wallow, and the big bull presently woke.

"Go get me some water," the big bull said, and the young woman, rising, plucked a horn from her husband's head and proceeded to the wallow, where her father roughly seized her arm. "No, no!" she warned. "They will follow and kill us both. We must wait until he returns to sleep, when I'll come and we'll slip away."

She filled the horn and walked back with it to her husband, who drank but one swallow and sniffed. "There is a person close by," said he. He sipped and sniffed again; then stood up and bellowed. What a fearful sound!

Up stood all the bulls. They raised their short tails and shook them, tossed their great heads, and bellowed back; then pawed the dirt, rushed about in all directions, and finally, heading for the wallow, trampled to death that poor Indian who had come to seek his daughter: hooked him with their horns and again trampled him with their hoofs, until not even the smallest particle of his body remained to be seen. The daughter was screaming, "Oh, my father, my father!" And her face was streaming with tears.

"Aha!" said the bull harshly. "So you're mourning for your father! And so now, perhaps, you will understand how it is and has always been with us. We have seen our mothers, fathers, all our relatives, killed and butchered by your people. But I shall have pity on you and give you just one chance. If you can bring your father back to life, you and he can return to your people."

The unhappy girl, turning to the magpie, begged him to search the trampled mud for some little portion of her father's body; which he did, again pecking about in the wallow until his long beak came up with a joint of the man's backbone. The young woman placed this on the ground carefully and, covering it with her robe, sang a certain song. Not long, and it could be seen that there was a man beneath the robe. She lifted a corner. It was her father, not yet alive. She let the corner down, resumed her song, and when she next took the robe away he was breathing. Her father stood up, and the magpie, delighted, flew round and round with a marvelous clatter. The buffalo were astounded.

"We have seen strange things today," said the big bull to the others of his herd. "The man we trampled to death is again alive. The people's power is strong."

He turned to the young woman. "Now, before you and your father go, we shall teach you our own dance and song, which you are never to forget." For these were to be the magical means by which the buffalo killed by the people in the future would be restored to life, as the man killed by the buffalo had been restored.

All the buffalo danced; and, as befitted the dance of such great beasts, the song was slow and solemn, the step ponderous and deliberate. And when the dance was ended, the big bull said, "Now go to your home and do not forget what you have seen. Teach this dance and song to your people. The sacred object of the rite is to be a bull's head and buffalo robe: all who dance the bulls are to wear a bull's head and buffalo robe when they perform."

It is amazing how many of the painted figures of the great paleolithic caves take on new life when viewed in the light of such tales of the recent hunting races. One cannot be certain, of course, that the references suggested are altogether correct. However, that the main ideas were much the same is almost certainly true. And among these we may number that of the animals killed as being willing victims, that of the ceremonies of their invocation as representing a mystic covenant between the animal world and the human, and that of song and dance as being the vehicles of the magical force of such ceremonies; further, the concept of each species of the animal world as a kind of multiplied individual, having as its seed or essential monad a semi-human, semi-animal, magically potent Master Animal; and the idea related to this, of there being no such thing as death, material bodies being merely costumes put on by otherwise invisible monadic entities, which can pass back and forth from an invisible other world into this, as though through an intangible wall; the notions, also, of marriages between human beings and beasts, of commerce and conversations between beasts and men in ancient times, and of specific covenanting episodes in those times from which the rites and customs of the peoples were derived; the notion of the magical power of such rites, and the idea that, to retain their power, they must be held true to their first and founding form—even the slightest deviation destroying their spell.

So much, then, for the mythic world of the primitive hunters. Dwelling mainly on great grazing lands, where the spectacle of nature is of a broadly spreading earth covered over by an azure dome touching down on distant horizons and the dominant image of life is of animal societies moving about in that spacious room, those nomadic tribes, living by killing, have been generally of a warlike character. Supported and protected by the hunting skills and battle courage of their males, they are dominated necessarily by a masculine psychology, male-oriented mythology, and appreciation of individual valor.

In tropical jungles, on the other hand, an altogether different order of nature prevails, and, accordingly, of psychology and mythology as well. For the dominant spectacle there is of teeming vegetable life with all else more hidden than seen. Above is a leafy upper world inhabited by winged screeching birds; below, a heavy cover of leaves, beneath which serpents, scorpions, and many other mortal dangers lurk. There is no distant clean horizon, but an ever-continuing tangle of trunks and leafage in all directions wherein solitary adventure is perilous. The village compound is relatively stable, earthbound, nourished on plant food gathered or cultivated mainly by the women; and the male psyche is consequently in bad case. For even the primary *psychological* task for the young male of achieving separation from dependency on the mother is hardly possible in a world where all the essential work is being attended to, on every hand, by completely efficient females.

It is therefore among tropical tribes that the wonderful institution originated of the men's secret society, where no women are allowed, and where curious symbolic games flattering the masculine zeal for achievement can be enjoyed in security, safe away from Mother's governing eye. In those zones, furthermore, the common sight of rotting vegetation giving rise to new green shoots seems to have inspired a mythology of death as the giver of life; whence the hideous idea followed that the way to increase life is to increase death. The result has been, for millenniums, a general rage of sacrifice through the whole tropical belt of our planet, quite in contrast to the comparatively childish ceremonies of animal-worship and -appeasement of the hunters of the great plains: brutal human as well as animal sacrifices, highly symbolic in detail; sacrifices also of fruits of the field, of the firstborn, of widows on their husbands' graves, and finally of entire courts together with their kings. The mythic theme of the Willing Victim has become associated here with the image of a primordial being that in the beginning offered itself to be slain, dismembered, and buried; and from whose buried parts then arose the food plants by which the lives of the people are sustained.

In the Polynesian Cook Islands there is an amusing local variant of this general myth in the legend of a maiden named Hina (Moon) who enjoyed bathing in a certain pool. A great big eel, one day, swam past and touched her. This occurred again, day after day, until, on one occasion, it threw off its eel costume and a beautiful youth, Te Tuna (the Eel), stood before her, whom she accepted as her lover. Thereafter he would visit her in human form, but become an eel when he swam away, until one day he announced that the time had come for him to leave forever. He would pay her one more visit, arriving in his eel form in a great flood of water, when she should cut off his head and bury it. And so indeed he came. And Hina did exactly as she was told. And every day thereafter she visited the place of the buried head, until a green sprout appeared that grew into a beautiful tree, which in the course of time produced fruits. Those were the first coconuts; and every nut, when husked, still shows the eyes and face of Hina's lover.

Art As A Revelation

JOSEPH CAMPBELL

Joseph Campbell, in this selection, identifies the changes in the mythologic thinking of different cultures and explains the particular associations that have been made with such symbols as the bear, fire, and women. By explaining the influence of the environment on the mythology of a particular culture, Campbell helps the reader to understand how particularities of myth and mythical perspectives are necessarily transformed.

PALEOLITHIC ROCK PAINTINGS

That an abrupt expansion of human consciousness occurred toward the close of the last glacial age, some 30,000 years ago, is evident in the appearance at that time in southwestern France and northern Spain of the earliest known works of visual art—both the rock paintings of the vast temple caves of the men's hunting ceremonials, and the numerous nude female figurines that have been found associated chiefly with dwelling sites. Evidence of the skulls suggests that an evolutionary advance from the mentality of Neanderthal to that of Cro-Magnon Man ("archaic" to "modern" *Homo sapiens*) is what unleashed this *creative explosion*—to borrow a term from a recent study of the period by John E. Pfeiffer.

According to Pfeiffer's interpretation of the phenomenon: There had been at that time an increase in population density with an associated social problem of conflict control, to which (in Pfeiffer's words) impressive institutions and occasions had to be addressed incorporating symbolic figurations, through which the regulations of a corpus of socially constructive rituals were pictorially encoded for storage and transmission through generations—which may be all very well as far, at least, as the sociological aspect of the remarkable occasion is concerned. The question remains, however, as to the originating source and sense of those affective symbolic figurations, which, when thus socially encoded by way of a "creative explosion" of literally thousands of magnificent masterworks of an unprecedented form of human expression, remained for some 20,000 years in force (*c.* 30,000 to 10,000 B.C.) as the founding mythological revelation for unnumbered generations of hunting tribes, whose lives and way of life were utterly dependent, not only on the physical support, but also on the spiritual instruction, of the animals thus displayed and celebrated in their subterranean chambers of initiation and invocation.

André Leroi-Gourhan, the present leading authority on the cavern art of Paleolithic Europe, has found through a comparative study and computerized count of the animals depicted in the caves, that the number of species represented is much lower than the number of species known to have existed at the time. "Paleolithic artists," he writes, "did not portray just any animal, but animals of certain species, and these did not necessarily play a part in their daily life. . . . The main actors," he discovered, "are the horse and the bison, the animals next in importance being the hinds, the mammoths, the oxen, the ibexes, and the stags. . . . Bears, lions, and rhinoceroces play an important part, but as a rule there is only one representation of each per cave, and they are by no means represented in every cave."

Such, then, were the animals selected by the Paleolithic master artists from the bounty of their environment for depiction in the galleries of their subterranean corridors and chambers, as being in some way significant of a mystic dimension of their landscape perceived by the eye of the mind, not the eyes of the physical look of things. One may compare their number to that of the masquerade of metaphorical birds and beasts selected by the Kwakiutl of the North American North Pacific Coast from the bounty of their environment, as shown in the photograph by Edward Curtis here reproduced, to be evoked to life in seasonal dances.

As Leroi-Gourhan discovered further in this comparative study of the painted caves, certain animals turned up next to each other "too often for such associations to be explained only as chance": oxen or bison next to horses, for example, or bisons next to mammoths. "The fundamental principle," he concluded, "is that of pairing. . . . Starting with the earliest figures, one has the impression of being faced with a system polished in the course of time—not unlike the older religions of our world, wherein there are male and female divinities whose actions do not overtly allude to sexual reproduction, but whose male and female qualities are indispensably complementary."

Moreover, as he also found, there is an architectural order in the distribution of the symbolic beasts and signs throughout the geographical as well as historical range of this earliest known, recorded testament to a system of ideas. "What constituted for Paleolithic men the special heart and core of the caves," he discovered, "is clearly the panels in the central part, dominated by animals from the male category and male signs. The entrance to the sanctuary, usually a narrow part of the cave, is decorated with male symbols, either animals or signs; the back of the cave, often a narrow tunnel, is decorated with the same signs, reinforced by horned men and the rarer animals (cave lion or rhinoceros). Although crowded with images this framework is quite simple; yet it leaves us completely in the dark concerning what we should like to know about the rites, and let us say, about an underlying metaphysics. However, it rules out any simplistic idea concerning the religious system of Paleolithic men."

This important statement from an unimpeachable authority confirms and validates the impression which immediately overtakes even the ill prepared visitor to such a sanctuary of Stone Age faith and hope as, say, the magnificently conceived grotto of Lascaux: which is, namely, of an ordered system of metaphorical reflections preserved from an age beyond our horizon of time in the pictorial script of this truly amazing Stone Age testament. It is certainly not a mere mindless arrangement of accurately observed animal forms, expertly delineated by a school of accomplished artists striving for decorative optical effects. The sense of an intelligible metaphorical statement is incontestable.

Comparably, on entering such a relatively modern sacred space as, for example, the cathedral of Notre Dame de Paris, or the temple in Nara, Japan, of the Great Sun Buddha, Vairochana, the visitor, the stranger from afar, will immediately know that he has entered an area where everything is not only metaphorical, but intelligibly related to everything else. And the reader of the present volume, likewise, on viewing the photograph by Edward Curtis, will surely have realized that the array of bizarre physical shapes there brought together and arranged for the camera's eye adumbrates a mythological order visible only to the eye of the informed mind, signifying a way of understanding and relating to the universe.

The universe viewed by the physical eye alone is but an array of shapes, some alive, some not, all changing either rapidly or slowly, which though linked along chains of cause and effect, are seen as distinct and separate from each other. In this phenomenal field, an Aristotelian logic prevails: *a* is not *not-a.* Moreover, except for the miracle of fire (and today, the electric bulb), the observed forms are not self-luminous. They are visible only by daylight to the daylight mind, when awake and outwardly cognitive, attentive to objects apprehended as apart from, and other than itself. As quoted by Ananda K. Coomaraswamy from a sermon by the medieval Rhineland mystic and theologian Meister Eckhart: "Subtract the mind, and the eye is open to no purpose."

By night, however, when the sun has set, the mind turns inward and, together with its universe, which is now a reflex of itself, "doth change Into something rich and strange." The forms now beheld are self-luminous and in definition ambiguous, unsubstantial yet insuppressibly affective. For *not-a* is now indistinguishable from *a.* The beholding mind and the objects beheld are of the same, non-dual, dreaming consciousness and of an intimately suggestive yet elusive import of some kind.

As defined, classified and described in the Indian Upanishads (first millennium B.C.), these contrasting states of consciousness in waking and in dream are known, respectively, as *Vaiśvānara,* meaning "Common to all Men," and *Taijasa,* "Originating from and consisting of *tejas* ('light')." They are identified, furthermore, with the first and second elements of the sacred syllable AUM or OM (*a* and *u* being in Sanskrit pronounced together as *o*), the concluding element, M, being then identified with the state of deep dreamless sleep.

As described in the *Māṇḍūkya Upanishad*:

1. "OM! This perishable sound is all. As further explained: it is the Past, the Present, the Future, all that has become, is becoming, and is yet to be. Moreover, whatever transcends this three-fold Time: that also is OM.
2. "For verily, this all is *brahman* (the Imperishable). One's Self, too, is *brahman.* The Self (*ātman*) is of four conditions:
3. "The Waking State (*jāgarita-sthāna*), outwardly cognitive, experiencing the gross enjoyments common to all men (*sthūlabhug-vaiśvānara*) by way of the body, its senses, organs of action, vital energies, and aspects of mind: this is the first condition of the Self.
4. "The Dreaming State (*svapna-sthāna*), inwardly cognitive, experiencing in exquisite solitude luminous enjoyments (*praviviktabhuk-taijasa*) by way of the body, its senses, organs of action, vital energies, and aspects of mind. This is the second condition.

5. "Where one asleep has no desires whatsoever, sees no dream whatsoever: that is deep sleep. The Deep Sleep State (*sushupta-sthāna*) is at one with itself (*ekīb-hūta*), an involuted cognitional mass (*prajñānaghana*), alone (*eva*), ensheathed in bliss (*ānandamaya*), enjoying bliss (*ānandabhuj*), the portal to knowledge [of the other two states]. Of the Knower (*prājña*) this is the third condition.
6. "This, verily, is the all-knowing Lord of all, Indweller and Controller; the Matrix (*yoni*) of all; the Beginning and End (*prabhavāpyayau*) of all beings.
7. "What is thought of as the fourth condition of the Self—cognitive neither inward, outward, nor of the intermediate state; not a uniform cognitional mass; neither knowing nor unknowing (*naprajñam-nāprajñam*); unseen (*adṛshṭa*); detached (*avyavahārya*); incomprehensible (*agrāhya*); without characteristics (*alakshaṇa*); beyond thought (*acintya*); indescribable (*avyapadeśya*); the essence of the assurance of a state of identity with the Self (*ekātmya-pratyaya-sāra*); the cessation of manifestation (*prapañcopaśama*); at peace (*śānta*); felicitous (*śiva*) and non-dual (*advaita*)—that is the Self (*ātman*) to be realized.
8. "This then is the Self (*ātman*) in the sense of the elements of the imperishable syllable. The elements are the four conditions and the four conditions are the elements: namely, A, U, M [and the Fourth, the Silence before, after, and supporting the sounding of the syllable OM].
9. "The Waking State, the Common to all Men, is the A-sound, the first element, for its comprehensiveness and priority. He who knows this obtains verily all his desires and becomes preeminent.
10. "The Dreaming State, Self-Luminous, is the U sound, the second element, for its eminence and intermediacy. He who knows this becomes verily a fountain of knowledge equal to any. In his family no one is born ignorant of *brahman*.
11. "The Deep Sleep State, of Intuitive Wisdom (*prājña*), is the M sound, the third element, as being the measure (*miti*), the termination or quenching (*apīti*) wherein all become one. He who knows this measures and comprehends all in himself.
12. "Transcendent, unsounded, beyond action (*avyavahārya*), is the Fourth, of no phenomenal existence, supremely blissful and without a second. Thus OM is indeed *ātman*, the Self. He who knows this enters through the Self into the Self—yea, he who know this."

Sigmund Freud, treating of the dreaming state in the early masterwork by which his reputation was established, *Die Traumdeutung* (*The Interpretation of Dreams*, published 1899 but dated 1900), distinguishes two sources of the impulses contributing to the instigation of dreams:

(1) wish-impulses originating in waking consciousness (*Cs.*) but carried into the dream state by way of what he termed the "Preconscious" (*Pcs.*), and then

(2) wish-impulses arising from the Unconscious (*Ucs.*) system.

"In general, I am of the opinion," he states, "that unfulfilled wishes of the day are insufficient to produce a dream in adults. I will readily admit that the wish-impulses originating in consciousness contribute to the instigation of dreams, but they probably do no more. The dream would not occur if the preconscious wish were not reinforced from another source.

"That source is the unconscious. I believe that *the conscious wish becomes effective in exciting a dream only when it succeeds in arousing a similar unconscious wish which reinforces it.* . . . I believe that these unconscious wishes are always active and ready to express themselves whenever they find an opportunity of allying themselves with an impulse from consciousness, and transferring their own greater intensity to the lesser intensity of the latter. It must, therefore, seem that the conscious wish alone has been realized in the dream; but a slight peculiarity in the form of the dream will put us on the track of the powerful ally from the unconscious. . . .

"The thought-impulses continued into sleep may be divided into the following groups:

1. Those which have not been completed during the day owing to some accidental cause.
2. Those which have been left uncompleted because our mental powers have failed us, i.e., unsolved problems.
3. Those which have been turned back and suppressed during the day. This is reinforced by a powerful fourth group:
4. Those which have been excited in our *Ucs.* during the day by the workings of the *Pcs.*; and finally we may add a fifth, consisting of:
5. The indifferent impressions of the day, which have therefore been left unsettled.

"We need not underrate the psychic intensities introduced into sleep by these residues of the day's waking life, especially those emanating from the group of the unsolved issues. . . . I cannot say what change is produced in the *Pcs.* system by the state of sleep, but there is no doubt that the psychological characteristics of sleep are to be sought mainly in the cathectic changes occuring in just this system. . . . On the other hand, I have found nothing in the psychology of dreams to warrant the assumption that sleep produces any but secondary changes in the conditions of the *Ucs.* system. Hence, for the nocturnal excitations in the *Pcs.* there remains no other path than that taken by the wish-excitations from the *Ucs.*; they must seek reinforcement from the *Ucs.*, and follow the detours of the unconscious excitations.

"But what is the relation of the preconscious day-residues to the dream? There is no doubt that they penetrate abundantly into the dream; that they utilize the dream-content to obtrude themselves upon consciousness even during the night; indeed, they sometimes even dominate the dream-content, and compel it to continue the work of the day; it is also certain that the day-residues may just as well have any other character as that of wishes."

And so, returning to India with this 19th-century European insight into the nature of what Freud called the "dream work" in mind, we find that the Vedantic theologian Sankara (*c.* A.D. 700–750), treating, in his commentary on *Māṇḍūkya Upanishad* verse 4, of the intermediate Dream State between Waking and Deep Dreamless Sleep, was already fully aware of the interplay there of (1) the preconscious day-residues, and (2) what in India is recognized as the "inner light" (*taijasa*) of the Self.

"Waking consciousness," we read in Sankara's discussion, "being associated with many manners of relationship and seemingly conscious of objects as external (although in reality they are nothing but states of mind), leaves in the mind corresponding impressions. That the mind in dream, maintaining none of the external connections, yet possessed of these impressions left upon it by the waking consciousness, like a piece of canvas with pictures painted upon it, should experience the dream state as though it were a waking state, is due to its being under the influence of ignorance (*avidyā*), desire (*kāma*), and impulse to action (*karma*). Thus it is said in the *Bṛihadāranyaka Upanishad* (*c.* ninth century B.C.):

"'When, on falling asleep, one takes along the material (*mātrā*) of this all-containing world, tears it apart and builds it up, oneself, and by one's own light dreams: that person becomes self-illuminated.

"'There are no chariots there, no spans, no roads. Yet he projects from himself chariots, spans, and roads. There are no blisses there, no pleasures, no delights. Yet he projects from himself, blisses, pleasures, and delights. There are no tanks there, no lotus-pools, no streams. Yet he projects from himself tanks, lotus-pools, and streams. For he is a creator.

"'To this point there are the following verses:

"'Striking down in sleep what is bodily,
Sleepless, he looks down upon the sleeping senses.
Having taken to himself light, he has returned to his own place:
That Golden Person (*hiraṇmaya purusha*), the Lone Wild Gander
(*ekahaṁsa*).

Guarding with his breath his low nest,
Out of that next the Immortal goes forth.
He goes wherever he pleases, that Immortal,
The Golden Person, the Unique Wild Gander.

In the State of Sleep, soaring high and low,
A god, he puts forth for himself innumerable forms:
Now, as it were, enjoying pleasure with women;
Now, as it were, laughing, or beholding even terrifying sights.

People see his pleasure ground;
Him no one sees at all.'"

Just what overpowering transpersonal experiences and associated thoughts informed the minds and moved the practiced hands of the artists of the Late Paleolithic, we shall of course never directly know, since no word of whatever speech may have been theirs has come down to us. From the Bushmen of the South African Kalahari Desert, however, whose rock art has been practiced (apparently without interruption) since the close of the last glacial age, verbal accounts have been recorded, not only of the mythology, but also of the out-of-body experiences in trance state, from which their spirited representations have been taken of an intelligible sphere, known to the eye of the mind, unseen by the light of day.

Certainly one of the most vivid accounts of such a visionary adventure yet recorded from any quarter of the globe is that from the report of an experienced Kung trance dancer, as delivered to Marguerite Anne Biesele scarcely more than a decade ago. Bearing in mind the image of the Self as the Lone Wild Gander in flight, let us here recall the circumstances of that remarkable out-of-body adventure, which is such, apparently, as occurs nearly every night among the Bushmen of the Kalahari.

A little cluster of women, who have lighted a large fire and are seated on the ground, has begun intoning a wordless chant while steadily clapping time. Eventually, a few of the men stray in behind them, to circle with short heavy stamps in a single line, which turns about, from time to time, to circle the other way. Their rhythms are complex, built into 5- and 7-beat phrases, and their body postures tight, hunched forward, arms close to the sides, slightly flexed. Others join the round, and as the night runs on, those approaching trance begin to concentrate intently. A climax strikes when one or another of them breaks and passes into the state known as half-death, while his spirit flies and climbs along threads of spider silk to the sky.

"When people sing," declared Marguerite Biesele's trance dancer, "I dance. I enter the earth. I go in at a place like a place where people drink water. I travel a long way, very far. When I emerge, I am already climbing. I'm climbing threads, the threads that lie over there in the south. I climb one and leave it, then I climb another one. Then I leave it and climb another. . . . And when you arrive at God's place you make yourself small. You have become small. You come in small to God's place. You do what you have to do there. Then you return to where everyone is, and you hide your face. You hide your face so you won't see anything. You come and come and come and finally you enter your body again. All the people who have stayed behind are waiting for you—they fear you. You enter, enter the earth, and you return to enter the skin of your body . . . And you say 'he-e-e-e!' That is the sound of your return to your body. Then you begin to sing. The *ntum*-masters are there around. They take powder and blow it—Phew! Phew!—in your face. They take hold of your head and blow about the sides of your face. This is how you manage to be alive again. Friends, if they don't do that to you, you die. . . . You just die and are dead. Friends, this is what it does, this *ntum* that I do, this *ntum* here that I dance."

The term *ntum* has been explained as a supernatural power that in the Mythological Age of the Beginning was put by the creator god into a number of things: medicine songs, ostrich eggs, certain plants and fruits, the sun, falling stars, rain, bees, honey, giraffes, aardvarks, blood, redwing partridges, and fires made in certain situations; also into certain persons, who may function as medicine men and healers.

Ntum varies in force in the various things it informs: beneficent in some, always strong, but in some things dangerously so. *Ntum* is so strong in the great god Gauwa that should he approach an ordinary mortal, like a lightning bolt his *ntum* would kill the man. The Kung call *ntum* a "death thing"; also, a "fight."

As discussed at length in my chapter on the Bushman trance dance and its mythic ground, the supreme occasion for the activation of *ntum* is the trance dance. The exertion of the tirelessly circling dancers heats their medicine power, which, as reported in the writings of Lorna Marshall and Richard B. Lee, they experience as a physical substance in the pit of the stomach. The women's singing, the men say, "awakens their hearts," and eventually their portion of *ntum* becomes so hot that it boils. "The men say it boils up their spinal columns into their heads, and is so strong when it does this," Lorna Marshall reports, "that it overcomes them and they lose their senses."

One of the very greatest Indian Godmen of the last century, Sri Ramakrishna (1836–1886), once described to a circle of his devotees the physical sensation of the unfolding ascent up his spine of the charge of psychic energy known in India as the *Kuṇḍalinī*, the "Coiled Serpent Power." In Ramakrishna's words:

"Sometimes the Spiritual Current rises through the spine, crawling like an ant. Sometimes, in *samādhi* [ecstasy in full trance], the soul swims joyfully in the ocean of divine ecstasy, like a fish. Sometimes, when I lie down on my side, I feel the Spiritual Current pushing me like a monkey and playing with me joyfully. I remain still. The Current, like a monkey, suddenly with one jump reaches the *Sahasrāra* [the supreme spiritual center, envisioned as a thousand-petalled lotus at the crown of the head]. That is why you see me jump with a start. Sometimes, again, the Spiritual Current rises like a bird hopping from one branch to another. The place where it rests feels like fire. . . . Sometimes the Spiritual Current moves up like a snake. Going in a zigzag way, at last it reaches the head and I go into samadhi. A man's spiritual consciousness is not awakened until his Kundalini is aroused."

In the Indian Vedantic manner of speech, such a consummate holy man as was Ramakrishna is known as a *Paramahaṁsa*, "Paramount of Supreme Wild Gander," in recognition of his having realized in life the ultimate Vedantic goal of complete self-identification with the Self; which is to say, with that metaphorical "Lone Wild Gander" whose invisible flight throughout the world of his creation has been described in the above-quoted lines from the *Bṛihadāranyaka Upanishad.*

It is relevant to recall at this point, that in Bushman rock paintings the released spirits of the trance dancers, as well as of the dead, are represented as flying antelope men, called in the literature, *alites*, or "flying bucks." Also to be noticed again is the shamanic figure pictured in the most sacred recess of the cavern of Lascaux, lying in trance before the vision of an eviscerated bull, wearing a bird mask, and with the figure of a bird on the top of what appears to be a shamanic staff at his side.

The image of a bird in flight is a well-nigh universal sign of the Holy Spirit, whether as in the Christian figure of the Holy Ghost descending in the form of a dove, both at the moment of Mary's conception of the Word, and again, at the time of the baptism of her Son, when the heavens opened "and he saw the Spirit of God descending like a dove and alighting on him" (Matthew 3:16); or as in the metaphorical Indian figure of the Lone Wild Gander (*ekahaṁsa*), as signifying the non-dual ground of all being (*brahman*) which is in each and every thing the one enlivening Self (*ātman*).

> "Striking down in sleep what is bodily,
> Sleepless, he looks down upon the sleeping senses."

Such is the experience of the knowing subject released in trance to the raptures of the untrammeled eye of vision. Suddenly unloaded of the material weight of commitments to the field of waking consciousness:

> "A god, he puts forth for himself innumerable forms."

The leading reference of the Upanishad here is, of course, to the Self (*ātman*) as identical with the sole, non-dual ground of all being and becoming (*brahman*), putting forth the phenomenal universe in a continuous act of creation. Also implied, however, as recognized in Sankara's application of the passage, is a reference to the microcosmic aspect of the Self, as the indwelling immortal part of each and every living being, enjoying in each in the Dreaming State a creative life analogous to that of the macrocosmic creator. The two creations, though analogous, are not exactly comparable, since the field of manifestation of the microcosmic Self is confined to the mental sphere of the dreamer, whereas that of the macrocosmic Creator extends to the grossly visible pleasure ground of the daylight consciousness that is "Common to All Men." Moreover, the chariots, spans and roads, blisses, pleasures and delights, envisioned by the microcosmic Self are not original to its inward field of exquisite manifestations, but have been imprinted, carried and recollected from an outward field of already manifest gross phenomena. And still further, the inward eye of the mind by which those forms are beheld is not of the immortal, creative, universally indwelling Self (*ātman*), but of an intervening, historically conditioned ego (*aham̐-kāra*: "making the sound I"), which is not only bounded by ignorance (*avidyā*) of their true nature, but also attached to them with desires (*kāma*) and impulses to action (*karma*).

Freud's name for the same ground and source of all human willing is the *id*, the "It," which he characterized (as summarized by his authorized translator A. A. Brill), as "an unorganized chaotic mentality . . . the sole aim of which is the gratification of all needs, the alleviation of hunger, self-preservation and love, the preservation of the species. . . . As the child grows older," Brill's explication continues, "that part of the id which comes in contact with the environment through the senses learns to know the inexorable reality of the outer world and becomes modified into what Freud calls the ego. This ego, possessing awareness of the environment, henceforth strives to curb the lawless id tendencies whenever they attempt to assert themselves incompatibly." A neurosis may result from "a conflict between the ego and the id," or from "a conflict between the ego and the super-ego [the internalized parental warnings and ethical laws of a local society]." In a psychosis, the illness results from "a conflict between the ego and the outer world."

"Psychoanalysis," we are informed in Freud's classic essay on *The Ego and the Id* (published 1923), "is an instrument to assist the ego in its progressive conquest of the id." "The id is completely amoral, the ego is striving to become moral; the super-ego can become hypermoral and as ferocious then as the id."

In contrast, according to Sankara's view and in general throughout the philosophical and yogic systems of India, no functional conflict of this kind between id and ego is recognized. Freud's id corresponds to the Indian reincarnating will to life, the so-called *jīva* (compare Latin *vīvo*, "I live, am alive, have life"; English *vivacious* and Greek *bios*, "life"), that chaotic bundle of motivating desires (*kāma*), impulses to action (*karma*), and ignorance of the true nature of being (*avidyā*) which "puts on bodies and puts them off as a man puts on and puts off clothing" (*Bhagavad Gita* 2.22). Freud's super-ego is the Indian *dharma:* the body of local laws and customs that has become internalized as constituting the dreamer's sense of his duties to society; and Freud's ego, understood as that part of the id which has become conscious of "the inexorable reality of the outer world," is known in Sanskrit, also, as "ego," *aham*—except that in India its commitment to what Freud regarded as "the inexorable reality of the outer world" is recognized as itself the maintaining cause of that state of ignorance (*avidyā*) of the *jīva* which it is the function of yoga and philosophy to illuminate and dissolve. The compulsive force of *dharma* will then also be annulled. And so, in diametric contrast to the Freudian formula of "where there is id there shall be ego," is Sankara's ideal, as stated in his Vedantic classic, "The Crest Jewel of Discrimination":

"Shedding the activities of ego and its adjuncts, abandoning all attachments in realization of the Highest Truth (*paramārtha*), be released from dualities through enjoyment of the bliss of the Self (*ātman*) and rest in *brahman*; for thou hast there attained thy full Self-Nature unqualified (*pūrṇātmanā brahmaṇi nirvikalpa*).

In contrast to this opening of ego (*aham*) to the light and force of its own larger reality in the Self (ātman), the psychoanalytic Freudian approach to the interpretation of dreams, and therewith of mythology and the arts, remains enclosed in a tight horizon of egocentric day-residues from both the recent and most distant past. For not only is the Preconscious (*Pcs.*) motivated by wish-impulses from the Waking State (*jāgarita-sthāna*), but so too is the Unconscious (*Ucs.*), since, as diagnosed by Freud, this occult source of all spiritual afflictions is but a manifold of libidinous desires and fears repressed from infancy and reflecting the individual's earliest Waking-State experiences of a mother and a father (*Oedipus Complex*). Freudian psychotherapy, therefore, is fixated in the past, its method being to uncover, not the zone of infinite light from which the energies of the psyche derive, but those suppressed infantile memories of specific traumatic scenes and events through which the light has been, so to say, deflected and occluded, the ideal of ego-enlightenment being of adjustment to the inexorable demands of both the local moral order (*super-ego*) and the "outer world."

Carl G. Jung, in his approach to the interpretation of dreams and the imagery of myth, recognized two distinct spheres of relevancy, as functional of two orders or levels of the unconscious: (1) a historically and biographically conditioned *personal* unconscious, corresponding to the Freudian id-to-ego-bounded system, and (2) a transpersonal, instinctive, biologically grounded *collective* unconscious, matching Sankara's understanding of the force of the Wild Gander, the macrocosmic atman, in the sphere of consciousness of Deep Dreamless Sleep (*sushupta-sthāna*).

"There are present," states Jung, "in every individual, besides his personal memories, the great 'primordial' images, as Jacob Burckhardt once aptly called them, the inherited powers of human imagination as it has been since time immemorial. The fact of this inheritance explains the truly amazing phenomenon that certain motifs from myths and legends repeat themselves the world over in identical forms. It also explains why it is that our mental patients can reproduce exactly the same images and associations that are known to us from the old texts."

"The collective unconscious is a part of the psyche," he states further, "which can be negatively distinguished from a personal unconscious by the fact that it does not, like the latter, owe its existence to personal experience and consequently is not a personal acquisiton. While the personal unconscious is made up essentially of contents which have at one time been conscious but which have disappeared from consciousness through having been forgotten or repressed, the contents of the collective unconscious have never been in consciousness and therefore have never been individually acquired, but owe their existence exclusively to heredity. Whereas the personal unconscious consists for the most part of *complexes*, the content of the collective unconscious is made up essentially of *archetypes*.

"The concept of the archetype, which is an indispensable correlate of the idea of the collective unconscious, indicates the existence of definite forms in the psyche which seem to be present always and everywhere. Mythological research calls them 'motifs'; in the psychology of primitives they correspond to Levy-Bruhl's concept of 'representations collectives,' and in the field of comparative religion they have been defined by Hubert and Mauss as 'categories of the imagination.' Adolf Bastian long ago called them 'elementary' or 'primordian thoughts.' From these references it should be clear enough that my idea of the archetype—literally a pre-existent form—does not stand alone but is something that is recognized and named in other fields of knowledge."

And with respect, then, to the imagery specifically of the Dream State (*svapnasthāna*): "The inner image," states Jung, "is a complex structure made up of the most varied material from the most varied sources. It is no conglomerate, however, but a homogeneous product with a meaning of its own. The image is a *condensed expression of the psychic situation as a whole*, and not merely, nor even predominantly, of unconscious content pure and simple. It undoubtedly does express unconscious contents, but not the whole of them, only those that are momentarily constellated. This constellation is the result of the spontaneous activity of the unconscious on the one hand and of the momentary conscious situation on the other, which always stimulates the activity of relevant subliminal material and at the same time inhibits the irrelevant. Accordingly the image is an expression of the unconscious as well as the conscious situation of the moment. The interpretation of its meaning, therefore, can start neither from the conscious alone nor from the unconscious alone, but only from their reciprocal relationship.

"I call the image *primordial* when it possesses an archaic character. I speak of its archaic character when the image is in striking accord with familiar mythological motifs. It then expresses material primarily derived from the *collective unconscious*, and indicates at the same time that the factors influencing the conscious situation of the moment are *collective* rather than personal. A *personal* image has neither an archaic character nor a collective significance, but expresses contents of the *personal unconscious* and a personally conditioned conscious situation.

"The primordial image, elsewhere also termed *archetype*, is always collective, i.e., it is at least common to entire peoples or epochs. In all probability the most important mythological motifs are common to all times and races; I have, in fact, been able to demonstrate a whole series of motifs from Greek mythology in the dreams and fantasies of purebred Negroes suffering from mental disorders."

The transformations of consciousness that occur when the mind, disengaged from its normal state of attachment to forms of the past, is through identification with luminous apparitions of the intermediate dreaming state transported toward a realization of non-duality in infinite being, are in the Upanishad termed, respectively, A. *Viśva* or *Vaiśvānara*, "Pervasive, Prevailing," or "Common to all Men"; U. *Taijasa*, "Originating from and Consisting of *tejas*, 'Light'"; and M. *Prajña*, the "Knower." As stated by the sage Guadapada (*c.* A.D. seventh century): "*Viśva* is he who cognizes with the senses; *Taijasa*, he who cognizes in the mind; *Prajña*, he who constitutes the infinite space (*ākāśa*) in the heart. Thus the one Self (*ātman*) is [experienced as] threefold in the body."

The condition of *Prājña* has been described in the Upanishad as "ensheathed in bliss" (*ānandamaya*), "enjoying bliss" (*ānandabhuj*). Hence, when the mind disengaged from the "gross enjoyments" of the Waking State (*jāgaritasthāna*), transported by way of the luminous enjoyments of the Dreaming State (*svapnasthāna*), approaches in exquisite solitude the condition of *Prājña*, in nondual identification with the *ātman*, it is overtaken by a surpassing rapture which is of the order of a trance-revelation. And it typically is out of ecstatic transports of this kind that works of visionary arts derive.

Unit II

Where People Lived Legends: American Indian Myths

Mythogenesis

JOSEPH CAMPBELL

In this selection, Joseph Campbell is concerned with the architecture of myth. He relates a story by Black Elk, an Oglala Sioux medicine man, to show how the tale's timeless, placeless theme can also be found in Greek mythology and in the Old and New Testaments. Thereby, he demonstrates that such myths and the metaphors on which they are based, are transformed through time, but the underlining structures of the myth remain constant.

AN AMERICAN INDIAN LEGEND

If an authority on architecture, looking at the buildings of New York, were to observe that most of the older ones were of brick, then, viewing the ruins of ancient Mesopotamia, remarked that the buildings were all of brick, and finally, visiting Ceylon, made the point that many of the early temples were of brick, should we say that this man had an eye for architecture? It is true that bricks appear in many parts of the world; true, also, that a study might be made of the differences between the bricks of Ceylon, those of ancient Sumer, those, say of the Roman aqueducts still standing in southern France, and those of the city of New York. However, these observations about brick are not all that we should like to hear about the architecture of the cities of the world.

Now let me suggest a problem in the architecture of myth.

Early one morning, long ago, two Sioux Indians with their bows and arrows were out hunting on the North American plains; and as they were standing on a hill, peering about for game, they saw in the distance something coming toward them in a strange and wonderful manner. When the mysterious thing drew nearer, they perceived that it was a very beautiful woman, dressed in white buckskin, bearing a bundle on her back, and one of the men immediately became lustful. He told his friend of his desire, but the other rebuked him, warning that this surely was no ordinary woman. She had come close now and, setting her bundle down, called to the first to approach her. When he did so, he and she were covered suddenly by a cloud and when this lifted there was the woman alone, with the man nothing but bones at her feet, being eaten by terrible snakes. "Behold what you see!" she said to the other. "Now go tell your people to prepare a large ceremonial lodge for my coming. I wish to announce to them something of great importance."

The young man returned quickly to his camp; and the chief, whose name was Standing Hollow Horn, had several tepees taken down, sewn together, and made into a ceremonial lodge. Such a lodge has twenty-eight poles, of which the central pole, the main support, is compared to the Great Spirit, Wakan Tanka, the supporter of the universe. The others represent aspects of creation; for the lodge itself is a likeness of the universe.

"If you add four sevens," said the old warrior priest, Black Elk, from whom this legend was derived, "you get twenty-eight. The moon lives twenty-eight days and this is our month. Each of these days of the month represents something sacred to us: two of the days represent the Great Spirit; two are for Mother Earth; four are for the four winds; one is for the Spotted Eagle; one for the sun; and one for the moon; one is for the Morning Star; and four are for the four ages; seven for our seven great rites; one is for the buffalo; one for the fire; one for the water; one for the rock; and finally, one is for the two-legged people. If you add all these days up you will see that they come to twenty-eight. You should know also that the buffalo has twenty-eight ribs, and that in our war bonnets we usually wear twenty-eight feathers. You see, there is a significance for everything, and these are the things that are good for men to know and to remember."

This wonderful old Oglala Sioux priest explained to the young scholar, Joseph Epes Brown, who had come to the Pine Ridge Reservation (South Dakota) expressly to gain knowledge first hand of the mystic dimension of American Indian mythology, the image of the man consumed by snakes. "Any man who is attached to the senses and things of this world," he said, "is one who lives in ignorance and is being consumed by the snakes that represent his own passions."

One is reminded by this Indian image of the Greek legend of the young hunter Actaeon, who, in quest of game, following a forest stream to its source, discovered there the goddess Artemis bathing, perfectly naked, in a pool: and when she saw that he looked lustfully upon her, she transformed him into a stag that was then pursued by his own hounds, torn to pieces, and consumed. And not only are the two legends comparable, but the old Sioux priest's interpretation of his own accords with the sense of the Greek. Moreover, his interpretation of the symbolism of the ceremonial lodge suggests a number of themes familiar to us as well: so that we are moved to wonder what the explanation of these accords might be.

When the people of Black Elk's legend had made their large ceremonial lodge that was symbolically a counterpart of the universe, they all gathered within it, extremely excited, wondering who the mysterious woman could be and what she wished to say to them. Suddenly, she appeared in the door, which was facing east, and proceeded sunwise around the central pillar: south, west, north, and again east. "For is not the south the source of life?" the old teller of the tale explained. "And does not man advance from there toward the setting sun of his life? Does he not then arrive, if he lives, at the source of light and understanding, which is the east? And does he not return to where he began, to his second childhood, there to give back his life to all life, and his flesh to the earth whence it came? The more you think about this," he suggested, "the more meaning you will see in it."

This sturdy old son of the American earth, now nearly blind, had been born in the early eighteen-sixties, fought as a young man at the battle of the Little Bighorn and the battle of Wounded Knee, had known the great chiefs, Sitting Bull, Crazy Horse, Red Cloud, and American Horse, and at the time of this retelling of his legend, the winter of 1947–1948, was Keeper of the Sacred Pipe. He had received the treasured talisman, together with its legend, from the earlier Keeper, Elk Head, who had prophesied at that time that as long as the pipe was used and its legend known, the Oglala Sioux would live, but as soon as the legend was forgotten, they would lose their center and perish.

The pipe and its legend, then, were of indeterminate age, anonymous, practically timeless. Yet actually, more closely regarded, they could not have been much older than a couple of hundred years; for the Oglala Sioux did not migrate to the plains and become buffalo hunters there until the end of the seventeenth century—1680 or so. They had formerly been a forest folk of the Upper Mississippi, a region of lakes and marshes, where they traveled in birchbark canoes. And yet, as already remarked, although we may never have heard before of the Sacred Pipe of the Oglala Sioux, or have numbered the buffalo's ribs, all the elements of this myth are curiously familiar. We recognize the bricks, so to say, though the way in which they have been put together is surprising.

The ceremonial lodge is exactly comparable to a temple, oriented to the quarters, with its central pole symbolic of the world axis. "We have established here the center of the earth," the nearly blind old medicine man explained to his attentive listener, "and this center, which in reality is everywhere, is the dwelling place of Wakan Tanka." But this figure of the "center" that is "everywhere" is a counterpart, exactly, of that of the hermetic twelfth-century "Book of the Twenty-four Philosophers," from which Nicholas Cusanus and a number of other distinguished European thinkers—Alan de Lille, Rabelais, Giordano Bruno, Pascal, and Voltaire, for example—derived their definition of God as "an intelligible sphere, whose center is everywhere and circumference nowhere." It is amazing indeed to catch the echo of such a metaphysical statement as this from the lips of an absolutely illiterate old Sioux, living out his fading day as the guardian of an aboriginal Amerindian fetish and its myth.

What are we to think of all these coincidences? Whence come these timeless, placeless themes?

Shall we join our voice to those who write of a great Perennial Philosophy, which, from time out of mind, has been the one, eternally true wisdom of the human race, revealed somehow from on high? How came this, then, with all its symbols, to the Sioux? Or shall we seek our answer, rather, in some psychological theory, like many of the most distinguished nineteenth-century ethnologists—Bastian, for instance, Tylor, and Frazer—attributing such cross-cultural accords to "the effect," as Frazer put it, "of similar causes acting alike on the similar constitution of the human mind in different countries and under different skies"? Do such images, that is to say, take form naturally in the psyche? Can they be assumed and even expected to appear spontaneously, in dreams, in visions, in mythological figurations, any place on earth, wherever man has made his home?

Or must it be said, on the contrary, that, since mythological orders—like architectural orders—serve specific, historically conditioned cultural functions, where any two can be shown to be homologous they must be assumed to be historically related? Can the Greeks and the Sioux, that is to say, be supposed to have received any part or parts of their mythological heritages from a common source?

Or, finally, shall we simply set aside this whole question of shared motifs (whether religiously, psychologically, or historically explained) as unworthy of a scientist's speculation, since—as a number of important field-anthropologists now hold—myths and rituals are functions of local social orders, hence meaningless out of context and, consequently, not to be abstracted and compared across cultural lines? Comparisons of that sort, enjoyed by dilettantes and amateurs, are, according to this view, neither significant nor of interest to a properly tutored scientific intelligence.

Let us look further, with an uncommitted eye, and try to judge for ourselves.

We note in Black Elk's commentary to his legend the formula four times seven, giving twenty-eight supports of the universe, the numbers four and seven being standard symbols of totality in the iconographies of both the Orient and the Occident; and this game of sacred numbers itself is a shared trait worth remarking. One of the twenty-eight supports is in the center of the great tepee, as the axis, the pivot of the universe. The number surrounding it then is twenty-seven: three times nine: three times three times three. One thinks of the psychologist C. G. Jung's numerous discussions of the symbolism of the four and three. One thinks of the nine choirs of angels (three times three) that surround and celebrate the central throne of the Trinity. Three is the number of time: past, present, and future; four is the number of space: east, south, west, and north. Space (four) and time (three) constitute the field—the universe—in which all phenomenal forms become manifest and disappear. (Another way would be to think of the vertical—above, here, and below—plus the four directions: which again yields three plus four.) The number four recurs in the "sacred turn," the clockwise circumambulation of the center, which is here associated not only with the four directions but also with the life stages of the individual, so that the symbolism is applied as well to the microcosm as to the macrocosm: the two being tied by the related number twenty-eight to the cycle of the moon, which dies and is resurrected and is accordingly a sign of the cycle of renewal.

Furthermore, as we are told, the buffalo has twenty-eight ribs and is therefore himself a counterpart of the moon—and the universe. Do not the buffaloes return every year, miraculously renewed like the moon? We think of the Moon Bull of the archaic Near East, the animal-vehicle of Osiris, Tammuz, and, in India, of Shiva. The horns of the moon suggest those of the moon-god Sin, after whom Mount Sinai was named, so that it should represent the cosmic mountain at the center of the world, from the summit of which Moses descended. And his face then shone with horns of light, like the moon, so that when he stood before the people he had to wear a veil (Exodus 34:29–35)—like the veils of archaic kings who, for centuries before his time, had been revered as incarnations of the self-renewing power of the moon. At the foot of Mount Sinai, the High Priest Aaron was found conducting a festival of the Moon Bull in the image of a golden calf, which Moses angrily committed to the fire, ground to bits, mixed with water, and caused the people to drink, in a kind of communion meal (Exodus

32:1–20). And—remarkably—it was only after this sacrifice that, returning to the mountaintop (where the earth-goddess and heaven-god are joined in eternal connubium), he received the full assignment of the Law and promise of the Promised Land, not for himself—for he was now, himself, to become the sacrifice—but for the Holy Race.

In the mythology of Christ crucified, three days in the tomb and resurrected, there is implicit the same symbolism of the moon that is three days dark. The Sacrificial Lamb, the Sacrificial Bull, and the Cosmic Buffalo: their symbology was perfectly interpreted by the old Sioux medicine man, Black Elk, when he explained that the buffalo was symbolic of the universe in its temporal, lunar aspect, dying yet ever renewed, but also (in its twenty-eighth rib) of the Great Spirit, which is eternal, the center that is everywhere and around which all revolves.

Chief Standing Hollow Horn—as Black Elk recounted in his tale—was seated at the west of the lodge, the seat of honor, when the beautiful woman entered: because there he faced the door, the east, whence comes the light, namely wisdom, which a leader must possess. And the woman, arriving before him, lifted the bundle from her back and held it out to him with both hands.

"Behold this bundle," she said, "and always love it! It is *lela wakan* [very sacred], and you must treat it as such. No impure man should ever be allowed to see it; for within this bundle there is a very holy pipe. With this pipe you will, during the years to come, send your voices to Wakan Tanka, your Father and Grandfather."

She drew forth the pipe and with it a round stone, which she placed on the ground. Holding the pipe with its stem to the heavens, she said: "With this sacred pipe you will walk upon the Earth; for the Earth is your Grandmother and Mother, and she is sacred. Every step that is taken upon her should be as a prayer. The bowl of this pipe is of red stone; it is the Earth. Carved in the stone and facing the center is a buffalo calf, who represents all the quadrupeds who live upon your Mother. The stem of the pipe is of wood, and this represents all that grows upon the Earth. And these twelve feathers that hang here, where the stem fits into the bowl, are feathers of the Spotted Eagle. They represent that eagle and all the winged things of the air. When you smoke this pipe all these things join you, everything in the universe: all send their voices to Wakan Tanka, the Great Spirit, your Grandfather and Father. When you pray with this pipe you pray for and with all things."

The proper use of the pipe, as expounded by its guardian priest, required that it should be ceremonially identified with both the universe and oneself. From the fire in the center of the lodge, the fire symbolic of Wakan Tanka, an attendant with a split stick lifted a coal, which he placed before the Keeper of the Pipe. The latter, holding in his left hand the pipe, took with his right a pinch of a sacred herb, and elevating this to heaven four times, prayed:

"O Grandfather, Wakan Tanka, I send you, on this sacred day of yours, this fragrance that will reach to the heavens above. Within this herb is the Earth, this great island; within it is my Grandmother, my Mother, and all four-legged, winged, and two-legged creatures who walk in a holy [*wakan*] manner. The fragrance of this herb will cover the entire universe. O Wakan Tanka, be merciful to all."

The bowl of the pipe was then held over the burning aromatic herb in such a way that the fragrant smoke, entering, passed through the stem and out the end, toward heaven. Thus Wakan Tanka was the first to smoke, and the pipe, by that act, was purified. It was then filled with tobacco that had been offered in the six directions: to the west, north, east, and south, to heaven, and to earth. "In this manner," said the medicine man, "the whole universe is placed in the pipe." And, finally, the man who fills the pipe should identify it with himself. There is a prayer in which this identity is described. It runs as follows:

> These people had a pipe,
> Which they made to be their body.
>
> O my Friend, I have a pipe that I have made to be my body;
> If you also make it to be your body,
> You shall have a body free from all causes of death.
>
> Behold the joint of the neck, they said,
> *That* I have made to be the joint of my own neck.
>
> Behold the mouth of the pipe,
> *That* I have made to be my mouth.
>
> Behold the right side of the pipe,
> *That* I have made to be the right side of my body.
>
> Behold the spine of the pipe,
> *That* I have made to be my own spine.
>
> Behold the left side of the pipe,
> *That* I have made to be the left side of my body.
>
> Behold the hollow of the pipe,
> *That* I have made to be the hollow of my body. . . .
>
> Use the pipe as an offering in your supplications,
> And your prayers shall be readily granted.

This game, this holy game of purifying the pipe, expanding the pipe to include the universe, identifying oneself with the pipe, and igniting it in symbolic offering, is a ritual act of the kind represented, also, in Vedic Brahmanic ceremonials, where the altar and every implement of the sacrifice is identified allegorically with both the universe and the sacrificing individual. "He who is in the fire," we read in the Maitri Upanishad, for example, "he who is here in the heart, and he who is yonder in the sun: he is one." Likewise, in the great Chandogya: "Now the light that shines higher than this heaven, on the backs of all, on the backs of everything, in the highest worlds, than which there are no higher—verily, that is the same as this light which is here within a person. . . .

"One should reverence the mind as *brahman*.—Thus with reference to oneself. And now, with reference to the gods: One should reverence space as *brahman.*—That is the two-fold instruction with reference to oneself and with reference to the gods.

"That *brahman* has four quarters. One quarter is speech. One quarter is breath. One quarter is the eye. One quarter is the ear.—Thus with reference to oneself. And now, with reference to the gods: One quarter is Fire. One quarter is Wind. One quarter is the Sun. One quarter is the quarters of the sky.—That is the twofold instruction with reference to oneself and with reference to the gods."

Now, as we have heard, the feathers of the sacred pipe are those of the spotted eagle, which is the highest-flying bird in North America and therefore equivalent to the sun. Its feathers are the solar rays—and their number is twelve, the number, exactly, that we too associate with the cycle of the sun in the months of the solar year and twelve signs (three times four) of the zodiac. There is a verse in a sacred song of the Sioux: "The Spotted Eagle is coming to carry me away." Do we not think here of the Greek myth of Ganymede transported by Zeus, who came in the form of an eagle to carry him away? "Birds," declares Dr. Jung in one of his dissertations on the process of individuation, "are thoughts and flights of the mind. . . . The eagle denotes the heights . . . it is a well-known alchemistic symbol. Even the *lapis*, the *rebis* [the Philosopher's Stone], made out of two parts, and thus often hermaphroditic, as a coalescence of Sol and Luna, is frequently represented with wings, in this way standing for premonition—intuition. All these symbols in the last analysis depict that state of affairs that we call the Self, in its role of transcending consciousness."

Such a reading nicely accords with the part played by our North American spotted eagle in the rites of the Indian tribes. It explains, also, the wearing of eagle feathers. They are counterparts of the golden rays of a European crown. They are the rays of the spiritual sun, which the warrior typifies in his life. Furthermore—as we have been told—their number in the war bonnet is twenty-eight, the number of the lunar cycle of temporal death and renewal, so that here Sol and Luna have again been joined.

There can be no doubt whatsoever that this legend of the Sioux is fashioned of at least some of the same materials and thoughts as the great mythologies of the Old World—Europe, Africa, and Asia. The parallels both in sense and in imagery are too numerous and too subtle to be the consequence of mere accident. And we have not yet finished!

For when the holy woman, standing before Chief Standing Hollow Horn, had taught him how to use the pipe, she touched its bowl to the round stone that she had placed upon the ground. "With this pipe you will be bound," she said, "to all your relatives: your Grandfather and Father, your Grandmother and Mother."

The Great Spirit, the medicine man explained, is our Grandfather and Father; the Earth, our Grandmother and Mother. As Father and Mother, they are the producers of all things; as Grandfather and Grandmother, however, beyond our understanding. These suggest the two modes of considering God that Rudolf Otto, in *The Idea of the Holy*, has termed the "ineffable" and the "rational": the same, as Joseph Epes Brown points out in his commentary on Black Elk's rite, that are referred to in India as *nirguṇa* and *saguṇa brahman*, the "Absolute without Qualities" and the "Absolute with Qualities," respectively *That* beyond names, forms, and relationships, and *That* personified as "God."

"This round rock," the very beautiful woman continued, "which is made of the same red stone as the bowl of the pipe, your Father Wakan Tanka has also given to you. It is the Earth, your Grandmother and Mother, and it is where you will live and increase. This Earth which he has given to you is red, and the two-legged people who live upon the Earth are red; and the Great Spirit has also given to you a red day, and a red road."

"The 'red road,'" Joseph Brown explains, "is that which runs north and south and is the good or straight way, for to the Sioux the north is purity and the south is the source of life. . . . On the other hand, there is the 'blue' or 'black road' of the Sioux, which runs east and west and which is the path of error and destruction. He who travels on this path, Black Elk has said, is 'one who is distracted, who is ruled by his senses, and who lives for himself rather than for his people.'" The latter was the road followed by the man at the opening of the story, who was eaten by snakes. And so we notice now that even the ethical polarity that we recognize between the bird and serpent as allegoric of the winged flight of the spirit and the earth-bound commitment of the passions, has been suggested here, as well.

"These seven circles that you see upon the red stone," the woman said, "represent the seven rites in which the pipe will be used. . . . Be good to these gifts and to your people; for they are *wakan* [holy]. With this pipe the two-legged people will increase and there will come to them all that is good." She described the rites and then turned to leave. "I am leaving now, but I shall look back upon your people in every age; for, remember," she said, "in me there are four ages: and at the end I shall return."

Passing around the lodge, in the sunwise way, she left, but after walking a short distance, looked back toward the people and sat down. When she got up again, they were amazed to see that she had turned into a young red and brown buffalo calf. The calf walked a little distance, lay down and rolled, looked back at the people, and when she got up was a white buffalo. This buffalo walked a little distance, rolled upon the ground, and when it rose was black. The black buffalo walked away, and when it was far from the people it turned, bowed to each of the four directions, and disappeared over a hill.

The *wakan* woman had thus been the feminine aspect of the cosmic buffalo itself. She herself was the earthly red buffalo-calf represented on the pipe bowl, but also its mother, the white buffalo, and its grandmother, the black. She had gone to be restored to her eternal portion, having rendered to man those sacred thoughts and visible things by which he was to be joined to his own eternity, which is here and now, within him and all things, in this living world.

The Metaphorical Journey

JOSEPH CAMPBELL

In this selection, Joseph Campbell discusses the symbols of an important Navajo sand painting in terms of its impact on the entire society. He identifies universal symbols as distinguished from local and historical transformations of metaphorical images. He explains the extent to which myths have been transformed through time.

In the Navaho sand painting the bounded area is equivalent to the interior of a temple, an Earthly Paradise, where all forms are to be experienced, not in terms of practical relationships, threatening or desirable, evil or good, but as the manifestations of powers supporting the visible world and which, though not recognized in practical living, are everywhere immediately at hand and of one's own nature. The painting is used in a blessing ceremony, for healing, or to impart the courage and spiritual strength requisite to the endurance of some ordeal, or for the performance of some difficult task. The dry painting is of colored sand strewn (with amazing skill and speed) upon the dirt floor of a native dwelling, or hogan. Neighbors and friends have assembled, and for a period of one to five (in many other rituals, nine) continuous nights and days of chanting, prayer, and metaphorical acts, the patient or initiate is ceremonially identified in mind and heart and costume with the mythological protagonist of the relevant legend. He or she actually enters physically into the painting, not simply as the person whose friends and neighbors have solicitously assembled, but equally as a mythic figure engaged in an archetypal adventure of which everyone present knows the design. For it is the archetypal adventure of them all in the knowledge of their individual lives as grounded and participative in a beloved and everlasting pattern. Moreover, all the characters represented in the ceremonial are to be of the local landscape and experience, become mythologized, so that through a shared witnessing of the ceremonial the entire company is renewed in accord with the nature and beauty of their spiritually instructive world.

The conformity of the imagery of this particular sand painting to the sense and symbolized experiences of the yogic *sushumnā* is certainly astonishing, but not more so than many other concordancies in the myths and ritual arts of peoples across the world. The axial Great Corn Plant corresponds here to the *sushumnā*; the footprints represent a spiritual ascent along the mystic way known to the Navaho as the Pollen Path. There is a verse from a sacred chant:

In the house of life I wander
On the pollen path,
With a god of cloud I wander
To a holy place.

With a god ahead I wander
And a god behind.
In the house of life I wander
On the pollen path.

The Great Corn Plant's upper half is marked by a lightning flash, which immediately suggests the oriental *vajra* ("thunderbolt of enlightenment") of Hindu and Buddhist iconography. It strikes to the exact center of the way, which corresponds (if we count the leaf-marked stages, the tassel above, and the root below) exactly to the fourth chakra, *anāhata*, where the sound is heard that is not made by any two things striking together. In Navaho myths and legends the god known as Sun Bearer, Tsóhanoai, possesses lightning-arrows. His domicile is above, in the sky, and when his twin sons, conceived of him by an earthly virgin, Changing Woman, arrived to receive from him the power and weapons with which to rid the world of monsters, the solar power of which they then became possessed was so great that when they returned to earth it had to be modified by a deity known as Hastyéyalti (Talking God), maternal grandfather of the pantheon, in whom male and female powers are combined. In the sand painting the realm of operation of this second god is represented by a modified rainbow of two colors, red and blue (here reproduced brick red and blue-gray), symbolic respectively of solar energy and of water and the moon. The high place onto which the Twins descended from the heavenly house of their father to earth was the central mountain of the world (locally, Mt. Taylor in New Mexico; height, 11,302 feet) and the corresponding station in the painting is at Chakra 4, *anāhata*. Moreover, the root of the corn plant is threefold, like *Yuktatriveṇī* in the *mūlādhāra*. The pollen path commences lower right and is of two colors, like the modified rainbow (or like the moon, which is of both matter and light). But at the turn between the Spirit Bringers the path becomes yellow and single (like the fire of the blended breaths, exploding to ascend the *sushumnā*); after which, in a sacred way, to an accompaniment of chants and prayer, the sanctuary is entered.

The parallelism of the tantric visionary ascent of the *sushumnā* and the ritually controlled transit of the Navaho Pollen Path is not simple to explain. We know that in Tibetan tantric colleges such as Gyudto in Lhasa, a strictly controlled tradition of ritualized sand painting was practiced until the calamity of 1959. We know also that the Navaho are a people of Athapascan stock, from northwesternmost North America, who migrated to the Southwest some time around the twelfth century A.D. The Athapascans of the Canadian Northwest, however, do not practice sand painting. The Navaho seem to have acquired the art from the Pueblo tribes around whose villages they settled. Any relationship between the Pueblos of New Mexico and the great monasteries of Lhasa is impossible, not only to document, but even to imagine. We are thrown back, therefore, for the present at least, upon the academically unpopular *psychological* explanation of the undeniable parallelism of these two visionary journeys, one, of an ecstatic, immediately visionary kind, the other, by way of a tribal ceremonial.

That the experience of an ascent of the *sushumnā* is of a psychological order is hardly questionable. And that the visions are culturally conditioned (as are all visions and all dreams) is certain; for every petal of the envisioned lotuses, from the four petals of the *mūlādhāra* to the two of *ājñā* and thousands of *sahasrāra*, bears a letter of the Sanskrit alphabet. Yet descriptions of the lotuses by those who have experienced them carry the conviction of a reality, not indeed of a gross material (*sthūla*), but of a subtle (*sukshma*), dreamlike, visionary kind.

"They are formed of Consciousness," said Ramakrishna, "like a tree made of wax—the branches, twigs, fruits, and so forth all of wax." "One cannot see them with the physical eyes. One cannot take them out by cutting open the body."

"Just before my attaining this state of mind," he told his company of followers, "it had been revealed to me how the Kundalini is aroused, how the lotuses of the different centers blossom forth, and how all this culminates in samādhi. This is a very secret experience. I saw a boy twenty-two or twenty-three years old, exactly resembling me, enter the sushumnā nerve and commune with the lotuses, touching them with his tongue. He began with the center at the anus and passed through the centers at the sexual organ, navel, and so on. The different lotuses of those centers—four-petalled, six-petalled, ten-petalled, and so forth—had been drooping. At his touch they stood erect.

"When he reached the heart—I distinctly remember it—and communed with the lotus there, touching it with his tongue, the twelve-petalled lotus, which was hanging head down, stood erect and opened its petals. Then he came to the sixteen-petalled lotus in the throat and the two-petalled lotus in the forehead. And last of all, the thousand-petalled lotus in the head blossomed. Since then I have been in this state."

In the sand painting of the pollen path, the two colors of the "female" and the "male," lunar and solar powers, having become one on passing between the guardian Spirit Bringers at the entrance to the sanctuary, the path, which is now of the single color of pollen, runs to the base of the World Tree, where three roots or ways of entrance are confronted. Those to right and to left point separately to the separated Spirit Bringers, whose arms and hands gesture together toward the portal of the middle way. The lower half, then, of the axial corn plant, which is to represent the graded stages of an initiation, is found signalized by a rainbow of the same two colors that were already of the initiate's pollen path of virtue before its right turn to the Spirit Bringer's call.

Now a rainbow is an insubstantial apparition composed of both matter and light, formed opposite the sun by reflection and refraction of the sun's rays in drops of rain. Thus it is at once material and immaterial, lunar and solar, as the matched colors, blue and red, here suggest. One thinks of the words of Goethe's Faust at the end of the opening scene of *Faust*, Part II, where, bedded on a flowery turf, Faust wakes to an alpine sunrise and beholds across the valley a rainbow spread across the face of a mighty waterfall. The sun behind him being too powerful to contemplate directly, his eyes dwell, "with rapture ever growing," upon the beauty of the insubstantial image formed of the sun's light refracted in the whirling spray of the tumbling torrent, and he exclaims, "*So bleibe denn die Sonne mir im Rücken!* ['So let the sun remain then behind me!']. . . . *Am farbigen Abglanz haben wir das Leben* ['Our living is in the colorful reflection']"; for he is stationed at the crucial turning point of Chakra 4, *anāhata*;

where he must choose whether to pass, like an Indian yogi, to extinction in the full blast of the sun or to rest engaged in the field of action, held to the recognition of the play of the sun's light through all the forms given life by its power!

The rainbow of the sand painting ends exactly midway of the path. There the lightning strikes, and to right and to left two apparitions appear. These are of a spiritual messenger-fly known to the Navaho as Dontso, "Big Fly," and in his second aspect, "Little Wind." In art and myth this little figure holds a prominent place, and like everything else in Navaho mythology, he is a transformation into metaphor of an actual feature of the environment. There is, namely, a local tachinid fly, of the species *Hystricia pollinosa* van der Wulp, which has the habit of lighting on a person's shoulder or on the chest just in front of the shoulder. In this role it has been admitted to the pantheon as a bringer of news and guidance and as a messenger of the spirit. In its present, dual apparition, black and pale yellow, it represents separately the male and female energies, which in the rainbow are united. For both black and red, in the color system of this painting, are symbolic of male energy, while both yellow and blue are female. Black is sinister and threatening to life, yet also protects. The lightning flash here is black, and Black Wind (which here appears in the form of the black Dontso) is the power of the sun. Yellow represents, in contrast, fructification, pollen, and the powers of vegetation. Women originated from a yellow corn ear, and yellow is the color of an inexhaustible food bowl symbolizing sustenance. The black zigzag line of the lightning flash is single, whereas in the rainbow, where the male power has been tempered to the female power, the male is red and the female blue. Red is the color of danger, war, and sorcery, but also of their safeguards; for it is the color, as well, of blood, flesh, and nourishing meat. Blue is a color associated by the Navaho with the fructifying power of the earth, with water and with sky. Combined as here in a rainbow, the two are known as "The Sunray" and said to stand for "light rays emerging from a cloud when the sun is behind it . . ."

The cornstalk of this painting is blue, and upon the central corn tassel at its top there is a bird, also blue, symbolic, it is said, of dawn, happiness, and promise. The only ears of corn to be seen are at nodes corresponding (by analogy with *sushumnā*) to Chakras 3 and 4, *maṇipūra* and *anāhata*, both of which are associated with powers and realizations of the period, in a lifetime, of maturity: mastery and illumination, which may or may not be by coincidence.

For the artist to whom we owe the reproduction of this precious document of an orally communicated tradition that is rapidly disappearing did not receive from the Singer (the Medicine Man) from whose Blessing Rite it was drawn an interpretation of its symbols. We now have, however, an ample resource of well-collected materials authoritatively interpreted, assembled by scholars chiefly of the 1930s and 1940s (when the normal Navaho household was said to have comprised a father and mother, one child and two anthropologists), out of whose publications a substantially acceptable reading of such an iconographic work can be confidently reconstructed.

My own researches in the material were undertaken in those same years of the 1930s and 1940s in connection, partly, with my task of editing, for the opening of the Bollingen Series, Maud Oakes' collection of paintings from the war ceremonial of the old Navaho warrior and medicine man Jeff King (1852/60(?)–1964, now buried with honors in Arlington National Cemetery), "Where the Two Came to Their Father," published in 1943. At the same time I was at work with Henry Morton Robinson on *A Skeleton Key to Finnegans Wake* (1944), besides helping Swami Nikhilananda with his translation from Bengali of *The Gospel of Sri Ramakrishna* (1942), having just commenced the twelve-year task of editing for the Bollingen Series four publications from the posthumous papers of my deceased friend Heinrich Zimmer: *Myths and Symbols in Indian Art and Civilization, The King and the Corpse, Philosophies of India,* and *The Art of Indian Asia,* in 2 volumes.

My distinct impression throughout those years was that I was at work only on separate chapters of a single mythological epic of the human imagination. Moreover, the epic was the same as that to which I had been introduced twenty years before in my graduate studies at Columbia University in Anthropology and European romantic and medieval literatures; in Paris at the Sorbonne on the Provençal poets and Arthurian romance; and at the University of Munich in Sanskrit and Far Eastern Buddhist art. Adolf Bastian's theory of a concord throughout the mythologies of the world of certain *Elementargedanken,* "elementary ideas" was confirmed for me beyond question by these apparently far-flung researches. The first task of any systematic comparison of the myths and religions of mankind should therefore be (it seemed to me) to identify these universals (or, as C. G. Jung termed them, archetypes of the unconscious) and as far as possible to interpret them; and the second task then should be to recognize and interpret the various locally and historically conditioned transformations of the metaphorical images through which these universals have been rendered.

Since the archetypes, or elementary ideas, are not limited in their distributions by cultural or even linguistic boundaries, they cannot be defined as culturally determined. However, the local metaphors by which they have been everywhere connoted, the local ways of experiencing and applying their force, are indeed socially conditioned and defined. Bastian termed such local figurations "ethnic ideas," *Völkergedanken,* and Mircea Eliade has termed them "hierophanies" (from hieros-, "powerful, supernatural, holy, sacred," plus *phainein,* "to reveal, show, make known").

"The very dialectic of the sacred," Eliade declares, "tends to repeat a series of archetypes, so that a hierophany realized at a certain 'historical moment' is structurally equivalent to a hierophany a thousand years earlier or later." Furthermore, "hierophanies have the peculiarity of seeking to reveal the sacred in its totality, even if the human beings in whose consciousness the sacred 'shows itself' fasten upon only one aspect or one small part of it. In the most elementary hierophany *everything is declared.* The manifestation of the sacred in a stone or a tree is neither less mysterious nor less noble than its manifestation as a 'god.' The process of sacralizing reality is the same; the *forms* taken by the process in man's religious consciousness differ."

The Elementary Idea is grounded in the psyche; the Ethnic Idea through which it is rendered, in local geography, history, and society. A hierophany occurs when through some detail, whether of a local landscape, artifact, social custom, historical memory, or individual biography, a psychological archetype or elementary idea is reflected. The object so informed becomes thereby sacralized, or mythologized. Correspondingly, a religious *experience* will be realized when there is felt an immediate sense of *identification* with the revelation. The sense of a mere *relationship* is not the same. In popular cult the experience of relationship is frequently all that is intended. Thereby a sense of social solidarity may be rendered. Through identification, however, a transformation of character is effected.

METAPHORICAL IDENTIFICATION

In the Blessing Ceremony of the Navaho the psychosomatic healing of spiritual initiation is accomplished by means of an identification, ritually induced, of the patient or initiate with the mythological adventure of the pollen path in its threshold crossings into and through a sacred space and out into the world transformed. We have not been told by what specific prayers, chants, and metaphorical acts the adventure is conducted. The pictorial statement of the painting, however, indicates the transformative stages.

Footprints of white cornmeal mark the path of the initiate throughout. As already noticed, they approach along a road that is of the two symbolic colors of the male and female powers, fire and water, sunlight and cloud, which at the place of the Spirit Bringers abruptly blend to one golden yellow of the color of pollen. The same color of pollen then appears bordering the outspread sacred space, which in this way opens to a display of the indwelling powers of pollen.

There are no footprints on this part of the path. To be assumed is a threshold passage of the mind as it turns from secular anxieties, identifications, and expectations, to a game of make believe ("as if"; *als ob*, recalling Kant), assuming an intentionally metaphorical, mythological cast of hierophantic personifications. The Navaho have a number of procedures to effect this indispensable alteration of consciousness: sweat baths, solemn recitations of the names of the gods to be represented in the rite, heraldic face and body painting, with an investiture of ornaments such as the god with whom the candidate is to identify would be wearing. The ordeal is an act of sacrifice. The mind is to abandon forever the whole way of relating to life which is of the knowledge of the two powers of the path only as distinct from each other, red and blue. Beyond the exit gate, returning to the world, the path is to be no longer red and blue, but of the one color of pollen. The neighbors and friends who have gathered to witness the occasion will experience an exaltation, but then return to the world along the path by which they came; for their participation will have been, not of identification, but of a relationship—something comparable to the attendance by a Roman Catholic family at holy mass, whereas the initiate, nearly naked and decorated as a god, will have become identified with the adventure, as in the way of the words of Jesus reported in the Thomas Gospel: "Whoever drinks from My mouth shall become as I am and I myself will become he, and the hidden things shall be revealed to him."

When footprints reappear in the painting they are already in or on the cornstalk, and Dontso has appeared as at once female and male, yellow and black, in two aspects: the one as two and the two as one. Moreover, the path of blue and red is now beheld in the luminous form of the Sunray, no longer of solid matter, but of light and cloud.

In the Hindu ascent of the *sushumnā*, the chakras of the lower centers are identified, respectively, as of the elements Earth, Water, and Fire, the associated divinities being Brahmā the Creator, Vishnu as Preserver and Lover, and Shiva as the Destroyer of Obstructions to Illumination. The involved yogi becomes to such a degree absorbed in these morphogenetic archetypes that as the *kuṇḍalinī* ascends, the whole body below the center of concentration goes cold. As stated by Sir John Woodroffe: "There is one simple test whether the Shakti [the energy as *kuṇḍalinī*] is actually aroused. When she is aroused intense heat is felt at that spot, but when she leaves a particular center the part so left becomes as cold and apparently lifeless as a corpse. The progress upwards may thus be externally verified by others. When the Shakti (Power) has reached the upper brain (Sahasrāra) the whole body is cold and corpse-like; except the top of the skull, where some warmth is felt, this being the place where the static and kinetic aspects of Consciousness unite."

Kuṇḍalinī yoga practiced in this outright way, in other words, is not a game of "as if" and make believe, but an actual experience of psychological absorption in a metaphysical ground of some kind, a morphogenetic field that has not yet, as far as I know, been scientifically recognized in the West except by C. G. Jung and, lately, by the physicist Rupert Sheldrake, whose works, according to one concerned scientific reviewer, ought to be burned. Ramakrishna, who surely knew whereof he spoke, warned (as we have already learned), that this form of yoga, not only is extremely difficult, but also should not be practiced by the lovers of God, since (in his words): "To one who follows it even the divine play in the world becomes like a dream and appears unreal; his 'I' also vanishes." At Chakra 4, *anāhata*, where the sound OM is first heard that is not made by any two things striking together, the element into which the yogi is absorbed is Air (*vāyu*, the life-breath; *prāṇa*, *spiritus*, pure spirit); while at Chakra 5 he devolves into *ākāśa*, Space. "Boundless am I as Space!" exclaims the ancient sage, Ashṭāvakra. "The phenomenal world is like an empty jar [enclosing Space, which nevertheless is boundless]. Thus known, phenomenality need be neither renounced, accepted, nor destroyed."

In Dante's *Divina Commedia* the corresponding stage of spiritual exaltation extends through all but the last three cantos of the *Paradiso (I–XXX)*, telling of the visionary's ecstatic ascent, led by the spirit of Beatrice (Dante's Shakti), from the garden of the Earthly Paradise on the summit of Mount Purgatory, through all the ranges of space to the outermost sphere of the Primum Mobile, by which the heavens are revolved.

> A Light is thereabove which makes the Creator
> visible to every creature that has
> his peace only in beholding Him.
> It spreads so wide a circle that the
> circumference would be too large
> a girdle for the sun.

Its whole expanse is made by a ray
reflected from the summit of the Primum
Mobile, which therefrom takes its life and potency.

The comparable figure in the Indian system is of the *nāda*, the creative sound OM of the energy of the light, ever resounding in the *ākāśa:* beyond which "space," both in Dante's concluding cantos (*Paradiso XXXI–XXXIII*) and in the stations six and seven of the Indian *sushumnā*, the culminating experiences are, first, of the vision of one's image of God and, then, of a transcendent Light which is the energy of the living world.

In oriental art, the image, whether of the Buddha, of Brahmā, of Vishnu, or of Shiva, is normally shown seated or standing upon a lotus. In Dante's vision, the Trinity with the whole heavenly host is beheld within the corolla of a radiant, pure white rose. In the Navaho sand painting the final image above and beyond the enclosed white space of the pollen-framed ceremonial field is of a symbolic bird perched on the central tassel at the top of the sacred corn plant. Noticing that the tassels are three, as are also the roots of the plant in the earth below, one is reminded that in Dante's *Divina Commedia* his vision of a Trinity above is matched by his Satan in the bowels of the earth, whose head shows three faces: the left face black, the right between white and yellow, and the middle red, signifying, respectively, in Dante's intention, ignorance, hatred, and impotent rage, in express contrast to the wisdom, love, and omnipotence of the Godhead.

In the Old Norse image of the World Ash, Yggdrasil, upon which Othin hung himself for nine days and nights in a sacrifice of himself to Himself to gain the wisdom of the runes (the scripture of the gods), there is an eagle perched on the topmost branch, and beneath, a dragon gnawing at the roots, of which there are three. The bird is a universal symbol of the spirit and spiritual flight, as is the feather of spiritual power. In India the wild gander, *haṁsa*, is symbolic of the *ātman*, the Self, and such a perfected saint as Ramakrishna is known as a Paramahamsa, ("Paramount or Supreme Wild Gander"). Jesus at the moment of his baptism saw a dove descending from the heavens. Zeus approached Leda in the aspect of a swan. Serpents and dragons, in contrast, are of the earth, its dynamism, urges, and demonic wisdom.

"The Pharisees and the Scribes have received the keys of Knowledge," said the Gnostic Jesus to St. Thomas and the disciples, "they have hidden them. They did not enter, and they did not let those enter who wished. But you, become wise as serpents and innocent as doves."

For the two modes of consciousness are of the one life, whereas the cherubim and flaming sword have forbidden entrance, even of the "justified," to the earthly garden where the two become again one.

Accordingly, in Dante's vision, Heaven and Hell are still separate, and the lower power (ethically judged) has been eternally condemned. The descent of the spirit is a fall. The life-giving demon has become a devil. The axial tree of the universe, around which all revolves, that is to say, is still cut in two, as it was in Yahweh's Eden of the two trees, one, of the Knowledge of Good and Evil, and the other, of the Knowledge of Eternal Life. Whereas in the unreformed, primeval archetype of the World Tree, such as appears in the Old Norse Yggdrasil and in the Navaho Blue Corn Stalk the life-giving roots and the pollen-bearing flowerings, or tassels, are of a single, organically intact, mythological image.

In the metaphorical Promised Land or Earthly Paradise of the Navaho sand painting (*paradise*, from the Greek *paradeisos*, "enclosed park"; Old Persian *pairi*, "around," and *daēza*, "wall," giving *pairidaēza*, "enclosure"), the rainbow curve, springing red and blue from the root, covers the zone of the first three chakras to the node of the second fruiting, where lightning strikes, flashing from the summit. The white footprints of the ceremonial path do not identify the initiate with either the rainbow or the lightning, only with the middle way; which is to say, apparently, that there is to be no absorption of the individual in identifications either with the gross matter of the world or with the unembodied light of sheer spirit. Both are recognized, evidently, as attributes of the Pollen Path. But the ordeal is to hold to the way between; so that even on coming to and passing the station of the Bluebird (Paramahamsa), the path, now the color of pollen (which is the color by which the garden is enclosed), does not ascend to the sky, but turning abruptly again to the right, bears the footprints back to the dwelling from which they came.

How Many Sheep Will It Hold?

BARRE TOELKEN

In this selection, Barre Toelken encourages the reader to "see" as Native Americans "see" in order to understand that there are various alternative patterns of perceptions. These different perceptions are important because learning about Navajo religion requires the learning of a whole new set of concepts, codes, patterns and assumptions. To understand any particular mythology and its metaphors we must learn to look carefully at the underlining assumptions of the culture "before we can see."

There are some things that one knows already if he or she has read very much about the native Americans. One of the most important is that there is almost nothing that can be said about "the Indians" as a whole. Every tribe is different from every other in some respects, and similar in other respects, so that nearly everything one says normally has to be qualified by footnotes. What I am about to say here does not admit room for that. I propose, therefore, to give a few examples from the Navajo culture and make some small glances at other Indian cultures that I know a little bit about; that is simply a device to keep my observations from appearing as though they were meant to be generally applicable to Indians of the whole country.

It is estimated that there were up to 2,000 separate cultures in the Northern hemisphere before the advent of the white man. Many of these groups spoke mutually unintelligible languages. Anthropologists estimate that there were as many as eighty such languages in the Pacific Northwest alone. In terms of language and traditions, these cultures were very much separated from each other; and although they have been lumped into one category by whites ever since (and that is the source of some of our problems), any given Indian will have a few things in common with some other tribes and many things not in common with others. My generalizations are made with this in mind from the start. But one must start *somewhere* in an attempt to cope with the vast conceptual gulf which lies between Anglos in general and natives in general, for it is a chasm which has not often been bridged, especially in religious discussion.

I do not claim, either, to be one of those rare people who *have* succeeded in making the leap—an insider, a confidant, a friend of the Red Man's Council Fire—in short, one of those Tarzans even more rare in reality than one would conclude from their memoirs. But I did have the good fortune to be adopted by an old Navajo, Tsinaabą̣ą̣s Yazhi ("Little Wagon"), in southern Utah in the mid-fifties during the uranium rush. I moved in with his family, learned Navajo, and lived essentially a Navajo life for roughly two years. Of course I have gone back since then on every possible occasion to visit my family, although my adopted father is now dead, as is his wife and probably 50 percent of the people I knew in the fifties. If one has read the Navajo statistics he knows why. This is not intended to be a tale of woe, however; I simply want it understood that I was not a missionary among the Navajo. Nor was I an anthropologist, a teacher, a tourist, or any of the other things that sometimes cause people to come to know another group briefly and superficially. Although, indeed, at one time I had it in my mind to stay with them forever, it is probably because my culture did not train me to cope with almost daily confrontation with death that I was unable to do so. I learned much from them, and it is no exaggeration to say that a good part of my education was gained there. It was probably the most important part. "Culture shock" attended my return to the Anglo world even though I left the Navajos as "un-Navajo" as when I arrived.

With that for background, though, I think I can say something about how differently we see things, envision things, look at things, how dissimilarly different cultures try to process the world of reality, which, for many native American tribes, includes the world of religion. In Western culture, religion seems to occupy a niche reserved for the unreal, the Otherworld, a reference point that is reached only upon death or through the agency of the priest. Many native American tribes see religious experience as something that surrounds man all the time. In fact, my friends the Navajos would say that there is probably *nothing* that can be called nonreligious. To them, almost anything anyone is likely to do has some sort of religious significance, and many other tribes concur.

Procedurally, then, our problem is how to learn to talk about religion, even in preliminary ways, knowing perfectly well that in one society what is considered art may in another be considered religion, or that what is considered as health in one culture may be religion in another. Before we can proceed, in other words, we need to reexamine our categories, our "pigeonholes," in order to "see" things through someone else's set of patterns. This is the reason for the odd title: "Seeing with a Native Eye."

Through our study of linguistics and anthropology we have learned that different groups of people not only think in different ways, but that they often "see" things in different ways. Good scientific experiments can be provided, for example, to prove that if certain ideas are offered to people in patterns which they have not been taught to recognize, not only will they not understand them, they often will not even see them. We see things in "programmed" ways. Of course Professor Whorf was interested in demonstrating the pervasiveness of this theory with respect to language, and many anthropologists and linguists have had reservations about his theories. But the experimentation continues, and there is some interesting and strong evidence that a person will look right through something that he or she is not trained to see, and that different cultures train people in different ways. I will not get into the Jungian possibilities that we may be born with particularized codes as well; this is beyond my area of expertise. But it is clear that when we want to talk about native American religion, we want to try to see it as much as possible (if it is possible) with the "native eye." That is to say, if we talk about native American religions using the categories of Western religions, we are simply going to see what *we* already know is there. We will recognize certain kinds of experiences as religious, and we will cancel out others. To us, for example, dance may be an art form, or it may be a certain kind of kinesis. With certain native American tribes, dance may be the most religious act a person can perform. These differences are very significant; on the basis of this kind of cultural blindness, for example, Kluckhohn classified the Navajo coyote tales as "secular" primarily because they are humorous.

The subtitle of this paper comes from my adopted Navajo father. My first significant educational experience came when I was trying to educate him to what the outside world looked like. Here was an eighty- or ninety-year-old man in the 1950s who had never seen a paved road or a train; he had seen airplanes flying overhead and was afraid of them. He had seen almost nothing of what you and I experience as the "modern, advanced world." I decided I would try to cushion the shock for him by showing him pictures, and then I would invite him into town with me sometime when I went to Salt Lake City. I felt he needed some preparation for the kind of bombardment of the senses one experiences in the city after living out in the desert.

I showed him a two-page spread of the Empire State Building which appeared in *Life* that year. His question was, immediately, "How many sheep will it hold?" I had to admit that I didn't know, and that even if I did know, I couldn't count that high in Navajo; and I tried to show him how big a sheep might look if you held it up against one of those windows, but he was interested neither in my excuses nor in my intent to explain the size of the building. When I told him what it was for, he was shocked. The whole concept of so many people filed together in one big drawer—of course he would not have used those terms—was shocking to him. He felt that people who live so close together cannot live a very rich life, so he expected that whites would be found to be spiritually impoverished and personally very upset by living so close together. I tried to assure him that this was not so. Of course I was wrong. Little by little one learns.

The next episode in this stage of my learning occurred about six months later, when I was at the trading post and found a magazine with a picture of the latest jet bomber on it. I brought that to him to explain better what those things were that flew over all the time. He asked the same question in spite of the fact there were lots of little men standing around the plane and he could see very well how big it was. Again he said, "How many sheep will it carry?" I started to shrug him off as if he were simply plaguing me, when it became clear to me that what he was really asking was, "What is it good for in terms of something that I know to be valid and viable in the world?" (That, of course, is not his wording either.) In effect, he was saying that he was not willing even to try to understand the Empire State Building or the bomber unless I could give those particular sensations to him in some kind of patternings from which he could make some assessment. He was not really interested in how big they were, he was interested in what they were doing in the world. When I told him what the jet bomber was for, he became so outraged that he refused ever to go to town, and he died without ever having done so as far as I know. He said that he had heard many terrible things about the whites, but the idea of someone killing that many people by dropping the bomb and remaining so far out of reach that he was not in danger was just too much!

The only other thing that approached such outrage, by the way, was when I explained to him about the toilet facilities in white houses, and I mentioned indoor toilet functions. He could hardly believe that one. "They do that right in the house, right inside where everyone lives?" "No, no, you don't understand. There is a separate room for it." That was even worse—that there could be a special place for such things. A world so neatly categorized and put in boxes really bothered him, and he steadfastly refused to go visit it. At the time I thought he was being what we call primitive, backward—he was dragging his feet, refusing to understand the march of science and culture. What I "see" now is that, as a whole, he was simply unable to—it did not "compute" in the way we might put it today; he did not "see" what I meant. In turn, he was trying to call my attention to that fact, and I was not receiving the impression.

I bring these matters up not because they are warm reminiscences, but because difficulties in communicating religious ideas are parallel to these examples. When my adopted father asked, "How many sheep will it hold?" he was asking, "What is it doing here, how does it function? Where does it go? Why do such things occur in the world?" We might consider the Pueblo view that in the springtime Mother Earth is pregnant, and one does not mistreat her any more than one might mistreat a pregnant woman. When our technologists go and try to get Pueblo farmers to use steel plows in the spring, they are usually rebuffed. For us it is a technical idea—"Why don't you just use plows? You plow, and you get 'x' results from doing so." For the Pueblos this is meddling with a formal religious idea (in Edward Hall's terms). Using a plow, to borrow the Navajo phrase, "doesn't hold any sheep." In other words, it does not make sense in the way in which the world operates. It is against the way things really go. Some Pueblo folks still take the heels off their shoes, and sometime the shoes off their horses, during the spring. I once asked a Hopi whom I met in that country, "Do you mean to say, then, that if I kick the ground with my foot, it will botch everything up, so nothing will grow?" He said, "Well, I don't know whether that would happen or not, but it would just really show what kind of person you are."

One learns slowly that in many of these native religions, religion is viewed as embodying the reciprocal relationships between people and the sacred *processes* going on in the world. It may not involve a god. It may not be signified by praying or asking for favors, or doing what may "look" religious to people in our culture. For the Navajo, for example, almost *everything* is related to health. For us health is a medical issue. We may have a few home remedies, but for most big things we go to a doctor. A Navajo goes to the equivalent of a priest to get well because one needs not only medicine, the Navajo would say, but one needs to reestablish his relationship with the rhythms of nature. It is the ritual as well as the medicine which gets one back "in shape." The medicine may cure the symptoms, but it won't cure you. It does not put you back in step with the things, back in the natural cycles—this is a job for the "singer." Considering the strong psychological and spiritual role of such a person, it should not come as a surprise that it is on spiritual (magic?) grounds, not medicinal, that some medicinal materials are *not* used. For example, Pete Catches, a Sioux medicine man (who practices the Eagle "way" of the Sacred Pipe), knows about but will not employ abortion-producing plants, for such use would run counter to and thus impede the ritualistic function of the pipe ceremony, a good part of which is to help increase the live things in the world. In the reciprocative life pattern, death is not a proper ingredient.

I want to go a little further into this, because these patterns, these cycles, these reciprocations that we find so prominently in native American religions, are things which for our culture are not only puzzling but often considered absolutely insane. It is the conflict or incongruency in patterning that often impedes our understanding. Let me give a few examples of this patterning. In Western culture—I suppose in most of the technological cultures—there has been a tremendous stress on lineal patterning and lineal measurements, grid patterns, straight lines. I think one reason for this is that technological cultures have felt that it is not only desirable but even necessary to control nature. We know there are very few straight lines in nature. One of the ways people can tell if they are controlling nature is to see that it is put in straight lines—we have to put things "in order." And so we not only put our filing cases and our books in straight lines and alphabetical "order," we also put nature in straight lines and grid patterns—our streets, our houses, our acreage, our lives, our measurement of time and space, our preference for the shortest distance between two points, our extreme interest in being "on time."

Those who have read the works of Hall and other anthropologists on the anthropology of time and space are familiar with these ideas. Each culture has a kind of spatial system through which one knows by what he sees as he grows up how close he can stand to someone else, how he is to walk in public and in private, where his feet are supposed to fall, where things are supposed to go. These patterns show up in verbal expressions too—we have to "get things straightened out," "get things straight between us," make someone "toe the line." We also arrange classrooms and auditoriums in some sort of lineal order (other groups might want these to be arranged in a circle). To us, having things "in order" means lining things up, getting things in line. We talk about "getting straight with one another," looking straight into each other's eyes, being "straight shooters." We even talk about the "straight" people vs. the "groovy" people. Notice how we often depict someone who is crazy with a circular hand motion around the ear. Someone who does not speak clearly "talks in circles," or uses circuitous logic. We think of logic itself as being in straight lines: A plus B equals C. We look forward to the conclusion of things, we plan into the future, as though time were a sort of straight track along which we move toward certain predictable goals.

If one knows much about native Americans of almost any tribe, he realizes that I am choosing, intentionally, certain lineal and grid patterns which are virtually unmatched in native American patterns. We learn to find each other in the house or in the city by learning the intersection of straight lines—so many doors down the hallway is the kitchen, or the bathroom, and we are never to confuse them. We separate them. One does not cook in the bathroom—it is ludicrous to get them mixed up. We have it all neatly separated and categorized. For most native American groups, almost the reverse is true—things are brought together. Instead of separating into categories of this sort, family groups sit in circles, meetings are in circles, dances are often—not always—in circles, especially the dances intended to welcome and include people. With the exception of a few tribes such as the Pueblo peoples, who live in villages which have many straight lines, most of the tribes usually live (or lived) in round dwellings like the hogan of the Navajo, the tipi of the plains Indians, the igloo of the Eskimo. The Eastern Indians and some Northwestern tribes sometimes lived in long houses, but the families or clans sat in circles within.

There is, then, a "logical" tendency to recreate the pattern of the circle at every level of the culture, in religion as well as in social intercourse. I think the reason for it is that what makes sense, what "holds sheep" for many tribes, is the concept that reciprocation is at the heart of everything going on in the world. I have had Pueblo people tell me that what they are doing when they participate in rain dances or fertility dances is not asking help from the sky; rather, they are doing something which they characterize as a hemisphere which is brought together in conjunction with another hemisphere. It is a participation in a kind of interaction which I can only characterize as sacred reciprocation. It is a sense that everything always goes this way. We are always interacting, and if we refuse to interact, or if some taboo action has caused a break in this interaction, then disease or calamity comes about. It is assumed that reciprocation is the order of things, and so we will expect it to keep appearing in all forms.

I think that it makes anthropological and linguistic sense to say that any culture will represent things religiously, artistically, and otherwise, the way its members "see" things operating in the world. But here is where the trick comes in. When we from one culture start looking at the patterns of another culture, we will often see what *our* culture has trained us to see. If we look at a Navajo rug, for example, we are inclined to say that Navajos use many straight lines in their rugs. And yet if we talk to Navajos about weaving, the *gesture* we often see is a four-way back-and-forth movement; and they talk about the interaction within the pattern—a reciprocation. Most often the Navajo rug reciprocates its pattern from side to side and from end to end, creating mirror images. My adopted sister, who is a very fine weaver, always talks about this kind of balance. She says, "When I am thinking up these patterns, I am trying to spin something, and then I unspin it. It goes up this way and it comes down that way." And she uses circular hand gestures to illustrate. While we are trained to see the straight lines, and to think of the rug in terms of geometric patterns, she makes the geometrical necessities of weaving—up one, over one—fit a kind of circular logic about how nature works and about how man interacts with nature. If we are going to talk about her beliefs with respect to rugs, we need somehow to project ourselves into her circles.

Let me give a couple of other examples. These, by the way, are not intended to be representative, but are just some things that I have encountered. They are simply illustrative of the way a Navajo might explain things. There is a species of beads that one often finds in curio shops these days. They are called "ghost" beads by the whites, though I do not know any Navajos who call them that except when talking to whites (they feel they ought to phrase it the way the whites will understand it). The brown beads in these arrangements are the inside of the blue juniper berries, which the Navajo call literally "juniper's eyes." In the most preferred way of producing these necklaces, Navajos search to find where the small ground animals have hidden their supply of juniper seeds. Usually a small girl, sometimes a boy, will look for likely hiding places, scoop them all out when she finds them, and look for the seeds that have already been broken open, so as not to deprive the animals of food. She puts all the whole seeds back, and takes only the ones that have a hole in one end. She takes them home, cleans them, punches a hole in the other end with a needle, and strings them together. I do not know any Navajo in my family or among my acquaintances who ever goes without these beads on him somewhere, usually in his pocket.

My Navajo sister says that the reason these beads will prevent nightmares and keep one from getting lost in the dark is that they represent the partnership between the tree that gives its berries, the animals which gather them, and humans who pick them up (being careful not to deprive the animals of their food). It is a three-way partnership—plant, animal, and man. Thus, if you keep these beads on you and think about them, your mind, in its balance with nature, will tend to lead a healthy existence. If you are healthy by Navajo standards, you are participating properly in all the cycles of nature, and thus you will not have bad dreams. Bad dreams are a *sign* of being sick, and getting lost is a *sign* of being sick. So these beads are not for warding off sickness itself; rather, they are reminders of a frame of mind which is essentially cyclic, in the proper relationship with the rest of nature—a frame of mind necessary to the maintenance of health.

Again, using the weaving of rugs as an example, I want to explain the significance of the spindle and the yarn. The yarn comes from the sheep, of course. The Navajos explain the relationship there not in terms of the rug, the end product—

which, of course, is what our culture is interested in—but in terms of the relationship with the yarn and with the sheep, and with the spinning of the yarn, which has to be done in a certain direction because it goes along with everything else that is spinning. Everything for the Navajos is moving; an arbitrary term in English such as *east* is phrased in Navajo, "something round moves up regularly." When one spins the yarn, then, one does not just twist it to make string out of it; one twists it in the right direction (sunwise) with everything else (otherwise, the thread will ravel). Thus, the yarn itself becomes a further symbol of man's interaction with the animal on the one hand, and with the whole of the cosmos on the other. When one works with yarn one is working with something that remains a symbol of the cyclic or circular interaction with nature. Even the spindle can be seen as an agency of, or a focal point in, a religious view of man and nature.

In a recent experiment by an anthropologist and a moviemaker, some young Navajos were given cameras and encouraged to make their own movies. One girl made a movie called "Navajo Weaving." It lasts, as I recall, almost forty-five minutes, but there are only a few pictures of rugs in it. Most of the film is about people riding horseback, wandering out through the sagebrush, feeding the sheep, sometimes shearing them, sometimes following them through the desert, sometimes picking and digging the roots from which the dyes are made. Almost the entire film is made up of the things that the Navajo find important about making rugs: human interaction with nature. That is what rugmaking is for the Navajos. Something which for us is a secular craft or a technique is for these people a part or extension of the reciprocations embodied in religion.

Religious reciprocity extends even into the creation of the rug's design. My Navajo sister wove a rug for me as a gift, the kind which the traders call *yei* (*yei* means something like "the holy people"). The pattern in this particular rug is supposed to represent five lizard people. The two on opposite ends are the same color, and the next two inward the same color, and the one in the middle a distinctly different color. The middle one is the dividing line, so that the pattern reciprocates from end to end of the rug. When my sister gave it to me, she said, "These represent your five children." Of course I was moved to inquire of her why she should represent my five children as lizards (I had private ideas about why she might). I wondered what her reasoning was, and I certainly knew children are not "holy people"—far from it. She pointed out, "Your oldest and youngest are girls, and they are represented by the two opposite figures on each end. Then you have twin boys—they are the two white ones, because they are alike. Then there is another boy, who doesn't have a mate in your family, so he is the center point of the family, even though he isn't that in terms of age." She made the pattern reciprocate from one end to the other not only in terms of representing my family but in terms of color. All the dyes were from particular plants which were related in her mind to good health. Lizards represent longevity, and by making my children congruent with the lizard people, she was making a statement of, an embodiment of, their health and longevity. This is a wish that any Navajo might want to express, because, as noted above, health and longevity are central to Navajo religious concerns. If I knew more about the symbolic function of certain colors in the rug, or the use of the dye-producing plants in

Navajo medicine, I have no doubt that I would have still more to say about the religious expression intended therein.

Reciprocity is central to the production of many other Navajo items, especially so in the making of moccasins. My brother-in-law, Yellowman, when he goes hunting for skins to put on the body, tries to produce what the Navajos call "sacred deerskin." It is supposed to be produced from a deer whose hide is not punctured in the killing. If one wants the deer for meat, it can be simply shot (Yellowman, though he is in his early sixties, still hunts with a bow and arrow). But when he hunts deer for moccasins, or for cradle boards for his family (the deerskin helps to surround the baby), then he wants skin of the sacred kind. To obtain sacred deerskin in the old fashion, one runs the deer down until it is exhausted, and then smothers it to death.

It is done in this manner: one first gathers pollen, which he carries with him in a small pouch. He then gets the deer out into open country and jogs along behind it, following until it is totally exhausted. Deer run very rapidly for awhile but soon get tired. The man who is good at jogging can keep it up for some distance. Still, it is no easy job, as you can imagine if you have ever visited the desert of the Monument Valley area. When the deer is finally caught, he is thrown to the ground as gently as possible, his mouth and nose are held shut, and covered with a handful of pollen so that he may die breathing the sacred substance. And then—I am not sure how widespread this is with the Navajos—one sings to the deer as it is dying, and apologizes ritually for taking its life, explaining that he needs the skin for his family. The animal is skinned in a ritual way, and the rest of the deer is disposed of in a ritual manner (I do not feel free to divulge the particulars here).

The deerhide is brought home and tanned in the traditional way. The coloration is taken from particular kinds of herbs and from parts of the deer (including its brains). Then the moccasins are made by sewing the deerhide uppers together with cowhide soles. In many cases they are buried in wet sand until the person for whom they are designed comes by. He puts them on and wears them until they are dry. In so doing, of course, he wears his footprints into them. You can always tell when you have on someone else's moccasins, if that mistake should ever occur, because they hurt. Your own toe prints are in your own moccasins, for they have become part of you. It is no accident that the word for moccasin or shoe is *shi ke'*, "my shoe," which is exactly the same word for "my foot." Religiously speaking, what happens is that the deerskin becomes part of us, and this puts us in an interactive relationship with the deer. The whole event is ritualized, carried out in "proper" ways, because it falls into a formal religious category, not a mere craft. The moccasin is more than something to keep the foot warm and dry: it is symbolic of that sacred relation and interaction with the plants and the animals that the Navajo sees as so central to "reality."

Also central to Navajo religion is the restoration of health when it has been lost. The *hogan* is the round dwelling the Navajos live in. The fire is in the middle of the floor, and the door always faces east. One of the reasons for this, as my adopted father told me, was to make sure that people always live properly oriented to the world of nature. The door frames the rising sun at a certain time of the year. The only light that comes in is either through the smokehole on the top, or through the door, if it happens to be open. Healing rituals involving "sandpainting" are usually enacted inside the hogan, and are oriented to the four directions. When the patient takes his or her place on the sandpainting, ritually they are taking their place within the world of the "holy people," related to all the cycling and reciprocation of the universe. It is partly that orientation which cures one.

Yellowman still hunts for meat with bow and arrow. His arrowheads are made out of ordinary carpenters' nails pounded out between rocks, although he has a whole deerskin bag of stone arrowheads that he has picked up on the desert. When I asked him why he did not use those nice stone points on his own arrows, he looked at me very strangely. (I knew that the Navajos put them in the bottom of medicine containers when they are making medicine, but I thought that perhaps he knew how to make them himself given the proper kind of rock.) I asked whether he knew how the oldtimers used to make them. He looked at me as if I were absolutely insane. He finally answered, "Men don't make them at all; lizards make arrowheads." For him, stone arrowheads, such as one might find, are sacred items, and they fall into the same category as lizards, lightning, and corn pollen. Lizards, as I mentioned earlier, are related to long life and good health. When one finds a lizard or an arrowhead, he picks it up and holds it against the side of his arm or over his heart, the same places where pollen is placed during a ceremony. Clearly, stone arrowheads are for curing, not killing; or, more properly, they are for killing *diseases*. Thus, even arrowheads have to do with special sacred medicinal categories, not with the kinds of practical categories our culture might see. In other words, learning about Navajo religion and daily life requires the learning of a whole new set of concepts, codes, patterns, and assumptions.

A student of mine paraphrased an old proverb this way: "If I hadn't believed it I never would have seen it." This is essentially what I am saying about viewing religion in other cultures. Our usual approach is in terms of pictures, patterns, gestures, and attitudes that we already know how to see. For example, when some dance specialists went to Tucson a couple of years ago to watch the Yaqui Easter ceremonies, all they saw were the dances. They did not see that on a couple of occasions, several people very prominent in the ritual were simply sitting next to the altar for extremely long periods of time. I talked to almost every person at that conference, and only a few of them had seen those people sitting there. There wasn't any dancing going on there, and so the dance people weren't "seeing." And yet it was probably a very important part of the dance. I do not pretend to have understood this part, but the point is that the strangers had not even seen it; they were watching for what they as Anglos and dance specialists could recognize as dance steps. I would not accuse them of stupidity, ignorance, or narrowmindedness. Rather, they had not been taught to "read," to see other kinds of patternings than their own.

To complicate matters further, many tribes feel the real world is not one that is most easily seen, while the Western technological culture thinks of *this* as the real world, the one that *can* be seen and touched easily. To many native Americans the world that is real is the one we reach through special, religious means, the one we are taught to "see" and experience *via* ritual and sacred patterning.

Instead of demanding proof for the Otherworld, as the scientific mind does, many native Americans are likely to counter by demanding proof that *this* one exists in any real way, since, by itself, it is not ritualized.

What the different cultures are taught to see, and how they see it, are thus worlds apart (although not, I think, mutually exclusive). One culture looks for a meaning in the visible, one looks for a meaning beyond the visible. The "cues" are different because the referents and the connotations are different. Add to this basic incongruency the fact that the patterning of one is based on planning, manipulation, predictability, competition, and power, while the other is based in reciprocation, "flowering," response to situation, and cooperation—and who would be surprised to find that the actual symbols and meanings of the two religious modes will be perceived and expressed in quite contrastive forms? We must seek to understand the metaphor of the native American, and we must be willing to witness to the validity of its sacred function, or else we should not pretend to be discussing this religion. Before we can see, we must learn how to look.

The Shadow of a Vision Yonder

SAM D. GILL

In this selection, Sam D. Gill demonstrates how Native American people apprehend reality. In his discussion of the Navajo sand painting, he points out that reality is dynamic for the Native American and that the concepts of being and becoming are communicated through the changing use of symbols. Through his discussions of the Hopi use of the Kachina mask, he shows how ordinary materials when presented in the proper ritual context, are transformed into the sacred and thereby, given their meaning.

Several summers ago while my family and I were living with a Navajo family north of Tuba City, Arizona, I witnessed an ordinary social event that at the time I thought to be curious but of little consequence. Since then I have found occasion to reflect upon that event. From it I think I learned something about the Navajo way of life, even something about their religion, which is the subject I had gone there to pursue. I confess that I went to the Navajo reservation not very well prepared to do fieldwork. I had not done enough homework to afford me the clearest view of my contact with Navajo people. As a result, much of what I was to learn came to me through insightful flashbacks sometime after I had left Navajo country.

By midsummer we had become well enough acquainted with our Navajo family to be trusted with some of their work. I considered it an honor to be asked to help hoe the weeds in the cornfield they had planted in the valley below the beautiful mesa on which we lived. Being from a farming family in Kansas, I willingly accepted the invitation and replied that I would gladly hoe the corn. To my dismay our Navajo friends expressed alarm. I am sure they were considering how they could retract the invitation as they told me, "Oh no, we don't hoe the *corn,* we hoe the weeds!" I assured them that I really did know the difference between corn plants and the unwanted weeds and that it was just the way we described the job back home. It was simply a product of the peculiarity of my own language, not theirs.

After getting the younger children on their way with the sheep, we headed for the cornfield early the next morning. Under cautious eyes, I set about proving that I not only knew the difference between weeds and corn, but that I was no slouch with a hoe. Of course, I was never to know the extent of my success. The weight of my experience with the Navajo people is that their quiet dignity always prevails.

With my flashing hoe gradually slowing to match the ordered, rhythmic movement of the other hoes—the native hoes—I was relieved when late in the morning it was time to stop for lunch. Moving to the arbor or "shade," a small partly enclosed brush structure, we took lunch. Then we prepared to rest for several hours during the heat of the day. It was in the shade that I was to observe the event on which I want to reflect.

The shade was perhaps half a mile from a narrow dirt road. In that part of the country the traffic is not what one would call heavy. During the quiet rest period after lunch, I was aroused from my drowsiness a couple of times by a soft but excited discussion of whatever motor vehicle, usually a pickup truck, passed by on the road. I noticed that all present expressed interest in the traffic. They arose and peered through the open areas in the brush on the side of the shade facing the road. I recalled the many times I had driven up to a Navajo dwelling, finding absolutely no sign of human activity.

What surprised me was the response my friends made when one of the passing pickups turned off the road and headed toward our shade. Watching with rapt attention, my Navajo friends carefully timed it so that as the truck pulled up to the shade and stopped, every member of the family was actively occupied. The grandmother sat on the ground with her back to the entrance near the truck and began her spinning. The children played a game in the dirt of the shade floor. Others sat about, gazing across the landscape, always in a direction away from the truck. This directed all attention away from the presence of the visitors.

The visitors in the truck were Navajos and knew how to respond. They sat in the truck for some minutes. It seemed like a very long time to me. Then quietly, the man, his wife and young daughter left their truck to enter the edge of the shade. There they sat upon the ground. The man quietly restrained the eagerness of the little girl to play with the other children. Again some minutes passed while my family continued their spinning, playing, and gazing. Finally, the man spoke a few soft words to the grandmother, who gently, almost inaudibly, responded without turning her head toward him. In a few moments he spoke again. This quiet conversation continued for some time, then the visitors arose and moved about the shade, talking softly to each of us, including me, extending their hands for a handshake and speaking the Navajo greeting, "*yá'át'ééh.*" Next my friends arose and began to intermingle with the visitors. I was informed that they were going to the trading post some ten miles away to get water and supplies to prepare a meal for the visitors. The entire proceedings had taken more than a quarter of an hour.

The insight that has come to me through continued reflection is that the incident illustrates the "way" of Navajo religion. I had witnessed the performance of a formal ritual for purposes of establishing certain kinds of social relationships, in this instance, between Navajo families. The ritual reflects the quiet dignity and the patient and formal manner of the Navajo people. And by its simplicity it helps place the almost infinite complexity of Navajo ceremonies in a better perspective. It also gives clues regarding the nature of Navajo religion, wherein relationships are established or reestablished with the holy ones.

Notice that the situation had been carefully analyzed by my Navajo friends. They followed proper conduct with deliberateness and patience. This resulted in the successful establishing of a relationship between two families. Each party made some sign to show that it understood its obligations and was committing itself to their fulfillment. The guests offered their hands as a sign of their entrance into the relationship. My family proceeded to meet their first obligation of the relationship by offering a meal to the guests.

It is commonly observed that Navajo religion centers largely upon the rituals by which an individual who is suffering a malady is healed. The sufferer is attended to by an individual called a "singer." The "singer" directs the ritual activities and is responsible for knowing the songs, prayers, and the order of the ritual processes. A Navajo ceremony is not performed unless it is called for. But when it is called for, the family of the sufferer must arrange with a "singer" to perform the ceremony. This requires making a formal relationship through social and ritual acts not unlike those characteristic of the introduction to which I have made reference. In both settings the relationship is bonded through a formal sign. In this case the "singer" receives payment in material goods or in cash in return for performing the ceremony. David Aberle has analyzed this exchange and has convincingly shown that the "fee" is not really payment for services rendered, but is a sign of the establishment of a reciprocity relationship. The "singer" is thus obliged to respond by conducting the requested ceremonial.

In the performance of Navajo ceremonies, the observer is struck by the material insignificance of the ritual objects. He also cannot help but notice the extreme care and formality with which these objects are treated. A singer's medicine bundle consists of nothing more precious than an odd assortment of sticks, feathers, bags of colored sands and vegetal materials, rocks, and so on. These things appear so common, even crude, that it makes one wonder how they could have any religious significance. But in the context of ritual the same objects are carefully handled, described in song, explicated in prayer, and manipulated in ritual. Their significance is developed to such a magnitude that they infinitely surpass their material content in signifying that which the Navajo regards as being holy.

This is in keeping with the story of creation in Navajo mythology. The story utilizes common objects in describing the process of bringing life to the world. First Man, who directed the creation, had a medicine bundle containing bits of colored rocks, called "jewels." First Man carefully placed these "jewels" upon the floor of the creation hogan to designate the life forces of the things which were to be created. All life forms were represented in these mundane substances, and their distinguishing characteristics were understood to be exemplified in the shapes of the jewels. Furthermore, the relative place where each was laid on the floor of the creation hogan designated the place each was to occupy in the world, together with the relationship each was to have to all other living things. These material representations of life were then clothed in a layer of colored sands to represent the outward appearance they would have in the created world. After the preparation of this symbolic microcosm has been completed, prayers were uttered to transform the symbolic creation into the more visible everyday world of the Navajos. This is the way in which the Navajos conceive the process of the creation of their world. When creation was completed the world was beautiful. All things were formed and set in a place, and proper relationships existed between them.

In both the creation of the world and in the creation of the social relationships formalities dominate. The formal enactment of ritual brings things to their proper place and serves to interconnect them by establishing binding relationships. Ritual acts are understood to be essential to the establishing of proper relationships. Navajo life depends upon such relationships.

Scholarly interpretation doesn't always catch the significance of this. Frequently, the interpretation of Navajo religion has called attention to the performance of "magical" acts. They are called "magical" to indicate that there is no ordinary causal principle which connects the acts performed with the expected results. I would never want to dismiss the presence of mystery and magic in Navajo religion. Yet it seems to me that the more significant factor is the process by which the visions and great conceptions are communicated by the formal manipulation of mundane objects. Let me illustrate the difference. The most common scholarly interpretation of the sandpainting rite is that it contains a kind of magical osmosis. The sandpainting is prepared upon the floor of the ceremonial hogan, the patient enters and sits upon the sandpainting, and the singer applies sands from the figures represented in the sandpainting to the patient. At the conclusion of the rite, the sandpaintings are formally destroyed and removed from the ceremonial hogan. According to the "magical osmosis" explanation, the sandpainting is understood to absorb the illness, or the evil cause of the illness, taking it from the patient and replacing it with goodness from the sandpainting. This explanation focuses attention on the removal of the sands after the rite, for it resembles and builds upon similarities between this act and the removal of sands into which one vomits in emetic rites.

In my view, this "magical osmosis" interpretation is partial. I would propose instead that sandpainting rites are meaningful curing acts because of the

Navajos' recognition of the *performative* powers of symbolic representation. In preparing the sandpainting, the Navajos follow the precedent established in the processes of world creation. In Navajo mythology it is said that "in the beginning" the forces of life were set forth in material form by arranging common objects of several colors upon the floor of a ceremonial hogan. Thus, in physical representations using ordinary materials, Navajos express their conception of the profound nature of life. In a healing ceremony the sandpaintings are closely associated with the elements identified with the cause of the illness suffered. As is told in the mythology of each ceremonial, the sandpaintings are revealed to the mythic hero as he is being cured of an illness. In most cases the illness is due to the fact that something is out of its proper place. For example, a ghost who will not remain in its domain, a person who has made contact with the dead, a deity who has been angered or offended by a person who has trespassed or violated a taboo, or a witch who has gained power by being out of bounds. The causal agent rather than the illness suffered determines the nature of the ceremonial cure. The ritual presents the forces of life in the shape and relative places assigned to them by sacred history as recounted in the myths. The identification of the patient with the sandpainting by touching the sands of the parts of the body of the painted figures to the corresponding parts of the patient is a gesture of communicating *proper relationships* to the patient. This is very similar to the acts performed to place the forces of life represented on the floor of the creation hogan within the representations of the outward forms they were to take in the real world. And, as in the case of the process of world creation, the formal removal of the sandpaintings designates a transition from the world of ritual to the world thus represented. This transition illustrates that reality is dynamic and will not always be contained in symbolic form.

Relationships are central to the Navajo way of life. Life's interrelationships are not casual. They are the product of careful ritual prescription, which acts both to bind and reestablish a proper order of relationships. In the Navajo conception, life and good health are not so much a matter of substance—for all things belong to the earth—as they are a matter of form and place with respect to the rest of the created world. Each living thing has an identity, a proper place, and a way to be. This identity, place, and way must be honored and carefully maintained.

The Navajo way of life can be characterized at one level as a kind of symbolic formalism, although Navajos would not describe it in these terms. The Navajos' own appreciation of symbolization becomes particularly compelling in their belief about the curative power of the healing rites. Here the symbols presented are appreciated for having the power to cure physical illness; and the Navajo have in mind something quite different from our common reduction of their religion to a kind of primitive psychology. In the symbols of their religion they recognize a power to change the shape of things in the world, even when the materials which compose the symbols are mundane. This performative power stems from a religious tradition that takes form in ritual acts. Such acts make earthly elements into a vehicle disclosing the deepest forces of life.

Navajo symbols within sandpainting rites are comparable to the shaking of hands to seal a social relationship. Both of these acts reflect the same temperament. Both indicate the way in which Navajos apprehend reality. In both cases, mundane ingredients find deeper symbolic significance. There is nothing special in the handshake, for example. But in the context of the formal ritual of establishing relationships, handshaking performs an essential role by assuring each party of the acceptance of the privileges and obligations of the relationship. It marks transformation from a relationship discussed to a relationship established and made operative. Similarly, in Navajo sandpainting rites the substance of the colored sands is not as important as the shapes which they form. Properly prepared and used, the sandpainting has the power to cure. It reestablishes the patient with the forces of life on which his health and happiness depend. In this regard, one of the most important components of native American religions is the process by which concepts of being and becoming are represented and communicated through the use of symbols. I have cited one instance of this in Navajo religion. We can find the same phenomenon when we turn to Hopi culture.

I remember feeling confused when I learned that Hopi children witness an event which they find shocking and bitterly disappointing at the conclusion of their first religious initiation. I am referring to the conclusion of the initiation into the kachina cult which is composed of two societies, the Kachina Society and the Powamu Society. Formally, this initiation begins the religious life of all Hopi children, boys and girls alike. The event occurs as a part of the Bean Dance which concludes the annual celebration of Powamu, a late winter ceremonial to prepare for the agricultural cycle. The newly initiated children are escorted into a kiva, an underground ceremonial chamber, there to await the entrance of the kachinas, the masked dancers they have come to know as Hopi gods. Prior to this time, the already initiated go to great efforts to keep the children from discovering that kachinas are masked male members of their own village. Announcing that they are kachinas, the dancers enter the kiva where the children are eagerly awaiting them. But they appear for the first time to the new initiates without their masks. The children immediately recognize the identity of the dancing figures. Their response is shock, disappointment, and bitterness. Don Talayesva, an old Oraibi Hopi, recalled his feelings at the time of his initiation in his autobiography *Sun Chief*.

> When the Katcinas entered the kiva without masks, I had a great surprise. They were not spirits, but human beings. . . . I had been told all my life that the Katcinas were gods. I was especially shocked and angry when I saw all my uncles, fathers, and clan brothers dancing as Katcinas.

It would seem to me that this concluding event in the Powamu ceremonial leaves the children in a peculiarly unstable state as new initiates. I would have expected the purpose of the initiation to reveal clearly the full nature of the kachinas to the children. But it appears that the initiation rites accomplish only the destruction of the belief in the identity of the kachina figures as gods, as it was held by the children prior to initiation. I understand that Margaret Mead likened this event to the European-American child learning of the identity of Santa Claus, which is often accompanied by the same kind of bitter disappointment. There are surface similarities, but this is not a satisfactory explanation. Nor should we accept another scholarly interpretation that it is inevitable the children learn that kachinas are "not real gods, but men dressed as gods." We may find some force in this argument, since, as outside occasional observers, we can easily recognize that the kachina dancers are masked mortals. Even when we hear a Hopi say that when in donning the kachina mask he "becomes the kachina," we tend to offer a critical interpretation. Another older position dismisses the statement as primitive nonsense, regarding it as a product of a primitive mentality not skilled in precise distinctions. Then too, a play theory has been advanced. This position argues that the Hopi acts "as if" he were a kachina and makes the statement while he is so pretending. But all of these interpretations are found wanting.

Instead, it is important that we take seriously what the initiated Hopi says. We must recognize that he actually means what he says, that in putting on the kachina mask he really becomes a god. This is a clear statement on his part. It is in light of this statement that we should attempt to see how at the conclusion of the initiation the shadow cast upon the kachina figures serves to reveal to the children the true nature of the kachinas. The ceremony appears to be deliberately calculated to engender the disappointment the children feel.

A fuller review of the contextual events is necessary. Prior to the initiation into the kachina cult, the children largely under the age of ten are carefully guided into the development of a particular kind of relationship with the kachinas. The kachinas, who frequent the villages during only half of the year, have a wide range of contacts with the children. Many of them are kind and benevolent to the children, presenting them with gifts. Others are frightening ogres who discipline naughty children by threatening to eat them. And some are silly clowns who entertain the children with their antics. In all these contacts the uninitiated children are protected against seeing kachinas unmasked or the masks unoccupied. They are also guarded against hearing anything which might disclose the masked character of the kachina figures. The children are told that the kachinas are gods who come to the village from their homes far away to overlook and direct the affairs of the Hopi people. They are taught that they too will become kachinas when they die. Prior to the initiation events, the children grow to accept the familiar kachina figures as being exactly what they appear to be.

The perspective nurtured in the children is given its final stage of development in the kachina cult initiation rites. During the Powamu ceremonial to which the initiation rites are attached, the initiates are given special attention by the kachinas. They come into closer contact. The kachina give the children special gifts. They are instructed in kachina lore. All of this seems to be carefully calculated to intensify the shock the children will feel when they observe the unmasked appearance of the kachinas during the Bean Dance.

When the kachinas enter the kiva, in one sharp and sudden blow the expectations so carefully nurtured are forever shattered. For the moment only pain and bitterness take its place, as is evident in the statement of a Hopi woman quoted by Dorothy Eggan in her 1943 study of Hopi adjustment.

> I cried and cried into my sheepskin, that night, feeling I had been made a fool of. How could I ever watch the Kachinas dance again? I hated my parents and thought I would never believe what the old folks said, wondering if Gods had ever danced for the Hopi as they said and if people really lived after death.

But even with the disappointment life goes on, and the initiated is given the privilege to participate in religious events. In time, the initiated can enter other religious societies and enjoy expanded privileges of participation. But once initiated into the kachina cult, religious events can never again be viewed naively. Unforgettably clear to the children is the realization that some things are not what they appear to be. This realization precedes the appreciation of the full nature of reality.

Professor Alfonso Ortiz, anthropologist and Tewa Indian, has pointed out that in the Pueblo worldview "all things are thought to have two aspects, essence and matter." In the shadow cast by the destruction of their naiveté, the initiated children are made aware of the "essence" or sacrality of things they had until then seen only as "matter." Thus, the initiation serves to bring the children to the threshold of religious awareness and as a consequence initiates their religious lives. Once begun, the lifetime of the Hopi is a gradual progression in the acquisition of knowledge and the appreciation of the nature of "space, time, being, and becoming," to use the subtitle of Ortiz's book *The Tewa World.*

This brings us back to the question of truth regarding the Hopi statement that when one dons the kachina mask he becomes a kachina. Given the appreciation by the initiated Hopi of the full nature of reality in both its material and essential aspects, the truth of the statement can be more clearly understood. By donning the kachina mask, a Hopi gives life and action to the mask, thus making the kachina essence present in material form. Mircea Eliade illuminates this point in his book *The Sacred and the Profane*: "by manifesting the sacred, any object becomes *something else*, yet it continues to remain itself." The anomaly we observe in the Hopi statement that he becomes a kachina is but an expression of the paradox of sacredness; but, in this case, the sacred object is the Hopi himself. By wearing the kachina mask, the Hopi manifests the sacred. He becomes the sacred kachina, yet continues to be himself. We, as uninitiated outsiders, observe only the material form. The spirit, or essence, of the kachina is present as well, but that can be perceived only by the initiated. The material presence without the spiritual is but mere impersonation—a dramatic performance, a work of art. The spiritual without the material remains unmanifest; it leaves no object for thought or speech or action. The spiritual must reside in some manifest form to be held in common by the community. The view, often taken, that the kachinas are "merely impersonations" fails to recognize the full religious nature of the kachina performances. It also fails to take into account the truth of the statement. If the kachinas are not present in both material and essential form, the events could scarcely be called religious.

Both Navajo and Hopi religions evidence an appreciation for the power of symbolization. Only through symbolization is the sacred manifest; the subtleties are many. On the one hand, the mundane materials which comprise religious symbols must never be taken as being more than the simple ordinary earthy elements they are. This fact is driven home in the disenchantment with the material appearance of the kachinas experienced by the children undergoing initiation. It is also evident in the example of the sandpainting rite from Navajo culture. On the other hand, the ordinary materials *when presented in the proper form* manifest the sacred. Both Navajo sandpaintings and Hopi kachinas have the power to order and affect the world in a very profound way.

I think that this deep appreciation for the process of the manifestation of the sacred is broadly held among native Americans. One of the best formulations of it I know is found in the wisdom of the Oglala Sioux, Black Elk, as told to John Neihardt in *Black Elk Speaks*. As a youth, Black Elk was the recipient of a remarkable vision which he looked to as a guide throughout his life. For many years he kept the vision to himself, fearing to tell others. But as time went on, he found rising within him an even greater fear. Part of the message given him was that he was to enact the vision in ritual form for the people to see. This was a common practice among the Dakota. An old medicine man from whom Black Elk sought guidance warned him that if the vision were not performed, something very bad would happen to him.

Under Black Elk's direction, preparations were immediately begun so that the vision could be enacted by the people. Black Elk recalls how he experienced the enactment of his vision.

> I looked about me and could see that what we then were doing was like a shadow cast upon the earth from yonder vision in the heavens, so bright it was and clear. I knew the real was yonder and the darkened dream of it was here.

There is a sense in which the Navajo sandpaintings, the Hopi kachina masks, and many other native American ritual acts share the properties of a shadow of a vision yonder. I have stressed that it is the form more than the substance that is important in manifesting the sacred forces of life. Certainly this is characteristic of shadows. Further, there is a "thinness" to ritual objects and acts which finds them to be real and meaningful only when cast in the light of "the yonder vision." This fragility is illustrated in the Navajo sandpaintings which are destroyed in the very acts by which they are of service. All of this is a constant reminder that the material symbols exist and are meaningful only in the degree to which they manifest the sacred. Were it not for these shadows cast by the vision yonder, American Indian religion would be confined to the experience of rarified mystical moments. The shadow may appear bright and clear as it did to Black Elk, or dark and foreboding as it does to Hopi kachina cult initiates, but the shadows integrate American Indian religion with a distinctive way of living and interpreting life.

In The Beginning

ALFONSO ORTIZ

Alfonso Ortiz is a social anthropologist and a native of the San Juan Pueblo, the largest of the surviving Tewa villages in New Mexico. In his writings he brings together his perspective as a social anthropologist and his experience and knowledge as a participant in the Pueblo culture. In this selection, Ortiz presents the "myth of origin and the early migrations of the Tewa."

The Tewa were living in *Sipofene* beneath Sandy Place Lake far to the north. The world under the lake was like this one, but it was dark. Supernaturals, men, and animals lived together at this time, and death was unknown. Among the supernaturals were the first mothers of all the Tewa, known as "Blue Corn Woman, near to summer," or the Summer mother, and "White Corn Maiden, near to ice," the Winter mother.

These mothers asked one of the men present to go forth and explore the way by which the people might leave the lake. Three times the man refused, but on the fourth request he agreed. He went out first to the north, but saw only mist and haze; then he went successively to the west, south, and east, but again saw only mist and haze. After each of these four ventures he reported to the corn mothers and the people that he had seen nothing, that the world above was still *ochu*, "green" or "unripe."

Next the mothers told him to go to the above. On his way he came upon an open place and saw all the *tsiwi* (predatory mammals and carrion-eating birds) gathered there. There were mountain lions and other species of cat; wolves, coyotes, and foxes; and vultures and crows. On seeing the man these animals rushed him, knocked him down, and scratched him badly. Then they spoke, telling him: "Get up! We are your friends." His wounds vanished immediately. The animals gave him a bow and arrows and a quiver, dressed him in buckskin, painted his face black, and tied the feathers of the carrion-eaters on his hair. Finally they told him: "You have been accepted. These things we have given you are what you shall use henceforth. Now you are ready to go."

When he returned to the people he came as Mountain Lion, or the Hunt chief. This is how the first Made person came into being. On approaching the place where the people awaited, he announced his arrival by calling out like a fox (*de*). This is his call. The people rejoiced, saying, "We have been accepted."

The Hunt chief then took an ear of white corn, handed it to one of the other men, and said, "You are to lead and care for all of the people during the summer." To another man he handed another ear of white corn and told him, "You shall lead and care for the people during the winter." This is how, according to the myth, the Summer and Winter chiefs were instituted. They joined the Hunt chief as Made People.

Among the people were also six pairs of brothers called *Towa é*, literally "persons." To the first pair, who were *blue*, the newly appointed chiefs said: "Now you shall go forth to the north and tell us what you see." They went with the older one in the lead, but could not walk very far because the earth was soft. All they saw was a mountain to the north. They returned and reported their observations to the people. Next the *yellow* pair were sent out to the west, followed by the *red* ones who went south, the *white* ones who went east, and the *dark* (*nuxu in*) ones who went to the zenith. Each successive pair returned and reported that the earth was still soft. The yellow, red and white brothers reported seeing mountains in each direction, while the dark pair who went above reported seeing *agoyo nuxu*, a large star in the eastern sky. The first four pairs each picked up some mud and slung it toward each of the cardinal directions, thereby creating four *tsin*, or flat-topped hills.

Finally, the last pair of *Towa é*, the *all-colored* (*tsege in*) ones of the nadir, were sent out. They found that the ground had hardened somewhat, and they saw a rainbow in the distance. When they returned and reported this, the people made preparations to leave the lake. The Summer chief led the way, but as he stepped on the earth it was still soft and he sank to his ankles in the mud. Then the Winter chief came forth, and as he stepped on the ground there was hoarfrost. The ground became hard, and the rest of the people followed. The original corn mothers and other supernaturals, the predatory mammals, and the carrion-eaters remained beneath the lake. The *Towa é*, or brothers, went to the mountains of the directions to stand watch over the people.

As the people started southward many began to fall ill. The Winter and Summer chiefs decided that they were not yet complete; something else was needed. All returned once again to the home under the lake, and there the Hunt chief opened up Summer chief's corn mother. He discovered that the hollow core was filled with pebbles, ashes, and cactus spines. The Hunt chief replaced these with seeds and declared that one among the people was "of a different breath," or a witch, for the items discovered in the corn mother were recognized as items of witchcraft. This, then, marked the beginning of witchcraft and other forms of evil. In order to combat these and to make the people well, the *Ke* (medicine man) was created as the fourth Made person. The people then started out once again.

Before they proceeded very far south, they all had to return to the lake three times more, because the chiefs felt they were still not complete. At each subsequent return the *Kossa* and *Kwirana* (clowns), the Scalp chief, and the *Kwiyoh* (Women's society) were instituted, in that order. The *Kossa* and *Kwirana* were created to entertain the people when they grew tired and unhappy, the Scalp chief to insure success in warfare, and the Women's society to care for the scalps and to assist the Scalp chief. The people at last felt they were complete, and prepared to proceed southward once again.

Before doing so, the Hunt chief divided the people between the Summer chief and the Winter chief. Those who were to follow the Summer chief would proceed south along the mountains on the west side of the Rio Grande. The Winter chief and his group would proceed along the mountains on the east side of the river. The Summer People, as the former group came to be called, subsisted by agriculture and by gathering wild plant foods, while the Winter People subsisted by hunting. Each group "took twelve steps" (made twelve stops) on this journey, and after each step they built a village and stayed for a day. "In that time one day was one year." Those who died along the way—for death was now known—were buried near the village and stones piled over the graves.

At the twelfth step the two groups rejoined and founded a village called *Posi,* near present day Ojo Caliente. The village grew and prospered, and the people remained there for a long while. In time, however, an epidemic struck and the elders decided to abandon the village. Six different groups left and founded the six Tewa villages we know today. San Juan was founded first, so it became the "mother pueblo" for the other five. Each of the six departing groups included both Winter and Summer people, so the chiefs and other Made People were replicated in each village. The origin and migration myth ends with an informant's observation:

> In the very beginning we were one people. Then we divided into Summer people and Winter people; in the end we came together again as we are today. But you can see we are still Summer people and Winter people.

Unit III

And We Washed Our Weapons in the Sea: Gods and Goddesses of the Neolithic Period

Archaeoastronomy and Its Components

ANTHONY F. AVENI

Anthony F. Aveni is known for his work in archeaoastronomy, a combination of astronomy and archaeological and ethnological research, especially in the area of pre-Colombian skywatching in the New World. In this selection, he indicates the uses of astronomy throughout the ancient world and explains how archaeologists today are coming to understand the complexity of that ancient use.

"... *Maya astronomy is too important to be left to the astronomers.*"
—*Sir Eric Thompson*

All developing civilizations exhibit a reverence for the sky and its contents. The cyclic movement of the sun, moon, planets, and stars represents a kind of perfection unattainable by mortals. The regular occurrence of sunrise and moonset provided the ancients with something dependable and orderly, a stable pillar to which their minds could be anchored.

Today we no longer have need of practical astronomy in our daily lives. Unlike our ancestors, we spend most of our time in a regulated climate with controlled lighting; we are detached almost totally from the natural environment. Technology has created an artificial backdrop against which we play out our lives. Any need we once had to watch carefully for celestial events has become lost. Who knows the time the sun rose today or the current phase of the moon? The clocks by which we pace our daily activities give us a distorted view of the dependence of real time periods upon circumstances transpiring in the heavens.

Though we may try, we cannot really appreciate the degree to which the minds of the ancients were preoccupied with astronomical pursuits. Modern science and technology have flavored our way of thought so as to rob us of any real sensitivity to the nature of the ancients' relation to the cosmos. The heavens touched nearly every aspect of their culture; consequently, we find ancient astronomy woven into myth, religion, and astrology. So great was the reliance of the ancients upon the sun and moon that they deified them. Representations of these luminaries adorned their temples as objects of worship and they were symbolized in sculpture and other works of art. The ancients followed the sun god wherever he went, marking his appearance and disappearance with great care. His return to a certain place on the horizon told them when to plant the crops, when the river would overflow its banks, or when the monsoon season would arrive. The planning and harvesting of crops could be regulated by celestial events. The important days of celebration and festivity could be marked effectively using the celestial calendar. Equipped with a knowledge of mathematics and a method of keeping records, the ancients could refine and expand their knowledge of positional astronomy. After several generations, with the advantage of a "written" record, they could learn to predict such celestial phenomena as eclipses well in advance. What a powerful advantage the priest-ruler would have over his followers with this bit of trickery in his repertoire.

We are continually amazed at the seemingly impossible accomplishments of our ancient ancestors. How did they erect the great pyramids, the statues on Easter Island, or the huge Olmec heads? In disbelief, some of us turn to extraterrestrial zoo keepers for the source of ancient wisdom and ability. According to one popular account, "the past teemed with unknown gods who visited the primeval earth in manned spaceships. Incredible technical achievements existed in the past. There is a mass of know-how which we have only partially rediscovered today."

Though the last part is substantially true, such a statement is uttered in total ignorance of the ways of ancient people. One of the goals of this text will be to show that the sophisticated astronomical and mathematical achievements of the people of ancient Mesoamerica followed logically in the evolutionary development of a civilization which intensely worshipped the heavens and steadfastly associated the phenomena they witnessed in the celestial environment with the course of human affairs.

Since the ancients expended considerable effort paying tribute to their celestial deities, we should not be surprised to find that, in many instances, astronomical principles played a role in the design of the ceremonial centers where they worshipped their gods. Stonehenge is perhaps the most famous example of an ancient structure believed to have served an astronomical function. In 1964, astronomer Gerald Hawkins wrote *Stonehenge Decoded,* thus rekindling an idea made popular at the end of the nineteenth century by Sir Norman Lockyer. Hawkins hypothesized that the megaliths standing for five thousand years on the plain of southern Great Britain constituted a calendar in stone, each component situated deliberately and precisely to align with astronomical events taking place along the local horizon. Detailed works by Alexander Thom and a cultural synthesis by Euan MacKie have since solidified the basis of our understanding of ancient megalithic astronomy and have elevated it to a level of scholarly respectability.

The Stonehenge controversy has been responsible for the resurgence of interest in the interdisciplinary field of astroarchaeology, a term first coined by Hawkins to encompass the study of the astronomical principles employed in ancient works of architecture and the elaboration of a methodology for the retrieval and quantitative analysis of astronomical alignment data. An alternate term, "archaeoastronomy," embodies the study of the extent and practice of astronomy among ancient civilizations. Such a definition fits the discipline which classicists call the "history of astronomy," except that the latter has dealt traditionally with the literate Western society and focuses largely on analyses of notational schemes in the Western style (i.e., ancient scriptures, Egyptian hieroglyphs, cuneiform tablets). Being somewhat less confined by tradition and often handicapped by the sparsity of a written record, archaeoastronomy has developed into a broader interdiscipline drawing upon the written as much as the archaeological and iconographic record. Consequently, discussions of astronomical symbolism and astronomical precision are often intermixed.

Though much emphasis has been placed on the megalithic sites in Europe, an increased interest has recently arisen in the study of astronomical building orientations in other parts of the world, particularly the Americas. Aerial photography has revealed that the remarkably straight lines etched across the Peruvian desert at Nazca continue up and down the steep sides of mountains for distances of several miles. Many of these lines are oriented to the rising positions of the sun at the solstices. Large figures at their intersection may have symbolized the constellations. In Central Mexico, the plan of the great ceremonial center of Teotihuacán seems to have been organized to harmonize with the positions of the sun and certain fundamental stars. Astronomical orientations have also been discovered in the Maya area of the peninsula of Yucatán. The so-called Group E structures at Uaxactún, Guatemala, represent the prototype of a series of sun watcher's stations found in that region. The Caracol at Chichén Itzá in Yucatán, an observatory in the shape of a round tower, contains horizontal sight tubes directed to positions of astronomical significance.

Anthropologists have become interested in studying relationships between the astronomical knowledge of civilizations of Mesoamerica and that of the native tribes of North America. Did cosmological ideas diffuse among the cultures and which concepts developed independently? Ceremonial mounds near St. Louis, Missouri, and in central Kansas probably functioned as solstice registers to mark the extreme positions of the rising sun. The Big Horn Medicine Wheel, a spoked wheel formed out of boulders in the mountains of Wyoming, also appears to have functioned as an astronomical observatory. Far to the south, the interconnected lines of the *ceque* system surrounding the ancient city of Cuzco in Peru may represent a calendar on the landscape which has astronomical, religious, and even political attributes.

In the Americas, a number of investigators from widely divergent fields have turned their attention to archaeoastronomical pursuits. As a result of the unique cooperation among them, there has been added to the literature an increasing body of evidence relating to the role of astronomy in the lives of the ancient people of this hemisphere. After a decade of progress on several fronts, it is time to begin the slow process of synthesis of the new material into the mainstream of human intellectual history.

This book is about the people of ancient Mexico and Central America and what we know of their system of astronomy. In studying them we have an enormous advantage over Thom, Hawkins, and their predecessors, for we know from the written record, the art, and the sculpture that the civilizations which developed in the New World before the arrival of Columbus were already highly advanced by the time of his arrival. Only within the last half-century have we begun to gain a full appreciation of the magnitude and sophistication of ancient New World civilizations. The ancient American calendric documents reveal that mathematics and astronomy were among their intellectual hallmarks; in fact, they were fanatically devoted to these disciplines. For them, time was an intricate natural system, each day being ticked off in a complex maze of endless cycles. But quite unlike our modern astronomy, the raison d'être of Mesoamerican, particularly Maya, astronomy was ritualistic and divinatory in nature.

To have accomplished as much as they did, the ancient Americans must have been keen observers of the heavens. Were they also brilliant theoreticians? To answer such a question we must assemble, all in one place, the material which is relevant to an objective assessment of the depth and extent of their astronomical knowledge. I have set such a goal in the production of this volume. In attempting to achieve it, I have necessarily ventured a few steps out of my own field in different directions in order to form canals between pools of material in disciplines usually regarded as unrelated. Any true interdisciplinary synthesis requires that such steps be taken. In cutting the path, I have made a special effort to tread softly, accepting the generous guidance of interested colleagues in allied fields.

Because an interdisciplinary approach to archaeoastronomy has developed, the serious scholar must become acquainted with certain segments of established fields which border upon it. What are these segments of knowledge? It seems clear that an understanding of basic positional astronomy is indispensable if one wishes to master the complexities of ancient astronomy. Maya archaeologist Sir Eric Thompson once suggested that one could understand Maya astronomy only by getting into the skin of the Maya priest-astronomer. In other words, a knowledge of the history and culture of the Native American people is vital to an understanding of their astronomical systems. Input from the archaeological discipline is important since it represents a large part of the record which survives. Pre-Columbian astronomy was strongly wedded to astrology and religion. Those of us trained in the modern sciences must be careful not to slant our view too much toward the present. We cannot assume that the Maya were always looking for the same celestial events which matter to us. Some astronomers, with a poor grasp of pre-Columbian thought, have made assertions about the Maya calendar which are strongly at odds with the facts gleaned from the anthropologists' studies.

Too often discussions of ancient astronomical systems have been couched in one-sided dialogue. The twentieth-century Western scientists are accused of fashioning their ancestors after their own image; they frame their arguments in the scientific jargon of their profession. As a result, the anthropologists either blindly accept their propositions out of awe and reverence for the complexity of their language and scientific method or refuse to consider the argument because they cannot comprehend the intricacies of positional astronomy delineated in tracts that were never intended for the nonscientific audience. Conversely, many outrageous astronomical statements have been uttered by untrained anthropologists, who, with a little understanding of elementary astronomy, could have carried their theories a long way.

The Prehistoric Goddess

ANNE L. BARSTOW

Anne L. Barstow teaches medieval history and women's religious history at SUNY College at Old Westbury, New York. She is the author of a feminist study Joan of Arc: Heretic, Mystic, Shaman *and* Married Priests and the Reforming Papacy: the Eleventh Century Debate. *In this article, Barstow provides the reader with insight into the earliest manifestations of the goddess, showing that the goddess is the deity of death as well as of life. She also shows that the attributes of the goddess and the predominance of female values and experience, although transformed by time, are still expressed by women today.*

Goddesses have been worshiped since earliest times, far longer than have male deities. Evidence of female figurines placed in sacred settings, as in circles of stones found on floors of caves, dates as far back as ca. 25,000 B.C. Traces of this worship have been found from Siberia to southern Africa, from the Indus to Ireland, and all over the New World as well. In caves, on mountaintops, at home altars, and in the earliest shrines, the goddess appeared, carved from stone, modeled from clay, etched in plaster.

Because these manifestations of the worship of a female deity begin long before recorded history, we call her the prehistoric goddess. She was the forerunner of the great goddess familiar to us from the written records of ancient Egypt and Mesopotamia, of ancient Greece and Rome. Because her worship antedates the invention of writing, we learn about her through archaeology. Who was she and why was she worshiped? The answer lies in archaeological records that reveal many richly varied and complex cults.

"She" is many goddesses: from the settings in which her likenesses are found we know that she was worshiped variously as the guardian of childbirth, the source of wisdom, the dispenser of healing, the Lady of the Beasts, the fount of prophecy, the spirit who presided over death. But preeminently she was the symbol of fertility, the guarantor of crops, animals, and humans. In this role she was the great mother, the earth mother, whose magical powers assured the food supply and the continuance of the human race.

Since evidence of her worship is found across the world, she must have been known by many names, of which our earliest written records give us hints—Cybele, Inanna, Isis. But despite her various names and her different purposes, she is a single goddess. The multiple roles created for her can be seen as different facets of one power, as expressions of one basic belief of these early Stone Age societies: that the female represented the principle of creativity and of power over both life and death. Prehistoric people expressed their deepest questions about life by constructing various cults that centered around a female deity.

What was she like? Were there common characteristics among the goddess figurines of the many cults over the centuries? Yes, despite differences in style, one can discern common traits. The goddess was faceless, as if to accentuate her universality, her ability to "stand for" the power of the female. Lacking feet, she appeared to come straight up out of the earth, with which she was identified. Unclothed, her very body seemed to have an efficacy. Often—but not always—she was big-breasted, and her hands were frequently placed under her breasts as if to display them. Many figurines show her entire body as ample, with huge breasts, belly, and buttocks, as if the very plenitude of her body would insure plentiful crops and herds. Sometimes she is pregnant, her enlarged belly emphasized by special markings. Sometimes she was sculpted nursing a child. Then again, she may be slender, the emphasis falling not on her procreative potency but on her sexual powers. In this case, her genitals received particular emphasis. In other representations she appears in a regal pose, often holding or supported by an animal or wearing jewelry, perhaps indicating that she was patroness of artisans. But regardless of manifestation or setting, it is clear that she was seen as a chief magical source of power, both spiritual and material.

I became interested in the prehistoric goddess when I first asked myself, "What would a religion created, at least in part, by women be like? What values would it express? What needs would it meet?" I knew that the Western religious tradition in which I had been raised, with its narrow patriarchalism, did not meet my spiritual needs, but I had no knowledge of alternative religious ideas. Although I was drawn to the ancient female figurines as expressions of female power, I could not appropriate them as meaningful symbols in my own religious life because they were from cultures totally alien to my own. The hunting and gathering groups and the early agricultural settlements that had produced the cults of the goddess were unknown to me. I discovered that I needed to understand the societies that had produced these female cults before I could appropriate the meaning of the prehistoric goddess for myself.

One must start with the question of women's roles in prehistoric cultures, a topic which, given the lack of textual evidence, is controversial among scholars. How much power did women wield in prehistoric societies, in the earlier hunting and gathering economies or the later agricultural settlements that emerged in the Neolithic period?

It has been argued widely that although artifact and myth may suggest that females held important roles in early societies, the actual power of women was illusory. It has been contended in addition that the projection of female images (by men, presumably) can be in itself a sign of women's subjugation. Whether woman is depicted as a sex object (Astarte or Eve) or as a magical virgin mother (the cult of Mary, for instance), the symbol serves as the projection of male needs, not as the expression of female experience or values.

Some scholars have agreed that women have never held dominant positions in any society, and thus have never been able to shape institutions to their own needs. I maintain that, although this analysis is true for many cultures, including our own, it may not apply to all early societies. In the first place, we simply do not know enough about the political organization of prehistoric groups to say for sure what was the balance of power between the genders. It seems premature, therefore, to close the door on the possibility that prehistoric women might have had the power to express themselves in an autonomous way in some areas of their lives.

It should be clear, however, that I am not talking about matriarchy. In acknowledging that matriarchy was a myth, we are free to ask a more realistic question about prehistoric societies: Did men and women perhaps relate in ways other than dominance/subjection? Male-dominant modes of social organization, after all, may not be the only types that humans have devised. Eleanor Leacock, for example, has shown that in some hunting and gathering societies women *share* political authority *when they have control over economic resources,* a point pertinent to the study of Neolithic women because the new wealth of that society was based on agriculture, domestication of animals, weaving, and pottery making—all activities associated in some degree with women's invention and control.

Our problem lies, at least in part, in the word dominance, for it is, in fact, foolish to judge this material by our common definitions of power. When we apply the usual Western concept of centralized power or of "power over" to tribal or early urban societies, we may well be led to ask the wrong questions. Power can be seen as power *with* rather than power over, and it can be used for competence and cooperation, rather than dominance and control. In reflecting on the dependence on nature that prehistorical people had to contend with, it occurred to me that a cooperative use of power may have been necessary, indeed crucial, for them: it is this different concept of power which I want to use. It is possible that neither patriarchy nor matriarchy describes the methods of early community control, and that these terms, which come from nineteenth-century comparative religious studies, are useless in analyzing many preliterate cultures.

However the issue of gender relations is decided, an understanding of the roles prehistoric women played is essential to an analysis of their religion, dominated as it was by powerful female imagery. We must learn what we can about their economic roles and their family arrangements, matters which varied according to the natural resources and level of technology of each society. For this reason, rather than surveying goddess cults over the centuries and vast geographical areas in which they were found, I will focus on the evidence for a female cult in one society, an early urban center of Anatolia that thrived during the sixth millennium B.C. This site, called Çatal Hüyük and rich in evidence of a female cult, was excavated by the British archaeologist James Mellaart. By studying its burial customs, its family organization, the sources of its wealth, its shrine decorations and carved figurines, we will be able to reconstruct much of the social, economic, and religious life in a Neolithic town. Then, and only then, can we answer whether it was possible for women there to assert themselves autonomously, that is, to express their spiritual needs directly through the public channels of power.

What was this culture like? To begin to answer this question, one must first picture a town built like a beehive: no streets, no large plaza, no palace; the homes were one-story abodes entered from the roof by a ladder and showed very little variation in wealth or possessions. Inside each one-room mud-brick house was a hearth and an oven. Every house had a large sleeping platform, always along the east wall, accentuated by wooden posts painted red; the skeletons of women and of some of the children were buried under this platform. Smaller platforms were scattered about the room in varying positions, children buried under some, men under others, but never children and men together. The houses were kept immaculately clean; refuse and sewage were disposed of in small courtyards.

This tightly constructed mass of housing was apparently a perfect defense: there is no evidence of invasion or violent destruction in the entire 1,000-year period of occupation.

The textiles and pottery found on the site are the earliest known and indicate a remarkably high level of specialization. Weaving of wool, flax, and grass; chipping and polishing of stone tools and weapons; and manufacture of beads, copper, lead jewelry, fine wooden vessels, and simple pottery were performed with such artistry that Mellaart assumes they were carried out by specially trained workers. Only gold and silver are missing from the materials he had hoped to find.

Trade was carried on far and wide in order to bring these raw materials to Çatal Hüyük. Timber was imported from the Taurus Mountains, shells from the Mediterranean, obsidian from the volcanic peaks near Çatal Hüyük, flint from south of the Taurus. But the wealth of these people lay primarily in their abundant farm produce. They ate plentifully from their crops of barley, peas, wheat, almonds, apples, and pistachios; they may have had yogurt and honey. They made wine from the hackberry and procured both milk and meat from domesticated sheep, cows, and goats.

Impressive as the economic production is, what is most extraordinary about this Neolothic town is that it has revealed a surprising number of shrines, one for every four or five homes in the area so far excavated. Because they are so numerous, they were most likely family shrines for extended family groups. We can identify them as sanctuaries by the ritual themes painted on their walls, their extravagant decorations of bulls' horns, bulls' heads, plaster goddess reliefs, cult statues, and evidence of elaborate burials. The practices of painting a handprint on sacred objects (the child's handprint on the breast of a goddess, for instance) and of destroying the sacred objects when a shrine fell into disuse offer conclusive proof that these buildings were set aside for a cultic purpose. However, there are no provisions for sacrifice, no altars or pits for blood or bones, such as later Bronze Age sites would lead us to expect. Although offerings were made to the deity—gifts of grain, stamp-seals, weapons, and votive figurines—there were no burnt offerings. Whatever the cult was, it was not bloody. What then was the religion that was celebrated here? Can we speculate without written records? Mellaart believes not only that we can speculate but that we must. He bases his decision on the wall paintings, which are the earliest-known frescoes painted on constructed walls. Admitting the difficulty of understanding these reliefs without an accompanying text, he maintains that we must interpret them, for if ever an early society tried to communicate through its art, this society tried, through these lively, natural, lifelike figures.

The earliest shrines have no goddesses, but are dominated by animal paintings and bulls' horns. Mellaart assumes that the hunt was then still an important food source and was seen as the great, magical source of power in the people's lives. Then, around 6200 B.C., the first goddess appears, in plaster outline on the wall, her legs spread wide, giving birth; below her, rows of plaster breasts, nipples painted red, are molded over animal skulls or jaws that protrude through the nipples. Already at her first appearance she is the deity of both life and death. She is the goddess of crops as well, as is suggested by the many statuettes of the goddess found in grain bins. Mellaart concurs with scholars who claim that women were the primary discoverers of agriculture and assumes therefore that women controlled this activity. He concludes that at Çatal Hüyük, agriculture was replacing hunting as the preeminent source of food and wealth, that the women controlled the new form of wealth and status, and that they introduced the new female religious images.

The goddess is faceless, usually naked, sometimes covered with a robe. She may be shown as a maiden (that is, not pregnant) who is running or dancing and whirling, her hair streaming behind her in the wind. Many times she is shown big-bellied with pregnancy, or actually giving birth. She is portrayed as the twin goddess, one of whom gives birth to an enormous bull. A human figure is pictured doing homage to a triple-goddess figure, the human covering her face as she approaches the deity. We may ask if the goddess shrines were special sanctuaries for women having difficulty in childbirth, for in one shrine, the skeletons of a woman and a baby were found buried together, and the body of a prematurely born child was carefully deposited nearby. One of the shrines is decorated entirely with floral patterns or textile designs, indicating that the goddess was the patroness of weaving. As the goddess of wild animals, she is shown flanked by leopards or holding leopard cubs.

As is true of many early goddesses, she is the deity of death as well as of life. Statues show her as a grim-faced goddess accompanied by a vulture. Many of the shrine wall paintings portray huge vultures attacking small headless human figures. In one such mural, painted above a row of four human skulls, the vulture in the painting appears to have human legs. Mellaart speculates from this that "priestesses" dressed themselves as vultures in order to preside over funeral rites. Because the care with which the dead were deposited indicates a belief in life after death, the goddess of death is in reality the goddess of another kind of life.

Mellaart is convinced that there was a priestly class at Çatal Hüyük, and that many of its practitioners were female. He observed that many abodes in what he calls the priestly quarter did not contain the usual equipment for weaving, spinning, reaping grain, or chipping stone. Lacking these standard tools, the occupants must have purchased essential items ready-made from the bazaar. He calls this professional priestly caste "elegant sophisticates."

Further evidence for a priesthood lies in the wall paintings, in which women and men dressed in leopard skins are shown chasing and dancing around deer and bulls. Mellaart assumes that only special persons could dress in rare leopard skins, and so concludes that some persons were set aside as priests and priestesses. His strongest evidence, however, comes from burials within shrines, where skeletons were buried with more care and with somewhat finer gifts than most of the home burials, although without indication of a wide differential in wealth. A number of the female skeletons are painted with red ochre (this does not occur outside shrines) are buried with costly obsidian mirrors. Males are accompanied by belt-fasteners, perhaps those born with the leopard-skin costumes. These burials represent a privileged class of people, possibly a priestly class. The majority of these shrine burials are of females.

Statues of male gods, found only outside shrines, show either a very young or a bearded male, usually riding on or standing beside bulls or leopards. But except for these figurines, the male is not depicted; he is represented only by bulls' or rams' horns. Mellaart assumes that the bull or ram was considered "a more impressive exponent of male fertility" than the human form. Another statue shows two scenes: on one side, two figures embracing, and on the other, a woman holding a child. Mellaart interprets the child as the product of the embrace and surmises that the population of Çatal Hüyük understood the role of the male in human conception. Although the male is represented in far fewer carvings (usually in connection with hunting) and although he appears to be strictly subordinate to the goddess, Mellaart believes that the male deity was still an object of "confidence, pride, and virility" at Çatal Hüyük. He may have been both the goddess's son and her consort, but he was not sacrificed—the dying god was probably not yet known in neolithic Anatolia. Mellaart concludes that men at Çatal Hüyük were figures still to be reckoned with, and that whatever forms of community control existed were shared by women and men.

Some statues show the goddess with a daughter, others show a son. Her daughter may be her successor; there may have been a religious hierarchy of females.

In the last period in which Çatal Hüyük was occupied, a shrine devoted to the hunt, decorated with hunt murals, was painted over with white paint. A bit later it was destroyed in order to build over it a shrine dedicated to agriculture and weaving, containing nine statuettes of the goddess. Mellaart speculates:

> Sometime in the fifty-eighth century B.C. agriculture finally triumphed over the age-old occupation of hunting and with it the power of the woman increased: this much is clear from the almost total disappearance of male statues in the cult, a process [observable elsewhere in Anatolia later.]

What conclusions, then, can be drawn about the roles of women in this society? Looking at women in the family, Mellaart interprets the family burial customs, in which the woman is always, the man never, buried under the main platform, to mean that the family centered around the woman, was matrilocal and probably matrilineal, and that women chose their mates. In the public sphere, the usual public leadership roles appear not yet to have existed: there is no evidence of central political organization, no major role for the military (no evidence of warfare whatever for a one-thousand-year period), and finally, no hunting ritual. Therefore, the positions of power that men usually hold did not exist. But in contrast to our lack of knowledge about male authority, we have evidence that women exercised certain kinds of power at Çatal Hüyük, because its chief source of wealth was agriculture and women were in charge of its development. They controlled the home and the economy, and eventually they molded the religion, incorporating into it their concerns about the fertility of crops and humans.

Mellaart makes a case for an impressive female status in this Neolithic society, a status that enabled women not necessarily to control the society but to express their own values and experiences in it. One of Mellaart's arguments for female creation of the religion is a bit farfetched. He posits that eroticism in art is "inevitably connected with male impulse and desire"—a dubious premise—and that because the art shows no erotic symbolism, no phallus or vulva, but instead stresses breasts, navel, and pregnant belly, it must have been commissioned by women. Whatever may have seemed erotic to the people of Çatal Hüyük, male or female, my own feelings are that the female representations convey a sense of dignity and great strength, more like objects of awe than sex objects. I agree with Mellaart, therefore, that the artwork does not appear to be erotic, and I would add that, with the exception of the ambiguous vultures, it does not relate to the demonic; it does not, therefore, express two of the themes commonly associated with the female in art. It is, rather, a celebration of fecundity and rebirth, and of the beauty and strength of textiles, animals, and women. And it is powerful.

More convincing is Mellaart's argument that Çatal Hüyük women created Çatal Hüyük religion by performing or controlling many of the economic tasks. In that way women gained authority in the community and became predominant in the priestly class. From this base they created the community's religion, a religion devoted to the conservation of life in all forms, devoted to the mysteries of birth and nourishment and life after death.

I know what I felt when I first saw the ruins of a shrine at Çatal Hüyük: the goddess figure above the rows of breasts and bulls' horns, her legs stretched wide, giving birth, was a symbol of life and creativity such as I had not seen in the Western church. But fertility symbols are no longer a sufficient image for twentieth-century women, just as the impressive agricultural society that produced those symbols no longer relates to our day. We cannot, therefore, go back to the ancient goddess cults.

But neither can we ignore these alternatives to the images of Western religions. Here is ample evidence that female deities were predominant in the religious lives of many people for millennia, and they raise the possibility that female values and experiences were expressed in the cults. They are thus an inspiration to women today who struggle to take their lives seriously (that is, gain economic independence) and to express their spirituality in new ways.

The Goddesses and Gods of Old Europe

MARIJA GIMBUTAS

Marija Gimbutas is professor of European archaeology at the University of California at Los Angeles and curator of Old World Archaeology at the UCLA Museum of Cultural History. In this selection, Gimbutas describes imagery associated with the goddess and shows how that imagery expresses the basic assumptions of the cultures in which it was found. By understanding the imagery of the goddess as reflected in various cultures, the reader can begin to understand how that imagery and the myths of the goddess have been transformed from the earliest times.

INTRODUCTION

The tradition of sculpture and painting encountered in Old Europe was transmitted from the Palaeolithic era. In art and mythical imagery it is not possible to draw a line between the two eras, Palaeolithic and Neolithic, just as it is not possible to draw a line between wild and domestic plants and animals. Much of the symbolism of the early agriculturists was taken over from the hunters and fishers. Such images as the fish, snake, bird, or horns are not Neolithic creations; they have roots in Palaeolithic times. And yet, the art and myths of the first farmers differed in inspiration and hence in form and content from those of the hunters and fishers.

Clay and stone figurines were being fashioned long before pottery was first made around 6500 B.C. The vast increase in sculptures in Neolithic times and the extent to which they departed from Palaeolithic types was not caused by technological innovations, but by the permanent settlement and growth of communities. A farming economy bound the villages to the soil, to the biological rhythms of the plants and animals upon which their existence wholly depended. Cyclical change, death and resurrection, were ascribed to the supernatural powers and in consequence special provision was made to protect the capricious life forces and assure their perpetuation. As early as the seventh millennium B.C. traits associated with the psychology and religion of the farmer are a characteristic feature of sculptural art. This art was not consciously imitative of natural forms but sought rather to express abstract conceptions.

About 30,000 miniature sculptures of clay, marble, bone, copper or gold are presently known from a total of some 3000 sites of the Neolithic and Chalcolithic era in southeastern Europe. Enormous quantities of ritual vessels, altars, sacrificial equipment, inscribed objects, clay models of temples, actual temples and pictorial paintings on vases or on the walls of shrines, already attest a genuine civilization.

The three millennia saw a progressive increase in stylistic diversity, producing ever greater variety of individual forms. Simultaneously, a more naturalistic expression of anatomical generalities gradually emancipated itself from an initial subordination to the symbolic purpose. The study of these more articulated sculptures, their ideograms and symbols and the highly developed vase painting enabled the author to distinguish the different types of goddesses and gods, their epiphanies, their devotees, and the cult scenes with which they were associated. Thus, it is possible to speak of a pantheon of gods, and to reconstruct the various costumes and masks, which throw much light on ritual drama and life as it was then lived.

Through the deciphering of stereotype images and signs with the help of quantitative and qualitative analyses it becomes clear that these early Europeans expressed their communal worship through the medium of the idol. In the miniature sculptures of Old Europe the emotions are made manifest in ritual drama involving many actors, both gods and worshippers. Much the same practice seems to have been current in Anatolia, Syria, Palestine and Mesopotamia in the corresponding periods, but only in southeastern Europe is such a quantity of figurines available for a comparative study.

The shrines, cult objects, magnificent painted and black pottery, costumes, elaborate religious ceremonialism, and a rich mythical imagery far more complex than was hitherto assumed, speak of a refined European culture and society. No longer can European Neolithic-Chalcolithic developments be summed up in the old axiom, *Ex oriente lux*.

When the magnificent treasures of the Minoan civilization were unravelled in the beginning of the twentieth century, Sir Arthur Evans wrote: "I venture to believe that the scientific study of Greek civilization is becoming less and less possible without taking into constant account that of the Minoan and Mycenaean world that went before it." While his remark was amply justified, the question of what went before the Minoan civilization remained to be posed. Now it is becoming less and less possible to understand the Minoan civilization without the study of the culture which preceded it. The study of this culture, to which I have applied the name "Old Europe," reveals new chronological dimensions and a new concept of the beginning of European civilization. It was not a single small legendary island claimed by the sea some 9000 years ago that gave rise to the fabulous civilization of Crete and the Cyclades, but a considerable part of Europe surrounded by the eastern Mediterranean, Aegean and Adriatic Seas. The many islands were an aid to navigation and facilitated communication with Anatolia, Levant and Mesopotamia. Fertile river valleys lured the first farmers deeper inland into the Balkan Peninsula and Danubian Europe. Old Europe is a product of hybridization of Mediterranean and Temperate southeast-European peoples and cultures.

European civilization between 6500 and 3500 B.C. was not a provincial reflection of Near Eastern civilization, absorbing its achievements through diffusion and periodic invasions, but a distinct culture developing a unique identity. Many aspects of this culture remain to be explored. One of the main purposes of this book is to present, as it were, the spiritual manifestations of Old Europe.

Mythical imagery of the prehistoric era tells us much about humanity—its concepts of the structure of the cosmos, of the beginning of the world and of human, plant and animal life, and also its struggle and relations with nature. It cannot be forgotten that through myth, images and symbols man comprehended and manifested his being.

Though profusely illustrated, this volume does not claim to present every aspect of the mythical imagery of Old Europe; the illustrations were selected from many thousands, with a view to showing the most representative examples and not just the most beautiful sculptures or vases. Basic information is derived from the systematically excavated sites, which are listed with full chronological details at the end of the book.

SCHEMATISM

Shorthand

In the earliest level of the Vinča mound, representing the Neolithic Starčevo complex, a ceramic figurine usually described as a "seated goddess with large buttocks and cylindrical neck" was found. For a female representation it has an extremely reduced form, with no distinction between head and torso, and only a cylindrical neck adjoining the buttocks. Its general shape suggests a bird but there is no indication of wings, beak or bird-legs. Even as a hybrid, perhaps half-woman and half-bird, it needlessly lacks naturalistic detail.

This means that we are confronted with the problem of determining the artist's ultimate intention. In the first place we must decide what the sculpture presents, its subject matter; beyond this, we must also try to understand its symbolic content, for only in this way can we hope to comprehend the psychosocial dynamic that inspired its production.

Satratigraphical evidence shows that this figurine dates from roughly 6000 B.C., and there are many like it in sites of the same period. Some figurines are even more reduced, rendering the merest outline of human or bird form. Excavation of Neolithic sites has yielded numerous 'bumpy' figurines, often little more than two centimetres long, which archaeologists classify only as indeterminate or ambiguous objects. Examined as isolated, individual pieces they remain enigmatic, their role unknown; but once we identify these miniatures as belonging to a single homogeneous group of figures, they can be recognized as vastly reduced versions of the larger "steatopygous" figurine-type which will be fully described in later chapters. With these and many larger figures lacking in detail, it is evident that the sculptor was not striving for aesthetic effects; he was producing sculptural "shorthand," an abstract symbolic conceptual art, images that were emblematic of the divine regardless of the extent of their schematization. The true meaning of the figures can best be sought in the more detailed, less abstracted figurines which reveal the naturalistic detail that betrays subject matter and so brings us closer to understanding the content of the work.

Sculptural "shorthand," unthinking and repetitive, illustrates the conservative nature of the tradition within which the sculptor worked; each culture translates its basic explanatory assumptions into equivalent form structures and creativity is only expressed in subtle variations from the socially prescribed norms. For the sociocultural historian it is more important to examine the conventional than the few and slight deviations from it, since his work is to comprehend the inherited and collective—rather than the individual—psyche.

The Neolithic Artist's Reality—Not a Physical Reality

Both figurine subject matter and the formal repetition of the collectively approved style give an insight into the content and purpose of figurine art. Art reveals man's mental response to his environment, for with it he attempts to interpret and subdue reality, to rationalize nature and give visual expression to his mythologizing explanatory concepts. The chaotic forms of nature, including the human form, are disciplined. While the Cycladic figurines of the third millennium B.C. are the most extremely geometricized, rigid constraint of this kind, though less marked, characterizes most of the groups of Old European Neolithic and Chalcolithic figures. The artist's reality is not a physical reality, though he endows the concept with a physical form, which is two-dimensional, constrained and repetitive. Supernatural powers were conceived as an explanatory device to induce an ordered experience of nature's irregularities. These powers were given form as masks, hybrid figures and animals, producing a symbolic, conceptual art not given to physical naturalism. The primary purpose was to transform and spiritualize the body and to surpass the elementary and corporeal.

It follows, then, that formal reduction should not be ascribed to the technical inability of the Neolithic artist to model in the round but to requirements dictated by deeply implanted concepts and beliefs. Nevertheless, since we are dealing with an art that has often been termed "primitive" in a partially pejorative sense, it is necessary to digress briefly in defence of the Neolithic sculptor's ability and to stress that he was not limited to unnaturalistic forms by the inadequacy of his manual skills, the nature of his raw materials or the lack of necessary techniques. In short, old European figurine art was the outcome of skilled craftsmanship, conforming to matured traditions.

The beginnings of pottery manufacture are blurred in the archaeological record, for the earliest clay vessels and artifacts were unbaked and have not survived. The earliest fired ceramics, including fine burnished and painted wares from the late seventh millenium B.C., are articulately modelled and reveal a complete mastery of ceramic technology. Stone and bone was finely carved and ground: Proto-Sesklo and Starčevo villagers in the Aegean area and central Balkans fashioned beautiful spoons of bone and painstakingly ground miniature stone ornaments such as perforated pendants and buttons. The serpentine toad from the site of Nea Nikomedeia in Macedonia is an outstanding work of art of the seventh millennium B.C.

Stone and bone sculptures are few compared to those of clay, but they show a like degree of stylization, though one might expect them to be, if anything, more schematic still. Two sculptures have been selected to demonstrate this: a typical Early Vinča clay figurine with a triangular masked head, bump for a nose, slanting incised eyes, stump-arms, projecting buttocks and naturalistically modelled breasts and navel; and the marble figurine from Gradac, also of the Early

Vinča period. The different raw materials do dictate a differing expression but the figures are alike in style and detail. Both comprise masked heads, arm stumps and inarticulate legs. Other marble sculptures are still more reduced, lacking all facial features. During the fifth millennium, carving in marble became more self-conscious and emancipated itself from the influence of clay-modelling. Bone figures were entirely schematic. A fifth-millennium example of a stylized human figure carved out of bone from a grave in the cemetery of Cernica near Bucharest is a case in point. Its head is broken. The two rounded protuberances apparently portray folded arms. The abdominal and pubic area is emphasized. Although drastically reduced, this little sculpture is probably a portrayal of a Great Goddess in a rigid position, standing in the nude with folded arms, a type encountered in graves throughout the Old European period and in the Cyclades of the third millennium B.C. Almost all of the known figurines of copper and gold are schematic, two-dimensional silhouettes of the human body, cut from a flat piece of material.

Throughout the seventh and sixth millennia B.C. figurine art was clearly dominated by abstract forms such as cylindrical pillar-like necks and a hybrid torso of female buttocks and a bird's body, but at the same time other quite different forms were produced, some of them strikingly naturalistic. An exceptional female figurine assigned to the Sesklo period in Thessaly sits in a relaxed position with her legs to one side, her hands resting on her thighs. In profile the nose is exaggerated and beaked but the head and body are naturalistically proportioned, dispensing with the pillar-like neck of earlier sculptures.

The Trend Towards More Naturalistic Sculpture in the Chalcolithic Era

The gradual trend toward more naturalistic sculpture can be traced in the Vinča statuary. The Vinča mound and other Vinča settlements provide a large group of figurines combining schematization of the upper part of the body with almost naturalistic modelling below. A sculpture from Selevac in central Yugoslavia provides a classic example in this series: the figure has exquisitely modelled abdomen and hips, the legs merging to provide a stable base. The head is schematized, pentagonal, with semi-globular plastic eyes: the arms are represented by perforated stumps. One of the most exquisite sculptures from the Vinča site is a perfectly proportioned squatting woman, unfortunately headless. Another remarkable Vinča sculpture, also headless, from the site of Fafos, depicts a man with knees drawn tightly to his chest, his hands placed on them and his back bent slightly forward. His life-like posture, with the exceptionally accurate modelling of the arms and the hands tightly grasping the knees, is unique in European art of c. 5000 B.C.

An exquisite rendering of the rounded parts of a female body, especially abdomen and buttocks, occurs occasionally in all parts of Old Europe. An extraordinary series of male sculptures, each individually seated on a stool, is distinguished for perfection in portrayal of the male body, particularly the slightly curving back.

The excavation of the Butmir site yielded several finely executed heads, remarkable for their realism; the conventional masked features are here replaced by a well-modelled forehead, eyebrows, nose, lips, chin and ears. Unmasked human heads modelled in the round occasionally occur in other cultural groups; even in the Cucuteni area, in which figurine art reached an extreme of schematic symbolism, a few naturalistically rendered human heads were discovered, with eyes, nostrils and mouth shown by impressed holes. Figurines with unmasked heads and human facial features comprise the rarest category of Neolithic and Chalcolithic sculptures.

The finest sculpture was certainly the product of exceptionally gifted members of society, though the varying intensity of individual motivation would also be reflected in the quality of the artifact. Nevertheless the cruder figurines which were the norm were no less rich in symbolic content.

Hekate and Artemis: Survival of the Old European Great Goddess in Ancient Greece and Western Anatolia

The question now arises as to what happened to the prehistoric goddess after the third millennium B.C. Did she disappear after the advent of the patriarchal Indo-European world or did she survive the dramatic change?

In Minoan (non-Indo-European) Crete the Great Goddess is seen in representations on frescoes, rings and seals. She is shown in association with bulls, or bull-horns, "double-axes" (butterflies), he-goats or lions. On a stamp from Knossos she appears as a lady of nature untamed on top of a mountain flanked by two lions and a male human worshipper. On a gold ring from Isopata near Knossos the butterfly- or bee-headed goddess is, as we have seen, surrounded by worshippers in festive garments wearing insect masks. She is represented on frescoes accompanied by worshippers, women or men in festive garments with upraised arms. Gigantic dogs portrayed on steatite vases or on cylinder seals are probably the companions of the same goddess. She or her animals, particularly the bulls, dominate the ritualistic scenes throughout the Palace period of Minoan Crete.

In Greece, as in India, the Great Goddess survived the superimposed Indo-European cultural horizon. As the predecessor of Anatolian and Greek Hekate-Artemis (related to Kubaba, Kybebe/Cybele) she lived through the Bronze Age, then through Classical Greece and even into later history in spite of transformations of her outer form and the many different names that were applied to her. The image of Hekate-Artemis of Caria, Lydia and Greece, based on descriptions of early Greek authors, vase paintings, and finds in actual sanctuaries dedicated to this multifunctional goddess, supplement and verify our understanding of the appearance and functions of the prehistoric goddess. Written sources pour blood into her veins of stone, clay, bone or gold.

In name and character she is a non-Greek, a non-Indo-European goddess. The name of Artemis is known from Greek, Lydian and Etruscan inscriptions and texts. Its antiquity is demonstrated by the appearance of the words *A-ti-mi-te* and *A-ti-mi-to*, the dative and genetive case of her name, on Linear B tablets from Pylos. *Hekate* (*Hekabe*) was Asiatic, not known to the Greeks in name. She was *Enodia* in Thessaly, perhaps an earlier name later replaced by Hekate. Whether Artemis and Hekate appear as two goddesses or as one, they both belong to the moon cycle. Hekate, gruesome and linked with death; Artemis, youthful and beautiful, reflecting the purity of untouched nature and linked with motherhood.

In Caria (western Turkey) Hekate was the primary goddess. Mysteries and games were performed in her sanctuary at Lagina. In Colophon, dogs were sacrificed to her and she herself could turn into a dog. West of Lagina was Zerynthos, from which Hekate derived her name of Zerynthia. In Samothrace, there was a cave called Zerynthos associated with Hekate. Dogs were sacrificed there and mysteries and orgiastic dances were performed. Hekate and her dogs are described as journeying over the graves of the dead and above the sacrificed blood. In the days of Aristophanes and Aischylos she is the mistress of the night road who leads travellers astray, of cross ways, of fate, and of the world of the dead, being known by both names, Hekate and Artemis. As Queen of the Ghosts, Hekate sweeps through the night followed by her baying hounds; as Enodia she is the guardian of crossroads and gates. Her sanctuaries stood at the gate to a hill-fort or at the entrance to a house. Pregnant women sacrificed to Enodia to ensure the goddess' help at birth. There is no mention in Aischylos of Hekate-Artemis assisting at birth. Clay figurines of the goddess in a seated position were sacrificed to her. A terracotta medallion found in the Athenian Agora portrays a triple-bodied Hekate-Artemis with stag and dog flanking her. She holds a torch, whip and bow-and-arrow. Sophocles in Antigone mentions Enodia as Persephone, the ruler of the dead. The torch of the goddess probably relates to the fertilizing power of the moon since Hekate's torches were carried around the freshly sown fields to promote their fertility. Statues of Roman Diana show her crowned with the crescent and carrying a raised torch. Hekate is responsible for lunacy and, on the positive side, is Giver of Vision.

The Lady of free and untamed nature and the Mother, protectress of weaklings, a divinity in whom the contrasting principles of virginity and motherhood are fused into the concept of a single goddess, was venerated in Greece, Lydia, Crete and Italy. She appears as Aretemis and under many local names: Diktynna, Pasiphae, Europa ("the wide-glancing one"), Britomartis ("the sweet virgin") in Crete, Laphria in Aetolia, Kallisto ("the beautiful") in Arkadia, or Agrotera ("the wild"), and Diana in Rome. She, 'the pure and strong one', was surrounded by nymphs, flanked by animals, and as huntress dominated the animal world. Games with bulls were among the rituals of this goddess. She was present everywhere in nature, above all in hills, forests, meadows, and fertile valleys, and often was therioform, appearing as a bear or doe. The Arkadian Kallisto, her companion and double, was said to have assumed the form of a bear. The stag is her standing attribute in plastic art; she is called 'stag-huntress' in the Homeric Hymns. Her companion Taygete was transformed into a doe and in the legend of the Alodae she herself assumes that form. Pausanias' records that in the temple of Despoina in Arcadia her statue was clothed with deer pelt. Near Colophon lay a small island sacred to Artemis to which it was believed pregnant does swam in order to bear their young. Well-bred Athenian girls of marriageable age danced as bears in honour of Artemis of Brauronia, and during rites of cult-initiation girls 'became' bears, *arktoi*. In paintings on vases, the worshippers of

Artemis wore animals masks while dancing. The girls and women of Lakedemonia performed orgiastic dances to glorify Artemis. Fat men, padded and masked as comic actors, participated in fertility dances for Artemis. We can recognize their ancestors in the masked and padded men of the Vinča culture of c. 5000 B.C.

Offerings to Artemis include phalli and all species of animals and fruits, for she was protector of all life, bestowing fertility on humans, animals and fields, and the sacrifice of any living thing was appropriate to her. According to the legend adapted by Sophocles, the most beautiful girl of the year, Iphigeneia, had to be sacrificed to the goddess. The possibility of human sacrifice is suggested by the fact that Artemis herself is called Iphigeneia in *Hermione and Ageira* and that human sacrifice was performed for Laphria who was also Artemis. Mutilated beasts, from which "a member was cut off," were sacrificed to Artemis in Boeotia, Euboea and Attica. In Asia Minor, in the great spring festival of Cybele, the shorn genitals of her priests were consecrated to her. He-goat and stag were particularly appreciated sacrificial animals, and also the hare, a moon animal. In nearly all shrines dedicated to Artemis spindle-whorls, loom weights, shuttles and *kallatoi* have been found and from inscriptions in sanctuaries it is known that woollen and linen clothes and threads wound on spools were offered as gifts to her. On Corinthian vases, Artemis and her priestesses are seen holding a spindle.

She appeared at births as the birth-giving goddess, Artemis Eileithyia ("Child-bearing"). Figurines in a seated position were sacrificed to her. Diana also presided over childbirth and was called "Opener of the Womb." As the Bear-Mother she nursed, reared and protected the newly born with the *pietas materna* of a bear. *Tithenidia,* the festival of nurses and nurslings in Sparta, honoured her name.

This goddess, as Otto poetically observed, mirrors the divine femininity of nature. Unlike the Earth-Mother who gives birth to all life, sustains it, and in the end receives it back into her bosom, she reflects the virginity of nature with its brilliance and wildness, with its guiltless purity and its strangeness. And yet, intertwined with this crystal-clear essence were the dark roots of savage nature. Artemis of Brauronia aroused madness. Her anger caused the death of women in childbirth.

It is no mere coincidence that the venerated goddess of the sixth and fifth centuries in Ancient Greece resembles the Goddess of Life and Death of the sixth and fifth millennia B.C. Mythical images last for many millennia. In her various manifestations—strong and beautiful Virgin, Bear-Mother, and Life-giver and Life-taker—the Great Goddess existed for at least five thousand years before the appearance of Classical Greek civilization. Village communities worship her to this day in the guise of the Virgin Mary. The concept of the goddess in bear shape was deeply ingrained in mythical thought through the millennia and survives in contemporary Crete as "Virgin Mary of the Bear." In the cave of Acrotiri near ancient Kydonia, a festival in honour of Panagia (Mary) Arkoudiotissa ('she of the bear') is celebrated on the second day of February. In European folk beliefs, she still moves within pregnant women in the shape of a wandering uterus or a toad. Each of her feminine aspects, virginity, birth-giving and motherhood, as well as her Terrible Mother aspect, is well represented in figurine art throughout the Neolithic and Chalcolithic eras of Old Europe.

CONCLUSIONS

In figurine art and pictorial painting the agricultural ancestors recreated their mythical world and the worship of their gods. Primordial events, principal personalities of the pantheon with their innumerable epiphanies, worshippers and participants in ritual ceremonies, all seem to have a life of their own in their various representations.

Myths and seasonal drama must have been enacted through the medium of the idol (the figurine), each with a different intention and with the invocation of appropriate divinities. The multiplicity of purpose and design is shown by sanctuaries, sacrifices, festive attire, masks, figures in dancing or leaping postures, musical instruments, shrine equipment, ladles and drinking cups, and other numerous and varied representations of objects and events which made up the context of these religious festivals. In making images of gods, worshippers and actors of the drama, man assured the cyclic returning and renewal of life. Many figurines were ex-votos and like the words of prayer were dedicated to the Great Goddess, the Bird or Snake Goddess, the Vegetation Goddess, or the Male God, a prototype of Dionysus, the daemon of vegetation.

Female snake, bird, egg, and fish played parts in creation myths and the female goddess was the creative principle. The Snake Goddess and Bird Goddess create the world, charge it with energy, and nourish the earth and its creatures with the life-giving element conceived as water. The waters of heaven and earth are under their control. The Great Goddess emerges miraculously out of death, out of the sacrificial bull, and in her body the new life begins. She is not the Earth, but a female human, capable of transforming herself into many living shapes, a doe, dog, toad, bee, butterfly, tree or pillar.

The task of sustaining life was the dominating motif in the mythical imagery of Old Europe, hence regeneration was one of the foremost manifestations. Naturally, the goddess who was responsible for the transformation from death to life became the central figure in the pantheon of gods. She, the Great Goddess, is associated with moon crescents, quadripartite designs and bull's horns, symbols of continuous creation and change. The mysterious transformation is most vividly expressed in her epiphany in the shape of a caterpillar, chrysalis and butterfly. Indeed, through this symbolism our ancestor proclaims that he believes in the beauty of young life. The ubiquity of phallic symbols connotes the glorification of the spontaneous life powers. Phallicism certainly had no obscene allusion; in the context of religious ritual it was a form of catharsis, not of symbolic procreation. There is no evidence that in Neolithic times mankind understood biological conception.

With the inception of agriculture, farming man began to observe the phenomena of the miraculous Earth more closely and more intensively than the previous hunter-fisher had done. A separate deity emerged, the Goddess of Vegetation, a symbol of the sacral nature of the seed and the sown field, whose ties with the Great Goddess are intimate.

Significantly, almost all Neolithic goddesses are composite images with an accumulation of traits from the pre-agricultural and agricultural eras. The water bird, deer, bear, fish, snake, toad, turtle, and the notion of hybridization of animal and man, were inherited from the Palaeolithic era and continued to serve as avatars of goddesses and gods. There was no such thing as a religion or mythical imagery newly created by agriculturists at the beginning of the food-producing period.

In Old Europe the world of myth was not polarized into female and male as it was among the Indo-European and many other nomadic and pastoral peoples of the steppes. Both principles were manifest side by side. The male divinity in the shape of a young man or a male animal appears to affirm and strengthen the forces of the creative and active female. Neither is subordinate to the other; by complementing one another, their power is doubled.

The central theme in re-enaction of myths obviously was the celebration of the birth of an infant. The baby as the symbol of a new life and the hope of survival is hugged by masked goddesses, Snake, Bird and Bear. Masked Nurses bearing a sack (the "hunch-back" figurines) seem to have played a role as protectresses of the child who later matures and becomes a young god. The male god, the primeval Dionysus is saturated with a meaning closely related to that of the Great Goddess in her aspect of the Virgin Nature Goddess and Vegetation Goddess. All are gods of nature's life cycle, concerned with the problem of death and regeneration, and all were worshipped as symbols of exuberant life.

The pantheon reflects a society dominated by the mother. The role of woman was not subject to that of a man, and much that was created between the inception of the Neolithic and the blossoming of the Minoan civilization was a result of that structure in which all resources of human nature, feminine and masculine, were utilized to the full as a creative force.

The Old European mythical imagery and religious practices were continued in Minoan Crete. The Minoan culture mirrors the same values, the same manual aptitude in artistic endeavour, the same glorification of the virgin beauty of life. The Old Europeans had taste and style—whimsical, imaginative and sophisticated; their culture was a worthy parent of the Minoan civilization.

Some scholars did in the past classify European prehistory and early history into matriarchal and patriarchal eras respectively. "The beginning of the psychological-matriarchal age," says Neumann, "is lost in the haze of prehistory, but its end at the dawn of our historical era unfolds magnificently before our eyes." It is then replaced by the patriarchal world with its different symbolism and its different values. This masculine world is that of the Indo-Europeans, which did not develop in Old Europe but was superimposed upon it. Two entirely different sets of mythical images met. Symbols of the masculine group replaced the images of Old Europe. Some of the old elements were fused together as a subsidiary of the new symbolic imagery, thus losing their original meaning. Some images persisted side by side, creating chaos in the former harmony. Through losses and additions new complexes of symbols developed which are best reflected in Greek mythology. One cannot always distinguish the traces of the old since they are transformed or distorted. And yet it is surprising how long the Old European mythical concepts have persisted. The study of mythical images provides one of the best proofs that the Old European world was not the proto-Indo-European world and that there was no direct and unobstructed line of development to the modern Europeans. The earliest European civilization was savagely destroyed by the patriarchal element and it never recovered, but its legacy lingered in the substratum which nourished further European cultural developments. The Old European creations were not lost; transformed, they enormously enriched the European psyche.

The teaching of Western civilization starts with the Greeks and rarely do people ask themselves what forces lay behind these beginnings. But European civilization was not created in the space of a few centuries; the roots are deeper—by six thousand years. That is to say, vestiges of the myths and artistic concepts of Old Europe, which endured from the seventh to the fourth millennium B.C. were transmitted to the modern Western world and became part of its cultural heritage.

Unit IV

Pharaoh's Rule: Egypt, the Exodus, and the Myth of Osiris

The Egyptian Gods

HENRI FRANKFORT

Henri Frankfort taught at the University of Chicago and the University of London. He also did extensive fieldwork in the Near East and led excavations in Iraq from 1924 to 1937 for the Oriental Institute of the University of Chicago. These excavations yielded much information about the early history of Babylonia from 4,000 to 2,000 B.C.E. In this selection, Frankfort provides an excellent background for the study of Egyptian mythology by discussing the gods and their symbols and their relationships to human problems in ancient Egypt.

THE GODS AND THEIR SYMBOLS

Religion as we Westerners know it derives its character and its unity from two circumstances: it centers on the revelation of a single god, and it contains a message which must be transmitted. The Torah, the Gospels, and Islam contain teachings sufficiently coherent for exposition. The Gospel and Islam must, moreover, be preached to the unconverted. In the whole of the ancient world there is only one religion with similar characteristics: the monotheistic cult of the sun introduced by the heretic Pharaoh Akhenaten. And Akhenaten was a heretic precisely in this: that he denied recognition to all but one god and attempted to convert those who thought otherwise. His attitude presents no problem to us; we acknowledge a conviction too deep for tolerance. But Egyptian religion was not exclusive. It recognized an unlimited number of gods. It possessed neither a central dogma nor a holy book. It could flourish without postulating one basic truth.

We find, then, in Egyptian religion a number of doctrines which strike us as contradictory; but it is sheer presumption to accuse the ancients of muddleheadedness on this score. In a recent book the reason and the meaning of this apparent confusion have been explained. I can only summarize the argument here in a few sentences. The ancients did not attempt to solve the ultimate problems confronting man by a single and coherent theory; that has been the method of approach since the time of the Greeks. Ancient thought—mythopoeic, "myth-making" thought—admitted side by side certain *limited* insights, which were held to be *simultaneously* valid, each in its own proper context, each corresponding to a definite avenue of approach. I have called this "multiplicity of approaches," and we shall find many examples of it as we proceed. At the moment I want to point out that this habit of thought agrees with the basic experience of polytheism.

Polytheism is sustained by man's experience of a universe alive from end to end. Powers confront man wherever he moves, and in the immediacy of these confrontations the question of their ultimate unity does not arise. There are many gods—one cannot know how many; a small handbook of Egyptian religion enumerates more than eighty. How, then, are they recognized? Here we may well use the evidence collected by anthropologists among living believers in polytheism. It appears that superhuman powers reveal themselves sometimes in a curiously accidental manner. For instance, a West African native is on an important expedition when he suddenly stumbles over a stone. He cries out: "Ha! Are you there?"—and takes the stone with him. The stone had, as it were, given a hint that it was powerful, and the Negro strengthened himself by taking possession of it. Under normal conditions he might not have taken notice of the obstacle that tripped him up, but the importance of the expedition had created the emotional tension which makes man receptive to signs of a supernatural order. Note that at the very moment that the stone reveals its imminent power, it acquires the quality of a person, for the native exclaims: "Are you there?" The next thing to observe is that such impact on a power in the outside world may be experienced as either of fleeting or of lasting significance. It will be of permanent significance especially when the community accepts it as valid and a cult is consequently established. For instance, an Ewe Negro enters the bush and finds there a piece of iron. Returning home he falls ill, and the priests explain that a divinity is manifesting its potency in the iron and that henceforth the village should worship it.

It seems to me that these examples throw some light upon the fact that Egypt knew a large number of gods and an astonishing variety of cult-objects. But our examples do not answer the question *why* certain experiences acquired a lasting significance (the iron) while others (the stone) did not. They show that we need not expect to be able to answer this question in relation to Egyptian cults, and we should do well to retain from our excursion among modern savages a certain skepticism as to the value of symbols and sacred objects as indicators of the meaning which the gods designated may have had for their worshipers. If many of the sacred objects seem devoid of mystery or meaning to us, it may well be that they were originally connected with the cult in a loose and accidental manner, mere adjuncts to an emotional reality from which the cult continued to draw its life but which we can neither recapture nor reconstruct. On the other hand, certain sacred objects possessed a deeper, a truly symbolical significance, and in such a case the relation between the deity and the object is capable of being understood. This is so, for instance, when the name of the goddess Isis is written as if it simply meant "throne," while she is also depicted with the throne as her distinctive attribute.

We know that many peoples consider the insignia of royalty to be charged with the superhuman power of kingship. Among these objects the throne occupies a special place: the prince who seats himself upon it at the coronation arises king. The throne "makes" the king—the term occurs in Egyptian texts— and so the throne, Isis, is the "mother" of the king. This expression might be viewed as a metaphor, but the evidence shows that it was not. The bond between the king and the throne was the intimate one between his person and the power which made him king. Now a power was not recognized objectively, as the result of an intellectual effort on the part of man. We have seen that the power reveals itself; it is recognized in the relationship of "I and thou;" it has the quality of a person.

The throne which "made" the king is comprehended as a mother, and thus it may be the object of profound and complex feelings. If we should try to resolve the complexity, we should merely be falsifying the evidence. We can neither say that Isis was originally the throne personified, nor that the throne acquired a transcendental quality because it was conceived as a mother. The two notions are fundamentally correlated, and mythopoeic thought expresses such a bond as identity. The throne made manifest a divine power which changed one of several princes into a king fit to rule. The awe felt before this manifestation of power became articulate in the adoration of the mother-goddess. There is no question of any evolution from a simpler to a complex notion. Complexity is of the essence of the relationship between man and deity. As early as the First Dynasty a Pharaoh calls himself "son of Isis." We do observe a historical development, but that concerns not the goddess but her cult. Originally Isis was significant only in her relation to the king; subsequently—and especially through the myth of Osiris, which we shall discuss later—she brought consolation to all men, and three thousand years after the first appearance of her name in Egypt, monuments were being erected to her throughout the Roman Empire, up to its very borders on the Rhine and the Danube.

Our discussion of Isis illustrates one possible relationshp between the gods and their symbols. Another possibility, as we have said already, is that the symbol lacks all deeper significance. This thought is especially disturbing because the symbols loom very large in our sources; in fact, they constitute in the case of many a god all—or almost all—we know about him. Moreover, since symbols are definite and distinct, they offer a delusive hold to modern research. And so one talks glibly of fetishes, sungods, ram-gods, falcon-gods, and so forth—as if the precision of those terms had any reference to the gods themselves! We must not generalize in this manner. Sometimes the symbol tells us something about a god, sometimes it does not; and mostly the evidence on hand does not allow us to decide one way or the other. But in any case the use of classificatory or generic terms in connection with the gods bars the road to understanding; for this can only be reached, if at all, by a circumspect interpretation of each individual case.

SACRED ANIMALS AND OTHERNESS

There is one generic term which is most difficult to avoid when we discuss Egyptian religion. That is the word "animal-gods." It should not be used, as we shall show in a moment. But we must admit—and the Greek, Roman, and early Christian writers too were struck by the fact—that animals play an altogether unusual role in Egyptian religion. We cannot evade the issue by referring back to what we said a moment ago, namely, that the origin of cults is beyond our ken and that we shall never know how certain gods came to be associated with certain animals. There are too many gods showing such an association and their cult is too widespread for us to pretend to understand Egyptian religion without at least a tentative explanation of this its most baffling, most persistent, and to us most alien feature.

It is wrong to say that the worship of animals is a survival from a primitive stratum of Egyptian religion. This view is often encountered and is supported by some plausible arguments. It is said that these cults are often of purely local significance; that they sometimes center on quite insignificant creatures like the centipede or the toad; and that we must therefore place the sacred animals on a par with certain sacred objects, like the crossed arrows of the goddess Neith, and consider all these symbols as mere emblems of—and means of promoting—tribal unity. Some scholars have even interpreted them as totems. But the characteristic features of totemism, such as the claim of descent from the totem, its sacrifice for a ceremonial feast of the clan, and exogamy, can not be found in Egyptian sources. Moreover, any treatment of the sacred animals which stresses their local or political significance at the expense of their religious importance flies in the face of the evidence. It is undeniable that there is something altogether peculiar about the meaning which animals possessed for the Egyptians. Elsewhere, in Africa or North America, for example, it seems that either the terror of animal strength, or the strong bond, the mutual dependence of man and beast (in the case of cattle cults, for instance), explains animal worship. But in Egypt *the animal as such,* irrespective of its specific nature, seems to possess religious significance; and the significance was so great that even the mature speculation of later times rarely dispensed with animal forms in plastic or literary images referring to the gods.

But there was nothing metaphorical in the connection between god and animal in Egypt. It is not as if certain divine qualities were made articulate by the creature, in the way the eagle elucidates the character of Zeus. We observe, on the contrary, a strange link between divinity and actual beast, so that in times of decadence animal worship may gain a horrible concreteness. Then one finds mummified cats, dogs, falcons, bulls, crocodiles, and so forth, buried by the hundreds in vast cemeteries which fill the Egyptologist with painful embarrassment—for this, we must admit, is polytheism with a vengeance. Nevertheless, these are grotesque but significant symptoms of a characteristic trait in Egyptian religion.

To understand this trait, we should first realize that the relation between a god and his animal may vary greatly. If Horus is said to be a falcon whose eyes are sun and moon and whose breath is the cooling north wind, we may think that this was a mere image to describe an impressive god of the sky. But the god was depicted as a bird from the earliest times and was apparently believed to be manifest either in individual birds or in the species. Thoth was manifest in the moon, but also in the baboon and in the ibis, and we do not know whether any relations were thought to exist between these different symbols, and if so, what they were. The relation between the Mnevis bull and the sun-god Re, and between the Apis bull and the earth-god Ptah, was different again. Ptah was never depicted as a bull or believed to be incarnate in a bull; but the Apis bull was called "The living Apis, the herald of Ptah, who carries the truth upwards to him of the lovely face (Ptah)." The Mnevis bull bore a similar title in connection with Re. We have to deal here, moreover, not with a species considered sacred, but with one individual identified by certain marks, not as the incarnation, but as the divine servant of the god. Other deities were regularly imagined in animal shapes but even in their case the incarnation did not limit—it did not even define—their powers. Anubis, for instance, was most commonly shown as a reclining jackal but he was by no means a deified animal. Already in the earliest texts in which he is mentioned, he appears as the god of the desert cemeteries. He ensured proper burial and when mummification became common he counted as the master of embalmment. The god was depicted in papyri and reliefs with a human body and a jackal's head.

Such hybrid forms are common in Egyptian art and the usual evolutionary theory explains them as "transitional forms," intermediate between the "crude" cult of animals and the anthropomorphic gods of a more enlightened age. This theory ignores the fact that the earliest divine statues which have been preserved represent the god Min in human shape. Conversely, we find to the very end of Egypt's independence that gods were believed to be manifest in animals. The goddess Hathor appears, for instance, in late papyri and even in royal statues as a cow. Yet she was rendered already in the First Dynasty, on the Palette of Narmer, with a human face, cow's horns, and cow's ears. This early appearance of human features was to be expected, for a god is personified power, and personification need not, but easily may, call up the human image. In any case, the gods were not confined to a single mode of manifestation. We have seen that Thoth appeared as moon, baboon, and ibis. He was also depicted as an ibis-headed man. To speak here of a transitional form seems pointless. There was no need for a transition. The god appeared as he desired, in one of his known manifestations. On the other hand, there was a definite need to distinguish deities when they were depicted in human shape, and in such an array the ibis-headed figure identified Thoth. I suspect that the Egyptians did not intend their hybrid designs as renderings of an imagined reality at all and that we should not take the animal-headed gods at their face value. These designs were probably pictograms, not portraits. Hathor, usually depicted as a cow, a woman's face with cow's ears, or as a woman wearing a crown of cow's horns, appears very rarely as a cow-headed woman; the meaning would be: This is the goddess who is manifest in the cow. The animal-headed figures are quite unorganic and mechanical; it makes no difference whether a quadruped's head, an ibis' neck, or a snake's forepart emerge from the human shoulder. That again would be easily explained if they were only ideograms, and this interpretation is corroborated by the truly vital character of the few monsters invented by the Egyptians: Taurt, for instance, is convincing even though she is composed of incongruous parts: the head of a hippopotamus, the back and tail of a crocodile, the breasts of a woman, and the claws of a lion.

Our rapid survey of the various relationships between gods and animals in Egypt does not clarify the role of the latter. But the very absence of a general rule and the variety of the creatures involved suggests, it seems to me, that what in these relationships became articulate was an underlying religious awe felt before all animal life; in other words, it would seem that *animals as such* possessed religious significance for the Egyptians. Their attitude might well have arisen from a religious interpretation of the animals' *otherness*. A recognition of *otherness* is implied in all specifically religious feeling, as Otto has shown. We assume, then, that the Egyptian interpreted the nonhuman as superhuman, in particular when he saw it in animals—in their inarticulate wisdom, their certainty, their unhesitating achievement, and above all in their static reality. With animals the continual succession of generations brought no change; and this is not an abstract and far-fetched argument but something which suggested itself also to Keats for instance; in the "Ode to a Nightingale" he writes:

> Thou wast not born for death, immortal Bird!
> No hungry generations tread thee down;
> The voice I hear this passing night was heard
> In ancient days by emperor and clown . . .

The animals never change, and in this respect especially they would appear to share—in a degree unknown to man—the fundamental nature of creation. We shall see in the following chapters that the Egyptians viewed their living universe as a rhythmic movement contained within an unchanging whole. Even their social order reflected this view; in fact, it determined their outlook to such an extent that it can only be understood as an intuitive—and therefore binding—interpretation of the world order. Now humanity would not appear to exist in this manner; in human beings individual characteristics outbalance generic resemblances. But the animals exist in their unchanging species, following their predestined modes of life, irrespective of the replacement of individuals. Thus animal life would appear superhuman to the Egyptian in that it shared directly, patently, in the static life of the universe. For that reason recognition of the animals' *otherness* would be, for the Egyptian, recognition of the divine.

This interpretation of the animal cults of Egypt requires qualification in two respects; it depends, of course, on the strength which the vision of an unchanging universe can be proved to have possessed in Egypt, and it requires therefore the cumulative evidence of the subsequent chapters. And, furthermore, even if it is true that animals in general were capable of inspiring all Egyptians with a feeling of religious awe, that feeling assumed definite and different forms in each of the ensuing cults. Their variety is reflected in the relationships which were claimed to exist between gods and animals, whether individuals or whole species.

The Gods of the Egyptians

E. A. WALLIS BUDGE

E. A. Wallis Budge, one of the foremost Egyptologists of this century, was Keeper of the Egyptian and Assyrian Antiquities in the British Museum and author of an extensive study of the religious and mythological texts of Ancient Egypt. In this selection, he explains the relationship of the gods to nature and discusses the elements of magic that were prevalent in ancient Egyptian ritual.

THE GODS OF EGYPT

The Greek historian Herodotus affirms that the Egyptians were "beyond measure scrupulous in all matters appertaining to religion," and he made this statement after personal observation of the care which they displayed in the performance of religious ceremonies, the aim and object of which was to do honour to the gods, and of the obedience which they showed to the behests of the priests who transmitted to them commands which they declared to be, and which were accepted as, authentic revelations of the will of the gods. From the manner in which this writer speaks it is clear that he had no doubt about what he was saying, and that he was recording a conviction which had become settled in his mind. He was fully conscious that the Egyptians worshipped a large number of animals, and birds, and reptiles, with a seriousness and earnestness which must have filled the cultured Greek with astonishment, yet he was not moved to give expression to words of scorn as was Juvenal,[1] for Herodotus perceived that beneath the acts of apparently foolish and infatuated worship there existed a sincerity which betokend a firm and implicit belief which merited the respect of thinking men. It would be wrong to imagine that the Egyptians were the only people of antiquity who were scrupulous beyond measure in religious matters, for we know that the Babylonians, both Sumerian and Semitic, were devoted worshippers of their gods, and that they possessed a very old and complicated system of religion; but there is good reason for thinking that the Egyptians were more scrupulous than their neighbours in religious matters, and that they always bore the character of being an extremely religious nation. The evidence of the monuments of the Egyptians proves that from the earliest to the latest period of their history the observance of religious festivals and the performance of religious duties in connexion with the worship of the gods absorbed a very large part of the time and energies of the nation, and if we take into consideration the funeral ceremonies and services commemorative of the dead which were performed by them at the tombs, a casual visitor to Egypt who did not know how to look below the surface might be pardoned for declaring that the Egyptians were a nation of men who were wholly given up to the worship of beasts and the cult of the dead.

The Egyptians, however, acted in a perfectly logical manner, for they believed that they were a divine nation, and that they were ruled by kings who were themselves gods incarnate; their earliest kings, they asserted, were actually gods, who did not disdain to live upon earth, and to go about and up and down through it, and to mingle with men. Other ancient nations were content to believe that they had been brought into being by the power of their gods operating upon matter, but the Egyptians believed that they were the issue of the great God who created the universe, and that they were of directly divine origin. When the gods ceased to reign in their proper persons upon earth, they were succeeded by a series of demi-gods, who were in turn succeeded by the Manes, and these were duly followed by kings in whom was enshrined a divine nature with characteristic attributes. When the physical or natural body of a king died, the divine portion of his being, i.e., the spiritual body, returned to its original abode with the gods, and it was duly worshipped by men upon earth as a god and with the gods. This happy result was partly brought about by the performance of certain ceremonies, which were at first wholly magical, but later partly magical and partly religious, and by the recital of appropriate words uttered in the duly prescribed tone and manner, and by the keeping of festivals at the tombs at stated seasons when the appointed offerings were made, and the prayers for the welfare of the dead were said. From the earliest times the worship of the gods went hand in hand with the deification of dead kings and other royal personages, and the worship of departed monarchs from some aspects may be regarded as meritorious as the worship of the gods. From one point of view Egypt was as much a land of gods as of men, and the inhabitants of the country wherein the gods lived and moved naturally devoted a considerable portion of their time upon earth to the worship of divine beings and of their ancestors who had departed to the land of the gods. In the matter of religion, and all that appertains thereto, the Egyptians were a "peculiar people," and in all ages they have exhibited a tenacity of belief and a conservatism which distinguish them from all the other great nations of antiquity.

But the Egyptians were not only renowned for their devotion to religious observances, they were famous as much for the variety as for the number of their gods. Animals, birds, fishes, and reptiles were worshipped by them in all ages, but in addition to these they adored the great powers of nature as well as a large number of beings with which they peopled the heavens, the air, the earth, the sky, the sun, the moon, the stars, and the water. In the earliest times the predynastic Egyptians, in common with every half-savage people, believed that all the various operations of nature were the result of the actions of beings which were for the most part unfriendly to man. The inundation which rose too high and flooded the primitive village, and drowned their cattle, and destroyed their stock of grain, was regarded as the result of the working of an unfriendly and unseen power; and when the river rose just high enough to irrigate the land which had been prepared, they either thought that a friendly power, which was stronger than that which caused the destroying flood, had kept the hostile power in check, or that the spirit of the river was on that occasion pleased with them.

They believed in the existence of spirits of the air, and in spirits of mountain, and stream, and tree, and all these had to be propitiated with gifts, or cajoled and wheedled into bestowing their favour and protection upon their suppliants.

It is very unfortunate that the animals, and the spirits of natural objects, as well as the powers of nature, were all grouped together by the Egyptians and were described by the word NETERU; which, with considerable inexactness, we are obliged to translate by "gods." There is no doubt that at a very early period in their predynastic history the Egyptians distinguished between great gods and little gods, just as they did between friendly gods and hostile gods, but either their poverty of expression, or the inflexibility of their language, prevented them from making a distinction apparent in writing, and thus it happens that in dynastic times, when a lofty conception of monotheism prevailed among the priesthood, the scribe found himself obliged to call both God and the lowest of the beings that were supposed to possess some attribute of divinity by one and the same name, i.e., *Neter.* Other nations of antiquity found a way out of the difficulty of grouping all classes of divine beings by one name by inventing series of orders of angels, to each of which they gave names and assigned various duties in connexion with the service of the Deity. Thus in the Ḳur'ân it is said that God maketh the angels His messengers and that they are furnished with two or three, or four pairs of wings, according to their rank and importance; the archangel Gabriel is said to have been seen by Muḥammad the Prophet with six hundred pairs of wings! The duties of the angels, according to the Muḥammadans, were of various kinds. Thus nineteen angels are appointed to take charge of hell fire; eight are set apart to support God's throne on the Day of Judgment; several tear the souls of the wicked from their bodies with violence, and several take the souls of the righteous from their bodies with gentleness and kindness; two angels are ordered to accompany every man on earth, the one to write down his good actions and the other his evil deeds, and these will appear with him at the Day of Judgment, the one to lead him before the Judge, and the other to bear witness either for or against him. Muḥammadan theologians declare that the angels are created of a simple substance of light, and that they are endowed with life, and speech, and reason; they are incapable of sin, they have no carnal desire, they do not propagate their species, and they are not moved by the passions of wrath and anger; their obedience is absolute. Their meat is the celebrating of the glory of God, their drink is the proclaiming of His holiness, their conversation is the commemorating of God, and their pleasure is His worship. Curiously enough, some are said to have the form of animals. Four of the angels are Archangels, viz. Michael, Gabriel, Azrael, and Israfel, and they possess special powers, and special duties are assigned to them. These four are superior to all the human race, with the exception of the Prophets and Apostles, but the angelic nature is held to be inferior to human nature because all the angels were commanded to worship Adam. The above and many other characteristics might be cited in proof that the angels of the Muḥammadans possess much in common with the inferior gods of the Egyptians, and though many of the conceptions of the Arabs on this point were undoubtedly borrowed from the Hebrews and their writings, a great many must have descended to them from their own early ancestors.

Closely connected with these Muḥammadan theories, though much older, is the system of angels which was invented by the Syrians. In this we find the angels divided into nine classes and three orders, upper, middle, and lower. The upper order is composed of Cherubim, Seraphim, and Thrones; the middle order of Lords, Powers, and Rulers; and the lower order of Principalities, Archangels, and Angels. The middle order receives revelations from those above them, and the lower order are the ministers who wait upon created things. The highest and foremost among the angels is Gabriel, who is the mediator between God and His creation. The Archangels in this system are described as a "swift operative motion," which has dominion over every living thing except man; and the Angels are a motion which has spiritual knowledge of everything that is on earth and in heaven. The Syrians, like the Muḥammadans, borrowed largely from the writings of the Hebrews, in whose theological system angels played a very prominent part. In the Syrian system also the angels possess much in common with the inferior gods of the Egyptians.

The inferior gods of the Egyptians were supposed to suffer from many of the defects of mortal beings, and they were even thought to grow old and to die, and the same ideas about the angels were held by Muḥammadans and Hebrews. According to the former, the angels will perish when heaven, their abode, is made to pass away at the Day of Judgment. According to the latter, one of the two great classes of angels, i.e., those which were created on the fifth day of creation, is mortal; on the other hand, the angels which were created on the second day of creation endure for ever, and these may be fitly compared with the unfailing and unvarying powers of nature which were personified and worshipped by the Egyptians; of the angels which perish, some spring from fire, some from water, and some from wind. The angels are grouped into ten classes, i.e., the Erêlîm, the Îshîm, the Bĕnê Elôhîm, the Malachîm, the Ḥashmalîm, the Tarshîshîm, the Shishanîm, the Cherûbîm, the Ophannîm, and the Serâphîm; among these were divided all the duties connected with the ordering of the heavens and the earth, and they, according to their position and importance, became the interpreters of the Will of the Deity. A comparison of the passages in Rabbinic literature which describe these and similar matters connected with the angels, spirits, etc., of ancient Hebrew mythology with Egyptian texts shows that both the Egyptians and Jews possessed many ideas in common, and all the evidence goes to prove that the latter borrowed from the former in the earliest period.

In comparatively late historical times the Egyptians introduced into their company of gods a few deities from Western Asia, but these had no effect in modifying the general character either of their religion or of their worship. The subject of comparative Egyptian and Semitic mythology is one which has yet to be worked thoroughly, not because it would supply us with the original forms of Egyptian myths and legends, but because it would show what modifications such things underwent when adopted by Semitic peoples, or at least by peoples who had Semitic blood in their veins. Some would compare Egyptian and Semitic mythologies on the ground that the Egyptians and Semites were kinsfolk, but it must be quite clearly understood that this is pure assumption, and is only based on the statements of those who declare that the Egyptian and Semitic languages are akin. Others again have sought to explain the mythology of the Egyptians by appeals to Aryan mythology, and to illustrate the meanings of important Egyptian words in religious texts by means of Aryan etymologies, but the results are wholly unsatisfactory, and they only serve to show the futility of comparing the mythologies of two peoples of different race occupying quite different grades in the ladder of civilization. It cannot be too strongly insisted on that all the oldest gods of Egypt are of Egyptian origin, and that the fundamental religious beliefs of the Egyptians also are of Egyptian origin, and that both the gods and the beliefs date from predynastic times, and have nothing whatever to do with the Semites or Aryans of history.

Of the origin of the Egyptian of the Palaeolithic and early Neolithic Periods, we, of course, know nothing, but it is tolerably certain that the Egyptian of the latter part of the Neolithic Period was indigenous to North-East Africa, and that a very large number of the great gods worshipped by the dynastic Egyptian were worshipped also by his predecessor in predynastic times. The conquerors of the Egyptians of the Neolithic Period who, with good reason, have been assumed to come from the East and to have been more or less akin to the Proto-Semites, no doubt brought about certain modifications in the worship of those whom they had vanquished, but they could not have succeeded in abolishing the various gods in animal and other forms which were worshipped throughout the length and breadth of the country, for these continued to be venerated until the time of the Ptolemies.

We have at present no means of knowing how far the religious beliefs of the conquerors influenced the conquered peoples of Egypt, but viewed in the light of well-ascertained facts it seems tolerably certain that no great change took place in the views which the indigenous peoples held concerning their gods as the result of the invasion of foreigners, and that if any foreign gods were introduced into the company of indigenous, predynastic gods, they were either quickly assimilated to or wholly absorbed by them. Speaking generally, the gods of the Egyptians remained unchanged throughout all the various periods of the history of Egypt, and the minds of the people seem always to have had a tendency towards the maintenance of old forms of worship, and to the preservation of the ancient texts in which such forms were prescribed and old beliefs were enshrined. The Egyptians never forgot the ancient gods of the country, and it is typical of the spirit of conservatism which they displayed in most things that even in the Roman Period pious folk among them were buried with the same prayers and with the same ceremonies that had been employed at the burial of Egyptians nearly five thousand years before. The Egyptian of the Roman Period, like the Egyptian of the Early Empire, was content to think that his body would be received in the tomb by the jackal-headed Anubis; that the organs of his corruptible body would be presided over and guarded by animal-headed gods; that the reading of the pointer of the Great Scales, wherein his heart was weighed, would be made known by an ape to the ibis-headed scribe of the gods, whom we know by the name of Thoth; and that the beatified dead would be introduced to the god Osiris by a hawk-headed god called Horus, son of Isis, who in many respects was the counterpart of the god Ḥeru-ur, the oldest of all the gods of Egypt, whose type and symbol was the hawk. From first to last the indigenous Egyptian paid little heed to the events which happened outside his own country, and neither conquest nor invasion by foreign nations had any effect upon his personal belief. He continued to cultivate his land diligently, he worshipped the gods of his ancestors blindly, like them he spared no pains in making preparations for the preservation of his mummified body, and the heaven which he hoped to attain was fashioned according to old ideas of a fertile homestead, well stocked with cattle, where he would enjoy the company of his parents, and be able to worship the local gods whom he had adored upon earth. The priestly and upper classes certainly held views on these subjects which differed from those of the husbandman, but it is a significant fact that it was not the religion and mythology of the dynastic Egyptian, but that of the indigenous, predynastic Egyptian, with his animal gods and fantastic and half-savage beliefs, which strongly coloured the religion of the country in all periods of her history, and gave to her the characteristics which were regarded with astonishment and wonder by all the peoples who came in contact with the Egyptians.

The predynastic Egyptians in the earliest stages of their existence, like most savage and semi-savage peoples, believed that the sea, the earth, the air, and the sky were filled to overflowing with spirits, some of whom were engaged in carrying on the works of nature, and others in aiding or obstructing man in the course of his existence upon earth. Whatsoever happened in nature was attributed by them to the operations of a large number of spiritual beings, the life of whom was identical with the life of the great natural elements, and the existence of whom terminated with the destruction of the objects which they were supposed to animate. Such spirits, although invisible to mental eyes, were very real creatures in their minds, and to them they attributed all the passions which belong to man, and all his faculties and powers also. Everything in nature was inhabited by a spirit, and it was thought possible to endow a representation, or model, or figure of any object with a spirit or soul, provided a name was given to it; this spirit or soul lived in the drawing or figure until the object which it animated was broken or destroyed. The objects, both natural and artificial, which we consider to be inanimate were regarded by the predynastic Egyptians as animate, and in many respects they were thought to resemble man himself. The spirits who infested every part of the visible world were countless in forms, and they differed from each other in respect of power; the spirit that caused the Inundation of the Nile was greater than the one that lived in a canal, the spirit that made the sun to shine was more powerful than the one that governed the moon, and the spirit of a great tree was mightier than the one that animated an ear of corn or a blade of grass. The difference between the supposed powers of such spirits must have been distinguished at a very early period, and the half-savage inhabitants of Egypt must at the same time have made a sharp distinction between those whose operations were beneficial to them, and those whose actions brought upon them injury, loss, or death. It is easy to see how they might imagine that certain great natural objects were under the dominion of spirits who were capable of feeling wrath, or displeasure, and of making it manifest to man. Thus the spirit of the Nile would be regarded as beneficent and friendly when the waters of the river rose sufficiently during the period of the Inundation to ensure an abundant crop throughout the land; but when their rise was excessive, and they drowned the cattle and washed away the houses of the people, whether made of wattles or mud, or when they rose insufficiently and caused want and famine, the spirit of the Nile would be considered unfriendly and evil to man. An ample and sufficient Inundation was regarded as a sign that the spirit of the Nile was not displeased with man, but a destructive flood was a sure token of displeasure. The same feeling exists to this day in Egypt among the peasant-farmers, for several natives told me in 1899, the year of the lowest rise of the Nile of the nineteenth century,[2] that "Allah was angry with them, and would not let the water come"; and one man added that in all his life he had never before known Allah to be so angry with them.

The spirits which were always hostile or unfriendly towards man, and were regarded by the Egyptians as evil spirits, were identified with certain animals and reptiles, and traditions of some of these seem to have been preserved until the latest period of dynastic history. Āpep, the serpent-devil of mist, darkness, storm, and night, of whom more will be said later on, and his fiends, the "children of rebellion," were not the result of the imagination of the Egyptians in historic times, but their existence dates from the period when Egypt was overrun by mightly beasts, huge serpents, and noxious reptiles of all kinds. The great serpent of Egyptian mythology, which was indeed a formidable opponent of the Sun-god, had its prototype in some monster serpent on earth, of which tradition had preserved a record; and that this is no mere theory is proved by the fact that the remains of a serpent, which must have been of enormous size, have recently been found in the Fayyûm. The vertebræ are said to indicate that the creature to which they belonged was longer than the largest python known.[3] The allies of the great serpent-devil Āpep were as hostile to man as was their master to the Sun-god, and they were regarded with terror by the minds of those who had evolved them. On the other hand, there were numbers of spirits whose actions were friendly and beneficial to man, and some of these were supposed to do battle on his behalf against the evil spirits.

Thus at a very early period the predynastic Egyptian must have conceived the existence of a great company of spirits whose goodwill, or at all events whose inaction, could only be obtained by bribes, i.e., offerings, and cajolery and flattery; and of a second large company whose beneficent deeds to man he was wont to acknowledge and whose powerful help he was anxious to draw towards himself; and of a third company who were supposed to be occupied solely with making the sun, moon, and stars to shine, and the rivers and streams to flow, and the clouds to form and the rain to fall, and who, in fact, were always engaged in carrying out diligently the workings and evolutions of all natural things, both small and great. The spirits to whom in predynastic times the Egyptians ascribed a nature malicious or unfriendly towards man, and who were regarded much as modern nations have regarded goblins, hobgoblins, gnomes, trolls, elves, etc., developed in dynastic times into a corporate society, with aims, and intentions, and acts wholly evil, and with a government which was devised by the greatest and most evil of their number. To these, in process of time, were joined the spirits of evil men and women, and the prototype of hell was formed by assuming the existence of a place where evil spirits and their still more evil chiefs lived together. By the same process of imagination beneficent and friendly spirits were grouped together in one abode under the direction of rulers who were well disposed towards man, and this idea became the nucleus of the later conception of the heaven to which the souls of good men and women were supposed by the Egyptian to depart, after he had developed sufficiently to conceive the doctrine of immortality. The chiefs of the company of evil spirits subsequently became the powerful devils of historic times, and the rulers of the company of beneficent and good spirits became the gods; the spirits of the third company, i.e., the spirits of the powers of Nature, became the great cosmic gods of the dynastic Egyptians. The cult of this last class of spirits, or gods, differed in many ways from that of the spirits or gods who were supposed to be concerned entirely with the welfare of man, and in dynastic times there are abundant proofs of this in religious texts and compositions. In the hymns to the Sun-god, under whatsoever name he is worshipped, we find that the greatest wonder is expressed at his majesty and glory, and that he is apostrophised in terms which show forth the awe and fear of his devout adorer. His triumphant passage across the sky is described, the unfailing regularity of his rising and setting is mentioned, reference is made to the vast distance over which he passes in a moment of time, glory is duly ascribed to him for the great works which he performs in nature, and full recognition is given to him as the creator of men and animals, of birds and fish, of trees and plants, of reptiles, and of all created things; the praise of the god is full and sufficient, yet it is always that of a finite being who appears to be overwhelmed at the thought of the power and might of an apparently infinite being. The petitions lack the personal appeal which we find in the Egyptian's prayers to the man-god Osiris, and show that he regarded the two gods from entirely different points of view. It is impossible to say how early this distinction between the functions of the two gods was made, but it is certain that it is coeval with the beginnings of dynastic history, and that it was observed until very late times.

The element of magic, which is the oldest and most persistent characteristic of the worship of the gods and of the Egyptian religion, generally belongs to the period before this distinction was arrived at, and it is clear that it dates from the time when man thought that the good and evil spirits were beings who were not greatly different from himself, and who could be propitiated with gifts, and controlled by means of words of power and by the performance of ceremonies, and moved to action by hymns and addresses. This belief was present in the minds of the Egyptians in all ages of their history, and it exists in a modified form among the Muḥammadan Egyptians and Sûdânî men to this day. It is true that they proclaim vehemently that there is no god but God, and that Muḥammad is His Prophet, and that God's power is infinite and absolute, but they take care to guard the persons of themselves and their children from the Evil Eye and from the assaults of malicious and evil spirits, by means of amulets of all kinds as zealously now as their ancestors did in the days before the existence of God Who is One was conceived. The caravan men protect their camels from the Evil Eye of the spirits of the desert by fastening bright-coloured beads between the eyes of their beasts, and by means of long fringes which hang from their *mahlûfas*, or

The Osirianization of the Hereafter

JAMES HENRY BREASTED

James Henry Breasted (1865–1935), considered the foremost influence in introducing Americans to the culture of ancient Egypt, was founder in 1919 of the Oriental Institute at the University of Chicago. In this selection, Breasted discusses the rituals that are described in the Pyramid Texts, the purpose of which was to help the deceased enter the world beyond. Breasted's work enables the reader to become aware of the way in which the Osiris myth transformed the Horus myth and became syncretized with it.

Probably nothing in the life of the ancient Nile-dwellers commends them more appealingly to our sympathetic consideration than the fact that when the Osirian faith had once developed, it so readily caught the popular imagination as to spread rapidly among all classes. It thus came into active competition with the Solar faith of the court and state priesthoods. This was especially true of its doctrines of the after life, in the progress of which we can discern the gradual Osirianization of Egyptian religion, and especially of the Solar teaching regarding the hereafter.

There is nothing in the Osiris myth, nor in the character or later history of Osiris, to suggest a celestial hereafter. Indeed clear and unequivocal survivals from a period when he was hostile to the celestial and Solar dead are still discoverable in the Pyramid Texts. We recall the exorcisms intended to restrain Osiris and his kin from entering the pyramid, a Solar tomb, with evil intent. Again we find the dead king as a star in the sky, thus addressed: "Thou lookest down upon Osiris commanding the Glorious (= the dead). There thou standest, being far from him, (for) thou art not of them (the dead), thou belongest not among them." Likewise it is said of the Sun-god: "He has freed king Teti from Kherti, he has not given him to Osiris." It is perhaps due to an effort to overcome this difficulty that Horus, the son of Osiris, is represented as one "who puts not this Pepi over the dead, he puts him among the gods, he being divine." The prehistoric Osiris faith, probably local to the Delta, thus involved a forbidding hereafter which was dreaded and at the same time was opposed to celestial blessedness beyond. To be sure, the Heliopolitan group of gods, the Divine Ennead of that city, makes Osiris a child of Nut, the Sky-goddess. But his father was the Earth-god Geb, a very natural result of the character of Osiris as a Nile-god and a spirit of vegetable life, both of which in Egyptian belief came out of the earth. Moreover, the celestial destiny through Nut the Sky-goddess is not necessarily Osirian. It is found, along with the frequent and non-Osirian or even pre-Osirian co-ordination of Horus and Set, associated in the service of the dead. The appearance of these two together assisting the dead cannot be Osirian. To be protected and assisted by Nut, therefore, does not necessarily imply that she is doing this for the dead king, because he is identified with Osiris, her son. It is thus probable that as a

Sky-goddess intimately associated with Re, Nut's functions in the celestial life hereafter were originally Solar and at first not connected with the Osirian faith.

When Osiris migrated up the Nile from the Delta, we recall how he was identified with one of the old mortuary gods of the South, the "First of the Westerners" (Khenti-Amentiu), and his kingdom was conceived as situated in the West, or below the western horizon, where it merged into the Nether World. He became king of a realm of the dead below the earth, and hence his frequent title, "Lord of Dewat," the "Nether World," which occurs even in the Pyramid Texts. It is as lord of a subterranean kingdom of the dead that Osiris later appears.[1]

As there was nothing then in the myth of the offices of Osiris to carry him to the sky, so the simplest of the Osirian Utterances in the Pyramid Texts do not carry him thither. There are as many varying pictures of the Osirian destiny as in the Solar theology. We find the dead king as a mere messenger of Osiris announcing the prosperous issue and plentiful yield of the year, the harvest year, which is associated with Osiris. That group of incidents in the myth which proves to be especially available in the future career of the dead king is his relations with Horus, the son of Osiris, and the filial piety displayed by the son toward his father. We may find the dead king identified with Horus and marching forth in triumph from Buto, with his mother, Isis, before him and Nephthys behind him, while Upwawet opened the way for them. More often, however, the dead king does all that Osiris did, receiving heart and limbs as did Osiris, or becoming Osiris himself. This was the favorite belief of the Osiris faith. The king became Osiris and rose from the dead as Osiris did. This identity began at birth and is described in the Pyramid Texts with all the wonders and prodigies of a divine birth.

> The waters of life that are in the sky come;
> The waters of life that are in the earth come.
> The sky burns for thee,
> The earth trembles for thee,
> Before the divine birth.
> The two mountains divide,
> The god becomes,
> The god takes possession of his body.
> The two mountains divide,
> This king Neferkere becomes,
> This king Neferkere takes possession of his body.

Osiris as Nile is thus born between the two mountains of the eastern and western Nile shores, and in the same way, and as the same being, the king is born. Hence we find the king appearing elsewhere as the inundation. It is not the mere assumption of the form of Osiris, but complete identity with him, which is set forth in this doctrine of the Pyramid Texts. "As he (Osiris) lives, this king Unis lives; as he dies not, this king Unis dies not; as he perishes not, this king Unis perishes not." These asseverations are repeated over and over, and addressed to every god in the Ennead, that each may be called upon to witness their truth. Osiris himself under various names is adjured, "Thy body is the body of this king Unis, thy flesh is the flesh of this king Unis, thy bones are the bones of this king Unis." Thus the dead king receives the throne of Osiris, and becomes, like him, king of the dead. "Ho! king Neferkere (Pepi II)! How beautiful is this! How beautiful is this, which thy father Osiris has done for thee! He has given thee his throne, thou rulest those of the hidden places (the dead), thou leadest their august ones, all the glorious ones follow thee."[2]

The supreme boon which this identity of the king with Osiris assured the dead Pharaoh was the good offices of Horus, the personification of filial piety. All the pious attentions which Osiris had once enjoyed at the hands of his son Horus now likewise become the king's portion. The litigation which the myth recounts at Heliopolis is successfully met by the aid of Horus, as well as Thoth, and, like Osiris, the dead king receives the predicate "righteous of voice," or "justified," an epithet which was later construed as meaning "triumphant." Over and over again the resurrection of Osiris by Horus, and the restoration of his body, are likewise affirmed to be the king's privilege. "Horus collects for thee thy limbs that he may put thee together without any lack in thee." Horus then champions his cause, as he had done that of his father, till the dead king gains the supreme place as sovereign of all. "O Osiris king Teti, arise! Horus comes that he may reclaim thee from the gods. Horus loves thee, he has equipped thee with his eye. . . . Horus has opened for thee thy eye that thou mayest see with it. . . . The gods . . . they love thee. Isis and Nephthys have healed thee. Horus is not far from thee; thou art his ka. Thy face is gracious unto him. . . . Thou hast received the word of Horus, thou art satisfied therewith. Hearken unto Horus, he has caused the gods to serve thee. . . . Horus has found thee that there is profit for him in thee. Horus sends up to thee the gods; he has given them to thee that they may illuminate thy face. Horus has placed thee at the head of the gods. He has caused thee to take every crown. . . . Horus has seized for thee the gods. They escape not from thee, from the place where thou hast gone. Horus counts for thee the gods. They retreat not from thee, from the place which thou hast seized. . . . Horus avenged thee; it was not long till he avenged thee. Ho, Osirisking Teti! thou art a mighty god, there is no god like thee. Horus has given to thee his children that they might carry thee. He has given to thee all gods that they may serve thee, and thou have power over them." A long series of Utterances in the Pyramid Texts sets forth this championship of the dead king as Osiris by his son Horus. In all this there is little or no trace of the celestial destiny, or any indication of the place where the action occurs. Such incidents and such Utterances are appropriated from the Osirian theology and myth, with little or no change. But the Osirian doctrine of the hereafter, absorbed into these royal mortuary texts by the priesthood of Heliopolis, could not, in spite of its vigorous popularity, resist the prestige of the state (or Solar) theology. Even in the Osirian Utterances on the good offices of Horus just mentioned we twice find the dead king, although he is assumed to be Osiris, thus addressed: "Thou art a Glorious One (Y'ḫwty) in thy name of 'Horizon (Y'ḫt) from which Re comes forth.'" The Osirian hereafter was thus celestialized, as had been the Osirian theology when it was correlated with that of Heliopolis. We find the Sky-goddess Nut extending to the Osirian dead her protection and the privilege of entering her realm. Nut "takes him to the sky, she does not cast him down to the earth." The ancient hymn in praise of the Sky-goddess embedded in the Pyramid Texts has received an introduction, in which the king as Osiris is commended to her protection, and the hymn is broken up by petitions inserted at intervals craving a celestial destiny for the dead king, although this archaic hymn had originally no demonstrable connection with Osiris, and was, as far as any indication it contains is concerned, written before the priestly theology had made Osiris the son of the Sky-goddess. Similarly Anubis, the ancient mortuary god of Siut, "counts Osiris away from the gods belonging to the earth, to the gods dwelling in the sky;" and we find in the Pyramid Texts the anomalous ascent of Osiris to the sky: "The sky thunders (lit. speaks), earth trembles, for fear of thee, Osiris, when thou makest ascent. Ho, mother cows yonder! Ho, suckling mothers (cows) yonder! Go ye behind him, weep for him, hail him, acclaim him, when he makes ascent and goes to the sky among his brethren, the gods." His transition to the Solar and celestial destiny is effected in one passage by a piece of purely mortuary theologizing which represents Re as raising Osiris from the dead. Thus is Osiris celestialized until the Pyramid Texts even call him "lord of the sky," and represent him as ruling there. The departed Pharaoh is ferried over, the doors of the sky are opened for him, he passes all enemies as he goes, and he is announced to Osiris in the sky precisely as in the Solar theology. There he is welcomed by Osiris, and he joins the "Imperishable Stars, the followers of Osiris," just as in the Solar faith. In the same way he emerges as a god of primeval origin and elemental powers. "Thou bearest the sky in thy hand, thou layest down the earth with thy foot." Celestials and men acclaim the dead, even "thy wind is incense, thy north wind is smoke," say they.

While the Heliopolitan priests thus solarized and celestialized the Osirian mortuary doctrines, although they were essentially terrestrial in origin and character, these Solar theologians were in their turn unable to resist the powerful influence which the popularity of the Osirian faith brought to bear upon them. The Pyramid Texts were eventually Osirianized, and the steady progress of this process, exhibiting the course of the struggle between the Solar faith of the state temples and the popular beliefs of the Osirian religion thus discernible in the Pyramid Texts, is one of the most remarkable survivals from the early world, preserving as it does the earliest example of such a spiritual and intellectual conflict between state and popular religion. The dying Sun and the dying Osiris are here in competition. With the people the human Osiris makes the stronger appeal, and even the wealthy and subsidized priesthoods of the Solar religion could not withstand the power of this appeal. What we have opportunity to observe in the Pyramid Texts is specifically the gradual but irresistible intrusion of Osiris into the Solar doctrines of the hereafter and their resulting Osirianization.

Even on his coffin, preserved in the pyramid sepulchre, the departed king is called "Osiris, lord of Dewat." The Osirian influence is superficially evident in otherwise purely Solar Utterances of the Pyramid Texts where the Osirian editor has inserted the epithet "Osiris" before the king's name, so that we have "Osiris king Unis," or "Osiris king Pepi." This was at first so mechanically done that in the offering ritual it was placed only at the head of each Utterance. In the earliest of our five versions of the Pyramid Texts, that of Unis, we find "Osiris" inserted before the king's name wherever that name stands at the head of the Utterance, but not where it is found in the body of the text. Evidently the Osirian editor ran hastily and mechanically through the sections, inserting "Osiris" at the head of each one which began with the king's name, but not taking the trouble to go through each section seeking the king's name and to insert "Osiris" wherever necessary in the body of the text also.

In this way the whole Offering Ritual was Osirianized in Unis's pyramid, but the editor ceased this process of mechanical insertion at the end of the ritual. A similar method may be observed where the same Utterance happens to be preserved in two different pyramids, one exhibiting the mechanical insertion of "Osiris" before the king's name, while the other lacks such editing. This is especially significant where the content of the Utterances is purely Solar.

But the Osirianization of the Pyramid Texts involves more than such mechanical alteration of externals. We find one Utterance in its old Solar form, without a single reference to Osiris or to Osirian doctrine, side by side with the same Utterance in expanded form filled with Osirian elements. The traces of the Osirian editor's work are evident throughout, but they are interestingly demonstrable in a series of five stanzas each addressed to a different god, whose name begins the stanza. The last stanza of the five begins with *two* gods' names, however, the second being "Sekhem, son of Osiris," although in the apostrophe, which constitutes this fifth stanza, the two gods are addressed by pronouns in the *singular number!* It is evident that, like the other four stanzas, the fifth also began with the name of a single god, but that the Osirian editor has inserted the name of an Osirian god as a second name, forgetting to change the pronouns. The insertion is enhanced in significance by the fact that all five gods in these five stanzas are Solar gods, and the last one, after which the name of Osiris was inserted, is identified with Re.

The process was carried so far that it was sometimes applied to passages totally at variance with the Osirian doctrine. In the old Solar teaching we not infrequently find Horus and Set side by side on an equal basis, and both represented as engaged in some beneficent act for the dead. Now when the dead king is identified with Osiris, by the insertion of the name "Osiris" before that of the king, we are confronted by the extraordinary assumption that Set performs pious mortuary offices for Osiris, although the Osiris myth represents Set as mutilating the body of the dead Osiris and scattering his limbs far and wide. Thus an old purification ceremony in the presence of the gods and nobles of Heliopolis (and hence clearly Solar) represents the dead as cleansed by the spittle of Horus and the spittle of Set. This ceremony had, of course, nothing to do with the Osirian ritual, but when the ritual introducing this ceremony was Osirianized, we find 'King Osiris, this Pepi" inserted before the formula of purification, thus assuming that Osiris was purified by his arch-enemy, the foul Set! Similarly, Set may appear alone in old Solar Utterances on familiar and friendly terms with the dead king, so that the king may be addressed thus: "He calls to thee on the stairway of the sky; thou ascendest to the god; Set fraternizes with thee," even though the king has just been raised as Osiris from the dead!

The ladder leading to the sky was originally an element of the Solar faith. That it had nothing to do with Osiris is evident, among other things, from the fact that one version of the ladder episode represents it in charge of Set. The Osirianization of the ladder episode is clearly traceable in four versions of it, which are but variants of the same ancient original. The four represent a period of nearly a century, at least of some eighty-five years. In the oldest form preserved to us, in the pyramid of Unis, dating from the middle of the twenty-seventh century, the Utterance opens with the acclamation of the gods as Unis ascends. "'How beautiful to see, how satisfying to behold,' say the gods, 'when this god ascends to the sky, when Unis ascends to the sky. . . .' The gods in the sky and the gods on earth come to him; they make supports for Unis on their arm. Thou ascendest, O Unis, to the sky. Ascend upon it in this its name of 'Ladder.' The sky is given to Unis, the earth is given to him by Atum." Such is the essential substance of the Utterance.[3] The ladder here barely emerges and the climber is the Pharaoh himself, though Atum is prominent. A generation later, in the pyramid of Teti the ladder is more developed and the original climber is Atum, the Sun-god; but the Osirian goddesses, Isis and Nephthys, are introduced. Finally, in the pyramid of Pepi I, at least eighty-five years after that of Unis, the opening acclamation of the old gods as they behold the ascent of the Pharaoh is put into the mouths of Isis and Nephthys, and the climber has become Osiris. Thus Osiris has taken possession of the old Solar episode and appropriated the old Solar text. This has taken place in spite of embarrassing complications. In harmony with the common coordination of Horus and Set in the service of the dead, an old Solar doctrine represented them as assisting him at the ascent of the ladder which Re and Horus set up. But when the ascending king becomes Osiris, the editor seems quite unconscious of the incongruity, as Set, the mortal enemy and slayer of Osiris, assists him to reach his celestial abode!

NOTES

[1]The situation of Dewat is a difficult problem. As the Nile flows out of it, according to later texts, especially the Sun-hymns, and the common designation of the universe in the Empire is "sky, earth, and Dewat," it is evident that it was later understood to be the Nether World. In the Pyramid Texts it is evidently in the sky in a considerable number of passages. It can be understood in no other way in passages where it is parallel with "sky," like the following:

The sky conceived thee together with Orion;
Dewat bears thee together with Orion.

Or again:

Who voyages the sky with Orion,
Who sails Dewat with Osiros.

Similarly "Dewat seizes thy hand, (leads thee) to the place where Orion (= the sky); and Orion and Sothis in the "horizon" are encircled by Dewat. Here Dewat is in the horizon, and likewise we find the dead "descends among" the dwellers in Dewat after he has ascended to the sky. It was thus sufficiently accessible from the sky, so that the dead, after he ascended, bathed in the "lake of Dewat," and while in the sky he became a "glorious one dwelling in Dewat." When he has climbed the ladder of Re, Horus and Set take him to Dewat. It is parallel with *'kr*, where *'kr* is a variant of Geb, the earth, which carries it down to earth again. It might appear here that Dewat was a lower region of the sky, in the vicinity of the horizon, below which it also extended. It is notable that in the Coffin Texts of the Middle Kingdom there appears a "lower Dewat." The deceased says: "My place is in the barque of Re in the middle of lower Dewat." Dewat thus merged into the Nether World, with which it was ultimately identified, or, being originally the Nether World, it had its counterpart in the sky.

[2]There is little distinction between the passages where the dead king receives the throne of Osiris, because identified with him and others in which he receives it as the heir of Osiris. He may take it even from Horus, heir of Osiris.

[3]The brief intimation of a mysterious enemy plotting against the life of the king, appended at the end of the Utterance, is perhaps an intrusive Osirian reference; but it does not affect the clearly celestial and Solar character of the Utterance. It is omitted in Ut. 480, but appears more fully developed in the Osirianized Utterances 572 and 474, but in none of the Utterances to which it is appended is the name of Osiris mentioned, while the epithet which is employed, "Ymnw (Hidden one?) of the Wild Bull," is usually Solar.

Unit V

The Sacred Source: The Perennial Philosophy of the East

A History of Indian Philosophy

SURENDRANATH DASGUPTA

Surendranath Dasgupta is a world renowned modern interpreter of Indian philosophy to the English-speaking world. His five-volume study of Indian philosophy is the source for this selection on Samkhya. Samkhya philosophy is, perhaps, the earliest formal philosophical system developed in India, several centuries before the Common Era. Because of its analysis of the Soul (Purusha) and Matter (Prakriti), later philosophies took the concepts and terminology of Samkhya into account in working out fresh interpretations. In order to follow the discussions about any Hindu philosophical system, it is important to know the basic concepts of Samkhya.

THE SAṂKHYA AND THE YOGA DOCTRINE OF SOUL OR PURUṢA

The Sāṃkhya philosophy as we have it now admits two principles, souls and *prakṛti*, the root principle of matter. Souls are many, like the Jaina souls, but they are without parts and qualities. They do not contract or expand according as they occupy a smaller or a larger body, but are always all-pervasive, and are not contained in the bodies in which they are manifested. But the relation between body or rather the mind associated with it and soul is such that whatever mental phenomena happen in the mind are interpreted as the experience of its soul. The souls are many, and had it not been so (the Sāṃkhya argues) with the birth of one all would have been born and with the death of one all would have died.

The exact nature of soul is however very difficult of comprehension, and yet it is exactly this which one must thoroughly grasp in order to understand the Sāṃkhya philosophy. Unlike the Jaina soul possessing *anantajñāna*, *anantadarśana*, *anantasukha*, and *anantavīryya*, the Sāṃkhya soul is described as being devoid of any and every characteristic; but its nature is absolute pure consciousness (*cit*). The Sāṃkhya view differs from the Vedānta, firstly in this that it does not consider the soul to be of the nature of pure intelligence and bliss (*ānanda*). Bliss with Sāṃkhya is but another name for pleasure and as such it belongs to prakṛti and does not constitute the nature of soul; secondly, according to Vedānta the individual souls (*jīva*) are but illusory manifestations of one soul or pure consciousness the Brahman, but according to Sāṃkhya they are all real and many.

The most interesting feature of Sāṃkhya as of Vedānta is the analysis of knowledge. Sāṃkhya holds that our knowledge of things are mere ideational pictures or images. External things are indeed material, but the sense data and images of the mind, the coming and going of which is called knowledge, are also in some sense matter-stuff, since they are limited in their nature like the external things. The sense-data and images come and go, they are often the prototypes, or photographs of external things, and as such ought to be considered as in some sense material, but the matter of which these are composed is the subtlest. These images of the mind could not have appeared as conscious, if there were no separate principles of consciousness in connection with which the whole conscious plane could be interpreted as the experience of a person. We know that the Upaniṣads consider the soul or ātman as pure and infinite consciousness, distinct from the forms of knowledge, the ideas, and the images. In our ordinary ways of mental analysis we do not detect that beneath the forms of knowledge there is some other principle which has no change, no form, but which is like a light which illumines the mute, pictorial forms which the mind assumes. The self is nothing but this light. We all speak of our "self" but we have no mental picture of the self as we have of other things, yet in all our knowledge we seem to know our self. The Jains had said that the soul was veiled by karma matter, and every act of knowledge meant only the partial removal of the veil. Sāṃkhya says that the self cannot be found as an image of knowledge, but that is because it is a distinct, transcendent principle, whose real nature as such is behind or beyond the subtle matter of knowledge. Our cognitions, so far as they are mere forms or images, are merely compositions or complexes of subtle mind-substance, and thus are like a sheet of painted canvas immersed in darkness; as the canvas gets prints from outside and moves, the pictures appear one by one before the light and are illuminated. So it is with our knowledge. The special characteristic of self is that it is like a light, without which all knowledge would be blind. Form and motion are the characteristics of matter, and so far as knowledge is mere limited form and movement it is the same as matter; but there is some other principle which enlivens these knowledge-forms, by virtue of which they become conscious. This principle of consciousness (*cit*) cannot indeed be separately perceived *per se*, but the presence of this principle in all our forms of knowledge is distinctly indicated by inference. This principle of consciousness has no motion, no form, no quality, no impurity[1]. The movement of the knowledge-stuff takes place in relation to it, so that it is illuminated as consciousness by it, and produces the appearance of itself as undergoing all changes of knowledge and experiences of pleasure and pain. Each item of knowledge so far as it is an image or a picture of some sort is but a subtle knowledge-stuff which has been illumined by the principle of consciousness, but so far as each item of knowledge carries with it the awakening or the enlivening of consciousness, it is the manifestation of the principle of consciousness. Knowledge-revelation is not the unveiling or revelation of a particular part of the self, as the Jains supposed, but it is a revelation of the self only so far as knowledge is pure awakening, pure enlivening, pure consciousness. So far as the content of knowledge or the image is concerned, it is not the revelation of self but is the blind knowledge-stuff.

The Buddhists had analysed knowledge into its diverse constituent parts, and had held that the coming together of these brought about the conscious states. This coming together was to them the point of the illusory notion of self, since this unity or coming together was not a permanent thing but a momentary collocation. With Sāṃkhya however the self, the pure *cit*, is neither illusory nor an abstraction; it is concrete but transcendent. Coming into touch with it gives unity to all the movements of the knowledge-composites of subtle stuff, which would otherwise have remained aimless and unintelligent. It is by coming into connection with this principle of intelligence that they are interpreted as the systematic and coherent experience of a person, and may thus be said to be intelligized. Intelligizing means the expression and interpretation of the events or the happenings of knowledge in connection with a person, so as to make them a system of experience. This principle of intelligence is called puruṣa. There is a separate puruṣa in Sāṃkhya for each individual, and it is of the nature of pure intelligence. The Vedānta ātman however is different from the Sāṃkhya puruṣa in this that it is one and is of the nature of pure intelligence, pure being, and pure bliss. It alone is the reality and by illusory māyā it appears as many.

THOUGHT AND MATTER

A question naturally arises, that if the knowledge forms are made up of some sort of stuff as the objective forms of matter are, why then should the puruṣa illuminate it and not external material objects. The answer that Sāṃkhya gives is that the knowledge-complexes are certainly different from external objects in this, that they are far subtler and have a preponderance of a special quality of plasticity and translucence (*sattva*), which resembles the light of puruṣa, and is thus fit for reflecting and absorbing the light of the puruṣa. The two principal characteristics of external gross matter are mass and energy. But it has also the other characteristic of allowing itself to be photographed by our mind; this thought-photograph of matter has again the special privilege of being so translucent as to be able to catch the reflection of the *cit*—the super-translucent transcendent principle of intelligence. The fundamental characteristic of external gross matter is its mass; energy is common to both gross matter and the subtle thought-stuff. But mass is at its lowest minimum in thought-stuff, whereas the capacity of translucence, or what may be otherwise designated as the intelligence-stuff, is at its highest in thought-stuff. But if the gross matter had none of the characteristics of translucence that thought possesses, it could not have made itself an object of thought; for thought transforms itself into the shape, colour, and other characteristics of the thing which has been made its object. Thought could not have copied the matter, if the matter did not possess some of the essential substances of which the copy was made up. But this plastic entity (*sattva*) which is so predominant in thought is at its lowest limit of subordination in matter. Similarly mass is not noticed in thought, but some such notions as are associated with mass may be discernible in thought; thus the images of thought are limited, separate, have movement, and have more or less clear cut forms. The images do not extend in space, but they can represent space. The translucent and plastic element of thought (*sattva*) in association with movement (*rajas*) would have resulted in a simultaneous revelation of all objects; it is on account of mass or tendency of obstruction (*tamas*) that knowledge proceeds from image to image and discloses things in a successive manner. The buddhi (thought-stuff) holds within it all knowledge immersed as it were in utter darkness, and actual knowledge comes before our view as though by the removal of the darkness or veil, by the reflection of the light of the puruṣa. This characteristic of knowledge, that all its stores are hidden as if lost at any moment, and only one picture or idea comes at a time to the arena of revelation, demonstrates that in knowledge there is a factor of obstruction which manifests itself in its full actuality in gross matter as mass. Thus both thought and gross matter are made up of three elements, a plasticity of intelligence-stuff (*sattva*), energy-stuff (*rajas*), and mass-stuff (*tamas*), or the factor of obstruction. Of these the last two are predominant in gross matter and the first two in thought.

FEELINGS, THE ULTIMATE SUBSTANCES

Another question that arises in this connection is the position of feeling in such an analysis of thought and matter. Sāṃkhya holds that the three characteristic constituents that we have analyzed just now are feeling substances. Feeling is the most interesting side of our consciousness. It is in our feelings that we think of our thoughts as being parts of ourselves. If we should analyze any percept into the crude and undeveloped sensations of which it is composed at the first moment of its apperance, it comes more as a shock than as an image, and we find that it is felt more as a feeling mass than as an image. Even in our ordinary life the elements which precede an act of knowledge are probably mere feelings. As we go lower down the scale of evolution the automatic actions and relations of matter are concomitant with crude manifestations of feeling which never rise to the level of knowledge. The lower the scale of evolution the less is the keenness of feeling, till at last there comes a stage where matter-complexes do not give rise to feeling reactions but to mere physical reactions. Feelings thus mark the earliest track of consciousness, whether we look at it from the point of view of evolution or of the genesis of consciousness in ordinary life. What we call matter-complexes become at a certain stage feeling-complexes and what we call feeling-complexes at a certain stage of descent sink into mere matter-complexes with matter reaction. The feelings are therefore the things-in-themselves, the ultimate substances of which consciousness and gross matter are made up. Ordinarily a difficulty might be felt in taking feelings to be the ultimate substances of which gross matter and thought are made up; for we are more accustomed to take feelings as being merely subjective, but if we remember the Sāṃkhya analysis, we find that it holds that thought and matter are but two different modifications of certain subtle substances which are in essence but three types of feeling entities. The three principal characteristics of thought and matter that we have noticed in the preceding section are but the manifestations of three types of feeling substances. There is the class of feelings that we call the sorrowful, there is another class of feelings that we call pleasurable, and there is still another class which is neither sorrowful nor pleasurable, but is one of ignorance, depression (*viṣāda*) or dullness. Thus corresponding to these three types of manifestations as pleasure, pain, and dullness, and materially as shining (*prakāśa*), energy (*pravṛtti*), obstruction (*niyama*), there are three types of feeling-substances which must be regarded as the ultimate things which make up all the diverse kinds of gross matter and thought by their varying modifications.

THE GUṆAS

These three types of ultimate subtle entities are technically called *guṇa* in Sāṃkhya philosophy. Guṇa in Sanskrit has three meanings, namely (1) quality, (2) rope, (3) not primary. These entities, however, are substances and not mere qualities. But it may be mentioned in this connection that in Sāṃkhya philosophy there is no separate existence of qualities; it holds that each and every unit of quality is but a unit of substance. What we call quality is but a particular manifestation or appearance of a subtle entity. Things do not possess quality, but quality signifies merely the manner in which a substance reacts; any object we see seems to possess many qualities, but the Sāṃkhya holds that corresponding to each and every new unit of quality, however fine and subtle it may be, there is a corresponding subtle entity, the reaction of which is interpreted by us as a quality. This is true not only of qualities of external objects but also of mental qualities as well. These ultimate entities were thus called guṇas probably to suggest that they are the entities which by their various modifications manifest themselves as guṇas or qualities. These subtle entities may also be called guṇas in the sense of ropes because they are like ropes by which the soul is chained down as if it were to thought and matter. These may also be called guṇas as things of secondary importance, because though permanent and indestructible, they continually suffer modifications and changes by their mutual groupings and regroupings, and thus not primarily and unalterably constant like the souls (*puruṣa*). Moreover the object of the world process being the enjoyment and salvation of the puruṣas, the matter-principle could not naturally be regarded as being of primary importance. But in whatever senses we may be inclined to justify the name guṇa as applied to these subtle entities, it should be borne in mind that they are substantive entities or subtle substances and not abstract qualities. These guṇas are infinite in number, but in accordance with their three main characteristics as described above they have been arranged in three classes or types called *sattva* (intelligence-stuff), *rajas* (energy-stuff) and *tamas* (mass-stuff). An infinite number of subtle substances which agree in certain characteristics of self-shining or plasticity are called the *sattva-guṇas* and those which behave as units of activity are called the *rajo-guṇas* and those which behave as factors of obstruction, mass or materiality are called *tamo-guṇas*. These subtle guṇa substances are united in different proportions (e.g. a larger number of sattva substances with a lesser number of rajas or tamas, or a larger number of tamas substances with a smaller number of rajas and sattva substances and so on in varying proportions), and as a result of this, different substances with different qualities come into being. Though attached to one another when united in different proportions, they mutually act and react upon one another, and thus by their combined resultant produce new characters, qualities and substances.

There is however one and only one stage in which the guṇas are not compounded in varying proportions. In this state each of the guṇa substances is opposed by each of the other guṇa substances, and thus by their equal mutual opposition create an equilibrium, in which none of the characters of the guṇas manifest themselves. This is a state which is so absolutely devoid of all characteristics that it is absolutely incoherent, indeterminate, and indefinite. It is a qualitiless simple homogeneity. It is a state of being which is as it were non-being. This state of the mutual equilibrium of the guṇas is called prakṛti[1]. This is a state which cannot be said either to exist or to non-exist for it serves no purpose, but it is hypothetically the mother of all things. This is however the earliest stage, by the breaking of which, later on, all modifications take place.

NOTES

[1]It is important to note that Sāṃkhya has two terms to denote the two aspects involved in knowledge, viz. the relating element of awareness as such (*cit*), and the content (*buddhi*) which is the form of the mind-stuff representing the sense-data and the image. Cognition takes place by the reflection of the former in the latter.

The Legacy of the Indus Civilization

F. R. ALLCHIN

F. R. Allchin, a well-known student of Indian religion, specializes in interpreting the earliest period of Indian civilization, linked to the Indus Valley located in the northwestern part of the Indian subcontinent. There in the third millennium B.C.E.*, the two cities of Mohenjodaro and Harappa were built and became the main centers of a widespread culture. The following selection discusses the recent interpretation of the Indus Valley culture, both as regards its origin and the legacy it left to the subsequent history of the region that is today's India and Pakistan.*

PART ONE

Perhaps the most significant discoveries of the past two decades have been those relating to the spread of peasant settlements in the Indus Valley in Pre-Harappan (or as one should prefer to call them, Early Indus) and even earlier Neolithic times. Among the excavated sites of this group are Amri, Kot Diji, Gumla, Harappa and Kalibangan. In the north an extension to the foothills is indicated by the discovery of Sarai Khola near Taxila. The development must have resulted from the initial exploitation of the rich flood plain of the Indus and its tributaries, and must have spread outward with the expansion of the population this produced. It seems to have been accompanied throughout by a tendency toward cultural convergence, and it provides the unequivocal basis in terms of human, technical and cultural resources on which the succeeding Mature stage of Indus Civilization is based. From the present point of view the maturity of the civilization must be seen as the climax of an organic developmental process starting in the early Indus times, if not yet earlier. Acceptance of this model suggests profoundly important analogies between it and the luminous words of Robert Redfield respecting the special character of Indian Civilization as indigenous, having "developed out of the precivilized people of that very culture, converting them into the peasant half of the same culture-civilization," so that "the continuity with their own native civilization has persisted," peasant tradition affects the doctrine of the learned, and there is continuing interaction between the learned and folk levels. In short, while there is no reason to neglect the possibility that the Indus Civilization arose as a partial result of stimuli applied from outside, either from the uplands to the west, or more distantly from the Persian Gulf or Mesopotamia in the Early Dynastic or Sargonid times, or from Central Asia, and that these stimuli may have involved the arrival of men as well as ideas, its actual emergence must be seen primarily as a dynamic socio-economic process, taking place on Indian soil, and not as something implanted from outside. This is profoundly significant for subsequent developments, since it implies that the legacy of the Indus Civilization may be sought in the life style of the common people of India and Pakistan, as much as in the learned tradition. This is exemplified by such things as the identity of ploughing patterns in the fields of Pre-Harappan Kalibangan and of the modern peasant population of the region; or by the direct analogy of one of the distinctive types of terracotta model carts from Mohenjodaro to a type which today survives only in upper Sind. These things can only mean that there is a direct and unbroken craft tradition or tradition of agricultural practice linking the two periods, and this in turn implies a continuum of population and a direct and unbroken rural life style from the Early Indus times forward.

Another important set of discoveries has provided evidence that the hunter-gatherers, using a stone technology and generally spoken of as Mesolithic, or Late Stone Age, had spread widely across the Subcontinent long before the emergence of the Indus Civilization. The current excavations of the French Archaeological Mission at Mehrgarh near Sibi have added a new perspective to the knowledge of the antecedents of the Indus Civilization, and may be expected in particular to throw very important light on the earliest stages of the development of settled agriculture in the Indian Subcontinent. This in turn is likely to contribute to the understanding of relations of the first agricultural communities to such groups of hunter-gatherers. Further, it is beginning to emerge that, at least in those places where evidence is available, groups of these people continued to live predominantly as stone-using hunter-gatherers, sometimes driven into areas of relative isolation, long after the use of metals had become common in the more advanced communities. It is now apparent that even before the emergence of the Indus Civilization in many regions of South Asia peasant or pastoral communities had appeared, and that in some cases these groups may have enjoyed social, economic or political relations of one kind or another with the Pre-Harappan settlements. How far both these and the settled communities arose as the result of the age-old tendency of peoples to move into the Subcontinent from the less hospitable lands to the west or north, and how far they arose by a process of local evolution from among the existing tribal populations has yet to be established. Probably both tendencies played their parts. But it is in such early settlements that the first localized cultural characters of some of the regions can be distinguished. The coexistence of groups at different socioeconomic and cultural levels, in close association with each other, often over long periods of time, may already at this stage be clearly distinguished. Both these tendencies were to play a significant role in the subsequent lifestyle of the regions of the Subcontinent. Thus they deserve to be borne in mind while considering the transmission of the legacy of the Indus Civilization.

It has been suggested that already during the Early Indus stage there was a period of rapid expansion of settled population throughout the Indus Valley and that it was accompanied by an outward spread towards less densely settled but attractive areas. This process may have involved both the establishment of new settlements in regions hitherto largely unsettled (i.e., populated by groups of hunters and collectors, or by primarily nomadic or seminomadic pastoralists); and the conquest or colonization of areas which already had a settled agricultural population. In some cases it may have involved the establishment of new settlements among the settlements of the already existing regional culture. Such variations in the pattern of culture contact are likely to be recognizable in the archaeological record, once their hypothetical existence is admitted. At a certain point this process triggered off changes resulting in the formation of cities, and the development of a new set of socioeconomic relations—the Indus Civilization. One must expect the same processes of growth and spread to have continued thereafter, and hence one would expect to find an outward spread of the Mature Urban Culture into areas hitherto peopled by tribal and/or peasant communities. Evidence to support this hypothesis appears to be forthcoming in several regions. In Saurashtra and Gujarat, Harappan settlements seem to be dispersed among those of a regional peasant culture, and this must be presumed to have led to contacts between the two at various levels. In East Punjab, in the Drishadvati and Sarasvati Valleys, and even further east in the Ganga/Yamuna *Doab*, there are somewhat similar indications. Dr. Suraj Bhan has reported sites related to the Early Indus stage at Kalibangan, and their continued occupation alongside sites which may more properly be called Harappan. This suggests the sort of culture contact situations to be found in the areas in which this spread was taking place. It is interesting to notice how the Rajasthan desert seems to have acted as a barrier to expansion, even though there was already a population of tribal people of mixed economy (as at Bagor) and perhaps also of agricultural settlements in parts of the region, and finds of Mesolithic tools suggest that such people were widely distributed there. One may expect that they enjoyed some sort of contacts with the Harappans prospecting for ores or raw materials and trading with them in such things as metal tools.

Within the area embraced by the Early and Mature Indus Civilization two opposing tendencies may be noticed: one convergent and the other divergent.

The first is the more prominent: It has long been apparent that one of the concomitants of the change from the Early to Mature Indus Cultures was the establishment of an extraordinary degree of cultural uniformity over a vast area. This convergent tendency is indeed already clearly visible during the Early Indus Period, and as this fact becomes more apparent it makes the change of style from Early to Mature Indus times all the more remarkable. Sir John Marshall discussed this aspect at length, and others have generally agreed with him. What was involved in terms of the population as a whole can only be partly guessed at, but it seems that the convergent tendency reached a new height with the growth of Mohenjodaro as a city, and therefore that the tendency may well go back to the foundation of that city whenever it may have been, perhaps towards the end of the second quarter of the third millennium. Presumably it would have more or less coincided with the first development of a full system of writing and the manufacture of inscribed seals. This seems to have been the signal for a rapid and wholesale diffusion of traits recognizable in the archaeological record, suggesting the spread or imposition of a common lifestyle which in time extended throughout the entire Harappan Culture region. The means for this spread remain unclear. That it was assisted by an unprecedented extension of internal trade in all manner of raw materials and commodities may be safely inferred, and that it witnessed a distribution of the specialist craft products of the cities is probable. Among trade goods one may cite the stone blades which appear to have been obtained and manufactured at such centers as the vast factories at Rohri, the similar indications of centralized manufacture of various classes of metal objects, of shell bangles and carnelian beads. An aspect of the mechanism by which this trade was carried on has been discussed by my wife (B. Allchin 1979). It seems certain that the emergence of urbanism must have also involved the extension of a single unifying socioeconomic pattern, including government and administration, and of a common pattern of beliefs and ideology. In sum it must have witnessed the promulgation of what one may call the Indus lifestyle, with all that went with it. The fact that the geographical confines of this culture region embraced an area far greater than that of any other of the great civilizations of the third millennium makes the process all the more remarkable. It may also be stated with certainty that the lifestyle incorporated not only popular matters but also the learned or 'great' tradition of the Indus Civilization. It is from among these things that scholars have found, or at least believed that they found, all manner of traits ancestral to those of later, even of modern, Indian civilization.

However, while recognizing the convergent tendency in the Indus Civilization, one must not neglect to notice the indications of regionalism or cultural divergence within the greater Indus system, leading to what may even be seen as separate culture provinces. One would expect there to be several kinds of divergence. First, the very real differences which already begin to emerge during the Neolithic stage, and later within the provinces of the early Indus Culture regions between such sites as Amri, Kot Diji, Gumla or Kalibangan, evidently continued to a certain extent, even after the imposition of the Indus style, reflecting no doubt these earlier differences. Second, even during the comparatively short life of the Mature Indus Civilization one may expect that there would have been a further general tendency towards separate development in certain respects in the different provinces, due to all sorts of possible causes. The divergences could in part reflect local reactions against the convergent tendency. A third possible cause could be the arrival in a given region of new groups who established some sort of power over, or relationship with, the existing population and proceeded in one way or another to influence the final stages of the Indus Culture therein. This aspect is crucially important in terms of legacy, and it must be considered further. A fairly circumscribed regional development appears to be represented by the Cemetery H Phase at Harappa and probably at related sites (at Bara and sites of East Punjab, in Bahawalpur, etc.). One is inclined to follow Vats in seeing this phase as a final stage in the Indus Culture representing the arrival of some sort of "Aryan" invaders in the region and their interaction with the existing population, leading to a degree of cultural synthesis. Another set of data which suggests something of the same kind is provided by the apparent contrast between certain sites at which stone or terracotta *linga* are reported, notably at Mohenjodaro, at Harappa, and at Surkotada, and other sites (or perhaps a phase at some sites) at which "ritual" fire altars are reported. Such altars are expressly absent at Surkotada, and also apparently at Mohenjodaro and Harappa, but are reported at Kalibangan from the beginning of the Harappan Period. Here several types of hearth are found; among them one is distinguished from the normal domestic types by the excavators. The "ritual" hearth, with a brick or clay "pillar" in its center occurs in three locations: on the top of a brick platform in the walled brick "Citadel" enclosure a row of seven were found, associated with a brick pit containing animal bones and with a well; single examples of the same type occur in small rooms in domestic houses, perhaps used for domestic rituals; and several more were found in a square brick enclosure outside the east wall of the lower city. These three contexts suggest that fire rituals formed a part of the religious life of the town, at a civic, domestic and popular level. They are also highly suggestive of an Indo-Iranian, if not more specifically Indo-Aryan, element in the culture of the period covered by these excavations.

Thus, if enough evidence of this sort were forthcoming, the regional divergences might in some instances be associated with the meeting of indigenous Indus populations and intrusive Indo-Iranian or Aryan elements, and with some sort of resulting cultural synthesis. This would clearly be of great significance for any discussion of legacy. It is of course wholly possible that similar intrusive groups may also have moved into the southern Indus provinces, but if they did so they seem to have coincided with a more or less complete extinction of the centralized urban authority rather than with its late stages, and their contribution to the legacy may be expected to have been therefore mainly at the village level. Thus there is likely to have been a major difference between the situation in the northern and southern parts of the Harappan Culture region.

One must now consider certain aspects of the civilization which resulted from this transformation of existing peasant and tribal social elements. The access of new data has already revealed much and doubtless much more awaits both discovery and analysis, but some aspects remain, and are long likely to, elusive for the archaeologist. Thus while one is learning more about the town planning, and such things as house plans, plant and animal foods, technology and crafts, one still has only rudimentary indications as to the meaning of all these in terms of social and economic relations. Similarly, while there are seals and art objects which reveal a clearly defined body of symbols and myths, there is still surprisingly little definite knowledge of the religious beliefs or ideology of their makers, although here one is beginning to gain an important new dimension from finds in the Early Indus Cultures. The reading of the script will almost certainly throw new light on trade and economics, and perhaps marginally on ideology, but even so the absence of longer inscriptions must mean that many topics will remain essentially speculative. This point is made mainly because it is just in these fields, in economic and social relations, in religious beliefs and ideology, that an important part of the Harappan legacy is likely to be most strongly evident in later Indian culture, and if one cannot positively identify such things at their source, how far is one entitled to speak of the legacy at all? This prompts one again to stress the importance of the framing of general hypotheses, as without them the contexts of individual facts, or groups of data, may be difficult to establish.

PART TWO

There is still much room for divergent views about the end of the Indus Civilization, but as this event provides, in a sense, with a terminus for consideration of the legacy, as opposed to the nature of the civilization itself, one must try to establish a satisfactory general hypothesis. The civilization has been thought of as a social, economic and cultural phenomenon produced as a consequence of the build up of population on the fertile plains of the Indus and the Punjab. It involved a delicate balance of internal relations between cities, towns and villages, and of external relations with neighboring peasant societies and with more distant urban societies. The end of the civilization probably arose from some major upsetting of this balance. This could have been produced by a variety of causes, acting either singly or in combination. It is possible, although there is as yet little supporting evidence, that there was a deterioration of climate; but as the main food production depended upon exploiting the Indus river's flood plain inundation this is unlikely to have been sufficient cause. It is possible too that there was a "wearing out" of the land, due to overcultivation; but in the light of the enduring fertility of the soils of the Indian Subcontinent over subsequent millennia of intensive cultivation this too seems unlikely. One accepts Lambrick's 1967 demonstration of the implausibility of the theory that Mohenjodaro was engulfed in a vast flood, and one is inclined to agree with his well-argued case that there may have been disastrous changes in the course of the Indus, resulting in the desiccation of areas which were essential for the feeding of the city's population, as a more likely cause of the end of that city. Such an event would lead to depredations by tribesmen from the nearby hills, and might well have brought about the desertion of the city and of outlying settlements. But would it have been a sufficient shock to upset the whole balance? This is a more difficult question, even if it is assumed, as it is more and more, that Mohenjodaro was in some way the "epicenter" around which the whole structure was held in balance. Would the attackers have been Aryan? This one cannot tell, but both in terms of the probable date of the event, and of the history of the dispersal of the Indo-European language family, there is no inherent impossibility in such a thing. If this was the state in Sind, what of the Punjab? Here too there are suggestions that there may have been major changes in the channels of rivers, due in part no doubt to tectonic events; but there is no very clear evidence that any of these coincided with the end of Harappa and Kalibangan. Moreover as has been seen at both sites there are suggestions that there may have been a period of coexistence of the population with conquering "Aryan" elements. Nevertheless, whatever may have been the cause or causes of the end of the civilization, what is of primary concern is that at a certain point in time it came to an end.

One must now proceed to consider "what was lost" and "what survived." Clearly the postulated central power and authority, together with whatever administrative machinery it possessed, must have been the first to go; and with them the economic organization and the highly organized trade or exchange of goods. All of these would seem to have followed upon the abandonment of the urban nucleus at Mohenjodaro, if this were the primary cause of the breakdown. But there need not, indeed there is most unlikely to have been, a comparable or simultaneous abandonment at all the other centers. Even if there is a marked and abrupt break in the material culture, this need not indicate desertion of a site, followed by reoccupation, but it may indicate no more than a withdrawal of the centrally imposed "urban" uniformity, and a return to (even reemergence of) the regional peasant styles. In some cases the break may indicate the arrival of conquerors, and the imposition of new elements upon the existing style. The uniformity of the Indus period is replaced by a whole series of local culture patterns; and at the socioeconomic level the breakdown of the centrally imposed authority must have been marked.

These things one may expect to have disappeared in their entirety. But what was retained or at least partly retained? First, many of the crafts and technical skills which were flourishing also at village level, and for which there would have been a continuing demand, would have persisted; while certain specialized urban or luxury crafts, including seal making, would have disappeared. Some parts of the urban lifestyle may have partly survived. For example, writing and the uniform system of weights and measures would almost certainly have gone as coherent systems, but must surely have left certain signs or convenient units of weight or measure in use. Many domestic aspects of the Indus lifestyle, the house plans, disposition of water supply, hearth and kitchen types, attention to bathing, etc. would survive in the settlements, and much could have been absorbed by newly arrived barbarian conquerors. There would be a very wide survival of traits pertaining to ideology and religious belief, particularly of those which were in common acceptance and which involved domestic practice.

The religious beliefs of the Indus Civilization would have been maintained in several ways: first, in the cities there must have been a class of specialist exponents or priests (who may also have constituted, or been closely associated with, the administrative group). Such people would be among those more likely to intermarry with a new ruling class, or in other ways to win for themselves positions of power or influence in a new order. Such a pattern was often witnessed much later, during the centuries of early Muslim conquest. Thus they would find themselves in a position to maintain an important body of their own beliefs, and to pass it on to their children, if not to their conquerors. At the popular level, tradition must also have been passed on within the family, in much the way as it has continued to be in India down to modern times. Through these two channels the cult of sacred places, rivers or trees, of sacred animals, and of symbols or myths would have survived; as too might a large part of the cosmology, philosophy and other parts of the learned tradition, even after the end of the cities. The strength and maintenance of the tradition would be greatly enhanced if, as one is inclined to believe, the initial period of "Aryan" settlement in the north coincided with the survival, at least for a time, of more or less full urban life under foreign rule, and with a situation such as that which one has surmised the Cemetery H Phase to have witnessed. This period of coexistence would have provided an opportunity for the priests or administrators to acquire the language of the conquerors and to have begun integrating the ideologies of the two groups. Here too the rapid acquisition first of Persian, later of English, by higher castes may suggest an appropriate model from recent times. Thus from the meeting of the two a new amalgam, an *Indian or Indo-Aryan cultural tradition was born*. This hypothesis does not demand "armies" of invaders. One would expect rather small bands, whose horses made them relatively mobile, and who may often have achieved whatever power they acquired by means other than open warfare. It is still not possible to decide at what precise stage groups ancestral to the authors of the *Rigveda* arrived, but one believes that there was a period during which there was a general restlessness, and it may well be that a whole series of waves drifted into the Indus Valley and the Punjab over succeeding centuries. This, after all, had been a pattern which continued through the historical period also. It is not even necessary that all the groups should have been Indo-Aryan speaking, or even Indo-Iranian. But it may be imagined that the closer were the ties of language and ancestry between such groups, the more marked would have been their own solidarity, and the polarity between them and the indigenous population.

Thus one can see that the survival and onward transmission of the Harappan legacy must have been at several different levels and of several different kinds. First, a widespread survival of the way of life among the common people particularly in the villages, in each of the main areas of settlement; and associated with this there would have been the survival of a series of little traditions in the several culture regions into which the peasant societies of Post-Harappan times devolved. With the removal of the urban authority the difference between the structure of village societies within and without the confines of the civilization would have been considerably reduced, and roughly similar structures would appear throughout. At the same time there would be a tendency for older, distinctive, culture traits to reemerge regionally. Hence, in all these regions one may expect the Harappan legacy to be passed down at the folk level, and to spread with the continuing expansion of the peasant society. Indeed this was probably the time when the village assumed the dominant role in Indian society which it has continued to occupy henceforward; so that while cities may have come and gone, the villages have survived with their own special Indian life-style. But at this level the surviving elements would be mainly those appropriate for the folk or village society, and many others of distinctly urban character would tend to disappear.

A second kind of transmission of the legacy must have been at the level of the great or learned tradition and would presumably have been much more restricted geographically, being mainly confined to those areas in which there was already a synthesis of Indus urban and "Aryan" ruling elements, during the later stages of the civilization. This sort of transmission probably developed in the Punjab and spread eastwards with the expansion of population and settlements into the *Doab* and Ganges Valley. Already by the time of a compilation of the Vedic *Samhitas* the process must have been providing an increasingly distinct element of the ideology, which one may now begin to call Indian, or culturally Indo-Aryan, as distinct from either Indus or Aryan. This is not the place to discuss the interesting and important question of which among the several groups of possible Aryans constituted the first arrivals in the Indus culture region; whether they were "Pre-Vedic" or "Non-Sanskritic," or "Proto-Rigvedic." This matter has recently been discussed by Dr. Parpola and I have touched on it in another paper (F. Allchin in press). But one has long been of the view that they must be regarded as culturally at least the direct ancestors of the Vedic Aryans. In these areas of course the transmission at the folk or village level also took place, and this in turn would have facilitated at a later date the secondary spread of the great tradition to other regions sharing the legacy at those levels. This hypothesis does not altogether preclude a similar survival of elements of the great tradition in other regions, notably in the South, in Sind or Saurashtra and beyond, but one believes that there it would be relatively much less powerful than in the North.

If the legacy thus transmitted was partly at the level of the great tradition, it is unlikely that it was done without considerable attenuation. The use of writing seems to have vanished, indeed one does not know how far it was used in Harappan times for purposes other than narrowly commercial, and probably much else with it. But the newly emerging Indian tradition must have received continuing enrichment from the folk level, and much may have survived at that level, to be later reabsorbed into the learned tradition.

The period following the end of the Harappan cities was one of continuing eastward expansion of Indo-Aryan Culture, now associated with the cultivation of rice and with an unprecedented growth of population. One may expect that already in the East (and for that matter the South) there were distinctive peasant societies in existence, each with its own cultural tradition, and thus the spread of the Indo-Aryan great tradition would have coincided with its encountering them. These factors led to the rapid expansion of settlements in the upper, middle and lower Ganges Valley and paved the way to the reemergence of cities there. These, like their Indus predecessors, were the products of their social and economic bases. But it is important to note that they were not the centers for the emergence of the early Indian tradition. This tradition was there before the cities, in the shape of the Vedic *Samhitas*, the accompanying schools of exegesis, and all that went with them; and also at the level of a more or less related series of little traditions, transmitted at the folk level. The new cities produced, however, a profound, even traumatic, reformulation of received tradition and ideology, and witnessed the development of Buddhism, Jainism and the other new "city" religious movements, notably Vaishnavism. But throughout this reformulation the prior existence of a tradition, from which to borrow and against which to react, can be clearly recognized. Thus while the life style of the Gangetic cities is also in many ways new, it embodies an incalculably large element which is very old, and which survived in one way or another from the earlier cities of the Indus.

The hypothesis advanced must be tested against the available data. For example, the presence of *lingas* or of an iconographic type suggestive of Śiva-Paśupati in the Indus cities, has often been seen as anticipating the later "emergence" of Rudra-Śiva in the Vedic-late Vedic literature. One may now postulate that Śiva-Mahādeva was a central concept of the Indus religion, which survived in both the great and folk traditions and developed as the process of the Indianization advanced. A problematic gulf appears to separate the narrow Indo-Iranian, polytheistic ideology postulated by philologists as that of the early Vedic hymns, from both the mature "Indian" character of the "late" hymns of the first and tenth *Mandalas* of the *Rigveda*, and their remarkably constant interpretation in Indian tradition thereafter. Although the gulf may be partly illusory, it undoubtedly exists. Just how early this shift began would depend, in terms of this hypothesis, upon the date at which the cultural synthesis of the two groups began. It is also worth considering whether the decline of the *Asuras* and the rise to eminence of the *Devas*, which seems to be happening in the body of the *Rigveda*, and to be looked back on as something already complete in such hymns as X.124, may indicate not so much a divergence of beliefs among separated Indo-Iranian groups, but rather—by this time—the process of Indianization in action.

One does not propose to anticipate objections to this hypothesis, but one is well aware that at more than one point the data are not available from which even a probable conclusion may be drawn, and where therefore it is possible to propose various alternative hypotheses. One such point concerns the moment when "Aryan" influence first began to exert itself on the Indus Civilization. For instance, one possible version would be that it was the arrival of Indo-Aryan speakers which provided the initial stimulus needed to tip the scales towards city life, and thus that the whole Indus Civilization from the start may have had a dominant Aryan strain. This hypothesis, attractive as in some ways it is, can only be sustained in the face of formidable objections, but it must not be too lightly dismissed. At the other extreme it is possible to argue that there was a final and irrevocable gap between the end of the cities and the arrival of the ancestors of the authors of the *Rigveda*. This raises almost insuperable problems of interpretation, not least in terms of the transmission of the Harappan legacy. Thus one is led to prefer a hypothesis which lies somewhere between the two extremes, that is, bringing the first Aryans into contact with the still flourishing Mature Indus society.

To sum up the main points of this essay—The Indus Civilization arose on Indian soil as an organic process: it was not primarily superimposed from outside, even if external stimuli may have contributed. Because of this there was already the necessary basis of continuity between the peasant and urban communities to permit the sort of persistence of the lifestyle which Robert Redfield remarked. An outward spread of peasant cultures from the Indus system had already begun in Pre-Harappan times, and the lifestyle spread with the continuation of that process both during the Mature Indus Civilization and after. Within the Early Indus and Mature Indus Civilization the tendency towards cultural convergence implies the emergence of a central ideology and learned tradition, and this one may call the Indus great tradition. In the north of the region there was an appearance of Indo-Aryan speaking people even during the life of the civilization, and this permits one to postulate a degree of synthesis between the exponents of the Indus great tradition and those of the arriving conquerors. This process is of enormous significance in terms of the onward transmission of the legacy, and of the translation of the Indus tradition into a unified Indian or culturally Indo-Aryan tradition. The end of the Indus Civilization appears to have been brought about by an upsetting of the delicate balance which maintained its social and economic life, and was probably linked with the abandonment at Mohenjodaro. The Indus legacy survived and was passed on most widely at the folk or village level, in almost all regions, while the learned tradition mainly survived in the Punjab, whence it spread eastwards with the spread of settlements in Post-Harappan times. The surviving tradition, an amalgam of Indus and Aryan elements was already active before the reemergence of cities in the Ganges Valley and in North India more generally during the first millennium B.C., and served as the ideological basis upon which the cities produced their own distinctive ideology. Therefore, to paraphrase an old saying: "if you seek a legacy, look about you."

Revelation

ELIOT DEUTSCH and J. A. B. VAN BUITENEN

Eliot Deutsch is a professor of Indian philosophy at the University of Hawaii. J. A. B. van Buitenen was professor of Sanskrit at the University of Chicago. Both have had an interest in the philosophical traditions of the classical period of Indian religion. The excerpts in this selection are from India's ancient scripture called the Veda. The Veda is called Shruti, or revelation, because its hymns and discourses were believed to be revealed to the ancient teachers, or Rishis, who transmitted them to later generations. These excerpts trace the early development of the idea that the goal of human life is the heroic quest for perfection, freedom, and perfect happiness.

Since a source book should avoid presenting sources in a controversial manner, the reader is urged to consult, e.g., Franklin Edgerton's translation in *Beginnings of Indian Philosophy,* Hume's in *Thirteen Principal Upaniṣads,* Radhakrishnan's in *The Principal Upaniṣads,* to quote the more accessible ones, for further reference.

1. There was Śvetaketu, the grandson of Aruṇa. His father said to him, "Śvetaketu, you must make your studies. Surely no one of our family, my son, lives like a mere Brahmin by birth alone, without having studied."

   At the age of twelve he went to a teacher and after having studied all the Vedas, he returned at the age of twenty-four, haughty, proud of his learning and conceited.

   His father said to him: "Śvetaketu, now that you are so haughty, proud of your learning and conceited, did you chance to ask for that Instruction by which the unrevealed becomes revealed, the unthought thought, the unknown known?"

   "How does this Instruction go, sir?"

   "Like this for example: by a single lump of clay everything is known that is made of clay. 'Creating is seizing with Speech, the Name is Satyam,' namely clay.

   "Like this for instance: by one piece of copper ore everything is known that is made of copper. 'Creating is seizing with Speech, the Name is Satyam,' namely copper.

   "Like this for instance: by one nail-cutter everything is known that is iron. 'Creating is seizing with Speech, the Name is Satyam,' namely iron."

   "Certainly my honorable teachers did not know this. For if they had known, how could they have failed to tell me? Sir, you yourself must tell me!"

   "So I will, my son," he said.
2. "The *Existent* was here in the beginning, my son, alone and without a second. On this there are some who say, "The *Nonexistent* was here in

the beginning, alone and without a second. From that Nonexistent sprang the Existent.'

"But how could it really be so, my son?" he said. "How could what exists spring from what does not exist? On the contrary, my son, the *Existent* was here in the beginning, alone and without a second.

"It willed, 'I may be much, let me multiply.' It brought forth Fire. The Fire willed, 'I may be much, let me multiply.' It brought forth Water. Hence wherever a person is hot or sweats, water springs in that spot from fire.

"The Water willed, 'I may be much, let me multiply.' It brought forth Food. Hence wherever it rains, food becomes plentiful: from water indeed spring food and eatables in that spot."

3. "Of these beings indeed there are three ways of being born: it is born from an egg, it is born from a live being, it is born from a plant.

"This same deity willed, 'Why, I will create separate names-and-forms by entering entirely into these three deities with the living soul.

"'I will make each one of them triple.' This deity created separate names-and-forms by entering entirely into these three deities with the living soul.

"Each of them he made triple. Now learn from me how these three deities each became triple."

4. "The red color of fire is the Color of Fire, the white that of Water, the black that of Food. Thus fireness has departed from fire. 'Creating is seizing with Speech, the Name is Satyam,' namely the Three Colors.

"The red color of the sun is the Color of Fire, the white that of Water, the black that of Food. Thus sunness has departed from the sun. 'Creating is seizing with Speech, the Name is Satyam,' namely the Three Colors.

"The red color of the moon is the Color of Fire, the white that of Water, the black that of Food. Thus moonness has departed from the moon. 'Creating is seizing with Speech, the Name is Satyam,' namely the Three Colors.

"The red color of lightning is the Color of Fire, the white that of Water, the black that of Food. Thus lightningness has departed from lightning. 'Creating is seizing with Speech, the Name is Satyam,' namely the Three Colors.

"As they knew this, the ancients of the great halls and of great learning said, 'Now no one can quote us anything that is unrevealed, unthought, unknown,' for they knew it by these Three Colors.

"If something was more or less red, they knew it for the Color of Fire; if it was more or less white, they knew it for the Color of Water; if it was more or less black, they knew it for the Color of Food.

"If something was not quite known, they knew it for a combination of these three deities. Now learn from me, my son, how these three deities each become triple on reaching the person."

5. "The food that is eaten is divided into three: the most solid element becomes excrement, the middle one flesh, the finest one mind.

"The water that is drunk is divided into three: the most solid element becomes urine, the middle one blood, the finest one breath.

"The fire that is consumed is divided into three: the most solid element becomes bone, the middle one marrow, the finest one speech.

"For the mind, my son, consists in Food, the breath consists in Water, the speech consists in Fire."

"Sir, instruct me further."

"So I will, my son," he said.

6. "The fineness of milk which is being churned rises upward, my son, and that becomes butter.

"In the same way, my son, the fineness of the food that is eaten rises upward, and that becomes the mind.

"The fineness of the water that is drunk rises upward, my son, and that becomes the breath.

"The fineness of the fire that is consumed rises upward, my son, and that becomes speech.

"For the mind, my son, consists in Food, the breath consists in Water, the speech consists in Fire."

"Sir, instruct me further."

"So I will, my son," he said.

7. "Man consists of sixteen parts, my son. Do not eat for fifteen days. Drink water as you please. The breath will not be destroyed if one drinks, as it consists in Water."

He did not eat for fifteen days. Then he came back to him. "What should I say, sir?"

"Lines from the *Ṛgveda*, the *Yajurveda* and the *Sāmaveda*, my son."

"They do not come back to me, sir."

He said to him, "Just as of a big piled-up fire only one ember may be left, the size of a firefly, and the fire does not burn much thereafter with this ember, thus of your sixteen parts one part is left and with that you do not remember the Vedas. Eat. Afterwards you will learn from me."

He ate. Then he returned to him, and whatever Veda he asked, he responded completely. He said to him, "Just as one ember, the size of a firefly, that remains of a big piled-up fire will blaze up when it is stacked with straw and the fire will burn high thereafter with this ember, so, my son, one of your sixteen parts remained. It was stacked with food and it blazed forth, and with it you now remember the Vedas. For the mind consists in Food, my son, the breath in Water, speech in Fire." This he learnt from him, from him.

8. Uddālaka son of Aruṇa said to his son Śvetaketu, "Learn from me the doctrine of the sleep. When a man literally 'sleeps' [*svapiti*], then he has merged with Existent. He has 'entered the self' [*svamapītaḥ*], that is why they say that he 'sleeps'. For he has entered the self.

"Just as a bird which is tied to a string may fly hither and thither without finding a resting place elsewhere and perches on the stick to which it is tied, likewise the mind may fly hither and thither without finding a resting place elsewhere and perches on the breath. For the breath is the perch of the mind, my son.

"Learn from me hunger and thirst. When a man literally 'hungers' [*aśiśiṣati*], water conducts the food he eats. And just as we speak of a cow leader, a horse leader, a man leader, so we speak of water as 'food leader' [*aśanāyā*, but first: hunger]. You must know a shoot has sprung up there, my son. This shoot will not lack a root.

"Where would this root be but in food? Thus indeed, my son, search by way of the food, which is a shoot, for the fire, its root. Search, my son, by way of the fire as a shoot, for the Existent, its root. All these creatures, my son, are rooted in the Existent, rest on the Existent, are based upon the Existent.

"And when a man literally 'thirsts' [*pipāsati*], fire conducts the liquid which is drunk. Just as we speak of a cow leader, a horse leader, a man leader, we speak of fire as 'water leader' [*udanyā*, but first: thirst]. You must know that a shoot has sprung up there, my son. This shoot will not lack a root.

"Where would this root be but in water? Search, my son, by way of the water as the shoot, for the fire, its root. Search, my son, by way of the fire as the shoot, for the Existent, its root. All these creatures, my son, are rooted in the Existent, rest on the Existent, are based upon the Existent. It has been said before how these three deities each become triple on reaching man. Of this man when he dies, my son, the speech merges in the breath, the breath in the Fire, the Fire in the supreme deity. That indeed is the very fineness by which all this is ensouled, it is the true one, it is the soul. *You are that,* Śvetaketu."

"Instruct me further, sir."

"So I will, my son," he said.

9. "Just as the bees prepare honey by collecting the juices of all manner of trees and bring the juice to one unity, and just as the juices no longer distinctly know that the one hails from this tree, the other from that one, likewise, my son, when all these creatures have merged with the Existent they do not know, realizing only that they have merged with the Existent.

"Whatever they are here on earth, tiger, lion, wolf, boar, worm, fly, gnat, or mosquito, they become that.

"It is this very fineness which ensouls all this world, it is the true one, it is the soul. *You are that,* Śvetaketu."

"Instruct me further, sir."

"So I will, my son," he said.

10. "The rivers of the east, my son, flow eastward, the rivers of the west flow westward. From ocean they merge into ocean, it becomes the same ocean. Just as they then no longer know that they are this river or that one, just so all these creatures, my son, know no more, realizing only when having come to the Existent that they have come to the Existent. Whatever they are here on earth, tiger, lion, wolf, boar, worm, fly, gnat or mosquito, they become that.

"It is this very fineness which ensouls all this world, it is the true one, it is the soul. *You are that,* Śvetaketu."

"Instruct me further, sir."

"So I will, my son," he said.

11. "If a man would strike this big tree at the root, my son, it would bleed but stay alive. If he struck it at the middle, it would bleed but stay alive. If he struck, it at the top, it would bleed but stay alive. Being entirely permeated by the living soul, it stands there happily drinking its food.

"If this life leaves one branch, it withers. If it leaves another branch, it withers. If it leaves a third branch, it withers. If it leaves the whole tree, the whole tree withers. Know that it is in this same way, my son," he said, "that this very body dies when deserted by this life, but this life itself does not die.

"This is the very fineness which ensouls all this world, it is the true one, it is the soul. *You are that,* Śvetaketu."

"Instruct me further, sir."

"So I will, my son," he said.

12. "Bring me a banyan fruit."

"Here it is, sir."

"Split it."

"It is split, sir."

"What do you see inside it?"

"A number of rather fine seeds, sir."

"Well, split one of them."

"It is split, sir."

"What do you see inside it?"

"Nothing, sir."

He said to him, "This very fineness that you no longer can make out, it is by virtue of this fineness that this banyan tree stands so big.

"Believe me, my son. It is this very fineness which ensouls all this world, it is the true one, it is the soul. *You are that,* Śvetaketu."

"Instruct me further, sir."

"So I will, my son," he said.

13. "Throw this salt in the water, and sit with me on the morrow." So he did. He said to him, "Well, bring me the salt that you threw in the water last night." He looked for it, but could not find it as it was dissolved.

"Well, taste the water on this side.—How does it taste?"

"Salty."

"Taste it in the middle.—How does it taste?"

"Salty."

"Taste it at the other end.—How does it taste?"

"Salty."

"Take a mouthful and sit with me." So he did.

"It is always the same."

He said to him, "You cannot make out what exists in it, yet it is there.

"It is this very fineness which ensouls all this world, it is the true one, it is the soul. *You are that,* Śvetaketu."

"Instruct me further, sir."

"So I will, my son," he said.

14. "Suppose they brought a man from the Gandhāra country, blindfolded, and let him loose in an uninhabited place beyond. The man, brought out and let loose with his blindfold on, would be turned around, to the east, north, west, and south.

"Then someone would take off his blindfold and tell him, 'Gandhāra is that way, go that way.' Being a wise man and clever, he would ask his way from village to village and thus reach Gandhāra. Thus in this world a man who has a teacher knows from him, 'So long will it take until I am free, then I shall reach it.'

"It is this very fineness which ensouls all this world, it is the true one, it is the soul. *You are that,* Śvetaketu."

"Instruct me further, sir."

"So I will, my son," he said.

15. "When a man is dying, his relatives crowd around him: 'Do you recognize me? Do you recognize me?' As long as his speech has not merged in his mind, his mind in his breath, his breath in Fire, and Fire in the supreme deity, he does recognize.

"But when his speech has merged in the mind, the mind in the breath, the breath in Fire, and Fire in the supreme deity, he no longer recognizes.

"It is this very fineness which ensouls all this world, it is the true one, it is the soul. *You are that,* Śvetaketu."

"Instruct me further, sir."

"So I will, my son," he said.

16. "They bring in a man with his hands tied, my son: 'He has stolen, he has committed a robbery. Heat the ax for him!' If he is the criminal he will make himself untrue. His protests being untrue, and covering himself with untruth, he seizes the heated ax. He is burnt, and then killed.

"If he is not the criminal, he makes himself true by this very fact. His protests being true, and covering himself with truth, he seizes the heated ax. He is not burnt, and then set free.

"Just as he is not burnt—that ensouls all this world, it is the true one, it is the soul. *You are that,* Śvetaketu."

This he knew from him, from him.

The Interpretation of Hindu Mythology

WENDY DONIGER O'FLAHERTY

Wendy Doniger O'Flaherty is the distinguished Mircea Eliade Professor of the History of Religions at the University of Chicago. She has written broadly on the Hindu system of myths. As she says, they deal "with every subject under the sun." How to make sense out of the great number and variety of these myths is the purpose of the following selection. The book from which this selection is taken is a study of the great Hindu god Shiva. Studying this selection will help the reader understand how the mythic imagination, which is akin to the collective unconscious, weaves its web of relations and meanings.

THE CHALLENGE OF HINDU MYTHOLOGY

"That which cannot be found here exists nowhere"—this was the boast of the author of the great Hindu epic, the *Mahābhārata*, and it is an excellent description of the Purāṇas, encyclopedic compendia of hundreds of thousands of Sanskrit verses dealing with every subject under the Indian sun. Here are found dynastic histories of obscure kings, detailed recipes for ritual offerings, hair-splitting philosophical arguments, tedious discussions of caste law, and, imbedded in the midst of all of this ("like a lotus in the mud," as the Hindus say), countless sublimely beautiful myths.

It is a time-consuming task to sift through the Purāṇas in search of the myths, but none of it is wasted labour, for the mud is as valuable as the lotus which it nourishes: the myths come alive only in the context of history, ritual, philosophy, and social law. Hidden somewhere in this maze is the key to the Hindu world view, vivid, startling, fascinating, and complex. The mythology of Śiva forms only a small part of the material of the Purāṇas, but it is an ideal model which reveals a pattern which pertains to the material as a whole. Śiva is not only an extremely important Hindu god; he is in many ways the most uniquely Indian god of them all, and the principles which emerge from an intensive study of his mythology lie at the very heart of Hinduism.

Can the mythology of Śiva be used to reveal a still more general, perhaps universal, truth? Questions of this sort have long tempted the student of mythology. It is an old maxim that we often find our home truths in foreign lands. The great Indologist, Heinrich Zimmer, once tried to explain his love for Indian myths by citing the Hassidic tale of the old Jew who travelled from Cracow to Prague only to learn from a young Christian soldier (who did not believe or understand the words that he spoke in jest and mockery) that a treasure of gold was buried under the stove in the old man's home, back in Cracow. As Zimmer remarks,

> Now the real treasure, to end our misery and trials, is never far away. . . . But there is the odd and persistent fact that it is only after a faithful journey to a distant region, a foreign country, a strange land, that the meaning of the inner voice that is to guide our quest can be revealed to us.

Yet the more reasonable goal—and the more rewarding—is simply to understand the myths *in situ,* to use methods which reveal the meanings that the Hindus saw in them, to enjoy them as the exotic and delightful creations that they are.

To extract these meanings without *reducing* the myths in any way is no simple task. The dilemma is at first complicated, but ultimately resolved, by the fact that there are many "meanings" in a Hindu myth: "Hindu mythology is much like a plum pudding. If you do not like the plums in the slice you have, or have been deprived of a favour, you may always cut another one. The first plum is the story itself, usually a rather good story, occasionally of the shaggy-dog variety but frequently with an immediately recognizable point on at least one level, which might be termed the narrative level. Closely related is the divine level, which concerns mythology as it used to be understood by scholars of the classics: the metaphorical struggles of divine powers and personalities. Above this is the cosmic level of the myth, the expression of universal laws and processes, of metaphysical principles and symbolic truths. And below it, shading off into folklore, is the human level, the search for meaning in human life, the problems of human society.

> No one meaning can be labelled the deepest or the truest. . . . The best words are ambiguous, and the more richly ambiguous the more suitable for the poet's or the myth-maker's job. Hence there is no end to the number of meanings which can be read into a good myth.

The various levels are simultaneous rather than alternative:

> Each level always refers to some other level, whichever way the myth is read. . . . We can only choose between various degrees of enlargement: each one reveals a level of organization which has no more than a relative truth and, while it lasts, excludes the perception of other levels.

In my analyses, I have certainly not exhausted all the meanings of each myth; I have not even mentioned all the meanings that I personally see in them. I have only discussed those themes in each myth which play a part in the basic schema of the corpus. This may seem a Procrustean method—to select one theme and then to maintain that it is uniquely important—but it is justified by the materials: I have selected that one theme because it *is* central. As one reads through the enormous mass of Purāṇic mythology, certain recurrent elements clearly emerge; myths which at first appeared obscure suddenly become obvious; one senses what the myths are about. No single myth contains the key, which is given only by the totality of variations. The pattern which they form involves a finite number of elements but will be seen to apply equally well to other (theoretically infinite) groups of myths, for the pattern is basic to Hindu thought. The completed model reveals that the apparent contradictions in individual variants are merely incomplete views of the whole, like the varying opinions of the blind men grasping the different parts of an elephant ("It is a rope," "It is a wall," "It is a snake"): it is, in fact, unmistakably an elephant. Although the themes in terms of which I have chosen to analyse the Śaiva cycle are not the only themes present in the myths, they are certainly extremely significant. This is evident from the impressive frequency with which their patterns recur, the ease and simplicity with which they account for otherwise puzzling idiosyncracies in the texts, and the convincing number of explicit discussions (within the myths and in other Indian materials) of the problems which they represent.

A further note of caution must be voiced here on the subject of arbitrary selection. Not only are the myths and themes themselves selected, but the very episodes and words of the myths are necessarily slanted as well. The Sanskrit Purāṇas are extremely garrulous and digressive, and in order to include as wide a selection of myths as possible I have summarized rather than translated, omitting large bodies of material superfluous to the present study, such as hymns of praise, ritual instructions, detailed descriptions, and philosophical discourses. This extraneous material is not only unwieldy but would have tended to obscure the patterns which emerge from a more selective treatment. I have also omitted those portions of the text which seem hopelessly corrupt, and in some instances where the meaning seems quite clear in spite of the garbled text I have given the best sense I could make of it. I have not (knowingly) added anything that is not in the text, but I may have omitted in one version certain details that occur in another, thus inadvertently disguising an actual correspondence in the text. In other circumstances I may translate in similar words two sentences which differ in the wording of the originals, thus suggesting a correspondence which does not in fact exist. I have omitted nonrecurring proper names and I have standardized epithets throughout my translations. (Secondary sources have been quoted literally unless enclosed in brackets to indicate my own summary.) A general indication of the degree to which any particular citation has been compressed may be obtained by comparing the length of the English version with the length of the Sanskrit text as indicated in the bibliographic note. In taking these liberties I hope I have not distorted the meanings of the myth, and in so doing I take heart from the words of Claude Lévi-Strauss, who has said that while poetry may be what is lost in translation, "the mythical value of myths remains preserved through the worst translations."

These measures, which are particularly necessary in a method of analysis which relies upon numerous different versions of a myth, have been defended by Lévi-Strauss:

> To avoid making the demonstration too unwieldy, I had to decide which myths to use, to opt for certain versions, and in some measure to simplify the variants. Some people will accuse me of having adapted the subject matter of my inquiry to suit my own purposes.

He answers these objections by arguing that, although the selection is to some extent arbitrary, further incidents would only be variants of those selected.

> A certain stage of the undertaking having been reached, it becomes clear that its ideal object has acquired sufficient consistency and shape for some of its latent properties, and especially its existence as an object, to be definitely placed beyond all doubt.

That the selected themes are mutually reinforcing and fully integrated is taken as evidence of their significance. These criteria are necessary but not sufficient. The recurrence of the isolated themes in other myths, as well as in other texts, iconography, and ritual, is further proof of their importance.

THE CENTRAL PARADOX OF ŚAIVA MYTHOLOGY

The wise applicability of the recurrent supplementary themes from the Purāṇic corpus is evident from the role that they play in one of the enduring problems of Hindu mythology, the paradox of Śiva the erotic ascetic. This problem has often been noted and by now is sometimes accepted as a matter of fact: the great ascetic is the god of the phallus (the *liṅga*). The *meaning* of this paradox, however, has never been properly explored.

The character of Śiva has always been an enigma to Western scholars. Only a small portion of the corpus of ancient Śaiva mythology has been translated from the Sanskrit; with this inadequate representation, it is not surprising that the mythology of Śiva was considered contradictory and paradoxical by scholars who saw only the two ends of the spectrum. Śiva the Creator and Destroyer, Life and Death, the *coincidentia oppositorum*—this much was accepted as consistent with Indian metaphysical thought, and the apparent sexual ambiguity of the god was regarded as simply one more aspect of a basically ambiguous character or a result of the chance historical assimilation of two opposing strains, a process well known in Indian religion. In the absence of critical editions of the Śaiva Purāṇas (medieval Sanskrit texts containing numerous myths), the problem was never properly considered, and the very fact of its paradoxical nature was taken as an accepted quality of Śaiva thought, a property upon which further speculation could be based.

At the beginning of the nineteenth century, the Abbé J. A. Dubois described with shocked disbelief the seemingly contradictory concept of sexuality exemplified by the forest-dwelling ascetics:

> By one of those contradictions which abound in Hindu books, side by side with the account of the punishments inflicted on a hermit for his inability to conquer his sensual passions, we find, related with expressions of enthusiasm and admiration, the feats of debauchery ascribed to some of their *munis* [ascetic sages]—feats that lasted without interruption for thousands of years; and (burlesque idea!) it is to their pious asceticism that they are said to owe this unquenchable virility.

This very "burlesque idea" is the core of the nature of Śiva, who is the god of ascetics, but out of its mythological context it could have no significance. Nor has contemporary scholarship found a satisfactory solution to the enduring enigma of Śiva: "Permanently ithyphallic, yet perpetually chaste: how is one to explain such a phenomenon?"

The problem was intensified by uncertainties regarding Śiva's place in the historical development of Hinduism. Failure to connect him with the Vedic gods Indra, Prajāpati, and Agni led to the assumption that the sexual elements of his cult were "non-Āryan" or at least non-Vedic, and obvious correspondences between Śaiva myths and Tantric cult led some scholars to seek the origins of Śiva's sexual ambiguity in this comparatively late development. Yet what is striking about the problem is that it extends from the period of the Vedas (*c.* 1200 B.C.) and even earlier, from the prehistoric civilization of the Indus Valley (*c.* 2000 B.C.), through the development of Tantrism, to the religion of present-day India.

The ancient Hindus themselves attempted to explain the Śaiva phenomenon. A Sanskrit poem dating from perhaps A.D. 900 muses upon Śiva:

> If he is naked what need then has he of the bow?
> If armed with bow then why the holy ashes?
> If smeared with ashes what needs he with a woman?
> Or if with her, then how can he hate Love?

In the Purāṇas, the nature of Śiva is often a source of worry for gods and mortals who become involved with him. When Himālaya learns that his daughter is to marry Śiva, he says, "It is said that Śiva lives without any attachments and that he performs asceticism all alone. How then can he interrupt his trance to marry?" Explicit reasons for Śiva's behaviour are given in the course of the myths, but the metaphysical arguments are both secondary and subsequent to the story. If philosophy could express the problem, there would be no need for the myth to mediate between the two opposed facets; the myth takes over where philosophy proves inadequate. Śiva himself is said to be troubled by the ambivalence in his character, for when Kāma, the god of desire, wounds him, shattering his trance and stirring his desire, Śiva muses, 'I dwell ever in asceticism. How is it then that I am enchanted by Pārvatī?' Only involvement in the eternal cycle of the myth can reveal—even to the god himself—the answer to this question.

The Resolution within the Texts

The solution is not an arbitrary construction of armchair scholarship, meaningless to the creators and preservers of the myths. Throughout Hindu mythology, even from the time of the Vedas, the so-called opposing strands of Śiva's nature have been resolved and accepted as aspects of one nature. They *may* be separated in certain contexts, and are frequently confused and misunderstood even by the tellers of the tales, but in every age there have been notable examples of satisfactory resolution. The Śiva of Brahmin philosophy is predominantly ascetic; the Śiva of Tantric cult is predominantly sexual. But even in each of these, elements of the contrasting nature are present, and in the myths—which form a bridge between rational philosophy and irrational cult—Śiva appears far more often in his dual aspect than in either one or the other.

As early as the Atharva Veda hymn to the *brahmacārin* (a young student who has undertaken a vow of chastity), there is a detailed description of a sage who has been identified with Śiva himself, the great *brahmacārin* but also the great *liṅga*-bearer, who spills his seed upon the earth. The first explicit reference to Śiva's ambiguous sexuality appears in the *Mahābhārata* (*c.* 300 B.C.), in a hymn in praise of Śiva:

> Whose semen was offered as an oblation into the mouth of Agni, and whose semen was made into a golden mountain? Who else can be said to be a naked *brahmacārin* with his vital seed drawn up? Who else shares half his body with his wife and has been able to subjugate Kāma?

The seed spilt creatively and contained in chastity, the ultimate act of desire and the conquest of desire—the essence of Śaiva mythology is in this passage.

Statements of resolution persist throughout the Purāṇas. Śiva says that if he marries, his wife must be a *yoginī* (female ascetic) when he does yoga, and a lustful mistress (*kāminī*) when he is full of desire. The sage Nārada describes Śiva:

> On Kailāsa mountain, Śiva lives as a naked yogi. His wife Pārvatī is the most beautiful woman in the universe, capable of bewitching even the best of yogis. Though Śiva is the enemy of Kāma and is without passion, he is her slave when he makes love to her.

And constantly, in less explicit ways, the two aspects of the god are equated or interchanged. It is said that Śiva is a great yogi, but in his meditation he fixes his mind upon his wife. Devī (the Goddess, who becomes incarnate as Satī and Pārvatī) says:

> Ever since I killed myself, Śiva has thought of me constantly, unable to bear his separation from me. He wanders naked, and has become a yogi, abandoning his palace, wearing unconventional clothing. Miserable because of me, he has abandoned the highest pleasure that is born of desire. He is tortured by longing and can find no peace as he wanders everywhere, weeping and behaving like a lover in distress.

In this passage, the two roles are enacted simultaneously: Śiva wears the garb of a yogi, but his behaviour is that of a lover in separation. In fact, he is a yogi *because* he is a lover. This may be read as a contradiction but psychologically it is entirely logical. So completely are the roles of ascetic and lover combined that the myth-makers themselves often confuse them. The Seven Sages say to Pārvatī, "How can you enjoy the pleasures of the body with an ascetic [*yati*] like him, so terrifying and disgusting?" But in another version of the same text they say, "How can you enjoy the pleasures of the body with a husband [*pati*] like him, so terrifying and disgusting? The sense remains the same in both readings, for the two roles are being compared and in fact interchanged.

The Pine Forest sages say, "If we have served Śiva from our birth with *tapas* [asceticism], then let the *liṅga* of this libertine [Śiva in disguise] fall to the earth." Thus they swear by Śiva the ascetic in order to destroy Śiva the erotic, not realizing that the two are one. This dualism is implicit in other versions of the Pine Forest myth as well, for the sages use the *tapas* of Śiva (their fiery curse) against the lust of Śiva (his *liṅga*), and they must be taught the unity of the two powers. The contrasting aspects of Śiva are artfully combined in a late poem of praise which invokes him first as the three-eyed god (ascetic, the third eye having burnt Kāma), then as the clever dancer (erotic), then as the wandering beggar (ambiguous, as in the Pine Forest), then as the rider on the bull (erotic), the poison-eater (ascetic), and finally as the god whose *liṅga* is worshipped by yoga (ambiguous).

For the yogi himself, using Śiva as his model, the god might appear in either aspect according to the worshipper's need: "The yogi who thinks of Śiva as devoid of passion himself enjoys freedom from passion. The yogi who meditates upon Śiva as full of passion himself will certainly enjoy passion." Nor was this choice limited to the initiated; a popular hymn to Śiva in Orissa says, "He is the much beloved husband of Gaurī [Pārvatī] and the only object of adoration by the ascetic." This dualism is taken for granted even in a modern English novel based upon the Śiva myth:

> Śiva . . . had two simultaneous identities. There was Śiva austere on the Himalaya, rambling down granite causeways . . . and yet at the same time a young buck around town, boyish, eager, with a shock of black hair over his eyes, and a girl on either arm. For it is possible "to comprehend flowering and fading simultaneously . . ."

Thus it would seem that this ambiguity has been comprehended and accepted by men of various ages and beliefs, notwithstanding its apparent logical contradiction and the difficulties which arise when its implications are literally applied to an actual or mythological social situation.

The Iconic Resolution of the Paradox: The Ithyphallic Yogi

In many texts, Śiva is said to be ithyphallic (with an erect phallus), an image which would certainly seem to be unambiguous sexually but for its particularly Hindu connotations, which tie it to the world of asceticism as strongly as it is naturally related to the realm of eroticism. Of all the characteristics of Śiva, this is perhaps the most basically iconic; images and paintings of Śiva frequently incorporate aspects described in the myths, but the image of the erect phallus preceded the myths related about it. The study of the ambiguity of the image, therefore, begins with the study of its earliest iconic form.

Sir John Marshall noted in the prehistoric Indus Valley civilization a seal on which was depicted a male god whom he identified as a prototype of Śiva; the god, seated in a position of yoga, has an exposed, erect phallus. More recently, Sir Mortimer Wheeler has suggested that another Indus Valley figure, the dancing torso from Harappā, also apparently ithyphallic, may be a prototype of Śiva Nāṭarājā (Lord of the Dance). So specific an identification of cult cannot be made with certainty, but there is evidence in the Indus Valley of yogic practices as well as of the phallic worship mentioned by the Ṛg Veda itself as characteristic of the enemies of the Āryans. Thus even at this early time there is a connection between the postures of yoga and of sexuality. The interrelation of asceticism and desire in the medieval Purāṇas, then, cannot be explained by any historical synthesis, but must be accepted as a unified concept which has been central to Indian thought from prehistoric times.

Ithyphallic images are as early as any Śiva images, and the Gūḍimallam *liṅga*, an erect phallus on which an image of Śiva is carved, has been called the earliest known Hindu sculpture. Many images of ithyphallic yogis represent Lakulīśa ("The Lord of the Club"), an incarnation of Śiva. Śiva mounted upon his bull should be ithyphallic, according to one textbook of aesthetics, and "the end of the phallus must reach the limit of the navel." Similarly, Śiva is ithyphallic as Nāṭarājā, as the androgyne, and frequently when sitting in a yogic posture beside Pārvatī, whose hand sometimes touches his erect phallus. Even Gaṇeśa, the son of Śiva, has the *ūrdhvaliṅga* (erect phallus) when he dances the dance of death (*tāṇḍava*) in imitation of his father. It has been suggested that several of the bearded figures portrayed on the Śaiva temples at Khajuraho are yogis participating in ritual orgies, and that some of their more convoluted poses represent "sexo-yogic" attitudes or esoteric yogic *āsanas* (postures). If this interpretation is accepted, the domain of the ithyphallic yogi is widely extended, even within purely Śaiva limits.

The ambiguity of ithyphallicism is possible because, although the erect phallus is of course a sign of priapism, in Indian culture it is a symbol of chastity as well. Śiva is described as ithyphallic, particularly in the Pine Forest, and this condition is often equated with a state of chastity. "He is called *ūrdhvaliṅga* because the lowered *liṅga* sheds its seed, but not the raised *liṅga*." The basic Sanskrit expression for the practice of chastity is the drawing up of the seed (*ūrdhvaretas*), but, by synecdoche, the seed is often confused with the *liṅga* itself, which is "raised" in chastity. The raised seed is a natural image of chastity; only Pārvatī can transform Śiva from one whose seed is drawn up into one whose seed has fallen. His seed "falls" when he begets Skanda, yet he is described as *ūrdhvaretas* even at the moment when he gives Agni his seed. The commentator glosses this term as *ūrdhvagāmivīryah*, "with his seed moving upwards," which may describe either the drawing up of the seed in chastity or the motion of the seed shed in the absence of chastity. Thus Śiva is both the god whose seed is raised up and the god whose *liṅga* is raised up. Even without this confusion, the image of the erect phallus is in itself accepted as representative of chastity. When the seed is drawn up, Śiva is a pillar (*sthāṇu*) of chastity, yet the pillar is also the form of the erect *liṅga*: "It is in this form of the Lord of Yogins that he becomes Sthāṇu or of *liṅga* form." Since, in the context of the Hindu attitude toward sexual powers, Śiva's chastity is his power of eroticism, the erect phallus can represent both phases:

> In many of his icons he [Śiva] is ithyphallic; often he appears with his consort. At the same time he is the patron deity of yogis. . . . This is not inconsistent with his sexual vitality. For the source of the yogi's power is his own divine sexuality, conserved and concentrated by asceticism.

The ithyphallic condition has been attributed by some not to priapism but to the Tantric ritual of seminal retention. To a certain extent, this technique may be considered a manifestation of yogic chastity, but Śiva's raised *liṅga* is symbolic of the power to spill the seed as well as to retain it.

> Shiva, the god of eroticism, is also the master of the method by which the virile force may be sublimated and transformed into a mental force, an intellectual power. This method is called Yoga, and Shiva is the great yogi, the founder of Yoga. We see him represented as an ithyphallic yogi. . . . Assuming the various postures of Yoga, Shiva creates the different varieties of beings. . . . Then in the posture of realisation (*siddhāsana*) he reintegrates into himself all the universe which he has created. It is in this posture that he is most often represented. His erect phallus is swollen with all the potentialities of future creations.

The yogi here gathers up his creative powers, retaining the promise of procreation in the form of the erect phallus, the embodiment of creative *tapas*. The raised *liṅga* is the plastic expression of the belief that love and death, ecstasy and asceticism, are basically related.

For the image retains its primary, more natural significance; it may symbolize actual, as well as potential or sublimated eroticism. This is clear from the statues of the ithyphallic Śiva embraced by Pārvatī. The wives of the Pine Forest sages touch Śiva's erect phallus, which is adorned with red chalk and bright white charcoal or with many bracelets, the latter a characteristic to which Dakṣa particularly objects. In a tribal myth it is said that a woman found an amputated phallus, and, "thinking it to be Mahadeo's [Śiva's] *liṅga*, took it home and worshipped it. At night she used to take it to bed with her and use it for her pleasure." In a similar manner, a female figure carved on the temple at Konārak is obviously using a stone Śiva-*liṅga* as a dildo, an act which seems to be explicitly prohibited in the lawbooks. In a variant of the episode in which Śiva tempts Pārvatī to abandon her *tapas*, he sends a flood to wash away the sand *liṅga* which she is worshipping, and she protects the *liṅga* by embracing it and smothering it with her breasts. Here *liṅga* worship replaces the usual episode of *tapas* with an image of unmistakable eroticism. In this way, the image of the ithyphallic yogi, simultaneously representative of chastity and sexuality, retains its ambiguities in myth, icon, and cult.

PROBLEMS AND METHODS

The paradox thus clearly represented in Indian art and explicitly discussed in philosophical texts also underlies the great corpus of Śaiva mythology, which combines the image and the idea. When this central pivot is selected, it is possible to balance the subsidiary themes around it in such a way that they become mutually illuminating. At this point we must reconsider the problem of mythological analysis.

Various Methods of Mythological Analysis

Just as there are many meanings in a myth, many themes that may be extracted, so there are many theories of interpretation that have been or might be applied to Hindu myths: There is the nature interpretation of the nineteenth-century German school (Rudra is the storm; Brahmā's daughter is the dawn), and, in reaction to this oversimplification, the school which "tried to reduce the meaning of myths to a moralizing comment on the situation of mankind" (Rudra is death, Kāma is love). There is the metaphysical interpretation of the Hindu theologians (Indra is the soul; the gods are the objects of the senses) and the ritual interpretation of the Hindu priests (Dakṣa is the sacrificial goat) and of certain anthropologists (Dakṣa's goat-head is the totemistic symbol of the clan). There is the Euhemerist or historical theory, often favoured by Hindus as it lends a 'scientific' air to tales once scoffed at by Europeans (the battle between Indra and the serpent Vṛtra represents the Āryan conquest of Nāga tribes in 1500 B.C.), and the related Marxist view (the battle represents the supplanting of an agrarian economy by a martial proletariat). There is the etiological method (the myth explains why the eclipse of the moon takes place); the psychological interpretations of Freud, Jung, and Rank; the comparative mythology of Sir James Frazer; Stith Thompson's attempt to index the entire spectrum of world-wide mythological and folk motifs; the structural methods of Vladimir Propp and Lévi-Strauss; and the text-historical method, which 'explains' a myth by finding its earliest known sources.

The text-historical method is the one which has been most frequently applied to Indian mythology. Scholars attempted to discount apparently incongruous elements of a myth by tracing them back to originally unrelated "*Ur*-texts"; the persistence of such elements in an inappropriate context in the later myth was thus attributed to an accidental historical conflation of several myths. If, on the other hand, these elements could not be found in any *Ur*-text, they were ignored as modern accretions, irrelevant to the analysis of the 'essential' myth. This technique is generally unrewarding when applied to the Sanskrit sources, which are extremely difficult to date with any accuracy, and Madeleine Biardeau has commented on the unsuitability of the text-historical approach to Purāṇa material of this nature: "The approach of historical philology will never be suitable for an oral tradition, which has no essential reference to its historical origin." Moreover, the historical method is misleading even when it succeeds; for the question to ask is not where the disparate elements originated, but why they were put together, and why kept together. It is when the combination seems most contradictory and arbitrary that it is most rewarding to analyse it as a combination—for only a strong emotional bond can bridge a wide logical gap. Lévi-Strauss uses the image of the *bricoleur*, the handyman who uses whatever tools and scraps are at hand, to describe the method of the myth-maker who uses pre-constrained images and actions. The interesting question is not where the tools came from but why the *bricoleur* persists in combining certain tools in certain ways once he has them. One of the first Europeans to study Indian mythology complained, "Hindu mythology is . . . very confused and contradictory, and almost any two things may be mythologically assimilated," but this is not true. Only those things are persistently assimilated which have some basic relationship in the Hindu view.

For these reasons, among others, the structuralists have rejected the text-historical emphasis on chronology, and Lévi-Strauss treats all versions of a myth, including what are usually called interpretations, as equally relevant or authentic. I have used "versions" of the Śiva myths written by modern Hindu authors (R. K. Narayan, Nirad Chaudhuri), eighteenth-century European scholars, and even one contemporary Western novelist (David Stacton). As Mary Douglas has remarked of this technique, "This challenging idea is not merely for the fun of shocking the bourgeois mythologist out of his search for original versions." The Hindu myths are part of an oral tradition extending through thousands of years. No one teller of a myth would include all the details that he might have known, and so a later version may express an idea which has been current in the tradition for a long time. Sometimes quite late versions make explicit what is present in an obscure form in a very early version: "These later sequences are organized in schemata which are at the same time homologous to those which have been described and more explicit than them." Even when later versions seem to contradict earlier ones, the pattern of the earlier myth will often persist. In many cases the explicit meaning of the myth will contradict the pattern which comparison with other versions will reveal in it; in such instances, the pattern is usually more basic to the myth than the superficial meaning. Similarly, the explicit intention of a character may contradict his actual behaviour, and comparison with other versions usually reveals the action to be older than the intention. In this way the new wine of later versions will often reveal the shape of the old bottles, and *all* variants may be considered on an equal basis.

Thus the text-historical method is of limited value in determining the pattern of the myth as a whole. It is, however, extremely useful in illuminating the individual elements of that pattern, the themes and motifs. For example, no Vedic "*Ur*-text" exists to clarify the episode in which Agni interrupts Śiva's love-play in order to take away Śiva's seed. However, once having *isolated* the significant motifs by means of a number of modern variants, having established that Agni usually assumes the form of a bird to steal the seed, we may then come to *understand* the motif better in the light of the history of the Vedic symbolism of the bird and the seed.

I have followed a modified text-historical method in tracing the historical backgrounds of the Śiva myths in the Vedas and Epics and the subsequent development of the motifs within the Purāṇa myths. The reader may trace the history of each motif through the citations in the index of motifs, bearing in mind the fact that, although it is impossible to 'date' any particular Purāṇic text, it is reasonable to postulate several broad areas of Indian mythology: Ṛg Veda (*c.* 1200 B.C.), Brāhmaṇas and Atharva Veda (900 B.C.), Upaniṣads (700 B.C.), *Mahābhārata* (300 B.C.–A.D. 300), *Rāmāyaṇa* (200 B.C.–A.D. 200), early Purāṇas (*Brahmāṇḍa*, *Mārkaṇḍeya*, *Matsya*, *Vāyu*, and *Viṣṇu*, 300 B.C.–A.D. 500), middle Purāṇas (*Kūrma*, *Liṅga*, *Vāmana*, *Varāha*, *Agni*, *Bhāgavata*, *Brahmavaivarta*, *Saura*, *Skanda*, and *Devī*, A.D.500–1000), late Purāṇas (all others, A.D. 1000–1500), and modern Hindu texts. In analysing the meaning of the myths as a whole, however, I have ignored chronology.

Almost every one of the traditional methods is applicable to some portion of some myth, though none can explain them all:

> There is some truth in almost all theories—as that is primitive philosophy or science, that its inner meaning is sexual, agricultural or astrological, that it is a projection of unconscious psychic events, and that it is a consciously constructed system of allegories and parables. No one of these theories accounts for all myths, and yet I do not doubt that each accounts for some.

The mythologist, like the myth-maker himself, cannot be too proud to accept diverse scraps from dubious sources. It is interesting to take note of Freudian overtones in analysing certain parts of the Hindu cycle, such as the castration myths; Lévi-Strauss himself, though objecting to the widespread Jungian interpretation of mythological functions in absolute terms, nevertheless (according to Edmund Leach) assumes that a myth is a kind of collective dream expressing unconscious wishes. Similarly, one need not stubbornly refuse the assistance of the ritualists in analysing other myths (such as the destruction of Dakṣa's sacrifice). Each has its place.

It may seem perverse to ignore the work of Frazer when analysing myths with obvious parallels in other cultures (such as the myth of primeval incest, or the androgyne), yet I have done so almost without exception. There seems little point in drawing attention to the parallels in primitive or Greek myths; readers who are familiar with those materials may draw their own parallels, while to others such comparisons would be cumbersome and meaningless. Only in a few particularly striking and obscure cases have I indicated the correlations, where these might help to elucidate the *Hindu* meaning of the myth. I had originally planned to key my entire analysis to Stith Thompson's index of world folklore motifs, in order to enable folklorists to identify motifs of other cultures, but I soon abandoned this plan when I discovered that, even on a superficial level, well over a thousand motifs of his index occurred in my material. A more serious objection, however, lay in the misleading nature of this use of the index, for a great many of these motifs were in fact cited by Thompson as known only from Indian folklore; thus a list of motifs from my corpus which appeared in the Thompson index would often suggest, wrongly, that a motif of purely Indian occurrence was a "world-wide" motif. The final objection, however, was one of basic method and philosophy: such a random and minute identification of motifs added nothing to one's understanding of the myths. As Lévi-Strauss has often pointed out, the motifs mean nothing in any abstract sense until they are placed in the context of the myth, in a structural relationship with other motifs.

The prevalence of certain of the Śaiva motifs—such as fire and water—in non-Indian cultures obviates the possibility of explaining them *only* in terms of Indian beliefs. But patterns within the Indian context reveal a particular meaning that the motifs do not have in any other cultures and may even serve to clarify some meanings in those cultures. In general, it seems to me that the universal elements of the myth are precisely the *least* important elements, that the point of the myth is to be sought in those areas in which the Hindu myth diverges from the general pattern. Betty Heimann has expressed this opinion:

> India can only be explained through its own nature. . . . Seemingly corresponding details that invite comparison are essentially unrelated as they sprang from different sources, and their seeming similarity is actually incidental, a common basis being lacking.

Even Lévi-Strauss has granted that a myth can be properly analysed only in the context of the culture of the listeners and tellers:

> I claim the right to make use of any manifestation of the mental or social activities of the communities under consideration which seem likely to allow me, as the analysis proceeds, to complete or explain the myth. . . . You cannot make mythology understood by somebody of a different culture without teaching him the rules and particular traditions of that culture.

Indeed, purely structural, Lévi-Strauss-inspired analyses of Indian tribal mythology have yielded interesting results; the authors have tried to relate these myths to the Hindu ritual materials at their disposal, but had they been able to make use of the full range of Sanskrit myths upon which the tribal stories are based, their conclusions might have been more subtle, if not necessarily more sound.

The validity of the method of analysis utilizing non-mythological materials for background is further supported by the manner in which characters within the myths actually cite Hindu lawbooks and philosophical works; the student of the myth seems justified in emulating their example. Once the Hindu myths have been set forth and analysed in their own right, it is possible to use them as the basis of more universal theories of myth, proceeding from the particular to the general, just as the general theories serve to illuminate the particular myths.

Yet the myth cannot be explained by ethnography alone, for if this were so, the myth could not add anything to the knowledge available in the ethnography. Moreover, the ambiguous relationship between myth and reality (the fact that the myth sometimes expresses a real situation by describing the reversal of actual conditions) produces yet another logical circle, for "ethnographic observation must decide whether the image [of a myth] corresponds to the facts." For these reasons, ethnographic commentary, though necessary, is not sufficient, nor is it able to compensate for certain shortcomings of the structural approach: "We soon find that the ethnographic commentary merely underlines even more sharply those aspects of the myth which cannot be further interpreted by this 'semantic procedure'."

Variants and Multiforms

The main source of explanatory material for any myth is not in ethnography but in other related myths: "A myth derives its significance not from contemporary or archaic institutions of which it is a reflection, but from its relation to other myths within a transformation group." A myth must be analysed in terms of its relationship to the "total myth structure" of the culture: "Each myth taken separately exists as the limited application of a pattern, which is gradually revealed by the relation of reciprocal intelligibility discerned between several myths." This theory is particularly well adapted to the variants of Hindu myths:

> The orthodox Hindu consciousness has always found it . . . easy to accept wide variations in the texts. Any scientific study should first of all preserve these variations and determine the kind of socio-religious idea they conveyed to people.

The latent structure of the myths can only be grasped by working through a large corpus of myths which express in many different ways certain similarities of structure, the repetition itself emphasizing the form behind the slight variations. "The ultimate conclusion of the analysis is not that 'all the myths say the same thing' but that 'collectively the sum of what all the myths say is not expressly said by any of them.'" Thus each myth within the cycle is a variant of every other myth.

The very delimitations of the episodes and symbols can only be identified by the comparison of several versions: integral units are those which appear in one version but not in another, on the principle of minimal pairs. As in algebra, the more terms (characters, episodes, symbols) involved in a myth, the more "equations" (variant myths) required to isolate the terms:

> Divergence of sequences and themes is a fundamental characteristic of mythological thought, which manifests itself as an irradiation; by measuring the directions and angles of the rays, we are led to postulate their common origin, as an ideal point on which those deflected by the structure of the myth would have converged had they not started, precisely, from some other point and remained parallel throughout their entire course.

The Śaiva myths involve many terms, and many permutations of each, so that literally hundreds of myths, all interrelated through their mutual characters and episodes, are required to elucidate even a relatively short sequence.

Moreover, one version may explain a particular element that is unclear in another:

> If one aspect of a particular myth seems unintelligible, it can be legitimately dealt with, in the preliminary stage and on the hypothetical level, as a transformation of the homologous aspect of another myth, which has been linked with the same group for the sake of argument, and which lends itself more readily to interpretation.

Biardeau has applied this method to Hindu mythology, with considerable success:

> As our text sheds very little light on this point, we may look for some similar stories where the symbolic meaning would be made more explicit. What appears to many people as an unmanageable overgrowth of myths in epics and *puranas* is actually an invaluable source of information for a better

> understanding of each of them. . . . The regional variants are all authentic as long as the overall significance of the epic remains. . . . The major variations in the text are likely each to have its own significance fitting into the whole.

This technique employed by the mythologist is justified by the theory that it is simply the mirror image of the technique employed by the myth-maker in the first place. Leach compares a myth with a message transmitted over a great distance (in time and space) and therefore repeated several times, with different wordings, so that when the different versions are reunited "the mutual consistencies and inconsistencies will make it quite clear what is 'really' being said." The mythologist is merely reassembling what the culture as a whole has fragmented.

Repetition enables the mythologist not only to separate the discrete units but to distinguish the more important elements from the trivial. The essential themes in a myth, impossible to identify from a simple reading of one version, emerge upon consideration of a number of other versions of that myth in which, despite various changes and reversals, certain elements persist. What is important is what is repeated, reworked to fit different circumstances, transformed even to the point of apparent meaninglessness, but always retained. In this way an element which occupies a relatively small part of a particular myth may be shown, in the context of the mythology as a whole, to be at the heart of that myth.

Multiple variants have a special importance in the analysis of myths which, like the Śaiva cycle, deal with contradictions. Myths which contain an insoluble problem are particularly prone to proliferate into many versions, each striving toward an infinitely distant solution, no one version able to confess its failure outright. The perplexing point, the crux, is constantly reworked in a vain attempt to find an emotional or logical resolution. This is apparent on the simple linguistic level as well, where false readings, alternative phrases, and blatantly corrupt or incorrect Sanskrit terms betray a point which the myth-maker did not himself understand but was unwilling to omit altogether, knowing it to be somehow essential.

For all of the foregoing reasons, it should be clear why it was necessary to include what may at first glance appear to be an overwhelming number of almost identical versions of certain myths. In the first place, they are *not* identical, and it is the seemingly insignificant minor variation that holds the key to the isolation of a motif, which in turn contributes to the final reconstruction of the total symbolism of the cycle. Every version adds an essential detail, the significance of which only becomes apparent when the entire cycle has been analysed.

The use of multiple variants, and the analogy of an imperfectly transmitted message, are supported by the irrefutable evidence that the ancient Hindu myth-makers were in fact well aware of the existence of these variants. Although it is possible that no one person knew all the versions of a particular myth (a theoretically infinite amount of material), it is likely that each author was acquainted with at least more than one version, and consciousness of these variants is demonstrated explicitly from time to time in the Purāṇas. Many myths are cited by characters within other myths or by the Paurāṇika (the bard reciting the Purāṇa) as apt examples or parallel situations. A frequent device used to accommodate multiple versions of a myth is the reference to multiple aeons of cosmic development. The seven variants of the birth of the Maruts describe how in seven former eras the Maruts were born, each time in a different way; a similar series of different versions of the birth of Rudra, in different eras, is narrated in another text. One passage refers explicitly to this technique: "Because of the distinction between eras, the birth of Gaṇeśa is described in different ways." Another text reveals an even more overt awareness of the use of this technique: the bard recites a myth in which a sage forgives his enemies; the audience then interrupts, saying, "We heard it told differently. Let us tell you: the sage cursed them in anger. Explain this." The bard then replies, "That is true, but it happened in a different era. I will tell you." And he narrates the second version of the myth. Similar references are often made to incidents occurring in a series of incarnations.

In addition to these, there are frequent variations of a different sort within a single myth, where an entire episode is told and then retold in a slightly different form as an expansion of the original theme, producing a series of multiforms similar to the multiforms of episodes in different myths. Leach has pointed out the redundancies of this type in the Bible:

> It is common to all mythological systems that all important stories recur in several different versions. Man is created in Genesis (chapter I, verse 27) and then he is created all over again (II, 7). And, as if two first men were not enough, we also have Noah in chapter VIII. Likewise in the New Testament, why must there be four gospels each telling "the same" story yet sometimes flatly contradictory on details of fact? . . . As a result of redundancy, the believer can feel that, even when the details vary, each alternative version of a myth confirms his understanding and reinforces the essential meaning of all the others.

This is a common artifice used in the composition of oral poetry, and it is appropriate to the rather florid style of the Purāṇas, which tend constantly to incorporate additional details and incidental material. In this way, for example, one Paurāṇika was forced to invent an elaborate episode to act as a bridge in order to include two different versions of the way in which Pārvatī accumulated her ascetic power (*tapas*): after narrating the first version, the Purāṇa states that Śiva then appeared in the form of a water demon and threatened to devour a small child unless Pārvatī transferred her *tapas* to him; Pārvatī agreed and then had to start all over again to accumulate her *tapas*; (there then follows the second version, which the Paurāṇika was loath to omit). A similar example of multiforms within the myth may be seen in a version of the Pine Forest story which distributes three different variants of the visit of Śiva to the forest in three different visits, each incorporating certain new details. One Purāṇa gives two multiforms of the birth of Skanda from Śiva's seed, one following immediately after the other but transferred from Skanda to Gaṇeśa, Śiva's other son; another Purāṇa gives two different versions of the story of Jalandhara, one clearly conscious of the other.

By comparing these variants and multiforms it is possible to isolate the basic themes and motifs out of which further versions, when and if they might be found or created, would be built as well, for the number of possibilities is infinite. The so-called repetitions are never *exactly* the same; beneath the apparent symmetry of structure is a fluctuation of detail, like the variations in a Persian carpet. We never reach a point when 'all the variants' have been considered, yet we may establish a reasonably complete basic vocabulary:

> There is no real end to mythological analysis, no hidden unity to be grasped once the breaking-down process has been completed. Themes can be split up *ad infinitum*. Just when you think you have disentangled and separated them, you realize that they are knitting together again in response to the operation of unexpected affinities.

This resistance against logical fragmentation leads the mythologist from one variant to another. "Myths are translations of one another, and the only way you can understand a myth is to show how a translation of it is offered by a different myth." The final "reconstruction" of the myth, once it has been dissected into some of its components, is simply a re-reading of the corpus in terms of these basic units. Thus I have constructed a chart of the basic motifs (see below), arranged according to the levels on which they occur (motifs of structure, symbolic motifs, ascetic/erotic motifs, and motifs of the interrelationships of Śiva with other characters), and reintroduced these motifs (designated by number) in the margins of all the myths I have cited. Moreover, after treating the component motifs as they occur in the principle myths of the Śiva cycle, I have discussed more general themes of function and structure which apply to the mythology on a broader scale.

There are many myths dealing with each theme, and many themes within each myth. There is no way to begin with any "basic" myth or any 'basic' theme, for the entire corpus interlocks and feeds back so that the total fabric resembles a piece of chain-mail rather than the brachiated, family-tree structure sought by the text-historical analysts and some structuralists. I have, therefore, selected certain myths best suited to illustrate certain motifs, while indicating in the margin other motifs which also occur therein. At some point the reader will have to re-read myths cited before their component motifs have been discussed; the chicken must precede the egg, and vice versa.

The final "explanation" of the myth cycle is thus the cycle itself, reread with a richer awareness of at least some of the resonances and harmonies behind the flickering images. It is to be hoped that the reader will not be disappointed with this 'explanation,' but rather, as Lévi-Strauss suggests,

> will find himself carried toward that music which is to be found in myth and which, in the complete versions, is preserved not only with its harmony and rhythm but also with that hidden significance that I have sought so laboriously to bring to light, at the risk of depriving it of the power and majesty that cause such a violent emotional response when it is experienced in its original state, hidden away in the depths of a forest of images and signs and still fresh with a bewitching enchantment, since in that form at least nobody can claim to understand it.

The present analysis has been patched up and amended and revised in successive incarnations until it became obvious that, unchecked, it would continue to generate new variants, like a myth itself, "until the final deluge," as the Hindus say. Paul Valéry once wrote that a poem is never finished; it is merely abandoned. Perhaps this is also true of any work of research in the arts; certainly it is true of any analysis of a body of mythology. For my part, I have had to force myself to abandon this opus for the moment; but it is of course unfinished.

Hindu Myths: A Sourcebook Translated from the Sanskrit

WENDY DONIGER O'FLAHERTY

The selection that follows is from a collection of essays by Professor Doniger O'Flaherty, compiled to illustrate the wealth of Hindu myths. The subtle themes of eroticism and asceticism are here cast against the background of the universe's fertility, with its consequences of life, death, and rebirth, for which Shiva is the antidote, the symbol of the highest god, or of the universal principle of being beyond all forms.

FROM THE *VĀMANA PURĀṆA*

The twice-born sage Maṅkaṇaka, the mind-born son of Kaśyapa, set out to bathe in his bark garment, and the celestial nymphs, Rambhā and the others, who were pleasing to look upon, shining, affectionate, and flawless, bathed there with him. Then the sage, whose ascetic power was great, became excited and shed his seed in the water; he collected that seed in a pot, where it became divided into seven parts, from which were born seven sages who are known as the bands of the Maruts: Wind-speed, Wind-force, Wind-destroyer, Wind-circle, Wind-flame, Wind-seed, and Wind-disc, whose heroic power is great. These seven sons of the seer support the universe, moving and still.

Once, long ago, the Siddha Maṅkaṇaka was wounded in the hand by the tip of a blade of *kuśa* grass, and plant sap flowed from that wound—so I have heard. When he saw the plant sap, he was filled with joy, and he started to dance; and then everything that was moving or still started to dance; the universe started to dance, for it was bewitched by his energy. When Brahmā and the other gods and the sages rich in ascetic power saw this, they reported to the great god Śiva about the sage: "You should do something so that he does not dance, O god." When the Great God saw that the sage was filled with joy to excess, he spoke to him for the sake of the welfare of the gods, saying, "Best of twice-born sages, what is the reason that has occasioned this joy in you who are an ascetic stationed on the path of *dharma*?" "Why, brahmin," said the sage, "do you not see the plant sap flowing from my hand? When I saw it I began to dance with joy."

The god laughed at the sage who was deluded by passion, and he said to him, "I am not amazed, priest. Look at this," and when Bhava, the god of gods, of great lustre, had said this to the eminent sage, he struck his own thumb with the tip of his finger, and from that wound ashes shining like snow came forth. When the priest saw this he was ashamed, and he fell at Śiva's feet and said, "I think that you are none other than the noble god who holds the trident in his hand, the best in the universe moving and still, the Trident-bearer. Brahmā and the other gods appear to be dependent upon you, faultless one. You are the first of the gods, the great one who acts and causes others to act. By your grace all the gods rejoice and fear nothing." When the sage had thus praised the great god, he bowed and said, "O lord, by your favour let my ascetic power not be destroyed." Then the god was pleased, and he answered the sage, "Your ascetic power will increase a thousand-fold by my favour, O priest, and I will dwell in this hermitage with you for ever. Any man who bathes in the Saptasārasvata and worships me will find nothing impossible to obtain in this world and in the other world, but he will certainly go to the Sārasvata world and, by the grace of Śiva, he will obtain the highest place."

Hīnayāna and Mahāyāna

HEINRICH ZIMMER

In this selection, Heinrich Zimmer examines the characteristic difference between the Hinayana (or Theravada) and the Mahayana. The development of Buddhist thought in this exposition rests upon the problem created by the Buddha's assertion that there is no soul (anatta). The Buddhist metaphysicians of Hinayana and Mahayana had to translate that assertion into an explanation that would fit all the cases of human experience of self, mind, and Nirvana. The splits between Hinayana and Mahayana reflected the disagreements that arose in the attempt to interpret the meaning of "no soul." All Buddhists, however, agree that the goal of existence is the cessation of dukkha or suffering to reach emancipation or Nirvana. Geographical distance as well as disagreements on philosophical points have kept the two expressions of Buddhism apart.

The village life of India was little modified by the rise and fall of the dynasties. The conquerors—even the complacent Greeks—soon recognized the virtues of the native way of being civilized. Alexander took to himself, as guru, the Jaina saint Kalanos, whom he invited to fill the vacancy of his old boyhood tutor Aristotle; while under the Kuṣāna warrior-kings both Buddhist art and Buddhist philosophy moved into a new and richly documented period. The Hellenistic Buddhist sculpture of Gandhāra, as well as the more spiritual and vigorous contemporary native Jaina and Buddhist art of Mathurā, gives ample evidence that under the protection of their foreign overlords the Indian religious systems were continuing to evolve. And we have the testimony of tradition for the statement that at the great Buddhist council assembled by Kaniṣka (the "Fourth Buddhist Council," held according to some reports at Jalandhar in the eastern Punjab, according to others at Kuṇḍalavana in Kashmir), the representatives of no less than eighteen Buddhist sects were in attendance.

The authenticity of the reports of this council has been questioned. Nevertheless it is obvious from many trains of evidence that a critical shift of weight took place in Buddhist teachings about this time. The religious practices of bhakti, which were already evident in the popular art and royal edicts of Aśoka's reign, began to receive the mature support of Buddhist philosophers. A canonical Buddhist literature in Sanskrit (no longer Pāli, the language of the earlier canon treasured in Ceylon) dating from the period of Kaniṣka stands for the view (already represented in Buddhist popular art) that the Buddha is to be reverenced as a divine being, and furthermore that numerous Buddhas (Buddhas of the past, Buddhas of the future) assist the devotee in his attempt to realize the Buddhahood latent within him. For whereas the earlier orthodox view had represented individual enlightenment (arhatship) as the goal to be attained, and this only by means of a literal imitation of the rigorous world-renunciation of the historical princely monk, Gautama Śākyamuni, the newer teaching was that Buddhahood (the status of a World Redeemer) is man's proper end, and furthermore that *since all things in reality are Buddha-things, all things potentially and actually are Saviors of the World.*

"It is as if a certain man went away from his father and betook himself to some other place. He lives there in foreign parts for many years, twenty or thirty or forty or fifty. In the course of time the father becomes a great man, but the son is poor; seeking a livelihood, he roams in all directions." The father is unhappy, having no son, but one day, while sitting at the gate of his palace transacting great affairs, he beholds his son, poor and tattered. The son thinks: "Unexpectedly have I here fallen in with a king or grandee. People like me have nothing to do here; let me go; in the street of the poor I am likely to find food and clothing without much difficulty. Let me no longer tarry in this place, lest I should be taken to do forced labor or should incur some other injury." The father orders his son brought to him, but, before revealing his birth to him, employs him for some years at all kinds of work, first at the meanest kinds, and then at the most important. The father treats his son with paternal kindness, but the son, although he manages all his father's property, lives in a thatched cottage and believes himself to be poor. At last, when his education is completed, he learns the truth.

In the same way we are the sons of the Buddha, and the Buddha says to us today, "You are my sons." But, like the poor man, we had no idea of our dignity, no idea of our mission as future Buddhas. Thus the Buddha has made us reflect on inferior doctrines. We have applied ourselves to them, seeking as payment for our day's work only nirvāṇa, and finding that it is already ours. Meanwhile the Buddha has made us dispensers of the knowledge of the Buddhas, and we have preached it without desiring it for ourselves. At last, however, the Buddha has revealed to us that this knowledge is ours, and that we are Buddhas, like himself.

This is the doctrine that has been termed, somewhat complacently, "The Big Ferryboat" (*mahāyāna*), the ferry in which all may ride, in contrast to "The Little Ferryboat" (*hīnayāna*), the way of those lonely ones, "lights unto themselves," who steer the difficult strait of individual release. The Big Ferryboat, with its pantheons of Buddhas and Bodhisattvas, prayer-wheels, incense, gongs, and graven images, rosaries and muttered syllables, has been disparaged generally by modern Occidental critics as a vulgar popularization of the Buddha's doctrine furthered by the advent into the northwestern provinces of barbaric peoples (not the Greeks, of course, but the Śakas and Yueh-chi), and yet, if anything is clear it is that the entire meaning of the paradox implicit in the very idea of Buddhahood has here come to manifestation. Nāgārjuna (*c.* 200 A.D.), the founder of the Mādhyamika school of Buddhist philosophy, which is the supreme statement of the Māhayāna view, was by no means a vulgarizer but one of the subtlest metaphysicians the human race has yet produced. While Asaṅga and his brother Vasubandhu (*c.* 300 A.D.), the developers of the Yogācāra school of the Mahāyāna, likewise merit the respect of whatever thinker sets himself the task of really comprehending their rationalization of Nāgārjuna's doctrine of the Void.

And Asvaghoṣa, the haughty contemporary of Kaniṣka (*c.* 100 A.D.), can have been no truckler to barbarians, even though his epic of the life of the Buddha, *Buddhacarita,* is graced with many unmonkish charms.

Fa Hsien (*c.* 400 A.D.), a Chinese Buddhist pilgrim to whom we owe much of our knowledge of the classic Gupta period, states that he found four Buddhist philosophic systems fully developed in India. Two of these, the Mādhyamika and Yogācāra, represented the Mahāyāna, while the others—Vaibhāṣika and Sautrāntika—were of the older Hīnayāna school. Fa Hsien declares that at Pāṭaliputra, the ancient capital of King Aśoka, the Mahāyānists had one monastery and the Hīnayānists another, with some six or seven hundred monks between them. His account of the cities further west reveals that at Mathurā the Bodhisattvas Mañjuśrī and Avalokiteśvara received worship as divinities—which is a Mahāyāna feature. The texts that he brought home with him were of both persuasions. And we learn, furthermore, from a second Chinese pilgrim two centuries later, Hsüan Tsang (629–640), that the two schools were then still in combat with each other. Relations between Buddhism and Hinduism were peaceful, but between the Mahāyāna and the Hīnayāna scholastic debate and mutual abuse, on the verbal level, were at such a pitch that the Buddha himself, had he returned, must certainly have been compelled to cry out piteously, as eleven centuries before: "The Order is divided! The Order is divided!"

The seeds of the conflict are in the words of the Buddha as recorded in the canon. The unsystematized epistemological, metaphysical, and psychological implications implicit in the program of spiritual therapy prescribed by him whose own one-pointedness had carried him beyond the sphere of simple ratiocination rankled in the minds of those still caught in the nets of thought; or one might say perhaps, in the minds of those who had been originally Brāhmans (rather than Kṣatriyas, like Gautama) and were therefore more disposed than he to adhere to the ways of thought, the way of jñāna, the processes of *thinking* problems through. Many, it is clear, attained Enlightenment; their formulae are not the vain scrimshaw of unreposeful intellects, but profoundly inspired, original renditions in philosophical terms of the realization promised by the Buddha's cure. Thus we find that just as the Doctrine, when rendered according to the dispositions of the bhakta, yielded a Buddhist art of broad popular appeal, so when apprehended by the Brāhman intellect, its implications opened into the most wonderfully subtle systems of metaphysical philosophy. The Wheel of the Law indeed was turning—churning the whole nature of man to new fulfillment. The Mahāyāna, the big ferryboat, and the Hīnayāna, the little, whether side by side or far apart, have together carried the millions of the Orient through centuries of transformations, secure in the understanding that the Buddha, somehow, in the most intimate, dependable way—no matter what their path of approach to him—was their indestructible, all-embracing Refuge.

"The ocean of tears shed by each being, wandering through life after life, without beginning," we are told, "is vaster than the Four Oceans together.

"The bones of the bodies that a man has worn in countless births, when heaped together would form a hill far larger than the lofty summits of the mountains that strike his eyes with awe."

This is the beginning of the wisdom of release. All life is sorrowful—and there is no beginning or end to it in this vale of tears; yet how substantial, how deep, how real, is this universal sorrow?

A parable is given in the Pāli *Stanzas of the Sister Elders*. There was a mother, we read, who had lost six children; one remained, a daughter. But eventually this child too died, and the mother was disconsolate. The Buddha came to her; and he said: "Many hundreds of children have we buried, you and I, hosts of kindred, in the times that are gone. Do not lament for this dear little daughter; four and eighty thousand with the same name have been burned on the funeral pyre by you before. Which among them is the one for whom you mourn?"

Obviously, there is here implicit an ontological problem; for in what sense can the woman with all the children, the man with all the bones, or the being with all the tears be said to have lived through multitudes of lives, if we are to accept as truth the Buddha's fundamental teaching: "All things are without a self (*an-attā*)"?

The philosophers of the Hīnayāna schools sought to solve the difficulty by contending that the ego-process consists of a series (*santāna, santati*) of moments (*kṣaṇa*) of transient entities (*dharmas*). There is nothing that abides. Not only are all the particles of being perishable (*anitya*), but their duration is infinitesimally short. "All things are as brief as winks" (*yat sat tat kṣaṇikam*). Their springing into existence is almost their ceasing to be. Yet they follow each other in chains of cause and effect that are without beginning and will go on for eternity. These chains, made up of momentary dharmas, are what appear to others, and in some cases to themselves also, as individuals—gods, animals, oceans, men, stones, and trees. Every phenomenal being is to be regarded as such a flux of particles that are themselves ephemeral. Throughout the transformations of birth, growth, old age, death, and the endless chain of rebirths, the so-called individual is no more than the vortex of such a causal sequence—never quite what is was a moment ago or what it is just about to be, and yet not different either. The simile is given of the flame of a lamp. During the first, the middle, and the last watches of the night, the flame is neither the same flame nor a different flame.

Or the simile is given of a bride who was bought from her father by the payment of the bride price when she was a little girl. The buyer thought that when she grew up she would become his wife, but he then went forth on a business voyage, and was away for many years. The girl grew up, came to marriageable age and, according to Indian custom, had to be supplied with a husband. The father, coming to a difficult decision, determined to accept a second bride price from a second suitor, and this man then became the actual husband of the girl. Presently, however, the first returned; and he demanded that the wife, now a mature woman, should be returned to him. But the husband replied: "This woman whom I have married is not your bride. The little girl of long ago is not the same as this adult woman who is today my wife." And so it is indeed! Throughout the series of our existences, bounded by death and birth, just as through the differing stages of our present biography, we are both identical with and not identical with ourselves.

The problem, that is to say, was resolved in the schools of the Hīnayāna by the assertion that what are usually taken to be permanent entities do not exist. All that can be said to exist—really—are the dharmas: small and brief realities, which, when grouped in aggregates and chains of cause and effect, create an impression of pseudo-individuals. There is no thinker, there is only thought; no one to feel, but only feelings; no actor, but only visible actions. This is a doctrine of phenomenalism, maintaining the nonexistence of substances and individuals while insisting on the reality of infinitesimal units, of which the world illusion is said to be compounded. The soul, the individual, is no more than a complex of momentary entities; and these are the only reality. Nirvāṇa, the attainment of release, therefore, is simply the cessation of the thought that all these phenomenal effects that we behold and feel constitute the reality that they appear to constitute. The extinction of that wrong idea—which is the leading thought and error of our lives—breaks off the entrainment of falsely grounded hopes and anxieties, life-plans, desires, and resentments. The arhat, the enlightened one, is simply a vortex no longer deluded about itself or about the phenomenality of other names and forms: in him knowledge concerning the conditions of phenomenality has released what once regarded itself as an entity from the consequences of the general delusion. But as to what more might be said about nirvāṇa—this is something the Hīnayāna philosopher does not attempt to face. The term *nirvāṇa* is by him treated strictly as a negative; for as a positive it would fall immediately under the ban of his basic formula, *yat sat tat kṣaṇikam:* whatever is, is momentary—like a wink.

The once extremely widespread Hīnayāna group known as Sarvāstivādins, or "realists" (*sarva*, "all"; *asti*, "exist"; *vāda*, "saying"), distinguish seventy-five dharmas, or "categories," under which, as they declare, every thought and form of being can be comprised. All are "real with regard to their substance" (*dravyato santi*); i.e., all exist as substance even though as ephemeral manifestations they are ever-changing and utterly perishable. They exist in a continuous series of births, durations, and destructions, which terminates when true knowledge puts an end to the restless movements of this contingent process, dissolving it in the quietness of extinction (*nirvāṇa*). This theory marked the climax of the *earlier* Hīnayāna, and supplied the basis for the speculations of the *later* Hīnayāna schools.

The question as to how suffering can be experienced in this world when there are no egos in which the suffering comes to pass, is answered by the thinkers of the later schools in different ways. The Sautrāntikas, for example, attempt to explain the situation in terms of their basic contention that our processes of thought do not represent a direct picture of external reality but follow each other in a thought series of their own—under pressure, as it were, from without, but otherwise autonomously. On the basis of our internal experiences it is to be inferred that external objects exist; yet this does not mean that those objects correspond in character to the internal thoughts generated under their influence. There is no chain of causation from outside to inside by which the suffering is produced that is felt within. The pains of hell, delights of heaven, and mixed blessings of the world, are all equally thoughts, brought about by precedent mental causes, not by outer facts, each chain of thoughts representing the force—more or less enduring—of a particular kind of ignorance or enthrallment. What imagines itself to be a suffering ego is only the continuity of the suffering itself. The continuous sequence of similar mental dharmas goes on simply as a reflex of its own imperfection.

The Vaibhāṣikas, on the other hand, object to this doctrine as "contradictory chatter" (*viruddhā bhāṣā*). They regard it as absurd to speak of inferences about the outer world on the basis of a thought series not in touch with it. For them, the world is open to direct perception. Occasionally, external objects may be known to exist by inference, but as a rule perception is what reveals to us their existence. Hence the objects of experience are of two kinds: 1. sensible or perceived and 2. cogitable or inferred. The laws that govern the causal sequences in these two realms are different, and yet the two interact upon each other. Suffering, for example, represents an actual impingement of the outer world upon the world within, even though the normal inference that an enduring individual is what is undergoing the ordeal is a thought without basis in fact. For whatever the mental stream presents as external does indeed exist without—though not in the mode of substances and enduring entities, as is generally inferred.

According to both of these late Hīnayāna schools of reasoning, the aggregates of experience—whether external or internal—are ephemeral yet real. In the Mahāyāna, on the other hand, they are not even real. A metaphysical substratum of all phenomenality is admitted, but the entire sphere of phenomenality itself (whether mental or physical, inferential or perceived, long-enduring or of briefest moment) is regarded as without substance. The philosophers of the Mahāyāna liken the universe to a magical display, a mirage, a flash of lightning, or the ripple of waves on the sea. The sea itself, the reality beyond and within the rippling forms, cannot be measured in terms of the ripples. Comparably, the objects in the world are of one reality and in reality therefore one; but this reality is beyond description in terms of phenomenality. This one reality, in its ontological aspect, can be termed only *bhūta-tathatā*, "the suchness of beings, the essence of existence." In its relation to knowledge it is known as *bodhi* ("wisdom") and *nirvāṇa*.

That is to say, by treating positively the term and problem to which the Hīnayāna philosophers accorded only a negative description, the Mahāyāna schools transcended the comparatively naïve positivism of their associates and approached the nondualism of the contemporary Vedānta. The Hīnayāna thinkers, in short, never really faced the question of the degree or nature of their so-called "reality." Precisely in what sense were their dharmas and chains of causes and effects being declared to "exist"? How much was being said, after all, when it was insisted that the dharmas were ultimately "real"? The Mahāyāna philosophers, turning their attention to this question, distinguished three aspects in terms of which the reality of any object could be considered: (1) quintessence, (2) attributes, (3) activities. The *quintessence* of a jar is earth or clay; its *attributes*, the coarseness or fineness, fragility or strength, beauty or ugliness, etc. of the form; while its *activities* are the receiving, holding, and discharging of water. Both attributes and activities are subject to laws of change, but quintessence is absolutely indestructible. The waves of the sea may be high or low, but the water itself neither increases nor decreases. And so it is that though all things are born to die—whether as long-lived individuals, or as infinitesimal momentary particles—the quintessence of them all remains unchanged. The universe, the entire world of names and forms, has both its phenomenal and its enduring aspect, but only the latter, this substratum of all things, is what can be known as *bhūta-tathatā*, the "suchness or essence of existence."

According to the basic argument of this metaphysical philosophy (which takes its stand, so to say, in *nirvāṇa*) every dharma is *pratītya-samutpanna*, "dependent on others." It cannot be explained by reference either to itself or to something else, or by bringing the two sets of references into a relationship. Every system of notions terminates in inconsistencies and is therefore simply void. And yet, to assert that all is "non-existent" (*abhāva*) would not be proper either; for that would be only another act of dialectical reasoning, whereas true wisdom is neither an affirmation or a negation. "Nothing is abandoned, nothing annihilated (*na kasyacit prahānam, nāpi kasyacin nirodhaḥ*)."

The only final truth, then, is the void—an ineffable entity, the state of "being thus" (*tathatā*), which is realized as opposed to the ever-changing mirage of contingent notions. This, and this alone, is the absolute that persists throughout all space and time as the essence of things. Things in their fundamental nature cannot be named or explained; they cannot be discussed; they are beyond the range of perception; they are possessed of absolute sameness; they have no distinctive features; they are subject to neither transformation nor destruction. The realm of their reality is that of "absolute truth" (*paramārtha-tattva*), not of "relative truth" (*saṁvṛti-tattva*)—in other words, precisely what in the Upaniṣads is known as *brahman*; though the negative as opposed to the positive formulation here, in Buddhism, conduces to a different attitude and a different style of thought and teaching.

Nāgārjuna (second or third century A.D.) is regarded as the great master of this doctrine. He is described (in a biography that a Hindu sage named Kumārajīva translated into Chinese about A.D. 405) as a Brāhman of southern India, who, while a mere boy, studied the four Vedas and became adept in all sciences, including those of magic. In his earliest youth he made himself and three friends invisible so that they might slip into the royal seraglio, but the four had hardly begun to take advantage of their situation when they were apprehended. The friends were condemned to die, but Nāgārjuna was allowed to elect the other death, namely, that of the monastic vow.

In ninety days he studied and mastered the whole of the Buddhist Pāli canon. Then he proceeded northward, in quest of further knowledge, until he came to the Himalayas, where a monk of immense age committed to him the Mahāyāna sūtras; after which a serpent king (*nāgarāja*) disclosed an authentic commentary on those pages. All these sacred writings had been preserved in secret—so the story goes—for centuries. They were, in fact, authentic revelations of the doctrine, which the Buddha himself had regarded as too profound for his contemporaries and had therefore put into the keeping of competent guardians. Mankind had required literally hundreds of years of preliminary training (the training of the Hīnayāna) in preparation for this higher law. But now that the world was ready, Nāgārjuna was permitted to spread the final Buddhist teaching of "The Great Ferryboat" throughout the land of India. And he did so (so runs the legend) for three hundred years.

The foundation of the Mahāyāna is thus in legend ascribed to Nāgārjuna. The evidence, however, now seems to show that the basic principles were formulated well before his century. Though practically the whole of the so-called *Prajñā-pāramitā* literature is attributed to him, it now appears that some of these texts preceded, while others followed, his lifetime. And yet it is certain, in spite of the fabulous details of his biography, that Nāgārjuna was an actual character, and, moreover, a brilliant, crystallizing, and energizing philosophical spirit. Throughout northern India they still speak of him as "the Buddha without his characteristic marks." And the works ascribed to him are revered equally with "the sūtras from the Buddha's own mouth."

The Doctrine of No-Soul: *Anatta*

WALPOLA SRI RAHULA

Walpola Sri Rahula is a Buddhist–scholar monk from Sri Lanka who taught for many years as a professor at Northwestern University. This selection discusses anatta or "no soul" which is essential to the understanding of Buddhism, a religion that is unique in stating that there is no separate self or soul whose salvation is the goal and purpose of existence. The human Guatama (Buddha) understood the heroic quest for the secret to abolish rebirth and suffering for all sentient beings. When he experienced Nirvana, he also discovered the truth of "no soul."

What in general is suggested by Soul, Self, Ego, or to use the Sanskrit expression *Ātman*, is that in man there is a permanent, everlasting and absolute entity, which is the unchanging substance behind the changing phenomenal world. According to some religions, each individual has such a separate soul which is created by God, and which, finally after death, lives eternally either in hell or heaven, its destiny depending on the judgment of its creator. According to others, it goes through many lives till it is completely purified and becomes finally united with God or Brahman, Universal Soul or *Ātman*, from which it originally emanated. This soul or self in man is the thinker of thoughts, feeler of sensations, and receiver of rewards and punishments for all its actions good and bad. Such a conception is called the idea of self.

Buddhism stands unique in the history of human thought in denying the existence of such a Soul, Self, or *Ātman*. According to the teaching of the Buddha, the idea of self is an imaginary, false belief which has no corresponding reality, and it produces harmful thoughts of "me" and "mine," selfish desire, craving, attachment, hatred, ill-will, conceit, pride, egoism, and other defilements, impurities and problems. It is the source of all the troubles in the world from personal conflicts to wars between nations. In short, to this false view can be traced all the evil in the world.

Two ideas are psychologically deep-rooted in man: self-protection and self-preservation. For self-protection man has created God, on whom he depends for his own protection, safety and security, just as a child depends on its parent. For self-preservation man has conceived the idea of an immortal Soul or *Ātman*, which will live eternally. In his ignorance, weakness, fear, and desire, man needs these two things to console himself. Hence he clings to them deeply and fanatically.

The Buddha's teaching does not support this ignorance, weakness, fear, and desire, but aims at making man enlightened by removing and destroying them, striking at their very root. According to Buddhism, our ideas of God and Soul are false and empty. Though highly developed as theories, they are all the same extremely subtle mental projections, garbed in an intricate metaphysical and philosophical phraseology. These ideas are so deep-rooted in man, and so near and dear to him, that he does not wish to hear, nor does he want to understand, any teaching against them.

The Buddha knew this quite well. In fact, he said that his teaching was "against the current" (*paṭisotagāmi*), against man's selfish desires. Just four weeks after his Enlightenment, seated under a banyan tree, he thought to himself: "I have realized this Truth which is deep, difficult to see, difficult to understand . . . comprehensible only by the wise . . . Men who are overpowered by passions and surrounded by a mass of darkness cannot see this Truth, which is against the current, which is lofty, deep, subtle and hard to comprehend."

With these thoughts in his mind, the Buddha hesitated for a moment, whether it would not be in vain if he tried to explain to the world the Truth he had just realized. Then he compared the world to a lotus pond: In a lotus pond there are some lotuses still under water; there are others which have risen only up to the water level; there are still others which stand above water and are untouched by it. In the same way in this world, there are men at different levels of development. Some would understand the Truth. So the Buddha decided to teach it.

The doctrine of *Anatta* or No-Soul is the natural result of, or the corollary to, the analysis of the Five Aggregates and the teaching of Conditioned Genesis (*Paṭicca-samuppāda*).

We have seen earlier, in the discussion of the First Noble Truth (*Dukkha*), that what we call a being or an individual is composed of the Five Aggregates, and that when these are analysed and examined, there is nothing behind them which can be taken as "I," *Ātman*, or Self, or any unchanging abiding substance. That is the analytical method. The same result is arrived at through the doctrine of Conditioned Genesis which is the synthetical method, and according to this nothing in the world is absolute. Everything is conditioned, relative, and interdependent. This is the Buddhist theory of relativity.

Before we go into the question of *Anatta* proper, it is useful to have a brief idea of the Conditioned Genesis. The principle of this doctrine is given in a short formula of four lines:

> When this is, that is (*Imasmiṃ sati idaṃ hoti*);
> This arising, that arises (*Imassuppādā idaṃ uppajjati*);
> When this is not, that is not (*Imasmiṃ asati idaṃ na hoti*);
> This ceasing, that ceases (*Imassa nirodhā idaṃ nirujjhati*).[1]

On this principle of conditionality, relativity and interdependence, the whole existence and continuity of life and its cessation are explained in a detailed formula which is called *Paṭicca-samuppāda* "Conditioned Genesis," consisting of twelve factors:

1. Through ignorance are conditioned volitional actions or karma-formations (*Avijjāpaccayā saṃkhārā*).
2. Through volitional actions is conditioned consciousness (*Saṃkhārapaccayā viññāṇaṃ*).
3. Through consciousness are conditioned mental and physical phenomena (*Viññāṇapaccayā nāmarūpaṃ*).
4. Through mental and physical phenomena are conditioned the six faculties (i.e., five physical sense-organs and mind) (*Nāmarūpapaccayā saḷāyatanaṃ*).
5. Through the six faculties is conditioned (sensorial and mental) contact (*Saḷāyatanapaccayā phasso*).
6. Through (sensorial and mental) contact is conditioned sensation (*Phassapaccayā vedanā*).
7. Through sensation is conditioned desire, 'thirst' (*Vedanāpaccayā taṇhā*).
8. Through desire ('thirst') is conditioned clinging (*Taṇhāpaccayā upādānaṃ*).
9. Through clinging is conditioned the process of becoming (*Upādānapaccayā bhavo*).
10. Through the process of becoming is conditioned birth (*Bhavapaccayā jāti*).
11. Through birth are conditioned (12) decay, death, lamentation, pain, etc. (*Jātipaccayā jarāmaraṇaṃ . . .*).

This is how life arises, exists and continues. If we take this formula in its reverse order, we come to the cessation of the process:

Through the complete cessation of ignorance, volitional activities or karma-formations cease; through the cessation of volitional activities, consciousness ceases; . . . through the cessation of birth, decay, death, sorrow, etc., cease.

It should be clearly remembered that each of these factors is conditioned (*paṭiccasamuppanna*) as well as conditioning (*paṭicca saamuppāda*). Therefore they are all relative, interdependent and interconnected, and nothing is absolute or independent; hence no first cause is accepted by Buddhism as we have seen earlier. Conditioned Genesis should be considered as a circle, and not as a chain.

The question of Free Will has occupied an important place in Western thought and philosophy. But according to Conditioned Genesis, this question does not and cannot arise in Buddhist philosophy. If the whole of existence is relative, conditioned and interdependent, how can will alone be free? Will, like any other thought, is conditioned. So-called "freedom" itself is conditioned and relative. Such a conditioned and relative "Free Will" is not denied. There can be nothing absolutely free, physical or mental, as everything is interdependent and relative. If Free Will implies a will independent of conditions, independent of cause and effect, such a thing does not exist. How can a will, or anything for that matter, arise without conditions, away from cause and effect, when the whole of existence is conditioned and relative, and is within the law of cause and effect? Here again, the idea of Free Will is basically connected with the ideas of God, Soul, justice, reward and punishment. Not only is so-called free will not free, but even the very idea of Free Will is not free from conditions.

According to the doctrine of Conditioned Genesis, as well as according to the analysis of being into Five Aggregates, the idea of an abiding, immortal substance in man or outside, whether it is called *Ātman*, "I," Soul, Self, or Ego, is considered only a false belief, a mental projection. This is the Buddhist doctrine of *Anatta*, No-Soul or No-Self.

In order to avoid a confusion it should be mentioned here that there are two kinds of truths: conventional truth (*sammuti-sacca*, Skt. *saṃvṛti-satya*) and ultimate truth (*paramattha-sacca*, Skt. *paramārtha-satya*). When we use such expressions in our daily life as "I," "you," "being," "individual," etc., we do not lie because there is no self or being as such, but we speak a truth conforming to the convention of the world. But the ultimate truth is that there is no "I" or "being" in reality. As the *Mahāyāna-sūtrālaṅkāra* says: "A person (*pudgala*) should be mentioned as existing only in designation (*prajñapti*) (i.e., conventionally there is a being), but not in reality (or substance *dravya*)."

"The negation of an imperishable *Ātman* is the common characteristic of all dogmatic systems of the Lesser as well as the Great Vehicle, and, there is, therefore, no reason to assume that Buddhist tradition which is in complete agreement on this point has deviated from the Buddha's original teaching."

It is therefore curious that recently there should have been a vain attempt by a few scholars to smuggle the idea of self into the teaching of the Buddha, quite contrary to the spirit of Buddhism. These scholars respect, admire, and venerate the Buddha and his teaching. They look up to Buddhism. But they cannot imagine that the Buddha, whom they consider the most clear and profound thinker, could have denied the existence of an *Ātman* or Self which they need so much. They unconsciously seek the support of the Buddha for this need for eternal existence—of course not in a petty individual self with small s, but in the big Self with a capital S.

It is better to say frankly that one believes in an *Ātman* or Self. Or one may even say that the Buddha was totally wrong in denying the existence of an *Āt*-

man. But certainly it will not do for any one to try to introduce into Buddhism an idea which the Buddha never accepted, as far as we can see from the extant original texts.

Religions which believe in God and Soul make no secret of these two ideas; on the contrary, they proclaim them, constantly and repeatedly, in the most eloquent terms. If the Buddha had accepted these two ideas, so important in all religions, he certainly would have declared them publicly, as he had spoken about other things, and would not have left them hidden to be discovered only 25 centuries after his death.

People become nervous at the idea that through the Buddha's teaching of *Anatta,* the self they imagine they have is going to be destroyed. The Buddha was not unaware of this.

A bhikkhu once asked him: "Sir, is there a case where one is tormented when something permanent within oneself is not found?"

"Yes, bhikkhu, there is," answered the Buddha. "A man has the following view: 'The universe is that *Ātman,* I shall be that after death, permanent, abiding, ever-lasting, unchanging, and I shall exist as such for eternity.' He hears the Tathāgata or a disciple of his, preaching the doctrine aiming at the complete destruction of all speculative views . . . aiming at the extinction of 'thirst,' aiming at detachment, cessation, Nirvāṇa. Then that man thinks: 'I will be annihilated, I will be destroyed, I will be no more.' So he mourns, worries himself, laments, weeps, beating his breast, and becomes bewildered. Thus, O bhikkhu, there is a case where one is tormented when something permanent within oneself is not found."

Elsewhere the Buddha says: "O bhikkhus, this idea that I may not be, I may not have, is frightening to the uninstructed worldling."

Those who want to find a 'Self' in Buddhism argue as follows: It is true that the Buddha analyses being into matter, sensation, perception, mental formations, and consciousness, and says that none of these things is self. But he does not say that there is no self at all in man or anywhere else, apart from these aggregates.

This position is untenable for two reasons:

One is that, according to the Buddha's teaching, a being is composed only of these Five Aggregates, and nothing more. Nowhere has he said that there was anything more than these Five Aggregates in a being.

The second reason is that the Buddha denied categorically, in unequivocal terms, in more than one place, the existence of *Ātman,* Soul, Self, or Ego within man or without, or anywhere else in the universe.

NOTE

[1]To put it into a modern form:
When A is, B is;
A arising, B arises;
When A is not, B is not;
A ceasing, B ceases.

Some Buddhist Concepts of Kuan Yin

JOHN BLOFELD

John Blofeld, an Englishman who settled in the Far East as a young man before World War II, dedicated his life's work to studying Chinese religion and philosophy. This selection, based upon Blofeld's interviews with the Chinese, presents Chinese attitudes toward Kuan Yin. Kuan Yin is the counterpart of the male Avalokiteshvara, prominent in ancient times in India and continuously in Tibet. One of the most popular features of Chinese Buddhism is devotion to the female Bodhisattva Kuan Yin. The secret of Kuan Yin meditation is to find the Bodhisattva within oneself. The search for the Bodhisattva Kuan Yin's realm has been a primary goal of Chinese Buddhism for many centuries.

To the perfection of her merits,
Worshipping, we bow our heads
Lotus Sūtra

I have always been intrigued by those masterpieces of Chinese ivory carving comprising a large number of exquisitely carved balls revolving one within another. Being intricately decorated, the outer layers largely conceal those within; though one gazes long and hard, it is not easy to discern the innermost ball or even to distinguish clearly one middle layer from another. So it was with my perception of Kuan Yin, some of her more and less materialistic aspects seeming to be inextricably intertwined. This was especially true of the levels of understanding at which she gradually exchanges her goddess-like attributes for those of a Bodhisattva. The guise she wears for those who burn incense to her in wayside shrines and mountain grottoes or in the temples of the fisher-folk and boat-dwellers undergoes no startling change when she presents herself to the less erudite members of Chinese and Japanese Buddhist communities. True, they know her by her proper title, Bodhisattva, but some would be hard put to it to explain in what manner gods and Bodhisattvas differ. The Mahayana sutras chiefly prized by the Pure Land Sect to which many of her devotees belong do not easily yield their hidden meaning. Indeed, they resemble the profound tantric works revered by Tibetans in that, if taken literally in the absence of oral instruction from a Master, they may repel rather than attract most Western students of the Way, who may deem them too full of marvels to merit serious attention.

Were the approach taken here to understanding Kuan Yin's true significance to be made fully consistent, it would be necessary to set forth the Pure Land teaching and practices as they appear to the uninstructed and reserve their esoteric meaning for a later chapter. However, that is not quite how I came upon them myself; thanks to the guidance of some Chinese friends, I began to have at least a vague conception of the esoteric meaning while still near the outset of my studies and I have ordered my exposition accordingly. Like a fair number of other Western Buddhists, I began by shying away from what could be seen of Pure Land practice all about me in South China, being unable as yet to reconcile it with the Buddhist teaching familiar to me. Though its manifestations were always beautiful, I did not believe they had much relation to reality.

Briefly, the doctrine in its literal form is that the Buddha, foreseeing the onset of a decadent age (in the midst of which we now find ourselves today) and recognising how difficult it would be for the beings born in that age to pursue Enlightenment by the means he had expounded hitherto, compassionately presented a much easier way. He delivered some sutras (discourses) anent celestial Buddhas such as Amitābha and Bodhisattvas such as Kuan Yin, each of whom had mentally created a spiritual realm (Pure Land) wherein all beings who aspired to it earnestly could secure rebirth under conditions ideally suited to making progress towards Enlightenment. Profiting by the vast merits of the creator of their chosen Pure Land, such beings, however great their demerits, could easily attain rebirth in, for example, Kuan Yin's Potala and thus escape the dreary round of birth, suffering and death endlessly renewed which is samsara. Furthermore, those celestial Buddhas and Bodhisattvas were portrayed as wielding miraculous powers wherewith they could instantly avert danger or affliction from any being who called upon them with absolute sincerity; thus, even should one be kneeling beneath an executioner's sword already raised to strike, a single heart-felt cry to Kuan Yin Bodhisattva would cause the blade to fall shattered to the ground!

Well, this teaching, though couched in terms of the utmost beauty, struck me as too good to be true, too redolent of primitive conceptions of heaven and of fairy godmother tales like Cinderella. However, my Chinese friends advised me not to dismiss the teaching out of hand, but rather to seek its inner purport. Soon after my arrival in the East, I had struck up a warm friendship with a most unusual man, a Chinese physician and keen Buddhist about ten years my senior. I was immediately attracted to him for several reasons, not the least of which was his fondness for tradition as exemplified by his garb—instead of a Western-style business suit, a mode of dress already becoming common in Hong Kong, he wore a graceful silken robe surmounted by an old-fashioned skull-cap of stiff black satin with a tiny scarlet bobble at the crown. Being fond of the more esoteric forms of Buddhism, he had first mastered the Shingon form recently brought back to China from Japan and then embarked on a life-long study of the Vajrayana taught by Tibetan lamas; and, with regard to the Pure Land Sect and others, he retained the typically Chinese attitude which the Venerable T'ai Hsü once summed up in the words: "All the sects are like beads on one rosary." It was to Ta Hai, my physician friend, that I took some doubts about the Pure Land teaching which persisted despite my feeling a special affinity to Kuan Yin.

"Elder Brother, since having that odd experience, I have been thinking much about Kuan Yin," I said one day, handing him an opened copy of the Heart of Great Compassion Dhāranī Sūtra and pointing to the words: "*Should any being recite and cleave to the sacred Dhāranī of Great Compassion and yet not be reborn in my Buddhaland, I vow not to enter upon Supreme Enlightenment.*"

"As you know," I continued, "there are many passages in the sutras which state that one who calls upon her name with great sincerity or recites her mantra from his heart will surely be reborn in her sacred Potala and there be trained to achieve Enlightenment. Does that not strike you as too easy to be true? Elsewhere the sutras stress again and again that the seed of Enlightenment latent in every being must be watered by self-cultivation, that no teacher, human or divine, can do the work on our behalf. Granted that those descriptions of Pure Lands full of jewelled trees, gem-studded lotus pools and birds warbling the sacred teaching can be understood figuratively and putting aside the fact that such descriptions come oddly from the lips of people who share our belief in the Mind Only doctrine, there is another obstacle. How can you reconcile the need for self-power (*tzû-li*) with the Pure Land Sect's reliance on other-power (*t'a-li*)? It seems so very illogical."

Ta Hai laughed delightedly. "I think, Ah Jon, you are still foreign-devil-man and cannot learn to think like Chinese. Why you care about logical, not logical? Truth have plenty faces. As you see things, so things are. As you expect things, so things come. Why? Because your mind make them so. You dream long time of jewelled paradise, you surely take rebirth there. You think wisdom help you reach formless world, you surely take rebirth there. You have learnt to recite Heart Sūtra, yes? So you know very well 'Form is void and void is form; form not differ from void, void not differ from form.' Then why you worry these nonsense things? I and my friends tell you and tell you and tell you that appearances are all in mind. Why you not understand? Outside mind—nothing!"

"Yes, but—"

"Listen, Ah Jon. Pure Land teacher say fix mind on sacred name or speak sacred mantra many, many times, then your mind become still, yes? All obscurations disappear. That way, you know, plenty people get objectless awareness which is first step to Enlightenment. That is very good, no? So why you care *how* they get it? All of us Buddhists are looking for goal higher than man can see or imagine. You agree? Good. Suppose my picture of goal is dull and your picture seems to you much clearer, you know quite well that both those pictures must be a long, long way—a million miles and more—from truth picture. We go one million mile walk; you start one inch in front of me not help you much, Ah Jon. Your talk about *tzû-li* and *t'a-li* sound very good, very clever, very wise to you? To me, all nonsense! You want to study Buddha Dharma, you must study mind. Only mind is real, but now you try to put front door and back door on it! Self? Other? Inside? Outside? How can be? Some people look for Enlightenment in mind. Some people look for Bodhisattva. You find them different? Never can be! Why? Because whole universe live inside your bony skull. Nowhere else at all. Amitābha Buddha in your skull. Kuan Yin Bodhisattva in your skull. Ch'an (Zen) followers seek Enlightenment from mind. Pure Land followers seek it in Pure Land. What difference? Two thoughts; one Source. You are philosopher, so think one way. Your friend just love Kuan Yin, so think another. Different, yes! What difference? Two faint ideas, same shining truth.

"You seek stirring of compassion in mind, soon you find. Seek shining compassion-being like Kuan Yin, soon you find. Suppose you run out in street now, tell everybody must use self-power, not other-power; you think they understand? Or they stand there gaping? You ought to welcome compassionate Buddha's thousand ways of teaching thousand kinds of people."

I believe I learned more from Ta Hai than from any other man, but it was not all plain sailing. His English was worse than I have represented it here and my Cantonese no better at that time. We managed to tackle all kinds of abstruse subjects, helping out the words with gestures, drawings and Chinese written characters, but relying greatly on the kind of telepathic understanding by which close friends can surmount most language barriers. With others of our group it was often easier, as some spoke English very well indeed. Asked why many Pure Land followers seemed to prefer evoking Kuan Yin rather than Amitābha Buddha of whom she is but an emanation, one of them replied:

"Like you, they feel drawn to her. It is because of your nature. If you were a horse, you would be sure to invoke the Horse-Headed Hayagrīva, who is also Kuan Yin. If a lobster, you would choose a lobster deity, just as nagas invoke serpent divinities. Picturing compassion in the form of a lovely woman is a reasonable thing to do. Amitābha Buddha is compassion seen as a noble quality, shining and majestic; Kuan Yin is compassion seen as intimate and a counterpart of gentle pity. Not having many heads in the Indian manner nor necessarily sharing Amitābha's vastness, she appeals to humanists like you and me and fits in well with our Chinese conception of divinity."

From all of this I began drawing a conclusion that proved absolutely wrong, for when I voiced it they opposed it vehemently and I perceived that the waters we had entered were deeper than they seemed.

"Do you mean," I asked, "that Amitābha, Kuan Yin and their vows to succour sentient beings are really myths used to persuade unlearned people to concentrate upon their names and thus achieve one-pointedness of mind even though unable to perceive its proper purpose?"

Before the words were well out of my mouth, I realised I had uttered an enormity. They glanced at one another in consternation—not, I think, because like Jovians in a similar situation they expected a shower of thunderbolts to greet this blasphemy, but because they were aghast that their way of putting things had led me so far astray, or perhaps they were just bewildered by my obtuseness.

"Make no mistake," cried old Mr. Lao sharply, speaking in Cantonese in his eagerness to set the record straight. "The Buddhas, the Bodhisattvas and their vows are real. If you doubt it, you will be beyond their help!"

"But—"

"Look at this blackwood desk. Is it real, do you think?"

"Yes, of course—real in the limited sense that any phenomenon is real. You can see and touch it."

"Good. How about, say, justice? We are told, for example, that Britain's legal system ensures you people a greater measure of justice than is to be had in Hong Kong. Is justice real?"

"Ye-es. If you put it that way. Justice can be quantified to some extent and seen to exist in one place but not in another. It would make sense to say that all justice had been banished from such-and-such a country."

"Excellent. Though you cannot hurt your hand by banging it against justice, you do agree it is real. But *why* is it real? Because mind conceives it. If human beings were mindless entities like motor-cars, there could be no such thing as justice. Whatever the mind conceives thereby achieves reality. Suppose it were found that, after all, our national sage, Confucius, had no historical existence, would it alter the fact of his overwhelming influence on our country? Conceived of in the way he is, he would have become a reality and his reported words and actions, as the causes of great effects, would also have reality. Be sure that Kuan Yin's vow is real, that if you earnestly desire rebirth in her Potala paradise, you will take birth there."

Such were the conversations Ta Hai and his friends most enjoyed and great was my bewildered admiration of their wisdom. The questions and answers reported above may seem obscure unless one has a background knowledge of Mahayana Buddhism. The crux of the matter is as follows:

According to that teaching, it is not profitable to spend time on such questions as whether there was ever a beginning to the succession of universes that have been arising and reaching their end for innumerable aeons, or why sentient beings must revolve endlessly from life to life in this sad realm of samsara. What is needed is to direct one's attention to the *present,* thinking: "This is how things are; what is to be done about them?" It is taught that reality has two aspects—the realms of Void and form—but that, due to obscurations arising from primordial ignorance and from evil karma accumulated in past lives, we fail to *see* that nothing can exist independently of everything else, that all entities (including people) are transient, mutable, unsatisfying and *lacking in own-being.* It is the illusion of possessing an ego that leads to such obscurations as passion, lust and inordinate desire. From these in turn spring a longing for continued *individual* existence which keeps beings revolving in the round of birth and death, reaping over and over again sorrow, frustration, disappointment, grief, adversity. Could one but be rid of all ego-born delusions, he would see himself as a shadow pursuing shadows and eagerly seek an end to the round—not in extinction, but in Nirvāna, the glorious apotheosis in which illusory egos are no more. The way out lies within each sentient being's mind in the form of latent wisdom-compassion energy (Bodhi). This is so because so-called individual minds are not truly apart from Mind, the Plenum in which everything exists forever and forever in the form of '*no thing.*' When entities vanish, nothing is lost, for they had no ultimate existence in the first place, being nevertheless real because not divorced from the Void.

Liberation is achieved by Enlightenment, the fruit of transcending all ego-delusion. A powerful technique for attaining it is meditation (better called contemplation) which results in a turning over of the mind upon itself, the expulsion of obscurations and recognition of oneself and all beings as being wholly without self or anything describable as own-being. Thus the effort has to be *self-effort.* No wise guru, no high divinity can accomplish this revolution on one's behalf; it must occur of itself. Now comes the surprise and seeming contradiction. Adherents of the Pure Land School seek a way out through rebirth in a "Pure Land" where they can give themselves over to the great task of seeking Enlightenment under ideal conditions, there being no hindrances, but only powerfully favourable influences. Exoterically it is taught that the Buddha Amitābha (like some others such as Kuan Yin) vowed to save all beings who call upon him wholeheartedly by admitting them to a Pure Land, this being his compassionate means of assisting those weak in self-power or too ignorant to understand how to use it. Esoterically it is recognised that the Pure Land is no other than Pure Mind, the condition of all minds when purged of ego-born obscurations. But this distinction between the exoteric and esoteric understanding of the doctrine is not simple or sharp-cut. Even exceedingly erudite Buddhists such as Ta Hai hold that the various Pure Lands, including Kuan Yin's Potala, do in a sense exist as places, since mind has thus conceived them. This seemingly startling departure from logic is somewhat less puzzling if one accepts that *all* entities are mental creations, none ultimately more or less real than any other.

A devotional approach, like that of Pure Land followers, is currently out of favour in the West, being too reminiscent of the Christian and Jewish faiths which many people no longer find acceptable. Few Asians share this antipathy. Even Theravada (southern) Buddhism has a greater element of devotional practice than is generally recognised by Westerners. The same is true of Ch'an (Zen) Buddhism, many of whose most ardent followers (including the great Daisets Suzuki) have affirmed the validity of the Pure Land doctrine and regard Pure Land practices as a particularly efficient means of attaining Enlightenment. Even today, the Pure Land School has a much greater following among Japanese Buddhists than any other, as it had in China prior to the submergence of all religion beneath the waters of the Red flood. Under present circumstances, Pure Land practice may not be well suited to the West; nevertheless, its critics among Western Buddhists would do well to ponder the implications of Mahayana philosophy more deeply before dismissing Pure Land teachings, as they sometimes do, as being contrary to the spirit of traditional Buddhism. As Asian Buddhists have always understood, different kinds of people need to make widely different approaches to the same truth. This is possible because one is not dealing with *understanding*, which is to some extent governed by the rules of logic, but with a *practice* that, if properly performed, will achieve results however one may initially conceive of it. A man who presses an electric light switch will succeed in turning on the light, even if he happens to be under the impression that he is switching on the radio.

To return to Kuan Yin. I believe my friend's point about invoking the Bodhisattva embodying wisdom-compassion in a form well suited to a people's cultural traditions was a good one; but, in my opinion, there may also be another reason for the preference given to Kuan Yin by people whose beliefs are generally in line with those of the Pure Land School. The visualisation prescribed for meditators in the Amitayus Sūtra is very difficult to perform in comparison with the popular manner of meditating upon Kuan Yin. When engaged in the sutra-type meditation, one has to build up a complicated picture involving, for example, the mental creation of eight pools flowing into fourteen channels, each with the radiant colours of seven jewels; in each pool there are 600,000 lotuses, each with seven jewels and each possessing a girth twelve times the distance covered by an army in a day's march! Amitābha's height equals that same distance multiplied by the number of sand-grains in *sixty thousand million million* rivers each of them the size of the Ganges! It is written that all this, and very much else besides, has to be visualised as clearly as one sees his hand before his eyes! The mind boggles—which is just what is intended, for, as with Ch'an (Zen) koans, the purpose is to exhaust the mind to the point where it is jerked into a new dimension. When successfully performed, this type of visualisation leads to a sudden transformation of consciousness, thereby opening up new realms hitherto far beyond the uttermost bounds of perception. Still, the task *is* daunting. The simple contemplation of Kuan Yin described in the chapter on meditation may perhaps be less effective, but is certainly better suited to the limited competence of ordinary meditators.

This view of the matter was suggested to me by a pamphlet I discovered in the very embryonic library of the school-room for novices in the Hua T'ing Monastery near Kunming. Locally printed, it was the work of a strange-looking man who occasionally paid us visits. Carelessly dressed, much given to laughter at unexpected moments, he would have impressed me as slightly demented, but for his reputation for wisdom—a great many holy men in China's history are reputed to have made just such an impression of daftness on their more staid contemporaries. Unable to recall more than the general purport of the pamphlet, I have had to fill in the details from my imagination, but my version is faithful in spirit to the original.

It began with some information regarding the writer's identity, birth-place, family and so forth, and then proceeded on the following lines:

"My father and grandfather, Confucian scholars of the old school, looked on Buddhists with disfavour, believing them to delude people with lurid tales of magic. It was through my mother—an unusually highly literate woman from a village near Ta Li—that I came upon the profound doctrine known as 'the voidness of non-void.' Not that she herself had much interest in such profundities. A devoted follower of the Pure Land School, she cared nothing for metaphysics, but it was her custom to buy whatever Buddhist works the pedlar who supplied us happened to bring. As a small child I learned to recite the sacred formula, Hail to Amitābha Buddha! many hundreds of times a day, though always in secret for fear of my father's anger. My mother believed that one-pointed repetition of this formula was enough in itself to ensure liberation from the round of birth and death. Once when my classmates in the middle school heard me softly invoking Amitābha Buddha they jeered at me so heartily that, to win back their esteem, I took to bringing to school Buddhist works which they had to admit would tax the understanding of erudite scholars. Pretending I understood these works myself, I came to study them in all seriousness. I believed then that the Pure Land practice was suitable only for women, peasants and similarly ill-educated people, and had turned instead to such works as The Pure Consciousness Treatise, the Avatamsaka and Lankāvatāra Sūtras. They availed me nothing, their only effect being to disturb my mind, so I returned to invoking Amitābha Buddha, but not without reflecting smugly that my understanding of this practice was now at a 'higher level' than my mother's! How ignorant I must have seemed to her! 'Higher' and 'lower,' 'deep' and 'shallow,' what have these dualisms to do with knowledge, wisdom, understanding?

"Once I went to listen to a lecture on Pure Land contemplation by a famous Tripitaka Master who fired me with ambition to visualise scenes of unimaginable vastness. This, too, got me nowhere. Having painfully built up a huge and glittering background of immense, heavily bejewelled trees and lakes, I had to set about creating images of the Three Holy Ones. No sooner had I started on Amitābha Buddha than the background slipped away; starting on Kuan Yin Bodhisattva, I lost Amitābha; starting on Mahāsthāma Bodhisattva, I lost Kuan Yin. It was all beyond my power. Only conceit hindered me from going back to simple repetition of the sacred formula, which my mother had never for one day abandoned.

"One night I dreamed of being shipwrecked, of clinging to a spar in a furiously raging sea. Mountainous waves curved about me like writhing dragons until, at last, I was cast upon a shore of unearthly beauty. Overlooking the rocky coast, a hill of turquoise rose from a forest of jade that was watered by foaming cascades of milk-white purity. The wings of birds and insects had a jewelled sheen; the spotted deer had coats of white and crimson fur. How could I doubt that I had come upon the seagirt paradise, Potala? Awed, but joyous, I climbed swiftly towards the crest of the hill.

"I had been observed, for a young girl came running down the slope to greet me. Her charming little feet seemed scarcely to touch the rocks over which she sped. When she turned and signalled me to follow her, I had difficulty in keeping up and was irked to notice how torn she was between good manners and an urge to burst out laughing. On our reaching the mouth of a great turquoise cavern, she ran in and soon disappeared from view, leaving me to follow as best I could. We had come to this place by skirting a lake of gold-flecked blue, an arm of which ran into the cave, its blue water hidden beneath masses of pink and white lotus. Though no direct sunlight penetrated beyond the entrance, the cave was illuminated as though by bright sunshine and a delicate fragrance filled the air. In the centre was a throne-shaped rock. Though it had neither cushions nor occupant, I knew it for the Bodhisattva's own and, kneeling, bowed my head to the gleaming silver sand at its foot. As I did so, my name was spoken by a voice as melodious as the tinkling of jade ornaments, the syllables distinct and long drawn out.

"'Cheng-Li, when my vow was uttered many aeons ago, I thought I had made things simple. Why do you *strive*? Let go! The whole Mahayana Canon contains no greater wisdom than the wisdom of letting go. This is also called *dāna*, giving.'

"There came a sweetly joyous laugh, then silence. I knew I was now alone in that shining cave. Already the magical colours were fading into powder-fine coloured sparks that vanished one by one. Darkness followed and, stretching out my hand, I brushed it against the gauze curtains hung around my bed.

"Now I have done with sutras and pious practices. Day and night I recite the Bodhisattva's sacred name, rejoicing in the beauty of its sound. Not for me its recitation in multiples of a hundred and eight, as though it were a duty. Does the runner count his breaths or the poet his words, or the stream its ripples? You sentient beings who seek deliverance, why do you not—let go? When sad, let go of the cause for sadness. When wrathful, let go of the occasion of wrath. When covetous or lustful, let go of the object of desire. From moment to moment, be free of self. Where no self is, there can be no sorrow, no desire; no I to weep, no I to lust, no 'being' to die or be reborn. The winds of circumstance blow across emptiness. Whom can they harm?"

Like many writings of this kind, it concluded with verses conveying the essence of its meaning. I remember that they were beautiful and made much of the magical setting—the gold-flecked lotus pool, the turquoise mountain ringed with a forest of jade leaves and the "dragon-curving waves." The Chinese language lends itself to poetic descriptions of this sort and its monosyllabic character saves the verses from being heavy or ornate. The verses went on to epitomise the qualities pertaining to a Bodhi-mind or heart of compassion and, at the end, came some such lines as:

> Wrathful, banish thought of selfhood;
> Sad, let fall the cause of woe;
> Lustful, shed lust's mental object;
> Win all by simply letting go.

However, I am sure the original concluding verse was a great deal more arresting.

Having received little personal instruction from Tripitaka Masters of the Pure Land School, I am not confident of having grasped the profound inner meaning of its teachings. It does not do to conceive of Kuan Yin and her Potala in the materialistic terms acceptable to the unlettered, who fully expect to behold physical splendours when the Bodhisattva, in response to their frequent invocations, comes to succour them at the moment of death; but nor should one treat the Pure Land sutras as wholly allegorical, or suppose that the Pure Land practice is of value only until the devotee "enters the Potala" in the sense of recognising it to be his own mind purged of obscurations. One must avoid an over-materialistic concept on the one hand and a purely allegorical interpretation on the other. Were you to say that Kuan Yin and her Potala exist objectively, you would be scolded for talking nonsense; but claim that she is wholly a creation of your own mind and you will be taken to task for arrogance or laughingly reminded that the Bodhisattva existed a long time before you were born. Pehaps full understanding is a fruit not to be won without intensive Pure Land practice, for there is certainly no logical solution to the riddle.

A recent exposition of the main practice of the Pure Land Sect—sustained recitation of one of the devotional formulas—is to be found in the writings of the Venerable Hsüan Hua, Abbot of Gold Mountain Monastery, San Francisco. It does not solve the riddle just discussed, for the Master was not elucidating that point, but it does reveal that the purpose of reciting the sacred name is very different from that underlying most of the theistic practices with which it has been erroneously confused. Speaking of the recitation of Amitābha Buddha's name, he says: "If you maintain your recitation morning and night without stopping, you may recite to the point that you do not know you are walking when you walk, you feel no thirst when you are thirsty, you experience no hunger when you are hungry, you do not know you are cold in freezing weather, and you do not feel warmth when you are warm. People and *dharmas* (entities) are empty and you and Amitābha become one. 'Amitābha Buddha is me and I am Amitābha Buddha.' The two cannot be separated. Recite single-mindedly and sincerely without erroneous thoughts. Pay no attention to worldly concerns. When you do not know the time and do not know the day, you may arrive at a miraculous state." He also says: "Day and night we recite the Buddha's name and with each sound we think of Amitābha. The phrase 'Namo' means 'homage.' To whom are we paying homage? Ultimately we pay homage to ourselves! On the day when you entirely forget yourself, the Amitābha of your own nature will appear." These quotations are of course applicable to the similar invocation offered to Kuan Yin, which must not be mistaken for a crude endeavour to win a divinity's favour by flattery, but recognised as a powerful technique for banishing ego-born obscurations and coming face to face with—Mind.

Another illuminating saying by the same Tripitaka Master runs: "As I have told you many times, the Dharma-door of Buddha recitation is false, and so are (those of) dyana (Zen) meditation and the Teaching School, the Vinaya (Discipline) School and the Secret (Esoteric) School. You need only believe in it and false becomes true; if you do not believe, then the true becomes false. . . . Everything is created from mind alone." On the face of it, this seems absurd. How can the false become true just by believing it? Yet to the mystic it makes excellent sense, for he recognises that *any picture* of what lies beyond the range of conceptual thought is bound to be too poor an approximation to have intrinsic worth; therefore all ways of picturing the path and goal are of value only as *convenient stand-ins* for use until direct intuitive perception is attained. Viewing the matter thus, one can more easily understand wherein lies the efficacy of the rites for invoking Kuan Yin.

Yet, even should one concede this point in relation to recitation of the sacred name or of the Dhāranī of Great Compassion, he may have difficulty in accepting what is written in the Lotus Sūtra and also in the sutra expounding that dhāranī concerning Kuan Yin's power to save beings from individual perils such as shipwreck, fire, storm, wild beasts, devils and even litigation. It must of course occur to one that, were these powers real in a literal sense, then a small body of devout believers, of whom there are many, could have stopped the Japanese invasion of China or the subsequent advance of the Red Army by causing weapons to fall to pieces in the soldiers' hands! So, short of rejecting the claims made in those sutras, one is tempted to seek some less literal interpretation.

Do those brave words mean perhaps that calling upon Kuan Yin with one-pointed mind makes one impervious to cold, heat, hunger, thirst, etc., and leads to such total freedom from the bonds of "self," such perfect identification of the individual's mind with Mind, that litigation and death by shipwreck or in a tiger's maw are to be feared no more than dreams? This rationalisation is convenient, but it does not take account of there having been too many instances of people being literally saved from disaster in the nick of time by calling on Kuan Yin for one to be able to discount them all as fabrications. What then? Can it be that absolute faith conjures from within oneself such powerful reserves in the face of danger that seemingly miraculous escapes do in fact occur; or is something more mysterious, more "magical" involved?

Personally I like to think that the inner purport of the passages about Kuan Yin's saving powers may be, in part, as follows. Sustained contemplation of the Bodhisattva as the embodiment of pure compassion inevitably affects the devotee's whole being. Seeking no advantages for himself, delighting to put himself out for others when urged to do so, he comes in some ways to resemble the Taoist sages of old—men so ungreedy, so easily satisfied with simple joys, so loath to take offence or put themselves forward unless pressed, so far removed from every kind of aggressive behaviour and factionalism that they were able to pass their lives in serene obscurity. Attracting no unwelcome attentions from robbers, government authorities or policemen, making no enemies, harbouring no grudges—in short, causing not the least offence to humans, animals or ghosts, they lived from day to day untroubled by savage beasts or extortioners, safe from the prisoner's manacles and strangers to the glittering horror of the executioner's sword. These were the 'Immortals whom ice could not freeze nor sunbeams scorch.' Calamities rarely if ever came their way. But whether this interpretation presents more than a fraction of the sutras' true meaning, I do not know.

In a commentary on the *Dhāranī Sūtra,* the Venerable Hsüan Hua relates a typical story of Kuan Yin's saving power. A certain devotee of hers was passing the night at an inn where the landlord was in the habit of administering drugged wine to the wayfarers who sought his hospitality and stealing into their rooms by night to rob and even murder them. This particular guest, however, was too faithful a Buddhist to touch wine; not being drugged, he awoke to find the landlord advancing upon him knife upraised; but at that very moment came a heavy banging at the front door. Hurriedly withdrawing, the landlord opened up to find a burly policeman-like individual who politely asked him to convey to one of his guests—the very one he had been on the point of murdering—that an old friend of such-and-such a name desired him to drop by in the morning. Tremblingly discarding his fell intention, he delivered this message soon after it was light. The wayfarer, though he gave no sign, had no difficulty in recognising the syllables of the visitor's "name" as a quotation from the Dhāranī of Great Compassion. In other words, the policeman-like individual had either been Kuan Yin herself or a being sent by her to protect a good man who had long made a practice of reciting that dhāranī!

That the Venerable Hsüan Hua should in one place equate the celestial Buddhas and Bodhisattvas with the devotee's own mind and in another relate a factual story of Kuan Yin's saving power well illustrates the difficulty of arriving at a satisfactory conception of the nature of those celestial beings.

Famous Mahayana Parables: Burning House, Prodigal Son

EDWIN A. BURTT

Edwin A. Burtt was the Sage Professor of Philosophy at Cornell University. The parables from this selection reveal the underlying theory of Mahayana Buddhism which is that the Bodhisattva has the goal of leading all beings to emancipation. To accomplish this he or she is motivated by compassion, but must use skillful means to attract human beings away from their hypnotic absorption in the material world. The purpose of the Bodhisattva is to be the leader, the example, and the savior for others. The Mahayana myths of the Bodhisattva show a development over the earliest theories of Buddhism in which the Bodhisattva was apparently less prominent.

As indicated in the introduction to this section, the main purpose of some of the most popular and frequently quoted Mahayana parables was to meet the objection that according to the ancient sutras Buddha taught Theravada doctrine rather than the Mahayana ideal. This purpose is present in each of the following parables, although the point of major emphasis varies. The "burning house" develops the thought that although the Buddha had a far more wonderful truth to give to those who could understand it, he was not deceiving the minds of the beginners by teaching them Theravada ideas. Such teaching was all they could receive and it would lead them in the right direction. In the "prodigal son" it is the fear and distrust of the immature seekers that must gradually and patiently be overcome by the loving Buddha-Father, in order that they may become able fully to respond to his love and share in his compassionate service of others. The "rain cloud" is a beautiful hymn to the grace of the Buddha that gently spreads over all the world and is able to meet the need of every parched living soul, whatever that need may be.

These parables are taken from the Lotus of the Wonderful Law. *This sutra provides the basis for the distinctive philosophy of the Ti'en T'ai School of Chinese Buddhism and the Nichiren School in Japan.*

"The dull, who delight in petty rules,
Who are greedily attached to mortality,
Who have not, under countless Buddhas,
Walked the profound and mystic Way,
Who are harassed by all the sufferings—
To these I (at first) preach Nirvana.
Such is the expedient I employ
To lead them to Buddha-wisdom.
Not yet could I say to them,

'You all shall attain to Buddhahood,'
For the time had not yet arrived.
But now the very time has come
And I must preach the Great Vehicle. . . . "

"Have I not before said that the Buddhas, the World-honoured Ones, with a variety of reasonings, parables and terms, preach the Law as may be expedient, with the aim of final Perfect Enlightenment? All their teachings are for the purpose of transforming [their disciples into] bodhisattvas. But Sariputra! Let me again in a parable make this meaning still more clear, for intelligent people, through a parable, reach understanding."

PARABLE OF THE BURNING HOUSE

"Sariputra! Suppose, in a [certain] kingdom, city, or town, there is a great elder, old and worn, of boundless wealth, and possessing many fields, houses, slaves, and servants. His house is spacious and large, but it has only one door, and many people dwell in it, one hundred, two hundred, or even five hundred in number. Its halls and chambers are decayed and old, its walls crumbling down, the bases of its pillars rotten, the beams and roof-trees toppling and dangerous. On every side, at the same moment, fire suddenly starts and the house is in conflagration. The boys of the elder, say ten, twenty, or even thirty, are in the dwelling. The elder, on seeing this conflagration spring up on every side, is greatly startled and reflects thus: 'Though I am able to get safely out of the gate of this burning house, yet my boys in the burning house are pleasurably absorbed in amusements without apprehension, knowledge, surprise, or fear. Though the fire is pressing upon them and pain and suffering are instant, they do not mind or fear and have no impulse to escape.'

"Sariputra! This elder ponders thus: 'I am strong in my body and arms. Shall I get them out of the house by means of a flower-vessel, or a bench, or a table?' Again he ponders: 'This house has only one gate, which moreover is narrow and small. My children are young, knowing nothing as yet, and attached to their place of play; perchance they will fall into the fire and be burnt. I must speak to them on this dreadful matter [warning them] that the house is burning, and that they must come out instantly lest they are burnt and injured by the fire.' Having reflected thus, according to his thoughts, he calls to his children: 'Come out quickly, all of you!'

"Though their father, in his pity, lures and admonishes with kind words, yet the children, joyfully absorbed in their play, are unwilling to believe him and have neither surprise nor fear, nor any mind to escape; moreover, they do not know what is the fire [he means], or what the house, and what he means by being lost, but only run hither and thither in play, no more than glancing at their father. Then the elder reflects thus: 'This house is burning in a great conflagration. If I and my children do not get out at once, we shall certainly be burnt up by it. Let me now, by some expedient, cause my children to escape this disaster.' Knowing that to which each of his children is predisposed, and all the various attractive playthings and curiosities to which their natures will joyfully respond, the father tells them, saying: '[Here are] rare and precious things for your amusement—if you do not [come] and get them, you will be sorry for it afterwards. So many goat-carts, deer-carts, and bullock-carts are now outside the gate to play with. All of you come quickly out of this burning house, and I will give you whatever you want.' Thereupon the children, hearing of the attractive playthings mentioned by their father, and because they suit their wishes, every one eagerly, each pushing the other, and racing one against another, comes rushing out of the burning house.

"Then the elder, seeing that his children have safely escaped and are all in the square, sits down in the open, no longer embarrassed, but with a mind at ease and ecstatic with joy. Then each of the children says to the father: 'Father! Please now give us those playthings you promised us—goat-carts, deer-carts, and bullock-carts.' Sariputra! Then the elder gives to his children equally each a great cart, lofty and spacious, adorned with all the precious things, surrounded with railed seats, hung with bells on its four sides, and covered with curtains, splendidly decorated also with various rare and precious things, draped with strings of precious stones, hung with garlands of flowers, thickly spread with beautiful mats, and supplied with rosy pillows. It is yoked with white bullocks of pure [white] skin, of handsome appearance, and of great muscular power, which walk with even steps, and with the speed of the wind, and also it has many servants and followers to guard them. Wherefore? Because this great elder is of boundless wealth and all his various storehouses are full to overflowing. So he reflects thus: 'My possessions being boundless, I must not give my children inferior small carts. All these children are my sons, whom I love without partiality. Having such great carts made of the seven precious things, infinite in number, I should with equal mind bestow them on each one without discrimination. Wherefore? Because, were I to give them to the whole nation, these things of mine would not run short—how much less so to my children! Meanwhile, each of the children rides on his great cart, having received that which he had never before had and never expected to have.'

"Sariputra! What is your opinion? Has that elder, in [only] giving great carts of the precious substances to his children equally, been in any way guilty of falsehood?"

"No, World-honoured One!" says Sariputra. "That elder only caused his children to escape the disaster of fire and preserved their bodies alive—he committed no falsity. Why? He thus preserved their bodies alive, and in addition gave them the playthings they obtained; moreover, it was by his expedient that he saved them from that burning house! World-honoured One! Even if that elder did not give them one of the smallest carts, still he is not false. Wherefore? Because that elder from the first formed this intention, 'I will, by an expedient, cause my children to escape.' For this reason he is not false. How much less so seeing that, knowing his own boundless wealth and desiring to benefit his children, he gives them great carts equally!"

"Good! Good!" replies the Buddha to Sariputra. "It is even as you say. Sariputra! The Tathagata is also like this, for he is the Father of all worlds, who has forever entirely ended all fear, despondency, distress, ignorance, and enveloping darkness, and has perfected boundless knowledge, strength, and fearlessness. He is possessed of great supernatural power and wisdom-power, has completely attained the Paramitas[1] of adaptability and wisdom, and is the greatly merciful and greatly compassionate, ever tireless, ever seeking the good, and benefiting all beings. He is born in this triple world, the old decayed burning house, to save all living creatures from the fires of birth, age, disease, death, grief, suffering, foolishness, darkness, and the Three Poisons,[2] and teach them to obtain Perfect Enlightenment. He sees how all living creatures are scorched by the fires of birth, age, disease, death, grief, and sorrow, and suffer all kinds of distress by reason of the five desires and the greed of gain; and how, by reason of the attachments of desire and its pursuits, they now endure much suffering and hereafter will suffer in hell, or as animals or hungry spirits. Even if they are born in a heaven, or amongst men, there are all kinds of sufferings, such as the bitter straits of poverty, the bitterness of parting from loved ones, the bitterness of association with the detestable. Absorbed in these things, all living creatures rejoice and take their pleasure, while they neither apprehend, nor perceive, are neither alarmed, nor fear, and are without satiety, never seeking to escape, but, in the burning house of this triple world are running to and fro, and although they will meet with great suffering, count it not as cause for anxiety.

"Sariputra! The Buddha, having seen this, reflects thus: 'I am the Father of all creatures and must snatch them from suffering and give them the bliss of the infinite, boundless Buddha-wisdom for them to play with.'

"Sariputra! The Tathagata again reflects thus: 'If I only use supernatural power and wisdom, casting aside every tactful method, and extend to all living creatures the wisdom, power, and fearlessness of the Tathagata, the living creatures cannot by this method be saved. Wherefore? As long as all these creatures have not escaped birth, age, disease, death, grief, and suffering, but are being burnt in the burning house of the triple world, how can they understand the Buddha-wisdom?'

"Sariputra! Even as that elder, though with strength in body and arms, yet does not use it, but only by diligent tact, resolutely saves his children from the calamity of the burning house, and then gives each of them great carts adorned with precious things, so is it with the Tathagata. Though he has power and fearlessness, he does not use them, but only by his wise tact does he remove and save all living creatures from the burning house of the triple world, preaching the Three-Vehicles, viz. the Sravaka, Pratyekabuddha, and Buddha vehicles. And thus he speaks to them: 'Ye all! Delight not to dwell in the burning house of the triple world. Do not hanker after its crude forms, sounds, odours, flavours, and contacts. For if, through hankering, ye beget a love of it, then ye will be burnt by it. Get ye out with haste from the triple world and take to the Three-Vehicles, viz, the Sravaka, Pratyekabuddha, and Buddha Vehicles. I now give you my pledge for this, and it will never prove false. Be only diligent and zealous!' By these expedients does the Tathagata lure all creatures forth, and again speaks thus: 'Know ye! All these Three-Vehicles are praised by sages: in them you will be free and independent, without wanting to rely on aught else.

Riding in these Three-Vehicles, by means of perfect faculties, powers, perceptions, ways, concentrations, emancipations, and samadhis,[3] ye will, as a matter of course, be happy and gain infinite peace and joy.'

"Sariputra! If there are living beings who have a spirit of wisdom within and, following the Buddha, the World-honoured One, hear the Law, receive it in faith, and zealously make progress, desiring speedily to escape from the triple world and seeking Nirvana for themselves—this [type] is called the Sravaka-Vehicle, just as some of those children come out of the burning house for the sake of a goat-cart. If there are living beings who, following the Buddha, the World-honoured One, hear the Law, receive it in faith, and zealously make progress, seeking self-gained Wisdom, taking pleasure in becoming good and calm, and deeply versed in the causes and reasons of the laws—this type is called the Pratyekabuddha-Vehicle, just as some of those children come out of the burning house for the sake of a deer-cart. If there are living beings who, following the Buddha, the World-honoured One, hear the Law, receive it in faith, diligently practise and zealously advance, seeking the Complete Wisdom, Buddha-Wisdom, the Natural Wisdom, the Masterless Wisdom, and Tathagata knowledge, powers, and fearlessness, who take pity on and comfort innumerable creatures, benefit devas and men, and save all beings—this type is called the Great Vehicle. Because bodhisattvas seek this vehicle, they are named mahasattvas. They are like those children who come out of the burning house for the sake of a bullock-cart.

"Sariputra! Just as that elder, seeing his children get out of the burning house safely to a place free from fear, and pondering on his immeasurable wealth, gives each of his children a great cart, so also is it with the Tathagata. Being the Father of all living creatures, if he sees infinite thousands of kotis[4] of creatures, by the teaching of the Buddha, escape from the suffering of the triple world, from fearful and perilous paths, and gain the joys of Nirvana, the Tathagata then reflects thus: 'I possess infinite, boundless wisdom, power, fearlessness, and other Law-treasuries of Buddhas. All these living creatures are my children to whom I will equally give the Great Vehicle, so that none will gain an individual Nirvana, but all gain Nirvana by the same Nirvana as the Tathagata. All these living creatures who escape the triple world are given the playthings of Buddhas, viz. concentrations, emancipations, and so forth, all of the same pattern and of one kind, praised by sages and able to produce pure, supreme pleasure.

"Sariputra! Even as that elder at first attracted his children by the three carts, and afterwards gave them only a great cart magnificently adorned with precious things and supremely comfortable, yet that elder is not guilty of falsehood, so also is it with the Tathagata; there is no falsehood in first preaching Three-Vehicles to attract all living creatures and afterwards in saving them by the Great Vehicle only. Wherefore? Because the Tathagata possesses infinite wisdom, power, fearlessness, and the treasury of the laws, and is able to give all living creatures the Great Vehicle Law; but not all are able to receive it. Sariputra! For this reason know that Buddhas, by their adaptability, in the One Buddha-Vehicle define and expound the Three. . . ."

Though Buddhas, the World-honoured,
Convert by expedient methods,
Yet the living they convert
Are indeed all bodhisattvas.
To such as are of little wit,
Deeply attached to desire and passion,
The Buddha, for their sake,
Preaches the Truth about Suffering,
And all the living joyfully
Attain the unprecedented.
The Buddha's Truth about Suffering,
Is real without distinction.
Those living beings who
Know not the root of suffering,
Cling to the cause of suffering,
Unable to leave it a moment.
Again, for the sake of these,
He preaches the Truth with tact (saying):
"The cause of all suffering
Is rooted in desire."
If desire be extinguished,
Suffering has no foothold.
To annihilate suffering
Is called the Third Truth.
For the sake of the Truth of Extinction,
To cultivate oneself in the Way,
Forsaking all ties to suffering,
This is called Emancipation.
From what then have these people
Attained Emancipation?
Merely to depart from the false
They call Emancipation;
But in truth that is not yet
Complete Emancipation.
So Buddha declares: "These people
Have not reached real extinction;
Because these people have not yet
Gained the Supreme Way,
I am unwilling to declare
That they have attained extinction.
I am the King of the Law,
Absolute in regard to the Law,
For comforting all creatures
I appear in the world.
"Sariputra! This,
My final Seal of the Law,
Because of my desire to benefit
All the world, is now announced;
Wherever you may wander,

Do not carelessly proclaim it.
If there be hearers who joyfully
Receive it with deep obeisance,
You may know those people
Are Avinivartaniyah.[5]
If there be any who receive
This Sutra-Law in faith,
These people must already
Have seen Buddhas of past times,
Revered and worshipped them
And listened to this Law.
If there be any who are able
To believe in your preaching,
They must formerly have seen me,
And also have seen you,
And these bhikshu-monks,
As well as these bodhisattvas.
"This Law-Flower Sutra
Is addressed to the truly wise;
Men shallow of knowledge hearing it,
Go astray, not understanding.
All the Sravakas,
And Pratyekabuddhas,
Cannot by their powers
Attain unto this Sutra.
Sariputra!
Even you, into this Sutra
Enter only by faith,
Much more must the other Sravakas.
The other Sravakas,
By believing the Buddha's words,
Obediently follow this Sutra,
Not by wisdom they have of their own."

THE PRODIGAL SON AND THE SEEKING FATHER

"It is like a youth who, on attaining manhood, leaves his father and runs away. For long he dwells in some other country, ten, twenty, or fifty years. The older he grows, the more needy he becomes. Roaming about in all directions to seek clothing and food, he gradually wanders along till he unexpectedly approaches his native country. From the first the father searched for his son, but in vain, and meanwhile has settled in a certain city. His home becomes very rich; his goods and treasures are incalculable; gold, silver, lapis lazuli, corals, amber, crystal, and other gems so increase that his treasuries overflow; many others and slaves has he, retainers and attendants, and countless elephants, horses, carriages, animals to ride, and kine and sheep. His revenues and investments spread to other countries, and his traders and customers are many in the extreme.

"At this time, the poor son, wandering through village after village, and passing through countries and cities, at last reached the city where his father has settled. Always has the father been thinking of his son, yet, though he has been parted from him over fifty years, he has never spoken of the matter to any one, only pondering over it within himself and cherishing regret in his heart, as he reflects: 'Old and worn, I own much wealth; gold, silver, and jewels, granaries and treasuries overflowing; but I have no son. Some day my end will come and my wealth be scattered and lost, for there is no one to whom I can leave it.' Thus does he often think of his son, and earnestly repeats this reflection: 'If I could only get back my son and commit my wealth to him, how contented and happy should I be, with never a further anxiety!'

"World-honoured One! Meanwhile the poor son, hired for wages here and there, unexpectedly arrives at his father's house. Standing by the gate, he sees from afar his father seated on a lion-couch, his feet on a jewelled footstool, revered and surrounded by brahmanas, kshatriyas,[6] and citizens, and with strings of pearls, worth thousands and myraids, adorning his body; attendants and young slaves with white chowries wait upon him right and left; he is covered by a rich canopy from which hang streamers of flowers; perfume is sprinkled on the earth, all kinds of famous flowers are scattered around, and precious things are placed in rows for his acceptance or rejection. Such is his glory, and the honour of his dignity. The poor son, seeing his father possessed of such great power, is seized with fear, regretting that he has come to this place, and secretly reflects thus: 'This must be a king, or some one of royal rank; it is no place for me to obtain anything for the hire of my labour. I had better go to some poor hamlet, where there is a place for letting out my labour, and food and clothing are easier to get. If I tarry here long, I may suffer oppression and forced service.'

"Having reflected thus, he hastens away. Meanwhile, the rich elder on his lion-seat has recognized his son at first sight, and with great joy in his heart has also reflected: 'Now I have some one to whom my treasuries of wealth are to be made over. Always have I been thinking of this my son, with no means of seeing him; but suddenly he himself has come and my longing is satisfied. Though worn with years, I yearn for him as of old.'

"Instantly he dispatches his attendants to pursue him quickly and fetch him back. Thereupon the messengers hasten forth to seize him. The poor son, surprised and scared, loudly cries his complaint: 'I have committed no offence against you; why should I be arrested?' The messengers all the more hasten to lay hold of him and compel him to go back. Thereupon, the poor son, thinking within himself that though he is innocent yet he will be imprisoned, and that now he will surely die, is all the more terrified; he faints away and falls prostrate on the ground. The father, seeing this from afar, sends word to the messengers: 'I have no need for this man. Do not bring him by force. Sprinkle cold water on his face to restore him to consciousness and do not speak to him any further.' Wherefore? The father, knowing that his son's disposition is inferior, knowing that his own lordly position has caused distress to his son, yet convinced that he is his son, tactfully does not say to others: 'This is my son.'

"A messenger says to the son: 'I now set you free; go wherever you will.' The poor son is delighted, thus obtaining the unexpected. He rises from the ground and goes to a poor hamlet in search of food and clothing. Then the elder, desiring to attract his son, sets up a device. Secretly he sends two men, doleful and shabby in appearance, saying—'You go and visit that place and gently say to the poor man—"There is a place for you to work here; you will be given double wages." If the poor man agrees, bring him back and give him work. If he asks what work you wish him to do, then you may say to him—"We will hire you for scavenging, and we both also will work along with you".' Then the two messengers go in search of the poor son and, having found him, place before him the above proposal. Thereupon the poor son, having received his wages beforehand, joins with them in removing a dirt-heap.

"His father, beholding the son, is struck with compassion for, and wonder at, him. Another day he sees at a distance, through a window, his son's figure, gaunt, lean, and doleful, filthy and unclean with dirt and dust; thereupon he takes off his strings of jewels, his soft attire, and ornaments, and puts on a coarse, torn, and dirty garment, smears his body with dust, takes a dust-hod in his right hand, and with an appearance fear-inspiring says to the labourers: 'Get on with your work, don't be lazy.' By such a device he gets near to his son, to whom he afterwards says: 'Aye, my man, you stay and work here, do not go again elsewhere; I will increase your wages; give whatever you need, bowls, utensils, rice, wheat-flour, salt, vinegar, and so on; have no hesitation; besides, there is an old and worn-out servant whom you shall be given if you need him. Be at ease in your mind; I am, as it were, your father; do not be worried again. Wherefore? I am old and advanced in years, but you are young and vigorous; all the time you have been working, you have never been deceitful, lazy, angry or grumbling; I have never seen you, like the other laborourers, with such vices as these. From this time forth you shall be as my own begotten son.'

"Thereupon the elder gives him a new name and calls him a son. The poor son, though he rejoices at this happening, still thinks of himself as a humble hireling. For this reason, during twenty years he continues to be employed in scavenging. After this period, there grows mutual confidence between them, and he goes in and out and is at his ease, though his abode is still the original place.

"World-honoured One! Then the elder becomes ill and, knowing that he will die before long, says to the poor son: 'Now I possess abundance of gold, silver, and precious things, and my granaries and treasuries are full to overflowing. The quantities of these things, and the amounts which should be received and given, I want you to understand in detail. Such is my mind, and you must agree to this my wish. Wherefore? Because, now, I and you are of the same mind. Be increasingly careful so that there be no waste.'

"The poor son accepts his instruction and commands, and becomes acquainted with all the goods, gold, silver, and precious things, as well as all the granaries and treasuries, but has no idea of expecting to inherit so much as a meal, while his abode is still the original place and he is yet unable to abandon his sense of inferiority.

"After a short time has again passed, the father, knowing that his son's ideas have gradually been enlarged, his aspirations developed, and that he despises his previous state of mind, on seeing that his own end is approaching, commands his son to come, and gathers together his relatives, and the kings, ministers, kshatriyas, and citizens. When they are all assembled, he thereupon addresses them saying: 'Now, gentlemen, this is my son, begotten by me. It is over fifty years since, from a certain city, he left me and ran away to endure loneliness and misery. His former name was so-and-so and my name was so-and-so. At that time in that city I sought him sorrowfully. Suddenly in this place I met and regained him. This is really my son and I am really his father. Now all the wealth which I possess entirely belongs to my son, and all my previous disbursements and receipts are known by this son.'

"World-honoured One! When the poor son heard these words of his father, great was his joy at such unexpected news, and thus he thought: 'Without any mind for, or effort on my part, these treasures now come of themselves to me.'

"World-honoured One! The very rich elder is the Tathagata, and we are all as the Buddha's sons. The Tathagata has always declared that we are his sons. World-honoured One! Because of the three sufferings, in the midst of births and deaths we have borne all kinds of torments, being deluded and ignorant and enjoying our attachment to trifles. To-day the World-honoured One has caused us to ponder over and remove the dirt of all diverting discussions of inferior things. In these we have hitherto been diligent to make progress, and have got, as it were, a day's pay for our effort to reach Nirvana. Obtaining this, we greatly rejoiced and were contented, saying to ourselves: 'For our diligence and progress in the Buddha-law, what we have received is ample.' But the World-honoured One, knowing beforehand that our minds were attached to low desires and took delight in inferior things, let us go our own way, and did not discriminate for us, saying: 'You shall yet have possession of the treasure of Tathagata-knowledge.' The World-honoured One, in his tactfulness, told of the Tathagata-wisdom; but we, though following the Buddha and receiving a day's wage of Nirvana, deemed this a sufficient gain, never having a mind to seek after the Great Vehicle. We, also, have declared and expounded the Tathagata-wisdom to bodhisattvas, but in regard to this Great Vehicle we have never had a longing for it. Wherefore? The Buddha, knowing that our minds delighted in inferior things, by his tactfulness taught according to our capacity, but still we did not perceive that we were really Buddha-sons.

"Now we have just realized that the World-honoured One does not grudge even the Buddha-wisdom. Wherefore? From of old we are really sons of Buddha, but have only taken pleasure in minor matters; if we had had a mind to take pleasure in the Great, the Buddha would have preached the Great Vehicle Law to us. At length, in this Sutra, he preaches only the One Vehicle; and though formerly, in the presence of bodhisattvas, he spoke disparagingly of sravakas who were pleased with minor matters, yet the Buddha had in reality been instructing them in the Great Vehicle. Therefore we say that though we had no mind to hope or expect it, yet now the Great Treasure of the King of the Law has of itself come to us, and such things as Buddha-sons should obtain, we have all obtained."

NOTES

[1]Ideal perfections or virtues.
[2]Sensual longing, ignorance, desire for continued separate existence.
[3]Ecstatic contemplation or meditation.
[4]Huge numbers.
[5]Such as never return to the realm of mortal existence.
[6]Members of the second highest caste.

The Great Retirement

HENRY CLARKE WARREN

Translated from the Introduction to the Jātaka

Henry Clarke Warren's work was originally published in Harvard University's classic series, The Sacred Books of the East. *His excellent translations from the original Pali give us the opportunity to appreciate one of the primary sources on which the Buddhist tradition is based. The Buddha's "retirement" or "renunciation" was the turning point or "life crisis" in his life. In the process of working out this life crisis, Guatama (Buddha) discovered the basic principles of Buddhism. "The Great Retirement" is the complex event of the Buddha's decision to embark on his heroic quest.*

Now on a certain day the Future Buddha wished to go to the park, and told his charioteer to make ready the chariot. Accordingly the man brought out a sumptuous and elegant chariot, and adorning it richly, he harnessed to it four state-horses of the Sindhava breed, as white as the petals of the white lotus, and announced to the Future Buddha that everything was ready. And the Future Buddha mounted the chariot, which was like to a palace of the gods, and proceeded towards the park.

"The time for the enlightenment of prince Siddhattha draweth nigh," thought the gods; "we must show him a sign:" and they changed one of their number into a decrepit old man, broken-toothed, gray-haired, crooked and bent of body, leaning on a staff, and trembling, and showed him to the Future Buddha, but so that only he and the charioteer saw him.

Then said the Future Buddha to the charioteer, in the manner related in the Mahāpadāna,—

"Friend, pray, who is this man? Even his hair is not like that of other men." And when he heard the answer, he said, "Shame on birth, since to every one that is born old age must come." And agitated in heart, he thereupon returned and ascended his palace.

"Why has my son returned so quickly?" asked the king.

"Sire, he has seen an old man," was the reply; "and because he has seen an old man, he is about to retire from the world."

"Do you want to kill me, that you say such things? Quickly get ready some plays to be performed before my son. If we can but get him to enjoying pleasure, he will cease to think of retiring from the world." Then the king extended the guard to half a league in each direction.

Again, on a certain day, as the Future Buddha was going to the park, he saw a diseased man whom the gods had fashioned; and having again made inquiry, he returned, agitated in heart, and ascended his palace.

And the king made the same inquiry and gave the same orders as before; and again extending the guard, placed them for three quarters of a league around.

And again on a certain day, as the Future Buddha was going to the park, he saw a dead man whom the gods had fashioned; and having again made inquiry, he returned, agitated in heart, and ascended his palace.

And the king made the same inquiry and gave the same orders as before; and again extending the guard, placed them for a league around.

And again on a certain day, as the Future Buddha was going to the park, he saw a monk, carefully and decently clad, whom the gods had fashioned; and he asked his charioteer, "Pray, who is this man?"

Now although there was no Buddha in the world, and the charioteer had no knowledge of either monks or their good qualities, yet by the power of the gods he was inspired to say, "Sire, this is one who has retired from the world;" and he thereupon proceeded to sound the praises of retirement from the world. The thought of retiring from the world was a pleasing one to the Future Buddha, and this day he went on until he came to the park. The repeaters of the Dīgha, however, say that he went to the park after having seen all the Four Signs on one and the same day.

When he had disported himself there throughout the day, and had bathed in the royal pleasure-tank, he went at sunset and sat down on the royal resting-stone with the intention of adorning himself. Then gathered around him his attendants with diverse-colored cloths, many kinds and styles of ornaments, and with garlands, perfumes, and ointments. At that instant the throne on which Sakka was sitting grew hot. And Sakka, considering who it could be that was desirous of dislodging him, perceived that it was the time of the adornment of a Future Buddha. And addressing Vissakamma, he said,—

"My good Vissakamma, to-night, in the middle watch, prince Siddhattha will go forth on the Great Retirement, and this is his last adorning of himself. Go to the park, and adorn that eminent man with celestial ornaments."

"Very well," said Vissakamma, in assent; and came on the instant, by his superhuman power, into the presence of the Future Buddha. And assuming the guise of a barber, he took from the real barber the turban-cloth, and began to wind it round the Future Buddha's head; but as soon as the future Buddha felt the touch of his hand, he knew that it was no man, but a god.

Now once round his head took up a thousand cloths, and the fold was like to a circlet of precious stones; the second time round took another thousand cloths, and so on, until ten times round had taken up ten thousand cloths. Now let no one think, "How was it possible to use so many cloths on one small head?" for the very largest of them all had only the size of a sāma-creeper blossom, and the others that of Kutumbaka flowers. Thus the Future Buddha's head resembled a kuyyaka blossom twisted about with lotus filaments.

And having adorned himself with great richness,—while adepts in different kinds of tabors and tom-toms were showing their skill, and Brahmans with cries of victory and joy, and bards and poets with propitious words and shouts of praise saluted him,—he mounted his superbly decorated chariot.

At this juncture, Suddhodana the king, having heard that the mother of Rāhula had brought forth a son, sent a messenger, saying, "Announce the glad news to my son."

On hearing the message, the Future Buddha said, "An impediment [rāhula] has been born; a fetter has been born."

"What did my son say?" questioned the king; and when he had heard the answer, he said, "My grandson's name shall be prince Rāhula from this very day."

But the Future Buddha in his splendid chariot entered the city with a pomp and magnificence of glory that enraptured all minds. At the same moment Kisā Gotamī, a virgin of the warrior caste, ascended to the roof of her palace, and beheld the beauty and majesty of the Future Buddha, as he circumambulated the city; and in her pleasure and satisfaction at the sight, she burst forth into this song of joy:—

"Full happy now that mother is,
Full happy now that father is,
Full happy now that woman is,
Who owns this lord so glorious!"

On hearing this, the Future Buddha thought, "In beholding a handsome figure the heart of a mother attains Nirvana, the heart of a father attains Nirvana, the heart of a wife attains Nirvana. This is what she says. But wherein does Nirvana consist?" And to him, whose mind was already averse to passion, the answer came: "When the fire of lust is extinct, that is Nirvana; when the fires of hatred and infatuation are extinct, that is Nirvana; when pride, false belief, and all other passions and torments are extinct, that is Nirvana. She has taught me a good lesson. Certainly, Nirvana is what I am looking for. It behooves me this very day to quit the household life, and to retire from the world in quest of Nirvana.[1] I will send this lady a teacher's fee." And loosening from his neck a pearl necklace worth a hundred thousand pieces of money, he sent it to Kisā Gotamī. And great was her satisfaction at this, for she thought, "Prince Siddhattha has fallen in love with me, and has sent me a present."

And the Future Buddha entered his palace in great splendor, and lay on his couch of state. And straightway richly dressed women, skilled in all manner of dance and song, and beautiful as celestial nymphs, gathered about him with all kinds of musical instruments, and with dance, song, and music they endeavored to please him. But the Future Buddha's aversion to passion did not allow him to take pleasure in the spectacle, and he fell into a brief slumber. And the women, exclaiming, "He for whose sake we should perform has fallen asleep. Of what use is it to weary ourselves any longer?" threw their various instruments on the ground, and lay down. And the lamps fed with sweet-smelling oil continued to burn. And the Future Buddha awoke, and seating himself cross-legged on his couch, perceived these women lying asleep, with their musical instruments scattered about them on the floor,—some with their bodies wet with trickling phlegm and spittle; some grinding their teeth, and muttering and talking in their sleep; some with their mouths open; and some with their dress fallen apart so as plainly to disclose their loathsome nakedness. This great alteration in their appearance still further increased his aversion for sensual pleasures. To him that magnificent apartment, as splendid as the palace of Sakka, began to seem like a cemetery filled with dead bodies impaled and left to rot; and the three modes of existence appeared like houses all ablaze. And breathing forth the solemn utterance, "How oppressive and stifling is it all!" his mind turned ardently to retiring from the world. "It behooves me to go forth on the Great Retirement this very day," said he; and he arose from his couch, and coming near the door, called out,—

"Who's there?"

"Master, it is I, Channa," replied the courtier who had been sleeping with his head on the threshold.[2]

"I wish to go forth on the Great Retirement to-day. Saddle a horse for me."

"Yes, sire." And taking saddle and bridle with him, the courtier started for the stable. There, by the light of lamps fed with sweet-smelling oils, he perceived the mighty steed Kanthaka in his pleasant quarters, under a canopy of cloth beautified with a pattern of jasmine flowers. "This is the one for me to saddle to-day," thought he; and he saddled Kanthaka.

"He is drawing the girth very tight," thought Kanthaka, whilst he was being saddled; "it is not at all as on other days, when I am saddled for rides in the park and the like. It must be that to-day my master wishes to issue forth on the Great Retirement." And in his delight he neighed a loud neigh. And that neigh would have spread through the whole town, had not the gods stopped the sound, and suffered no one to hear it.

Now the Future Buddha, after he had sent Channa on his errand, thought to himself, "I will take just one look at my son;" and, rising from the couch on which he was sitting, he went to the suite of apartments occupied by the mother of Rāhula, and opened the door of her chamber. Within the chamber was burning a lamp fed with sweet-smelling oil, and the mother of Rāhula lay sleeping on a couch strewn deep with jasmine and other flowers, her hand resting on the head of her son. When the Future Buddha reached the threshold, he paused, and gazed at the two from where he stood.

"If I were to raise my wife's hand from off the child's head, and take him up, she would awake, and thus prevent my departure. I will first become a Buddha, and then come back and see my son." So saying, he descended from the palace.

Now that which is said in the Jātaka Commentary, "At that time Rāhula was seven days old," is not found in the other commentaries. Therefore the account above given is to be accepted.

When the Future Buddha had thus descended from the palace, he came near to his horse, and said,—

"My dear Kanthaka, save me now this one night; and then, when thanks to you I have become a Buddha, I will save the world of gods and men." And thereupon he vaulted upon Kanthaka's back.

Now Kanthaka was eighteen cubits long from his neck to his tail, and of corresponding height; he was strong and swift, and white all over like a polished conch-shell. If he neighed or stamped, the sound was so loud as to spread through the whole city; therefore the gods exerted their power, and muffled the sound of his neighing, so that no one heard it; and at every step he took they placed the palms of their hands under his feet.

The Future Buddha rode on the mighty back of the mighty steed, made Channa hold on by the tail, and so arrived at midnight at the great gate of the city.

Now the king, in order that the Future Buddha should not at any time go out of the city without his knowledge, had caused each of the two leaves of the gate to be made so heavy as to need a thousand men to move it. But the Future Buddha had a vigor and a strength that was equal, when reckoned in elephant-power, to the strength of ten thousand million elephants, and, reckoned in man-power, to the strength of a hundred thousand million men.

"If," thought he, "the gate does not open, I will straightway grip tight hold of Kanthaka with my thighs, and, seated as I am on Kanthaka's back, and with Channa holding on by the tail, I will leap up and carry them both with me over the wall, although its height be eighteen cubits."

"If," thought Channa, "the gate is not opened, I will place my master on my shoulder, and tucking Kanthaka under my arm by passing my right hand round him and under his belly, I will leap up and carry them both with me over the wall."

"If," thought Kanthaka, "the gate is not opened, with my master seated as he is on my back, and with Channa holding on by my tail, I will leap up and carry them both with me over the wall."

Now if the gate had not opened, verily one or another of these three persons would have accomplished that whereof he thought; but the divinity that inhabited the gate opened it for them.

At this moment came Māra,[3] with the intention of persuading the Future Buddha to turn back; and standing in the air, he said,—

"Sir, go not forth! For on the seventh day from now the wheel of empire will appear to you, and you shall rule over the four great continents and their two thousand attendant isles. Sir, turn back!"

"Who are you?"

"I am Vasavatti."

"Māra, I knew that the wheel of empire was on the point of appearing to me; but I do not wish for sovereignty. I am about to cause the ten thousand worlds to thunder with my becoming a Buddha."

"I shall catch you," thought Māra, "the very first time you have a lustful, malicious, or unkind thought." And, like an ever-present shadow, he followed after, ever on the watch for some slip.

Thus the Future Buddha, casting away with indifference a universal sovereignty already in his grasp,—spewing it out as if it were but phlegm,—departed from the city in great splendor on the full-moon day of the month Āsāḷhī,[4] when the moon was in Libra. And when he had gone out from the city, he became desirous of looking back at it; but no sooner had the thought arisen in his mind, than the broad earth, seeming to fear lest the Great Being might neglect to perform the act of looking back, split and turned round like a potter's wheel.[5] When the Future Buddha had stood a while facing the city and gazing upon it, and had indicated in that place the spot for the "Shrine of the Turning Back of Kanthaka," he turned Kanthaka in the direction in which he meant to go, and proceeded on his way in great honor and exceeding glory.

For they say the deities bore sixty thousand torches in front of him, and sixty thousand behind him, and sixty thousand on the right hand, and sixty thousand on the left hand. Other deities, standing on the rim of the world, bore torches past all numbering; and still other deities, as well as serpents and birds, accompanied him, and did him homage with heavenly perfumes, garlands, sandal-wood powder, and incense. And the sky was as full of coral flowers as it is of pouring water at the height of the rainy season. Celestial choruses were heard; and on every side bands of music played, some of eight instruments, and some of sixty,—sixty-eight hundred thousand instruments in all. It was as when the storm-clouds thunder on the sea, or when the ocean roars against the Yugandhara rocks.

Advancing in this glory, the Future Buddha in one night passed through three kingdoms, and at the end of thirty leagues he came to the river named Anomā.

But was this as far as the horse could go? Certainly not. For he was able to travel round the world from end to end, as it were round the rim of a wheel lying on its hub, and yet get back before breakfast and eat the food prepared for him. But on this occasion the fragrant garlands and other offerings which the gods and the serpents and the birds threw down upon him from the sky buried him up to his haunches; and as he was obliged to drag his body and cut his way through the tangled mass, he was greatly delayed. Hence it was that he went only thirty leagues.

And the Future Buddha, stopping on the river-bank, said to Channa,—

"What is the name of this river?"

"Sire, its name is Anomā [Illustrious]."

"And my retirement from the world shall also be called Anomā," replied the Future Buddha. Saying this, he gave the signal to his horse with his heel; and the horse sprang over the river, which had a breadth of eight usabhas,[6] and landed on the opposite bank. And the Future Buddha, dismounting and standing on the sandy beach that stretched away like a sheet of silver, said to Channa,—

"My good Channa, take these ornaments and Kanthaka and go home. I am about to retire from the world."

"Sire, I also will retire from the world."

Three times the Future Buddha refused him, saying, "It is not for you to retire from the world. Go now!" and made him take the ornaments and Kanthaka.

Next he thought, "These locks of mine are not suited to a monk; but there is no one fit to cut the hair of a Future Buddha. Therefore I will cut them off myself with my sword." And grasping a simitar with his right hand, he seized his top-knot with his left hand, and cut it off, together with the diadem. His hair thus became two finger-breadths in length, and curling to the right, lay close to his head. As long as he lived it remained of that length, and the beard was proportionate. And never again did he have to cut either hair or beard.

Then the Future Buddha seized hold of his top-knot and diadem, and threw them into the air, saying,—"If I am to become a Buddha, let them stay in the sky; but if not, let them fall to the ground."

The top-knot and jewelled turban mounted for a distance of a league into the air, and there came to a stop. And Sakka, the king of the gods, perceiving them with his divine eye, received them in an appropriate jewelled casket, and established it in the Heaven of the Thirty-three as the "Shrine of the Diadem."

> "His hair he cut, so sweet with many pleasant scents,
> This Chief of Men, and high impelled it towards the sky;
> And there god Vāsava, the god with thousand eyes,
> In golden casket caught it, bowing low his head."

Again the Future Buddha thought, "These garments of mine, made of Benares cloth, are not suited to a monk."

Now the Mahā-Brahma god, Ghatīkāra, who had been a friend of his in the time of the Buddha Kassapa, and whose affection for him had not grown old in the long interval since that Buddha, thought to himself,—

"To-day my friend has gone forth on the Great Retirement. I will bring him the requisites of a monk."

> "Robes, three in all, the bowl for alms,
> The razor, needle, and the belt,
> And water-strainer,—just these eight
> Are needed by th' ecstatic monk."

Taking the above eight requisites of a monk, he gave them to him.

When the Future Buddha had put on this most excellent vesture, the symbol of saintship and of retirement from the world, he dismissed Channa, saying,—

"Channa, go tell my father and my mother from me that I am well."

And Channa did obeisance to the Future Buddha; and keeping his right side towards him, he departed.

But Kanthaka, who had stood listening to the Future Buddha while he was conferring with Channa, was unable to bear his grief at the thought, "I shall never see my master any more." And as he passed out of sight, his heart burst, and he died, and was reborn in the Heaven of the Thirty-three as the god Kanthaka.

At first the grief of Channa had been but single; but now he was oppressed with a second sorrow in the death of Kanthaka, and came weeping and wailing to the city.

NOTES

1The Future Buddha puns upon the word "happy" in Kisā Gotamī's verses. The word in Pāli is *nibbuta*, and is in form a past passive participle of a verb which perhaps does not occur in Pāli in any finite form, but which appears in Sanskrit as *nirvr*. Now there is a Pāli verb of which the third person singular present indicative is *nibbāyati*, and from this verb is formed the verbal noun *nibbána* (Sanskrit, *Nirvāna*). *Nibbuta* is constantly made to do duty as past passive participle to this verb, so that what would be the true form (*nibbāta*) is never found. The Future Buddha therefore puns when he pretends that Kisā Gotamī was using *nibbuta* as the participle of *nibbāyati*, and was urging him to Nirvana.

The verb *nibbāyati* means "to be extinguished," as the flame of a candle; and, when used as a metaphysical term, refers to the fires of lust, desire, etc. And as when fire is extinguished coolness results (a consummation devoutly to be wished in a hot climate like India), the verb acquires the future meaning of "be assuaged," "become happy." And in like manner the verbal noun Nirvana (in Pāli *nibbāna*), meaning both literally and metaphorically "becoming extinguished," comes to stand for the *summum bonum*.

I add a retranslation of the passage, to show the punning meanings given by the Future Buddha to the words, *nibbuta*, *nibbāyati*, and Nirvana:—

"Nirvana hath that mother gained,
Nirvana hath that father gained,
Nirvana hath that woman gained,
Who owns this lord so glorious!"

On hearing this, the Future Buddha thought, "In beholding a handsome form the heart of a mother is made happy (*nibbāyati*), the heart of a father is made happy, the heart of a wife is made happy. This is what she says. But wherein does happiness (*nibbuta*) consist?" And to him whose mind was already averse to passion, the answer came: "When the fire of lust is assuaged (*nibbuta*), that is happiness (*nibbuta*); when the fires of hatred and infatuation are assuaged, that is happiness; when pride, false belief, and all other passions and torments are assuaged, that is happiness. She has taught me a good lesson. Certainly, happiness (Nirvana) is what I am looking for. It behooves me this very day to quit the household life and to retire from the world in quest of happiness. I will send this lady a teacher's fee."

2In India it is customary to hang doors at the height of about two feet from the ground for the sake of coolness and ventilation. The threshold is thus exposed even when the door is shut.

3The Buddhists recognize no real devil. Māra, the ruler of the sixth and highest heaven of sensual pleasure, approaches the nearest to our Satan. He stands for the pleasures of sense, and hence is The Buddha's natural enemy.

4About the first of July.

5I think the conception here is that a round portion of the earth, on which the Future Buddha stood, turned around like a modern railroad turn-table, thus detaching itself from the rest and turning the Future Buddha with it.

6An usabha is 140 cubits.

Unit VII

From Id to the Ego in the Orient: Kundalini Yoga Part I

The Way and the Apparent Eroticism of Tantrism

HERBERT V. GUENTHER

Herbert V. Guenther is a well-known authority on Tibetan Buddhism and a specialist in Buddhist Tantra. In this selection, he discusses a philosophy of sexuality that he feels is different from that of the West. Tantric views of sexuality are not bound by the reticence and disdain that characterize traditional Western views of eroticism. Tantrism was able to affirm the value of sexuality even in the framework of spiritual transcendence. Guenther offers a useful introduction to the refined erotic theory of Tantric Buddhism and perspective on the difference between Eastern and Western views of the subject. It is not uncommon in Indian discussions of eroticism to see the subject as an "heroic encounter" between lovers. The pursuit of erotic goals is in the universal mythic category of the Indian quest for freedom, perfection, and ultimate happiness.

The attempt to resolve the tension that exists between the feeling of frustration and the sense of fulfilment, between the fictions about man's being and the awareness of his Being, is termed "the Way." It is not an inert rod lying between two points, nor is it the favouring of one side in the dilemma that constitutes the human situation, but grounded in Being it is an exercise of regaining and staying with Being. In other words, it is the actualization of intrinsic awareness, Mind-as-such (*semsnyid*), together with or inseparable from value-being (*chos-kyi-sku*). As this is not the same as the ideas we may have about it, the "Way" is summarized in the statement:

> Free from the concepts of *maṇḍala* and (*gaṇa*) *cakra*,
> Of Karmamudrā and Jñānamudrā.

Padma dkap-po explains *maṇḍala* as the "bearer" (*rten*) of this or that psychic activity, *gaṇa-cakra* (*brten*), manifesting itself as "divine" forces (*lha*); Karmamudrā as a woman (*mo*) who yields pleasure containing the seed of frustration; and Jñānamudrā as a woman who yields a purer, though unstable, pleasure. He goes on to say:

> By attending to these facets alone we may be able to reach the Akaniṣṭha realm, the ultimate in sensuousness, but not the absolute, because not free from concretizations, we convert (the real) into un-knowing.

Obviously, our conceptualizations and concretizations of some pleasurable experience may provide a temporary escape, but an escape into sentimentality is not the solution of man's burning problem to find himself. In the same way, an intellectually induced suspension of all mental activity is no answer; nor is the problem solved by an essentially intellectual negativism, as advocated by the Prāsangikas. Sentimentality is compassion divorced from understanding, and the open dimension of Being divorced from all feeling becomes negativism. Therefore Saraha said:

> He who becomes involved with openness without compassion
> Will never set forth on the most excellent path.
> So also by attending to compassion alone
> He will stay in Saṃsāra, but not become free.

Against such one-sided efforts the following statement is directed:

> Do not negate, do not suspend (the mental working), do not find fault,
> Do not fix (the mind on something), do not evaluate, but just let be.

In other words, the way is not travelled by abrogating the ability to think, by destroying the inner continuity of one's being and by introducing a division where there is none, but by preserving the unitary character of Being. Again we may quote Saraha:

> He who can combine both (compassion and openness of Being),
> Stays neither in Saṁsāra nor in Nirvāṇa.

Moreover, apart from Being there is no other being that can serve as a way:

> Friend, since words falsify, give up this infatuation,
> And to whatever you become attached, give that up, too.
> Once you understand (the real), all turns out to be It;
> Nobody knows anything else but this.

But it is the tendency of our un-knowing to look for our Being where it cannot be. So Saraha declares:

> Where it is present
> There we do not see it.
> Still, the doctrinaires all explain the texts
> But do not understand the Buddha to be in (their) body.

Karma Phrin-las-pa explains this verse as referring to the "togetherness-awareness" that is present in and with every individual but is not recognized as such by him who is involved with his propositions. Such an individual, therefore, is unable to see Being as it is, but by looking outward he tries to understand what actually is within him. It is in his own body, speech, and mind, that the individual must understand Buddhahood to reside, though not in the manner of the body being a container, but as the representation, the embodiment of Buddhahood. Due to the fact that our concrete existence is an intricate pattern of interacting forces, not only can it be viewed from different angles, but even more so experienced on different levels, and since our individual life is our "Way," at every step it partakes of ritual and imagery. This can be seen from the following verse by Saraha and its explanation by Karma Phrin-las-pa:

> By eating and drinking and by enjoying copulation
> Forever and everywhere one fills the rounds.
> Thereby the world beyond is reached,
> And one goes away having crushed the head of infatuation under one's feet.

Karma Phrin-las-pa's interpretation is based on the importance which the tactile sense has for the relationship between man and his outer and inner environment and their corresponding evaluation as well as on the significance of aesthetic perception. We must never forget that man is in the world in the sense that it is through his body that there is for him the corporeality of things, and the interaction between the environment and its impressions on the tactile organs or the body-surface induces sensations of change and intensity in our physical condition. At the same time this experience of materiality and thereby of an objective reality gives way to a visual world picture which is much wider than the limitations imposed by the purely tactile experience, and this is meant by the use of the word "beyond," which must never be understood as the impossibility of there being a world other than the one we experience. However, there are wide-spread ramifications of the tactile sense, and the corresponding world experiences interlace man with the world or nature on the one hand and with the physiological side of his being on the other. It is this interlacing pattern with various focal points that is termed *risa*, which we can best translate by "pattern," "structure" and, in specific localizations as "focal points of experience." Because of the importance of the tactile sense which gives us immediate contact with the world surrounding us, and because of the fact that we are embodied beings, the cognition that is most highly valued is the aesthetic one, not the one that through its association with concepts introduces the painful separation of object and subject. Togetherness and separateness can best be illustrated by a reference to the place a work of art, particularly sculpture, has in either framework. While the conceptual framework was responsible, as far as our Western tradition is concerned, for removing the work of art from the space and time of our experience and locating it in an ideal space, thereby enabling the spectator to look at it coldly from a distance, in aesthetic perception the work of art remains alive; it calls out to be felt and touched, and each part of it is perceived as if it were for the moment all of the world, unique, desirable, perfect, not needing something other than itself in order to be itself. In this experience there is the warmth of closeness, not the coldness of distance. Instead of disrupting the unity of Being by separating and downgrading the instinctive side, as represented by the tactile experience, from the perceptual side which then becomes over-evaluated conceptually, the Tantric 'Way' attempts to preserve this unity of sensuousness and spirituality, the latter being essentially the former's value, by clarifying the various aspects. It is in this light that Karma Phrin-las-pa gives different interpretations of the above quoted verse by Saraha:

> Discussing the problem objectively: Having received the (necessary) empowerments, (the person) eats the meat (prepared for) the assemblage and drinks the beer (or other alcoholic beverage). Then he unites with his partner having the appropriate characteristics, by developing three ideas. In the act of the rubbing together of the two organs he concretely fixates and preserves the origination of the four kinds of delight in an ascending or descending manner as taught by the Guru, and thus forever fills the four focal points in his (existential) pattern by making the (*bodhicitta*) move downward or by forcing it to move upward. By such an experience he reaches a world-transcending Buddhahood experience. Stepping on the head of the worldly people who, not having received the empowerments, are deluded about the maturing effect and, not having received guidance, are deluded about the instruction, one crushes this delusion by (the above) non-delusive method and reaches the level of Buddhahood.
>
> Discussing the problem in terms of a subjective experience: He who follows the Mantrayoga, eats and drinks the five kinds of nectar in (what is a mixture of) the pure and impure; he unites the motility in (his existential) patterns with the *bodhicitta* and he steadies in his being the awareness of the four kinds of delight due to attending to the process of unification. Continuously attending to this experience he fills the focal points, that is, the pure in his body, with the awareness of absolute bliss. Thereby he attains a Nirvāṇa beyond this world. Stepping on the head of those who are deluded about the Mantrayāna method and crushing this delusion, he goes to a place superior to their status.
>
> Discussing the problem in terms of a mystical experience: 'Eating' means to know the world of appearance to be mind, through instruction in the meaning of "memory;" "drinking" means to know mind to be open, through instruction in the meaning of "non-memory;" through instruction in the meaning of 'unorigination' appearance and mind meet in one flavour and become united; and through instruction in the meaning of "transcendence," the self-validating intrinsic awareness rises as spontaneous joy; and by experiencing the ineffable, one forever and everywhere fills one's noetic being with original awareness, through an instruction which is like an uninterruptedly on-going effort; through this experience he goes to the world beyond.
>
> Discussing the problem from the viewpoint of ultimate Being: A follower of the Mahāmudrā teaching takes as his food the world of appearance rising incessantly in splendour, and has for his drink the open dimension (of Being) merging in the absoluteness of Being. By experiencing the unity and inseparability of appearance and openness of Being he is immediately aware with unsurpassable joy. Forever and everywhere making this experience in the above gradation he fills the rounds, i.e., the world of the knowable or the whole of appearance and possibility, with a spontaneous original awareness, and by this (feeling of) unity he goes to the world beyond.

This fourfold interpretation represents a growing awareness as a continuous process, in which ideas act as functions of unification rather than as separating agents. This, of course, places a different connotation on our concept of ideas which is mainly an instrument for perpetuating the gulf between subject and object and for preventing man from penetrating to his Being which is possible only through experience. According to the above fourfold interpretation, the experience "A" is understood by the experience "B," since "B" is of a higher order than "A." To speak, in the last analysis, of an identification of the cognizer with the cognized is another instance of "misplaced concreteness." What happens is the emergence of the feeling of unity. The idea as a vehicle of unification is indicated by gNyis-med Avadhūtipa who, in commenting on the first part of Saraha's verse, explains Karma Phrin-las-pa's cryptic "Three ideas." They are the idea that the body is a god, speech a mantra, and mind absolute Being. To see the body, by which the body as lived by me is meant, as a god is to appreciate it as a value in its own right; similarly speech as mantra is not empty talk, rather it is communication which does not depend on words with their conventional meaning in usage. Lastly, mind as absolute Being is not the absolutization of subjectivism, it is rather the cognitiveness of Being-as-such which expresses itself in and through the activity of our Mind.

Throughout Tantrism reference is made to the body as lived by me, perceiving, moving, acting, and so on. Taking this reference as our clue we can say that sexuality is itself a mode of being of the person in question, and is concretely interpreted in the stream of lived experience. A human being, whether man or woman, is in this world with his or her body and the body discloses itself as meaningful in its attitudes, gestures, and actions. As an embodied being man is embodied with a certain sex, and the sexuality of the body manifests itself in a variety of manners so that it is justifiable to say that sexuality expresses a human being's existence in the same way as his existence expresses his sexuality. Thus, if the body expresses Existence, it does so because the body actualizes it, and at the same time is its actualization. In other words, the body is not something external to my existence, but is its concrete realization and hence both "expression" and "the expressed." Another point to be noted is that the body discloses itself to my experience as being *mine* and somehow belonging to me who "lives" it. At the same time it is peculiarly ambiguous, and this ambiguity may be stated as follows: That body over there is simultaneously a woman herself and not herself; her sex presents me with her, and she as embodied presents me with her sex. In the same way, this body here is simultaneously a man himself and not himself; his sex presents her with himself, and he as embodied presents her with his sex. In terms of subject and object, each individual is both subject and object, but the individual is object in a special way, both for himself, as when I speak of *my* body, and for others as a mere body (to be manipulated and controlled). Although human beings are male and female and although sexuality is coextensive with life, sexuality cannot be reduced to Being-as-such, nor can the latter be reduced to sexuality. Hence sexuality is the dialectic of lived experience, in which I apprehend the other as subject or, to put it more cautiously, in which I should apprehend the other as a subject, which means to recognize the intrinsic value of the other, as indicated by the statement that in the realm of lived experience men and women are gods and goddesses. The failure to grasp the meaning of 'Being,' of 'body' and 'sexuality' has resulted in a thorough misunderstanding of Buddhist Tantrism. This is mainly due to the difference of 'climate' contributing to the development of ideas. Western civilization derives from the early Mediterranean slave societies with their attendant postulate of a celestial lawgiver who "legislates" for both human beings and non-human natural phenomena, and who "owns" the human beings as his chattels, just as a shepherd owns his flock and takes up an active attitude of command. Pastoral dominance, on the one hand, and among seafaring people, the unquestioning obedience to the one in command of a ship, on the other, greatly assisted the development of a "dominance" psychology which attempts to rationalize the crave for power, domination, and control. It aims not only at turning the other into an object to be used or misused at will, but also at making the other feel as an object in the eyes of the master or postulated super-power. This is, of course, impossible because an object, a slave, cannot give the recognition sought for by the master as only a subject can do so, and it is precisely the individual's subject character that the master cannot tolerate and that he attempts to deny. Inasmuch as Hindu Tantrism has been deeply influenced by the dominance psychology of the Sāṃkhya system, professing a dualism of *puruṣa* who is male, and of *prakṛti* who is female and who dances or stops dancing at the bidding of the Lord or *puruṣa*, this purely Hinduistic power mentality, so similar to the Western dominance psychology, was generalized and applied to all forms of Tantrism by writers who did not see or, due to their being steeped so much in dominance psychology, could not understand that the desire to realize Being is not the same as the craving for power. Hence Tantrism was equated with 'power.' And since *puruṣa* and *prakṛti* involved a sexual symbolism, which was concretized in the sense that the sexual act was the proof of one's masculinity, the paranoid Western conception about Tantrism resulted: it is the paranoid who is obsessed with his sexual potency and attempts to compel the object to come towards him (the *prakṛti* dances at the bidding of the *puruṣa*). He tries to make the other (*the* woman) responsible for the action of satisfying his needs. At the same time he identifies himself with his sexuality, and this identification becomes the basis for his idea of power, preferably of "omnipotence."

It is a fact that any dominance psychology inevitably destroys the individual as subject. Its dehumanizing force was keenly felt by those brought up in the Western world and so they turned to the "mysterious" East which was supposed to hold the key to their acquiring the powers that the Western institutions denied them. But exchange of one kind of dominance for another does not lead to the realization of Being; it remains a slave's dream of becoming a master.

There is another area in which the destruction of the individual as a living being is deeply felt, and where traditional Western religion fails and has always failed us. This is the feeling of sex, rigorously excluded from the realm of speech and thought and frowned upon in deeds. This exclusion, too, has a long history and is inextricably tied up with the contempt for and fear of the body. The official attitude has been and is in favour of continence, abstinence and asceticism having their root in fear, and while contempt might assist the official attitude, it more often has been in opposition to it, particularly in its aspect of defiance. Libertinism did not appear under the auspices of communion and joy, but under those of arrogance and contempt. To suffer from an obsessive fear of the body is perhaps not so different from a compulsive addiction to sex, be these addicts virility-provers or seductiveness-provers. The important point to note is that in all these cases sex is confined to only one dimension, sensual pleasure and exploitation, but the aesthetic experience of joy and through it the enrichment of one's Being is missed. The use of sex as an instrument of power distorts its function. Instead of being an experience of feeling for the partner, it becomes a manoeuvre to establish one's imaginary superiority. A man who sees himself as a sexual object will imagine himself as "the great lover," and a woman who sees herself as a sexual object believes in the irresistibility of her sex appeal. Both may feel repelled by their body, but they are convinced of its power.

Tantrism certainly is not on the side of asceticism, but it would be wrong to conclude that therefore it must of necessity advocate libertinism and that its appeal to Western man, reared in an atmosphere hostile to women, pleasure, and life, is due to the fact that Tantrism approves of women and of sex and, by implication, can serve as the moral justification for the sex addict's compulsion. It is true, Tantrism recognizes pleasure as valuable and positive, but much more than mere pleasure-seeking is involved. It is equally true that in its Hinduistic form it combines power with pleasure which essentially is appreciation and is meant to lead to aesthetic enjoyment, and so has a positive content, unlike Christianity which advocates the impotence of man, denounces pleasure and condemns its source, woman. Buddhist Tantrism dispenses with the idea of power, in which it sees a remnant of subjectivistic philosophy, and even goes beyond mere pleasure to the enjoyment of being and of enlightenment unattainable without woman.

> How can enlightenment be attained in this bodily existence
> Without thine incessant love, o lovely young girl?

Enlightenment is the name for a change in perspective, and Tantrism is the practical way of bringing about this change. It does not mean that in this change of perspective something is seen that others cannot see, but that things and, above all, persons are seen differently. This is clearly stated by Padma dkar-po:

> In attending to the vision of (seeing himself as) a god (*lha*) the yogi apprehends, not incorrectly, what ordinarily appears before his eyes; however, mentally he takes a firm pride in his being a god (*lha'i sku*). This is termed *adhiṣṭhānayoga*. In this term, *adhi* means "superior," and *ṣṭhāna* means "arrangement," "accomplishment," "adornment," hence "to be graced." A "superior accomplishment" is termed "superior feeling of reverence" from the root *adhimuñc*.

To see oneself as a god is to be aware of one's existence as valuable and as good; it is not deifying one's shortcomings which are the products of a limited and selfish vision and hence negative and evil. The emotional quality of this value-perception is the "feeling of reverence" which is not contradictory to exaltation or pride; the negative counterpart to pride is arrogance and to reverence, contempt and self-debasement. The attempt to see oneself as a god is not yet to be enlightened, but it enfeebles the negative view one takes of oneself. Only by taking a positive view of oneself can one truly be. The experience of really being is not only felt as blissful but is also an identity-experience. Here man has found himself and is no longer a "thing-of-and-in-the-world." In order to find himself man needs the "other" who is no intellectual abstraction, but part of himself, needed in order to be himself. Sahajayoginī Cintā, speaking of the state when one spontaneously is oneself, says:

> Here, in spontaneity which is non-dual and naturally pure, one's Being (*bdag-nyid*), in order that one may understand one's Being, manifests itself in the shape of man and woman.

The concrete "other" person, for me, is whoever enters my life-world and whom I accept as one accepting me in order to accept me or her as one who is willing to accept me as one accepting myself. This complex situation of the interaction of man and woman is termed Karmamudrā and Jñānamudrā, the one referring to the "without," the other to the "within," each of them representing an "encounter" that changes both partners.

Swāmi Muktānanda and the Enlightenment Through Śakti-Pāt

CHARLES S. J. WHITE

Charles S. J. White is a specialist in Hinduism and professor of religion at the American University. This selection illustrates the way that Tantric ideas and practices have been brought to bear on an interesting Hindu revitalization movement, created by a contemporary "saint" of Hinduism named Swami Muktananda. The version of Hinduism taught by Muktananda is called "Shakti-Pat" or the Way of Shakti. The methods employed by Muktananda in "awakening the Kundalini" of his disciples are supposed to be typical of Tantric masters' use of Siddhis or acquired paranormal powers. The appeal of Muktananda's "Way," especially during his lifetime, extended to Westerners as well as Indians.

The main burden of proof lying upon the cult of the Indian saint is in the area of psychic claims, as representing the unusual state in which the saint finds himself when he becomes "divine." The following passage from a biography of Sathya Sāi Bābā illustrates what we mean: "His [Sathya Sāi Bābā's] Omniscience and Omnipresence are revealed to everyone who meets Him in the Interview Room. He tells the visiting devotee what he has said, has done or has felt; to whom he has spoken, and on what; what he has feared and plotted, suffered and lost. If you want to consult Him on ten points, He will have answered them and more even before you ask! He might reveal what you actually experienced in your dreams, repeating the very words which in the dream you had heard Him say. He may even lay bare your history down to the minutest detail, and where there was sorrow and weakness, He will replace it with joy and strength."

Obviously, it is this kind of "proof" that draws to the feet of the saints the masses of Indians and, with the convenience of fast aerial transportation, increasing numbers of foreigners as well. In what follows I shall be discussing a relatively new cult in Maharashtra: that of the saint, Swāmi Muktānanda of Ganeshpuri—a small village about two or three hours distant from Bombay by car. As will be perceived, the saint who resides there is held to have achieved remarkable development, but his cult is different in most particulars from what I had earlier grasped about the saints stemming from Sāi Bābā of Shirdi. My study of Muktānanda had been inspired originally by the hope of enlarging the scope of analysis of detailed interconnections in the Sāi Bābā Movement; but I learned that there are a number of Maharashtrian saints, similar in some ways to Sāi Bābā of Shirdi (Muktānanda among them), whose practices derive from tantric sources. This could not be said, so specifically I believe, of the Sāi Bābā group.

It was interesting to discover, nevertheless, that the famous Sāi Bābā had inspired a number of his own disciples after his own death, through dreams, to go for help to a saint called Swāmi Nityānanda, the preceptor of Muktānanda. For example, I was told about a neighbor of Amma (the principal devotee of Muktānanda) who some years ago had developed gangrene in his leg from diabetes. In a dream, his wife saw Sāi Bābā of Shirdi, who told her to go to Ganeshpuri to a temple where Swāmi Nityānanda was accustomed to bathe in the naturally hot spring water of the tank and to follow his example. The man went with his wife, bathed as ordered, and, although the doctors had been preparing to amputate the leg, was healed. Such instances help establish a tie, however tenuous, between the two saintly *gaddis* of Sāi Bābā and Nityānanda.

DEFINITION

The Saint—The English word "saint" comes from the Latin *sanctus*, which in turn derives from a verb meaning "to make sacred, ordain, or establish." A saint, thus by definition, is related to the sacred or holy and exhibits to the world the same qualities associated with terms often used to epitomize the religious dimension of life in events and places. The saint is a focus of the sacred. To combine some of the terminologies accredited to the sacred or holy, we might expect that the saint would have unusual power and compelling attributes of the *mysterium tremendum et fascinans* in addition to permeating an environment with numinous qualities. The question remains, however, whether the term "saint" itself is sufficiently precise to exclude other possibilities. It seems likely that many religious figures—the priest, the prophet, the seer, the healer, not to mention the sorcerer, the medium or the magician, and the shaman and medicine man—have some characteristics in common with "saint" so defined. Moreover, we are dealing with a term that has traditional meanings in the Christian West.

The care which the Roman Catholic church has taken to safeguard the official processes leading to the recognition of sainthood among its members is well known. Yet the bases for final judgment are remarkably similar to those which act, it would appear, "automatically" in the appointment of Hindu saints. Take, for example, the question of miracles. The following incident is recorded in connection with Saint Thérèse of Lisieux shortly after her death:

> Amongst the youthful and critical seminarists of Bayeux . . . was a certain ill young abbé Anne, whose doctors diagnosed galloping consumption and gave him at most a few more days to live. A *neuvaine* to Thérèse was instantly begun (and this alone shows how high her reputation already stood), but apparently to no avail: after some days of intercession it was thought that the end would come during the night. Forewarned of the probability, the young man, who, for all his piety, had no wish to die, pressed a relic of the nun to his heart and passionately invoked her in silent prayer, saying that although he felt sure she was in Heaven, he was on earth where much work remained

> to be done, and that, in a word, she must cure him. An extraordinary change visibly taking place without delay in the patient, the doctors were summoned in haste and to their astonishment were obliged to declare that they found him completely restored to health. Lest any exaggeration should be suspected, it is as well to give the statement in the words of Lisieux' official chronicler in a work crowned by the French Academy, "The destroyed and ravaged lungs had been replaced by new lungs, carrying out their normal functions and about to revive the entire organism. A slight emaciation persists, which will disappear within a few days under a regularly assimilated diet." The miracle was so well authenticated, owing to the attendance and the testimony of the doctors, that it was later taken as one of the two test cases demanded.

For the cult of Thérèse, miracles substantiate the efficacy of her sainthood; the religiously interested investigator is drawn to considerations of the deeper meaning of the milieu in which such events are recorded as well as to reflect upon the development of religious practices or theology on the basis of such happenings.

In thinking about the situation in Hinduism, one observes that there the line grows indistinct at times between the gods and the saints. In the measure that the cult theory allows, most saints are considered to be divine and often receive public worship in the manner of divinities. Indeed, it is not beyond possibility that such great gods of Hinduism as Kṛṣṇa and Rāma in part arose from historical figures. The virtual deification of Gautama Sākyamuni is another striking example from India, and we notice parallels in the cases of the popular divinities of the Chinese.

No doubt the Christian church, at times the fierce guardian of the unique divinity of Jesus Christ, would have to provide both for the extraordinary phenomena that were identified with canonized persons and at the same time maintain the theological distinctions among saints in general, the Blessed Virgin Mary, and the Second Person of the Trinity. In Hinduism there is a good deal of theological justification for the divinization of historical persons. From Vedic statements onward, the theological expressions of Hinduism reinforce the concept that God or some profound ontological principle very deeply penetrates the manifested world. Popular stories repeat the theme that everything is God, as is everyone—although in the latter case, it is only with considerable effort that one becomes a realized person, that is, self-consciously divine.

It seems clear that a wide-ranging structure must encompass the concept of saint even when necessary distinctions are made. The saint is always an exemplary figure, and his status is never merely given in the normal course of religious life. The appearing of a saint is an unusual event, and the establishment of the claim of being one, however much longed for in the religious society, is subject to controls. These controls are such that full acceptance is delayed until the phenomenon can be completely grasped after the saint's death, even though preliminary acknowledgment is often given. The latter is due to the fact that the demands of the religious society upon these individuals are so pressing that they cannot be postponed once some evidence is available. Moreover, the term "saint," as contrasted with any other designations of religious proficiency, must refer to the most exalted capacity of all, recognizable in human, historical terms and in the religious cultures where human sensibility can express itself with the greatest degree of clarity and to a universal constituency. However, the term "saint" may not always be the most acceptable one to that constituency if it should prefer a definition indicating the presence of a divine being. It would appear to be difficult to maintain a functional difference between two such categories (as saint and god) if other criteria are more or less harmonious. Even though some religions must attempt to maintain such a distinction, we have noted, this need not cause theological problems for Hinduism and some other Eastern religions.

PHENOMENAL ASPECTS

As we have previously discussed, the saints of the Sāi Bābā Movement were a homogeneous group to the extent that they appeared to share certain elements in common in their cult theory or activity. For example, the god Dattātreya, familiar in the cult of saints in Maharashtra, figured in some aspects of the cult of Shirdi Sāi Bābā, Upasani Bābā, and Mātā Godāvarī (the first three in this series), but not in that of the fourth saint, Sathya Sāi Bābā. On the other hand Shirdi Sāi Bābā and Sathya Sāi Bābā made use of sacred ash, *vibhuti,* in the performance of miracles although the other two saints did not; and this practice linked them to the Kānphaṭa yogīs who have as their patron Dattātreya—and through them, at least in the case of Sāi Bābā of Shirdi, to the Muslim saint tradition. Thus we thought that a set of tautologous relations binds the members of this group into a kind of system, all of whose elements we cannot review here.

It is true that the saints of a given region are united by the many accidental circumstances of cultural life such as a common language, a lineage going back through hierarchies of preceptors (sometimes to medieval or ancient times), and a sectarian milieu. The homogeneity of Indian cultural life is another factor that must be taken into account in spite of regional differences, so that the scope of possible forces at work on any particular example of sainthood must include regional as well as national elements. This panoply of influences and connections becomes staggering to contemplate but no more so than that in other religious expressions whose structures we only glimpse with present techniques.

I first learned about Swāmi Muktānanda through a letter received from a friend (a New Zealander, a Cambridge graduate, and a candidate for the Ph.D. degree in an Indian university) who had received Śakti Pát initiation:

> One becomes aware that one has received the dīkshā in several ways, according to one's prārabdha. The Kuṇḍalinī shakti is awakened and immediately starts to work: one's body may shake violently, one may laugh or cry, get overjoyed or very depressed, develop a temporary illness, start performing yoga āsanas automatically, chant mantras (even without ever

> having known a word of Sanskrit—in one case I heard of, the man, an Englishman, actually wrote down Sanskrit shlokas tho he knew no Sanskrit, not even Devanagari), goes into trances, dances, sees visions, hears locutions, etc. It varies with everyone. In my case my body started to shake uncontrollably; I felt great upsurges of bhakti so that I felt as tho I had fallen in love, then I felt like a woman who had just realised she was pregnant with a new and divine life, but above all I felt a happiness I had not known in my whole life. Since that day I have never been depressed as I used to be. I have full confidence and hope for the future. After a few weeks I started to do a lot of very intense prāṇāyāma which gradually subsided into long kumbhakas. I found myself adopting some yogic mudras. One day I stood up and danced slowly and gracefully. All these things are called Kriyas and their onset and process is called nādī-shuddhi. Everything that happens, no matter how difficult it may seem (like breath retention) is to be allowed to run its course for they are directed by the shakti, the cit-shakti, who is supremely intelligent and forces one's body to do just those yogic and other exercises which are necessary for the purification of the sthūla sharīra. Purified, the cit-shakti moves inward and starts to operate on the level of the sūkshma-sharīra, according to one's prārabdha—the will of God—giving rise to visions, the hearing of the divya nāda dhvani, and other psychic phenomena and powers.

I subsequently returned to India and was able to gather materials for a preliminary description of Muktānanda's history and present activities. I have also taken this occasion to speculate on the noumenal area of his experience and influence. What follows is presented somewhat schematically to give a sense of the structure that one might expect to be able to extrapolate, for example, in a comparative study together with other similar data.

1. Muktānanda was born into a wealthy family in the vicinity of Mangalore in Mysore State. His mother was a pious woman who had made a pilgrimage to Dharmasthala to invoke the aid of a form of Śiva, known as Manjunāth Mahādev, to have a son. She was told afterward by a Sadhu to repeat the mantra, Om Nāmaḥ Śivāya, and her prayer would be granted. (The same mantra is used by the devotees of Muktānanda in the ashram at Ganespuri today to assist in the work of arousing the Kuṇḍalinī.)
2. The mother was taken in labor while she was at the washbasin under a tree in the compound of her house. The baby emerged so swiftly that she did not have time to catch it before it fell into the washbasin. The time was dawn on May 16, 1908, the full moon day of the month of Vaiśākh: very auspicious. They named him Kṛṣṇa.
3. The story goes that as a boy and youth he was stronger, more handsome, and more aggressive than his peers and bored with formal education but highly intelligent. "Nor did any theories or dogmas interest him; for he was one who could be convinced only by actual observation and direct experience." At fifteen, by chance at a festival, he met Swāmi Nityānanda who was afterward to direct the final stages of his spiritual development. Nityānanda embraced Kṛṣṇa and gently stroked his cheeks. He then strode away and they did not

meet again for many years. A pious home life had already aroused interest in a religious vocation, so six months after the encounter with Nityānanda Kṛṣṇa left to become a sadhu.

4. Unlike certain other saints in India his full realization did not occur until middle age, while the process leading up to it included instruction under several different gurus.

a) Among the saints, siddhas, and sadhus with whom he took training of some sort as a young man—including Sanskrit language, study of the scriptures, yoga practice, and other subjects—his first teacher was Siddharuddha Swāmi, from whom he received initiation as sannyasī and his religious name, Muktānanda. Thus his formal entrance into the ascetic life was under the direction of a monk trained in the disciplines of śakti. In 1929 he began to wander through India, staying here and there with various teachers. In all, during this period he mentions having met sixty great saints. Besides his first guru, among others there were "Popat Maharaj at Satana, Upasani Baba at Sakori, Swami Prakashananda at Gondal, Sitaram Bairagi at Dwarkabet, Bhagari Baba at Lasalgaon, Narayan Maharaj at Kedgaon, Mauni Baba at Chikhali, Chaitanya Swami at Paithan, Munsoji Baba at Varad, Prembhikshu at Jamnagar, Ramana Maharshi at Tiruvannamalai, Jagannath Baba at Ahmedabad and a saint at Howrah (Calcutta) who was known to subsist on stones." In the reports about his experiences, in the latter part of his first phase he received special attention from two saints who noted his impending realization. The one, Zipruanna, a naked ascetic who passed his days seated on a refuse heap, healed Muktānanda's incessant headaches by licking the latter's head and, through water that had been poured on Zipruanna's foot, helped Muktānanda to cure a woman in an advanced stage of tuberculosis. Of Muktānanda, Zipruanna said, "Your fame will touch the highest heaven." Another saint, Harigiri Bābā, said of Muktānanda shortly before the last and crucial stage of the sadhana began, "You have now to live in a palatial building. Cast away your ochre clothes and wear silken garments instead. You are no longer a *sannyasin*, but a *maharaja*. You shall not ask but only give."

b) Muktānanda has discussed in detail the special relationship that developed between himself and the saint known as Bhagavān Nityānanda. Nityānanda became for Muktānanda his guru, both in the sense that he received final initiation from him and in a very special manner when it became clear that Muktānanda was the "chosen disciple." The date of his initiation was August 14, 1957. On that occasion Nityānanda gave his own sandals to him—the closest physical objects to the lotus feet of the guru. The presentation of the initiatory symbol marked a turning point in the quality of Muktānanda's noumenal experience which now began to conform with that of the Siddha Paraṃparā, the adepts in Kuṇḍalinī. He writes of what happened: "[Nityānanda] looked into my eyes. Watching carefully, I saw a ray of light entering me from his pupils. It felt hot, like burning fever. Its light was dazzling, like that of a high-powered bulb. As that ray emanating from Lord Nityānanda's pupils penetrated mine, I was thrilled with amazement, joy and fear. I was beholding its color, and also chanting 'Guru Om.' It was a full unbroken beam of divine radiance. Its color kept changing from molten gold to saffron, to a shade deeper than the blue of a shining star. I stood utterly transfixed."

From shortly after the time that this experience occurred, Muktānanda lived away from Ganeshpuri in very intense *sādhana* until his realization was complete. After he returned to Ganeshpuri, when the devotees of Nityānanda were preparing a small temple to enshrine an image of their guru to worship after his passing, which they expected would be soon, on November 16, 1956 they were directed instead to install Muktānanda in the temple with ceremonies appropriate to a divinity. Thereafter, the two saints lived side by side in the ashram until Nityānanda's death on August 8, 1961. Since then Muktānanda has directed the ashram and greatly expanded both its size in respect to property holdings and its activities with increasing numbers of devotees.

5. In regard to the organization of the ashram and related subjects I have first-hand kowledge—having visited Ganeshpuri and Shri Gurudev Ashram in July of 1971 and having met Swāmi Muktānanda. Moreover, it was possible to engage in extended discussions with several of the devotees who had been initiated. The ashram has attracted people from all over India and among them large numbers of the elite of Bombay who come, particularly on week-ends, to have *darśan* of Muktānanda and to join temporarily in the ashram's religious life. There is also a considerable group of Western devotees, including several who have been living in the ashram for a year or more since it was established. The emotional atmosphere if this sanctuary for those who are experiencing Kuṇḍalinī awakening is no doubt very exalted. At the edge of the road with rice fields making a half-moon around it and everything, including the little jungle at a distance behind, washed and hearteningly green in the monsoon rain, to the observer the ashram gave a sense of cleanliness, order, and peace. Muktānanda rules as Guru Mahārāj and demands conformity to the horarium, which is somewhat similar to that of a monastic house in the West. Seeing him, listening to his remarks, hoping for some favorable sign bestowed on oneself constitute the framework of the relationship that the individual devotee develops with the guru.
6. We have already mentioned the "royal" style of Muktānanda. He has been described by some as "virile," and by that is meant that he has a commanding, masculine personality. He speaks directly and even coarsely at times to express his ideas, but on the other hand his erudition and skill in expression in written form are often commented on as well. In this he contrasts with Nityānanda who was illiterate; likewise, in his personal habits, Muktānanda favors silken *lūngīs* and has taken pains to make the ashram attractive with plants, an enclosed garden, a spacious meditation hall and dining room, and so on. In personal manner Muktānanda is often said to be "restless," and one observes this in his sudden appearances and disappearances around the grounds, his quick walk, the changing expressions on his face, and his penchant for unexpected excursions to holy places in the region or for calling the people to vary the routine of services during the day with some novel celebration. He usually wears dark glasses, and it is reported that he does so because he perceives

the spiritual quality of the world and of the persons who come to see him in varying degrees of a kind of light which hurts the physical eyes. He does not have the reputation of being able to produce phenomena, or at least he does not particularly cater to demands that he do such, although the psychic experiences which his followers say stem from his powers are remarkable enough.

7. There are several persons at the ashram who undoubtedly have a much closer bond with the guru than do others. Chief among these is a very bright woman, a former professor of Sanskrit, who serves as the leader of the woman devotees and is in charge of some of the other internal operations of the ashram. She is called Amma, or Mother, and has written about the tantric *sādhana* taught by Muktānanda. She has been at Ganeshpuri since before the time that the two saints lived together there in the fifties. Since she is a person of great personal refinement and charm, her allegiance to this saint is an instance of the testimony of the distinguished disciple to the worthiness of the master. A young, former English professor, Mr. Jain, who serves as Muktānanda's interpreter and has been in the ashram for several years, is also high in the ranking of the disciples. Among others of them one might mention an American girl, now called Uma, who met Muktānanda in New York City during his American tour and was able, through a combination of seemingly miraculous circumstances, to come to Ganeshpuri to live permanently. Indeed, I was able to talk with several young Western men and women, including married couples who felt called to be followers of Muktānanda and reside with him. Reports of the experiences of some of the disciples appear in *Shree Gurudev-Vani.*

8. As Muktānanda's reputation grows and he becomes known in wider and wider circles, no doubt he will be increasingly called upon to serve in the role of guru to society. If one might compare him with the somewhat more famous Sathya Sāi Bābā, one would say that his social role is likewise less well developed. It is difficult to completely clarify these aspects of the demands upon the Hindu saint, but they may be compared with the tasks that some of the clergy in Western countries, particularly famous prelates or evangelists, are called upon to perform.

   Beyond such activities, Muktānanda is a man who can capture the attention of the reading public. In his own right he has published in Hindi a spiritual memoir, entitled *Citśaki Vilās*, which was translated and abridged in the American version, *Guru*. Members of the ashram, but particularly Amma and Mr. Jain, are likewise skilled writers. In this regard, as compared with the majority of the saints one would find in India, Muktānanda may be thought of as more clearly employing his intellectual gifts to reach a literate public. This is true also in the question-and-answer series, published from transcriptions of conversations with the disciples.

9. As far as the further development of the cult is concerned, it is clear already that Muktānanda has been adopted as guru and avatar by a proliferating body of followers and that he will experience the fame of a national celebrity. Besides the testimonials of both Eastern and Western disciples appearing in ashram publications, independent writings of disciples attest to the master's

influence. There is a temple, Shree Gurudev Dhyan Mandir, in Johannesburg, South Africa, where devotions are held regularly in honor of Muktānanda; and his cult is observed in various places in India and the United States at the present time. It should be emphasized that this is happening while the saint is still alive.

NOUMENAL ASPECTS

Muktānanda in a unique manner has described his inner states both in his Hindi and English writings on the subject. As the devotees point out, what is unusual about Muktānanda's "way" is that, for the first time, the secret initiations and experiences of the Śakti Pāt, the yoga of the goddess, of the primordial energy of Śiva, is presented openly in a manner suitable to universal acceptance. We often speak of this kind of yoga as Kuṇḍalinī Yoga.

When we look through the reports of Muktānanda's and his disciples' experiences, we find that there is conformity to the technical theory. For instance, one might recall the passage quoted previously from the letter of my friend who experienced the initiation very emotionally. As regards its further physical effects, he goes on in the same letter to say the following:

> Now when I visit a temple, a samādhi or dargāh [Hindu or Muslim saint's tomb, respectively], I become acutely aware of the shakti present there, to the extent that my body shakes very violently. Near Ganeshpuri is a very old Devi temple—Vajreshvari—and when I visited it, I found myself doing the mūla-bandha. Since then I have been doing theuddi-yāha-bandha and the jālandhara-bandha at various times. I have found that the samādhi of Dyāneshvar at Ālandī is particularly powerful in this respect. As I approach the temple my body starts to shake and by the time I reach the sanctum sanctorum it is uncontrollable. Dyāneshvar is there for me. The pūjarī throws some water on me and my body twists and writhes even more—such is the power of the shakti.

A devotee writing about his and a friend's experiences after being initiated at the ashram mentions equally striking results.

> We sat for meditation after the recitation of the Vishnu Sahasra Nama. Within half an hour I could notice some movements in my body: I was swinging from left to right, and back and forth. My entire body was trembling as if I had received an electric shock or there was an earthquake. I felt a heaviness in my head and it touched the ground. I was in that condition for about an hour. I felt a kind of wheel moving or whirling in the stomach at high speed. I had some visions of Lord Ganapati and Lord Dattatreya. My eyes had become red and tears rolled down the cheeks. Even when I got up my body was trembling and I could not keep my balance. Shri Zarapakar could sit for about half an hour only and in that period he said he saw some colours, a flame, and visions of Lord Dattatreya, Swami Vivekananda and others. He also felt heaviness in his head especially between the two eye-brows. His eyeballs were sometimes rotating and then fixed between the two eyebrows.

One soon becomes aware that the underlying current of excitement at Ganshpuri is directly related to the hope of some who go there to be graced with initiation, to have these kinds of experiences, and perhaps to aspire after the "realization" which is at the basis of Muktānanda's claim to be a saint. During the course of the devotions, chanting, and meditations in the prayer hall in front of the shrine of Nityānanda, it was not unusual for an observer to witness the physical phenomena described above. Among both Eastern and Western devotees I saw some moving their arms, prostrating, revolving the head, even shrieking and falling to the ground in tears. I traveled and visited for a fortnight with an Indian friend who had received this initiation. As a consequence of it, he did automatic breath arrest in the manner of the Prāṇāyāma of Rājā Yoga. He had some control over it, but he asserted to me that it was nearly irresistible; he had become so blissful that he could hardly remain in normal consciousness. Indeed, he appeared to go into trance even in the middle of a conversation, while involuntarily holding his breath for an inordinately long time.

Of course it will occur to the reader to ask to what extent these expressions attest to any particularly real or new thing, or whether they are not hysterical or other types of psychologically abnormal states. The answer, naturally, cannot be given because we do not have a technique for analyzing psychological conditions in respect to spiritual realization. The whole recent trend in science would be to eliminate anything but "normal" states of mind—that is, to bend every effort to retain or return to a "normal" state; but apparently this would provide a hindrance to the physical and psychic transformations that are held to appear along the way to Śakti Pāt realization.

On the other hand, I had, as it were, anticipated the discovery of evidence in favor of the possible widespread character of these experiences when, on an earlier visit to India, I met Mr. Mark Sunder Rao at the Christian Institute for the Study of Religion and Society in Bangalore. Mr. Rao had converted to Christianity from Hinduism, but his most striking "Christian" experience occurred in a manner very reminiscent of what we have just been discussing. He presented me with a copy of the printed document from which I quote this passage:

> In 1951, after attending the episcopal consecration of Bishop Appasamy at Erode I spent a month at my sister's (a Brahmin) home. There I had the privilege of meeting an old friend of the family, a yogin. From 9 a.m. to 5 p.m. with intervals for refreshment and rest, we held conversations on spiritual life and studied together the Maharashtra saint Jnandev's work Amritānubhava. When it was time for me to return home, this friend expressed the wish that I should have the mystic vision and experience. Indeed he said that he coveted it for me because of my steadfast studentship and adhikāra. Within a few moments of his saying so I became aware of a strangely attractive mass of brilliant but cool light approaching me from an immense distance. Coming nearer it appeared to be a circle of radiant gold, blue and white, with the head of Jesus Christ at its centre. Approaching me at great speed this mass seemed to hit me between the eyebrows, penetrate through the skull to the rear brain, when I heard a sound like that of a camera click, and I lost consciousness of my surroundings while internally alert and aware of unutterable oneness with All and inexpressible joy . . .

> When I came to, there was the senior friend beaming with a smile and tearful eyes. He told me that his wish had been fulfilled and that we could part with joy.

After the initiatory experience described earlier, Muktānanda retired to a lonely place in central Maharashtra and underwent the final stages of the Śakti Pāt *sādhana.* It is true, he frequently found himself in a condition bordering madness or complete physical breakdown as a result of the psychic "tension" that brought him from level to level. He experienced the operation of the śakti upon him in a primarily visual way. His progress was measured by an evolution through visions of unearthly red, white, and even "black" light which bore with them ecstatic epiphanies of the Hindu deities as well as paranormal "precognitive" knowledge. He developed such classical yogic powers as the "Bindu-Bhed . . . in place of normal binary vision, the eyes attain a uniform field of perception . . . can see or visualize on all sides." The climax of all was the sight of the blue pearl, or *nīla bindu,* that Muktānanda's devotees hope to perceive through his help and that he asserts is the symbol or even the vehicle of the highest realization:

> while I was gazing at the tiny Blue Pearl, it began to expand in all directions, spreading its blue radiance. The entire region from the earth to the sky became irradiated. It was a Pearl no longer, having enlarged into the shining, sparkling infinite Light. This has been designated by the scriptural authors or the seers of the highest Truth as the conscious Light of Chiti. I actually saw the universe arise from this expanding Light like clouds of smoke from a fire. . . . I meditate these days also. But I feel certain that there is nothing more for me to see. The certitude that I have attained the highest arises spontaneously in meditation. For the extremely subtle, tranquil, all-pervasive, Conscious Blue Light that followed the three visions with the Blue Pearl has not altered or vanished and still bathes the external universe. Even with closed eyes I perceive it glowing and shimmering, softer than the soft, tenderer than the tender, subtler than the subtle. When, I open my eyes, I see the blue rays all around. Whenever I look at anyone, I first see the blue Light and then him. When I look at an object, I first see the honeyed subtle Conscious rays and then the object. Regardless of what occupies my mind, I perceive the universe within the lustrous mass of Light.

Tantrism

AGEHANANDA BHARATI

Agehananda Bharati, a professor of anthropology at Syracuse University, is a Western convert to Hinduism. In addition to the following selection, he has written a full-length book on Tantra. This selection is a historical survey of Tantra from its origins to the present with special discussions of such features as Tantric practices, differences between Hindu and Buddhist Tantrism, and Tantric art. Note that this selection is reprinted from the Abingdon Dictionary of Living Religions—*cross-references are to articles within that dictionary.*

TANTRISM tăn' trĭsm (H, B & Ja—Skt.; lit. "that which extends, spreads"). 1. In a general sense, a non-Vedic practice (*see* VEDAS), including rites open to women and persons not of the BRAHMIN caste. It also includes the worship of deities for the purpose of specific religious merit or wordly gain. 2. In a narrower but more popular sense, an esoteric, radical way to achieve MOKṢA (emancipation from rebirth and suffering). It includes a large body of scriptural and oral lore, parallel, and in HINDUISM and JAINISM, marginal to official scriptures. While orthodox Hindus view occult powers (*see* SIDDHI) as impediments to the quest for *mokṣa*, tantrics court them as proofs of progress. The orthodox view a temperate life-style as essential to the pursuit of YOGA, but tantrics cultivate the sensuous elements in their psychic makeup.

1. **Types of tantrism.** First is the clandestine, often eroticized version of yoga. In Tibetan Buddhism it is synonymous with the "diamond vehicle" (*see* VAJRAYĀNA) and is part of mainstream doctrine and practice. In Hinduism, however, it is peripheral and antagonistic to orthodoxy and orthopraxis. Second, tantrism on the Indian subcontinent is largely identified with shamanistic behavior, with sorcery and witchcraft. A tantric is a person who commands extrahuman forces for his own benefit or that of his clients. Third, a learned convention among Hindu scholars calls all non-Vedic practice tantric (*see* VEDIC HINDUISM), particularly domestic and women's rites. Practicing tantrics pay less attention to scripture than to personal transmission from their GURU and to psycho-experimental manipulations.
2. **Tantric literature.** Hindu tantrics are at pains to show that their texts are truly Vedic. Their apologetic centers almost entirely on asserting Vedic respectability, because mainstream Hindus kept attributing clandestine and nefarious actions to them. Since Buddhist tantric teachings were absorbed into the Vajrayāna canon, Tibetan tantrics never felt such need for legitimation. Hindu tantric texts are a distinct, less respected, and often censured corpus, having canonical status only for the rather small audience of learned

tantrics, few of whom are BRAHMINS. In fact, tantrism is overtly anti-brahmanical (though not anti-Vedic) and can be seen as a rebellion against Brahmin sacerdotalism.

3. **History of Tantrism.** Although the roots of this tradition are very ancient, the material was reduced to writing only later. The first tantric works were Buddhist, e.g. the *Mañjuśrīmūlakalpa* "radical institutions of MAÑJUŚRĪ" and the *Guhyasamājatantra* "tantra of secret association," both compiled between A.D. 300 and 600. The first authentic and extant Hindu tantric texts are the *Mahānirvaṇatantra* "tantra of the great liberation" of the eleventh century and the *Kaulāvalinirṇaya* "description of the garland of adepts" of approximately the fourteenth century. These texts derive much of their raw material from folk sources which may reach back into pre-Vedic times. The manipulation of chthonic powers, the psychodynamic experimentation underlying tantrism, is much closer in form and content to what many scholars now perceive as *Indian* contrasting with *Aryan* on the subcontinent. The robust, extrovert, philosophically naïve tenor of the Vedas, the books of the Aryans, differs sharply from tantric style and lore, tantric apologetic notwithstanding. Tantrism became strong in area of late or weak Aryan, Brahmanical penetration.

There is some support for the notion that the earliest era (*see* INDUS VALLEY CIVILIZATION) contained proto-tantric elements like ithyphallic representations of a fertility god. If there is a connection, the origins of tantrism could indeed be placed as far back as 2500 B.C. The early portions of the Vedas indicate strong opposition to phallic worship. In later sections, however, phallic ritual was accepted and established (*see* LINGA). By about 1000 B.C., phallic lore was apparently well established.

Since the tantric style originated with non-Aryan segments of the population, it remained identified with low-caste ritual. Tantrics courted, trained, and revered teachers of low-caste background who were barred from Vedic ceremony. This included women and non-Hinduized tribal groups.

In buddhist trantrism being female or of low caste was a positive qualification. In theory at least, Brahmins were disqualified from tantric apprenticeship, but most of the later commentators on tantric codes were Brahmins, particularly from Bengal, a region where even Brahmins are largely non-vegetarian and often given to the worship of mother-goddesses.

After the destruction of two main centers of Buddhist learning by a Muslim chieftain in the eleventh century, tantric Buddhism retained ecclesiastic status only in Tibet. A few pockets of Vajrayāna institutions probably survived into the fourteenth century in northeastern and southwestern India, after which time it was defunct in India.

Organized Hindu tantrism peaked between the ninth and the fourteenth centuries under feudal and royal patronage. The decline of sophisticated, scriptural tantrism coincided with and was accelerated by new rural cults of monotheistic devotionalism (*see* BHAKTI HINDUISM), which shared an anti-erotic, puritanical ideology. To the teachers and followers of these cults, tantrism epitomized all that was reprehensible, to be shunned and rejected primarily because of its sensuous overtones, and secondarily due to its being tied in the popular mind to magical practices.

4. **Tantric teachings.** The theologies of tantrism are not fundamentally different from those of mainline Hinduism, Buddhism, and Jainism; it was tantric practice which roused the ire of the orthodox.

   a) Hindu tantrism theologically largely overlaps with the monistic school (*see* ADVAITA), which postulates a single existent, BRAHMAN (neuter), defined as "being-consciousness-ecstasy," *saccidānanda*. Multiplicity is a delusion (*see* MĀYĀ) and the intuitive, irreversible realization of numerical oneness with Brahman implies liberation. The tantric seeker, however, conceptualizes Brahman as the union of the male and female principles; namely, SHIVA (benign) and ŚAKTI (energy). In this initial polarity the male stands for the quiescent, for cognition, and for wisdom. The female represents action, conation, and the energetic élan. While Shiva and *Śakti* belong to the general pantheon, they are here transmuted into the cosmic principles of cognition and action. "The universe arises through the copulation of Shiva and *Śakti*." The process is reversed in the experience of the individual adept who by applying the proper techniques realizes that even that cosmic duality is illusion and that the ultimate reality is the nondual Brahman. This knowledge dissolves the adept's ego and the impersonal Absolute shines forth. The successful practitioner, while identifying with Shiva if he is male and with *Śakti* if female, eventually transcends this partial, albeit divine duality to *be* the Absolute.

   b) Buddhist tantrism (*see* VAJRAYĀNA) postulates the exactly obverse polarity. Here the male principle is Buddhahood, active, outgoing, energetic. It is the means (Skt. *upāya*) of emancipation and compassion (Skt. *karuṇā*) as instantiated by the numerous Buddhas and BODHISATTVAS (*see* BUDDHISM) and visualized in sexual embrace with their female counterparts, which represent the quiescent, static, wisdom principle (Skt. *prajñā*). The two poles are symbolically juxtaposed as the "diamond" (Skt. *vajra*, hence Vajrayāna), conjoined with the "bell" (Skt. *ghaṇṭā*) representing the womb. The intuitive knowledge of their underlying oneness propels the aspirant from the experimentally fertile yet ephemeral *karuṇā-prajñā* model into the liberating, incontrovertible knowledge of momentariness and voidness (Skt. *śūnyatā*), again in line with mainstream Mahāyāna doctrines. The one who interiorizes this knowledge realizes that both worldly being (SAMSARA) *and* salvation (NIRVANA) have no essence, and that Nirvana and samsara are identical.

   c) Jaina tantric teachings differ in no way from the atomistic doctrines of the official texts (*see* JAINISM). The distinction between orthodox and tantric Jainism lies entirely in the tantrics' emphasis on meditations using MANTRAS, particularly those pertaining to the goddess Padmavatī, the tutelary deity of Jaina tantrics.

5. **Practice and meditation** are ranked far above scriptural and all other theological knowledge in all tantric schools. The basic difference between orthopractical yoga (*see* PATAÑJALI) and other mainstream Hindu, Buddhist, and

Jaina meditation on the one side, and tantric practice on the other, lies in what makes tantrism suspect to the orthodox: its full harnessing rather than the renunciation and rejection of the senses and its maximizing of the sensuous personality in contrast to the ascetic style of the official traditions.

Hindu tantrism of the "right hand" (*dakṣiṇācāra*) is coextensive with all nonbrahmanical ritualistic performance. The critical break with mainstream Hinduism occurs in the "left way" (*vāma-mārga*) practices. The aspirant must first master the usual techniques of physical control (*see* HAṬHA YOGA) before proceeding to the more esoteric meditations. "Haṭha Yoga is to tantric yoga what a B.A. is to a Ph.D. degree," a modern tantric explained. Next, the aspirant learns to raise the "dormant power" (*see* KUṆḌALINĪ) within himself or herself. It is here that the discipline diverges from orthodox yogic practice, which is a solitary procedure. The neophyte is initiated into a "circle" (*cakra*) of fellow aspirants consisting of an equal number of male and female disciples, guided by a male adept (*cakreśvara*) and his adept female consort (*cakreśvarī*). All male participants are designated as Shiva, all female ones as *Śakti*. The latter sit to the left of their male counterparts—hence "left-handed" tantra. The convening of a *cakra* is controlled by complex astrological and ritualistic preparations. The ritual commences with prolonged chanting of Vedic and tantric texts. Each participant then silently meditates on the special formula (*see* MANTRA) given to her or him by the guru, who may or may not be the *cakreśvara*, while the latter keeps chanting the requisite hymns. At this stage, the participants imbibe impressive quanitites of "victory" *vijayā*, a *sandhā*-term for *cannabis sativa* (i.e., marijuana) blended with sherbet and sweet milk. This introduces the core segment of the exercise, namely the seriatim use of the "five Ms" (*pañca-makāra*), which stand for the initial letters of the Sanskrit words for the main "ingredients" (i.e., fish [*matsya*], meat [*māṃsa*], parched grain or kidney bean believed to be aphrodisiacs [*mudrā*], liquor [*mada*], and finally sexual union [*maithuna*]). All these are highly stylized events with little leeway for innovation. During the last phase, the *Śaktis* place themselves astride their Shivas and initiate copulatory movement, in line with the doctrinal notion of woman as energy and man as quiescent. Since "Shiva without *Śakti* is (like a) corpse," no tantric male can achieve emancipation without being thus aided by a *Śakti*.

The key technique within this ritual is retention of semen during *maithuna*. It is in effect a ritualized *coitus reservatus* for the avowed purpose of achieving simultaneous control of mind, breath, and semen. This technique is seen as a shortcut, albeit a dangerous one, to *mokṣa*. Tantrics aver that this successfully controlled *maithuna* rushes the *kuṇḍalinī* upward into the "thousand-petalled lotus" atop the subtle body, merging the Shiva and *Śakti* principles. This explains, in part at least, the importance and the ubiquity of the phallic shape of the Shiva icon (*see* LIṄGA) as joined to the *pīṭha* "seat, vulva." Shiva must not be formally worshiped in any anthropomorphic form like other deities; the *liṅga-pīṭha* icon is mandatory (*see* YONI). Shiva is the tutelary god of all ascetics, the "vanquisher of Cupid." The contradiction is only apparent: the ithyphallic representation is nonpriapic. It implies complete control as retention of semen at the point of orgasm. The nonejaculatory union of the tantric adepts thus reenacts the cosmic resorptive union, just as regular coitus culminating in ejaculation replicates the priapic, procreative aspect of Shiva and *Śakti.*

6. **Buddhist tantric practice** (*see* VAJRAYĀNA) is based on similar psycho-experimental principles, but the actual copulation between the practitioners is less formalized. In Vajrayāna practice today the preliminary exercises take up a much larger portion than sexual congress; in fact, the latter element is now often eliminated. Meditation consists in increasingly complex visualizations of and gradual identification with divinities of the Vajrayāna pantheon. (*See* MANDALA.) The advanced practitioners interiorize the "honorable father and honorable mother" (Tibetan *yab yum*) imagery of the deities in sexual embrace. Where there is actual copulation, retention of semen is axiomatic: "having brought down the *vajra* into the lotus, let him not eject the knowledge mind." Such use of code or "intentional language" is a feature shared by Hindu and Buddhist tantrism. It serves as key terminology for the initiates and as a means to screen the teachings from outsiders. "Knowledge-mind" (*bodhicitta*) for example, is a code term for semen.

   Not part of the formal ritual, consumption of alcohol is accepted and even recommended by some meditation masters preceding the practice of visualization and interiorization.

7. **Places and calendars.** Neither Buddhist nor Hindu tantrics adhere to a specific calendar of festivals and celebrations—there are no tantric festivals *per se.* However, certain regional festivals connected with the worship of female deities are regarded as auspicious occasions for tantric practice. There is a very large number of such shrines on the subcontinent, but the majority of pilgrims are hardly aware of the site's tantric significance.

8. **Tantric art** is a term of recent Western origin. Yet is is safe to assume that most of the erotic imagery on Hindu shrines, especially in southern and southeastern India, is due to tantric inspiration. Muslim chiefs destroyed a large number of shrines in northern India, which accounts for the virtual absence of erotic sculpture in that area. Such representations were and are abominations to Muslims, Christian missionaries, and modern puritanical Hindus. Nepalese Hindu temples display an abundance of such imagery. The best known and most readily accessible shrines are KHAJURĀHO in central India, KONĀRAK, Purī, and Bhuvaneshvar in the eastern State of Orissa, and hundreds of shrines in the south.

   Tibetan and Nepalese Vajrayāna has generated and continues to produce tantric artifacts in quantity, and some of exquisite craftsmanship as well. Painted silk and paper scrolls (*thanka*), *yab yum* bronzes particularly from Nepal, and *vajra-ghaṇṭā* bell metal representations have entered the international market in quantity after the Chinese occupation of Tibet.

9. **Current trends** indicate a steady decline of Hindu tantrism and a strong revival of Buddhist tantra in the diaspora in North America and Western Europe. India's official culture being puritanical and its threshold of tolerance low, tantrism is virtually blacklisted by administrators. In the late 1950s there

were sporadic police actions against tantric centers like that of Pagli Baba, an eminent tantric in Orissa. Śrī Rajneesh, a psychology lecturer turned tantric guru, attracts large Indian and foreign audiences in Bombay. He concocted his own version of tantrism, incorporating enough general urban Hinduism to make it tolerated in spite of some protest. Yogi Bhajan, a Sikh customs officer turned guru in California, teaches a blend of hardline SIKHISM and *kuṇḍalinī* yoga. He distinguishes between "white" (acceptable) and "red" (wicked) tantrism, terms entirely of his own invention. Ananda Marg, whose leader was acquitted of a charge of murder of some renegade disciples, has a large, monastically garbed following in India and chapters on many North American college campuses and represents itself as tantrism. (*See* HINDUISM IN AMERICA 2.)

On the popular level, tantric practice has been observed and reported over the past three decades, and there is no reason to believe that its occurrences have significantly increased or decreased. Groups of male tantrics gather in remote forest areas and meet with women who seek the same powers or who have been persuaded to participate "for pleasure." The women divest themselves of their blouses (hence "way of the blouse" for this ritual), which they place on one heap. The men then approach, each of them picking one blouse. Each man then enters into a one-night liaison with the owner of the blouse. This guarantees randomness, which is part of the tantric notion that the "other woman" (*parā-strī,* i.e., either not the practitioner's wife, or another man's wife) is *parā-śakti* (the highest power), the prefix *parā* meaning both "other" and "highest." There is little chanting, except for some preparatory invocations in the vernacular language, interspersed with some Sanskrit mantras. There is no insistence on retention of semen during the act. The target of the rite is not liberation, but occult powers. Modern Hindus who hear about these events condemn them as "dirty" and "superstitious," and as things Hinduism must reject if it is to survive in this age of science.

The energetic missionary work of a number of learned Tibetan lamas in the West has, by contrast, created nuclei of serious Vajrayāna practice in Europe and America. The number of expatriate Tibetan monks is on the increase, adding leadership and direction to Occidentals who seek this experience, a thing which was totally impossible before Buddhism's forced exit from Tibet. No tantrics are left in Chinese-occupied Tibet, and the Tibetan refugees settled in India are bound to assume the prevailing Indian mores which frown upon esoteric practice.

The Synthesis of Yogas: *Haṭha*, *Rāja*, and *Tantra*

HARIDAS CHAUDHURI

Haridas Chaudhuri brought the teachings of Sri Aurobindo, a well-known modern interpreter of Tantrism in India, to the United States where he taught for many years. In this selection, Chaudhuri presents an interpretation of Tantra that conforms to many modern views of the application of religious theories to the solution of human problems. In both India and the West many modern seekers for religious values are motivated by the desire to live a life that is healthier, that liberates one from the anxieties brought about by the pressures of contemporary life. Chaudhuri's interpretation of Tantra is, therefore, "revisionist" to a certain extent. It offers an approach to Tantra that is not heavily driven by the esoteric and erotic tendencies that, in the classical Indian tradition itself, were regarded as fully appropriate only for a small group of individuals. Chaudhuri sees aspects of Tantra that are universally applicable.

Integral Yoga represents the crowning fulfilment of the traditional yoga systems of India. It takes note of their limitations and one-sided tendencies. And it incorporates their inherent truths in a higher synthesis.

Broadly speaking, there are six traditional systems of yoga; *Haṭha, Rāja, Tantra, Jñāna, Bhakti,* and *Karma.*

HAṬHA YOGA

Haṭhayoga is the system which starts with the body. Body and mind being closely interrelated, it aims at mastery over the mind. Control of nervous and vital energies produces control of mental functions. The mind-body complex being brought under perfect control, the in-dwelling spirit shines out and the higher self is realized.

Haṭha is derived from the roots, *ha* (sun) and *ṭha* (moon). *Haṭha* is the equalization and stabilization of the "sun breath" (i.e., the breath which flows through the right nostril) and the "moon breath" (i.e., the breath which flows through the left nostril). *Haṭha* also means violence, force. Through the regulation of the physiological processes, *Haṭhayoga* forcibly releases the dormant energies of human personality.

The principal steps of *Haṭhayoga* are *āsana* and *prāṇāyāma*. *Āsana* consists of certain bodily postures such as lotus posture, hero posture, head stand, shoulder stand, etc. They are designed to stimulate the glands, vitalize the body, and strengthen the nervous system. Purified and strengthened nerves are the most important pre-requisite of yogic practice.

Prāṇāyāma means control of the vital energy through breath-regulation. It aims at mastery over the vital forces which are operative in the body. Through control of breath and mobilization of vital forces, it endeavors to secure the release and free flow of the fundamental psycho-physical energy (*kuṇḍalinī*) latent in the human system. This root energy being dynamized, the individual is set on the path leading to his reintegration with the ultimate ground of existence.

It is believed that one who acquires success in *Haṭhayoga* gains supernormal powers. He enjoys vibrant health, youthfulness and longevity. He attains spiritual liberation and supernal bliss.

The chief merit of *Haṭhayoga* lies in its insistence upon the basic importance of the body. Various bodily postures and breathing exercises recommended in *Haṭhayoga* are very effective means of developing the body as a fit and strong instrument of higher spiritual living. Mystics who have neglected the bodily factor, have suffered immensely on the physical plane. They have suffered from disease and disability, and have met with premature death. Profound spiritual experiences put an inordinate strain on the nervous system. They often come with the impact of a rushing flood. Without a prior bodily training and nervous firmness, many mystics fail to stand that impact. They are carried off by waves of emotion; they sing, dance, cry, and roll on the floor, failing to convert the flood of emotion into calm creative energy. *Haṭhayoga* can prepare and fortify one against this kind of mishap.

The chief defect of *Haṭhayoga* lies in its over-emphasis upon the physical side of existence. The body is sometimes almost deified. Preoccupation with the body produces excessive self-concern. Acquisition of supernormal powers and the bliss of personal salvation loom large on the mental horizon. An indifference to the affairs of the world and the requirements of society is generated. Not much interest is left for higher cultural pursuits. The need for intellectual development is not sufficiently recognized. The danger of going astray through selfish appropriation of whatever unusual power is gained is rampant. The spectacle of *Haṭhayogis* making a vainglorious display in public of their extraordinary bodily control is not an uncommon sight. It is such misguided persons who have brought much disrepute upon the fair name of yoga.

RĀJA YOGA

Whereas *Haṭhayoga* starts with the body, *Rājayoga* starts with the mind. It works with the mental apparatus considered as a whole. It endeavours to achieve a complete cessation of all mental functions, so that the light of the indwelling spirit may shine out. It recommends no doubt the methods of bodily posture (*āsana*) and breath-control (*prāṇāyāma*), but it does not require the practice of them in their full elaborateness as developed by *Haṭhayoga*. On the contrary it adapts them to its central purpose of mental calmness, balance and equilibrium. Of the numerous forms of bodily posture it specially selects that which keeps the body motionless in the fittest and most comfortable position helpful to the practice of breath-control and meditation. It recommends breathing exercises with a view to harmonizing the vital forces of the body, so that the obstructive elements of ignorance, inertia and restlessness may be removed.

Prior to the practice of bodily posture and breath-control, *Rājayoga* stresses the need for adequate ethico-religious training. The powers of body and mind are likely to be abused if the right spiritual foundation has not already been laid. The ethico-religious training recommended in *Rājayoga* consists of two steps: moral discipline (*yama*) and religious observances (*niyama*).

Moral discipline includes the practice of non-violence (*ahiṁsā*), truthfulness (*satya*), non-stealing (*asteya*), control of the sexual impulse (*brahmacarya*), and abstinence from greed or avariciousness involving non-acceptance of unnecessary gifts from those whose motives are questionable (*aparigraha*). Religious observances include the practice of internal and external purity (*sauca*), contentment implying the principle of plain living and high thinking (*santoṣa*), endurance of hardship and adverse circumstances (*tapas*), devoted study of spiritually ennobling books (*swādhyāya*) and self-surrender to the Divine (*Iśwara-praṇidhāna*).

The fifth important step in *Rājayoga* is self-withdrawal (*pratyāhāra*). It is the withdrawal of the senses from their external objects. It is the act of transcending the natural world. It should not be misconstrued as absolute and final world negation. It is the methodological device of temporarily setting aside the world with a view to inquiring with sustained energy into the nature of the spirit. It corresponds to what Edmund Husserl has called "the phenomenological reduction." It is the method of putting into brackets the whole natural world, without paying any attention to the question whether the world is real or unreal and without using any "judgement that concerns spatio-temporal existence." The idea is to gain thereby full freedom and untrammelled energy in investigating the field of consciousness. When the question concerning the ontological status of the natural world is set aside, the contents and functions of consciousness can be observed in their essence as pure phenomena.

As an aspect of spiritual practice, *pratyāhāra* implies a shift in attention from the impulsive to the higher values of life. It signifies one's dissatisfaction with exclusive pre-occupation with material values or with the traditional and conventional mode of living. It involves what Plato has called "divine discontent." It involves a kind of "metaphysical rebellion" against man's condition in the universe. *Pratyāhāra* is one's readiness to plunge into the unchartered sea of deeper self-inquiry and critical investigation into the meaning of life. It is the disengagement of the self from unthinking attachment to the not-self. It aims at transcending the world of false identifications and illusory projections in a deep search for the unconditioned spiritual reality.

In order to achieve the ultimate goal of freedom, *pratyāhāra* has to be supplemented by three other processes, namely, concentration (*dhāraṇā*) meditation (*dhyāna*) and self-integration (*samādhi*).

Concentration is the focusing of all mental energies upon one object, one central idea, or one relevant truth. It releases latent energies of the psyche and marshalls all psychic forces in a definite direction. Meditation is the higher phase of concentration. It is the free and uninterrupted flow of thought in one direction, centering round a definite theme. That theme may be the self, or pure existence, or the supreme value. It purifies the inner being, thoroughly cleanses the mental apparatus, and removes all unconscious obstructions to the unitary functioning of personality. It prepares the ground for self-integration or existential self-awareness (*samādhi*).

At first, existential self-awareness takes place on the mental level. This is called *savikalpa samādhi*. At this stage a person sees his own image reflected clearly and distinctly, as if on a flawless mirror or on the tranquil and transparent water of a pool. His purified mind, emancipated from the drive of desire and the taint of ignorance, is such a mirror. But still what is seen here is only an image of the self, the self known objectively, not the subjective reality of the self. So one has to advance still farther. At the next higher stage, the mental level is transcended. All mental functioning comes to a stop. The mirror or the pool disappears. A man now knows himself just by being his own true self. It is no more his image in a mirror that is seen, but his innermost reality. This is called *nirvikalpa samādhi*. This unobstructed and unmediated self-abiding is the essence of spiritual liberation according to *Rājayoga*.

Now, integral yoga fully appreciates the perfection of the technique which has been elaborately developed by *Rājayoga*. But it points out that the methods of *Rājayoga* are tailored to the concept of static realization of the self in its pure transcendence. They are not quite adequate for the purpose of dynamic self-identification with the Divine immanently operative in history. They are not quite suitable for man's intelligent co-operation with the creative force of cosmic evolution. For the fulfilment of the latter purpose, an active dedication from the very beginning to the cosmic purpose of existence is imperatively necessary. *Rājayoga* emphasizes the method of mental tranquillization as a means of attaining static self-realization. The danger of life negation, even though not a necessary sequel, is present in this approach. According to integral yoga, simultaneously with the processes of inward self-purification and mental serenity, active participation in life is essential. Social, cultural and humanitarian activities pursued in a spirit of self-offering to the Divine are an indispensable adjunct to the inward processes of concentration and meditation.

TĀNTRIC YOGA

Tāntric yoga is also known as *Kuṇḍalinī* or *Kuṇḍalī* yoga. It has some noteworthy characteristics of its own.

Tāntric yoga is closely connected with the worship of God as the supreme Mother. The Divine has two inseparable aspects: the archetypal masculine (Śiva), the archetypal feminine (Śakti). Śiva is pure Being, timeless perfection, eternal wisdom, logos. Śakti is the power of Becoming, the creative energy of time, the joy and love of self-expression, eros. Śakti is the Divine Mother who mediates between Being and the flux of becoming, between the Absolute and the sphere of relativity, between eternal perfection and the ceaseless flow of time. On the one hand Śakti is the medium of manifestation of the infinite in the finite. On the other hand she is the medium of self-fulfilment of the finite in the infinite. So the most natural approach for those who desire perfection is to seek the help, guidance and grace of the dynamic Divine.

The world as the manifestation of energy is an unceasing process, a perpetual flow. Our life is movement and action. But all movements, acts and processes ultimately flow from the universal creative energy, Śakti. This universal energy cannot be blind and unconscious. Nor is it conscious in the way in which the human mind is conscious. It is infinitely superior to human consciousness in depth of insight and breadth of vision. Having created the human individual, Śakti enters into him and dwells within him as his main support and centre of gravity. This dynamic nucleus, the central psycho-physical power latent in man, is called *Kuṇḍalinī*, the coiled power (the serpent energy). The serpent while resting stays in coiled form, and while moving and acting it uncoils itself. Similarly, the creative energy has its dormant and dynamic, static and kinetic aspects. Different vital functions such as respiration, digestion, procreation, elimination, etc. are different modes of operation of the *Kuṇḍalinī*. Different mental functions such as perception, reflection, emotion, volition, etc. are also modes of manifestation of the *Kuṇḍalinī*. Science teaches us that in the structure of an atom, there are electrons or negative charges of electricity which move round a positive nucleus which is apparently static. Likewise, in the human organism there are numerous vital and mental functions which are centrally supported by the positive nucleus of the *Kuṇḍalinī*.

Tāntric yoga is the art of splitting the spiritual atom in man. It is the technique of releasing the pent-up energies of the human psyche. When the *Kuṇdalinī* is dynamized, the individual experiences a tremendous upsurge of energy from within. He feels it as the power of God working within him. He feels that he is being guided from within, with infinite patience and love, by the Divine Mother. A re-orientation of his outlook towards spiritual values takes place. A deep longing for the eternal leaps into flame. New vistas of thought are opened. Centres of extra-sensory perception are stimulated. The search for the ultimate is intensified.

According to *Tāntric* yoga true spiritual development begins with the awakening of the *Kuṇḍalinī*. Prior to this awakening, all ethical and religious practices are in the nature of self-preparation. They purify the heart of the individual and direct his attention to the spiritual destiny of life. After the *Kuṇḍalinī* is awakened, spiritual growth seems to be guided no more by the ego but by a deeper power within. Meditation becomes in a sense effortless and spontaneous. A process of deepening self-awareness and joyful self-expansion sets in. Unconscious motivations are gradually brought to light, and a genuine spirit of dedication to the Divine begins to permeate the whole being.

According to *Tāntric* yoga, the ultimate goal of spiritual effort is the union of the dynamic and static aspects of personality. We have noted that Tantra affirms the reality of God as the unity of timeless perfection (Śiva) and the dynamism of time (Śakti). Man who is an image of God is also essentially the unity of the power of becoming and the perfection of being. Through social, cultural, ethical and religious activities, man prepares himself for the fulfilment of his spiritual destiny. The practice of yoga, which involves self-energizing and self-transcending, advances him to a higher phase of spiritual growth. This process of growth is carried to perfection when the timeless dimension of existence is discovered. In *Kuṇḍalinī* yoga union with the timeless is believed to be total and complete. It is not simply union through the intellect or the heart. It is a kind of total psychophysical union. It is the union of one's total self-energy with the timeless ground of being. Symbolically, it is represented as a sort of mystic marriage (*mahāmaithuna*) between the feminine and the masculine aspects of personality—between the principles of basic energy and pure existence. A flood of delight is released from this mystic union. On all the levels of body and mind indescribable waves of joy are experienced.

Tāntric yoga is boldly affirmative in its methodological approach. Other yoga systems have laid much stress upon renunciation and desirelessness as essential aids to liberation. But *Tāntric* yoga affirms the need for intelligent and organized fulfilment of natural desires. In its view there is no basic antagonism between nature and spirit. Nature is the creative power of spirit in the objective sphere. Nobody therefore can enter the kingdom of spirit without first obtaining a passport from nature. Practice of austerity, asceticism and self-mortification is an insult to nature. It creates more difficulties than it can solve. By weakening the body and producing inner conflicts and tensions, it undermines balanced and healthy development. It is only by following the spirit of nature that one can swim with the current and capture the kingdom of heaven by storm.

Worship of the Divine Mother implies appreciation of the presence of profound wisdom in nature, both external and internal. There is a principle of cosmic intelligence operative in external nature. It controls the process of cosmic evolution. Similarly, there is deep wisdom inherent in man's inner nature, in his unconscious psyche. It secretly determines his inner evolution. If a person intelligently follows the bent of his own nature, his desires become more and more refined and lofty. Base desires gradually yield place to noble desires. Lower impulses are replaced by higher impulses. When a child's natural desire to play with toys is duly satisfied, it is soon outgrown yielding place to a keen interest in books or living playmates. When a man's natural desire for sex is lawfully satisfied, it gives rise to a growing interest in social welfare or humanitarian service. When his desire for enjoying the world is duly satisfied on the basis of intelligent self-organization, one day it gives rise to a deeper longing for Transcendence.

So *Tāntric* yoga prescribes what is called desireful prayer and worship (*sakāma upāsanā*). All natural desires are accepted as modes of manifestation of the creative spirit of nature. The problem is to organize them intelligently with a view to the maximum satisfaction and fulfilment of one's nature. There is divine sanction behind such self-fulfilment. One can also invoke divine blessings in such self-fulfilment. One places one's desires before God, and then, with God's sanction and sanctification, proceeds to fulfil them in a spirit of self-offering to the Divine. This brings about an increasing refinement and spiritual transformation of one's desire-nature. A constructive channelling of the libido towards the higher ends of existence takes place.

Tantra believes in the principle of 'like cures like.' When a person suffers from water in his ears, the doctor injects more water into his ear drums so that all the accumulated water comes out. When a person gets sick on account of a certain kind of poison inside his body, the doctor may prescribe for him the same kind of poison to be taken in the right dose in medicinal form. When a person falls down to the ground, it is with the support of the ground that he jumps to his feet again. Similarly, it is with the help of passion that the problem of passion can be solved. Sensuous desires and personal ambitions are usually condemned as impediments to spiritual progress. Cravings for delicacies, for sex, for stimulants, etc., as well as longings for wealth, social position, political power, etc., are often regarded by religion as temptations of the devil. But *Tantra* says: "All these desires ultimately come from the divine will. In the final analysis they are aids to the process of evolution and progress. The important thing is to fulfil them with that understanding in a spirit of co-operation with the creative force of evolution. The creative force of evolution is no other than the divine will immanent in the world process. The more a person co-operates with the evolutionary impetus, the more his desires are purged of the egotistic taint and are transmuted into the pure flame of aspiration for divine life." This is the underlying truth of the *Tāntric* theory of the five M's. Such ingredients as wine (*Madya*), meat (*Māṁsa*), fish (*Matsya*), parched cereal (*Mudrā*), and sexual union (*Maithuna*) are considered valuable aids to vigorous growth and development. They represent different modes of manifestation of energy. Those who can make profitable use of them, in union with the supreme creative power, towards the ends of harmonious self-development and social progress, belong to the heroic type of yogi.

It may be observed here that the affirmative approach of *Tāntric* yoga contains a precious element of truth. It is a protest against the extreme tendencies of asceticism, self-mortification, world and life negation, etc. It affirms nature as the dynamism of spirit. It affirms life as the diversified expression of transcendence. But it has often the tendency to carry the spirit of affirmation too far. Over-emphasis upon life affirmation may prove as misleading as over emphasis upon ascetic renunciation. The balanced spiritual ideal lies midway between these two extremes. When people develop a strong ascetic tendency, it is necessary to tell them about the positive significance of life, nature and society. On the other hand, when people become too affirmative in following the way of nature, it is desirable to tell them about the transcendent glory of the spirit. Otherwise, one may get lost in the labyrinth of desire and practise self-deception in the name of religion. That is why we find that a huge amount of malpractice came in course of time to be associated with *Tāntric* yoga. Sexual promiscuity is sometimes sanctioned as a mode of concerted power worship. Black magic puts on the religious mask. Ruthless slaughter of animals is approved as symbolic self-sacrifice to God.

Even though *Tāntric* yoga is affirmative in its method and approach, it differs little from asceticism with regard to the ultimate goal of spiritual effort. The *Tāntric* method of life affirmation is also designed oftener than not to statically blissful union with the formless consciousness of the eternal. It has been said: "When Kundali 'sleeps' man is awake to this world. When she 'awakes' he sleeps, that is, loses all consciousness of the world and enters his causal body. In Yoga he passes beyond to formless Consciousness." Thus here again we come across the ideal of static and transcendent realization inherent in traditional mysticism. At this point integral yoga wishes to remind us emphatically that static and formless consciousness is not the ultimate goal. The formless nontemporal is only one aspect of Being. The evolving world of endless forms is another no less important aspect of Being. Our goal is to join forces with the evolution of higher forms and values in union with the formless depth of Being.

The concept of total union with the eternal is a very significant contribution of *Tantra*. We have to be integrated with Being not only by way of contemplation or devotion or love. Our entire existence including the physical and the unconscious has to be lifted up to the thrilling touch of the eternal. The conscious and unconscious aspects of our being have to be unified. The static and dynamic aspects of our personality have to be harmonized. But this notion of total union could not be developed in *Tantra* to its furthest logical sequel on account of limitations imposed by medieval metaphysics. Medieval metaphysics conceived of the inmost essence of Being in terms of transcendence and eternity. It had no adequate comprehension of the ontological significance of evolution and history.

In conformity with the metaphysical outlook of the age, total union is envisaged in *Tantra* as the union of a man's entire personality with formless eternity. The central psycho-physical energy (*Kuṇḍalī*) is awakened so that with his whole being the yogi may experience the transcendent bliss of the absolute, having completely withdrawn himself from the world of form and change. Now, strictly speaking, this is not total union in the full sense of the term. The eternal is incomplete without the historical. Perfection is incomplete without evolution. The creative flow of time is an essential factor in the structure of eternity. So total union must imply that while we are anchored in the timeless foundation of Being, we have to act in our historical situations as dynamic centres of Being. The ultimate purpose of the transformation of our physical and vital nature is to prepare us for the supreme task of life. That task is to establish higher values in society and to manifest the glories of eternity in time.

Thus in integral yoga the notion of total union develops into integral union. It is the union of the total self with total reality. It implies union with the creative force of evolution as well as union with the immutable joy of eternity. Contact with the eternal brings supreme wisdom and joy and love into our being. But wisdom, joy and love cannot be divorced from action. To know God is to love God. To love God is to serve God. To glimpse the will of God is to act for the glory of God in the world. Such action is co-operation with the creative force of evolution, because the latter is another name for the will of God operative in the world process. Enlightened action is indeed of the very essence of human reality. Through such action we joyfully participate in the movement of time, anchored in the serenity of the timeless. So it may be stated that integral union is the union of wisdom, love and action. It is union with history as well as with transcendence.

Tantra

HEINRICH ZIMMER

The following selection from Heinrich Zimmer's major work on Indian philosophy was edited by Joseph Campbell, who, like Zimmer, was familiar with the special role played by the Bengali saint of the nineteenth century, Sri Ramakrishna, in the revitalization of Hinduism. Ramakrishna was the quintessential Hindu and yet he opened up Hinduism to its vocation to become a universal religion in a way that no one else did. He saw that the truths of religion are universal and that Hinduism was uniquely equipped to express that universalism because of its tradition of a multiplicity of "paths" to the final goal of spiritual emancipation.

WHO SEEKS NIRVĀṆA?

The later Buddhist change of attitude toward the final goal is paralleled exactly by the contemporary Hindu development. As we have seen, in Hīnayāna usage the term *bodhisattva* denoted a great being on the point of becoming a Buddha and so passing from time to nirvāṇa, an archetype of the Buddhist lay-initiate escaping from the world, whereas in the Mahāyāna the concept was translated into a time-reaffirming symbol of universal saviorship. Through renouncing Buddhahood the Bodhisattva made it clear that the task of mokṣa, "release, liberation, redemption from the vicissitudes of time," was not the highest good; in fact, that mokṣa is finally meaningless, saṁsāra and nirvāṇa being equally of the nature of śūnyatā, "emptiness, the void." In the same spirit the Hindu Tāntric initiate exclaims: "Who seeks nirvāṇa?" "What is gained by mokṣa?" "Water mingles with water."

This point of view is rendered in many of the conversations of Śrī Rāmakrishna with his lay disciples.

"Once upon a time," he told them one evening, "a sannyāsin entered the temple of Jagganāth. As he looked at the holy image he debated within himself whether God had a form or was formless. He passed his staff from left to right to feel whether it touched the image. The staff touched nothing. He understood that there was no image before him; he concluded that God was formless. Next he passed the staff from right to left. It touched the image. The sannȳasin understood that God had form. Thus he realized that God has form and, again, is formless."

"What is vijñāna?" he said on another occasion. "It is knowing God in a special way. The awareness and conviction that fire exists in wood is jñāna, knowledge. But to cook rice on that fire, eat the rice, and get nourishment from it is vijñāna. To know by one's inner experience that God exists is jñāna. But to talk to Him, to enjoy Him as Child, as Friend, as Master, as Beloved, is vijñāna. The realization that God alone has become the universe and all living beings is vijñāna."

And with respect to the ideal of becoming annihilate in Brahman, he would sometimes say, quoting the poet Rāmprasād, "I love to eat sugar, I do not want to become sugar."

The Mahāyāna Bodhisattva tastes unending saviorship by devoting himself with absolute selflessness to his teaching task in the vortex of the world; in the same spirit, the Hindu Tāntric initiate, by persevering in the dualistic attitude of devotion (*bhakti*), enjoys without cease the beatitude of the knowledge of the omnipresence of the Goddess.

"The Divine Mother revealed to me in the Kālmi temple that it was She who had become everything," Śrī Rāmakrishna told his friends. "She showed me that everything was full of Consciousness. The Image was Consciousness, the altar was Consciousness, the water-vessels were Consciousness, the doorsill was Consciousness, the marble floor was Consciousness—all was Consciousness. I found everything inside the room soaked, as it were, in Bliss—the Bliss of Satcidānanda.[1] I saw a wicked man in front of the Kālī temple; but in him also I saw the Power of the Divine Mother vibrating. That was why I fed a cat with the food that was to be offered to the Divine Mother."

"The jñānī, sticking to the path of knowledge," he explained again, "always reasons about the Reality, saying, 'Not this, not this.' Brahman is neither 'this' nor 'that'; It is neither the universe nor its living beings. Reasoning in this way, the mind becomes steady. Then it disappears and the aspirant goes into samādhi. This is the Knowledge of Brahman. It is the unwavering conviction of the jñāī that Brahman alone is real and the world illusory, like a dream. What Brahman is cannot be described. One cannot even say that Brahman is a Person. This is the opinion of the jñānīs, the followers of Vedānta philosophy.

"But the bhaktas accept all the states of consciousness. They take the waking state to be real also. They don't think the world to be illusory, like a dream. They say that the universe is a manifestation of God's power and glory. God has created all these—sky, stars, moon, sun, mountains, ocean, men, animals. They constitute His glory. He is within us, in our hearts. Again, He is outside. The most advanced devotees say that He Himself has become all this—the twenty-four cosmic principles, the universe, and all living beings. The devotee of God wants to eat sugar, not to become sugar. (*All laugh.*)

"Do you know how the lover of God feels?" Rāmakrishna continued. "His attitude is 'O God, Thou are the Master, and I am Thy servant. Thou are the Mother, and I am Thy child.' Or again: 'Thou art my Father and Mother. Thou art the Whole, and I am a part.' He doesn't like to say, 'I am Brahman.'

"The yogī seeks to realize the Paramātman, the Supreme Soul. His idea is the union of the embodied soul and the Supreme Soul. He withdraws his mind from sense-objects and tries to concentrate it on the Paramātman. Therefore, during the first stage of his spiritual discipline, he retires into solitude and with undivided attention practices meditation in a fixed posture.

"But the Reality is one and the same. The difference is only in name. He who is Brahman is verily Ātman, and again, He is the Bhagavān, the Blessed Lord. He is Brahman to the followers of the path of knowledge, Paramātman to the yogīs, and Bhagavān to the lovers of God.

"The jñānīs, who adhere to the nondualistic philosophy of Vedānta, say that the acts of creation, preservation, and destruction, the universe itself and all its living beings, are the manifestations of Śakti, the Divine Power. If you reason it out, you will realize that all these are as illusory as a dream. Brahman alone is the Reality, and all else is unreal. Even this very Śakti is unsubstantial, like a dream.

"But though you reason all your life, unless you are established in samādhi, you cannot go beyond the jurisdiction of Śakti. Even when you say, 'I am meditating,' or 'I am contemplating,' still you are moving in the realm of Śakti, within Its power.

"Thus Brahman and Śakti are identical. If you accept the one, you must accept the other. It is like fire and its power to burn. If you see the fire, you must recognize its power to burn also. You cannot think of fire without its power to burn, nor can you think of the power to burn without fire. You cannot conceive of the sun's rays without the sun, nor can you conceive of the sun without its rays.

"What is milk like? Oh, you say, it is something white. You cannot think of the milk without the whiteness, and again, you cannot think of the whiteness without the milk.

"Thus one cannot think of Brahman without Śakti, or of Śakti without Brahman. One cannot think of the Absolute without the Relative, or of the Relative without the Absolute.

"The Primordial Power is ever at play. She is creating, preserving, and destroying in play, as it were. This Power is called Kālī. Kālī is verily Brahman, and Brahman is verily Kālī. It is one and the same Reality. When we think of It as inactive, that is to say, not engaged in the acts of creation, preservation, and destruction, then we call It Brahman. But when It engages in these activities, then we call It Kālī or Śakti. The Reality is one and the same; the difference is in name and form."

This introductory exposition of the Tāntric point of view was given on the deck of a little excursion-steamer, sailing up and down the Ganges, one beautiful autumn afternoon in 1882. Keshab Chandra Sen (1838–84), the distinguished leader of the semi-Hindu, semi-Christian Brāhmo Samāj, had come with a number of his following, to visit Śrī Rāmakrishna at Dakshineswar, a suburb of the modern city of Calcutta, where the saintly teacher was serving as priest in a temple dedicated to the Black Goddess, Kālī. Keshab was a modern, occidentalized Hindu gentleman, with a cosmopolitan outlook, and a sāttvic, humanistic, progressive religious philosophy—not unlike that of his New England contemporary, the Transcendentalist (and student of the *Bhagavad Gītā*), Ralph Waldo Emerson. Rāmakrishna, on the other hand, was a thorough Hindu—intentionally ignorant of English, nurtured in the traditions of his motherland, long-practiced in the techniques of introverted contemplation, and filled with the experience of God. The coming together of these two religious leaders was a meeting of the modern, timely India and the timeless—the modern consciousness of India with the half-forgotten divine symbols of its own unconscious. Noteworthy, moreover, is the fact that on this occasion the teacher was not the Western-educated, tailored gentleman, who had been entertained in London by the Queen, but the yogī in his loincloth, speaking of the traditional Indian Gods out of his own direct experience.

KESHAB (*with a smile*): "Describe to us, sir, in how many ways Kālī, the Divine Mother, sports in this world."

ŚRĪ RĀMAKRISHNA (*also with a smile*): "Oh, She plays in different ways. It is She alone who is known as Mahā-Kālī ["The Great Black One"], Nitya-Kālī ["The Everlasting Black One"], Śmaśāna-Kālī ["Kālī of the Cremation Ground"], Rakṣā-Kālī ["Goblin Kzmalī"], and Śyāmā-Kālī ["Dark Kālī"]. Mahā-Kālī and Nitya-Kālī are mentioned in the Tantra Philosophy. When there were neither the creation, nor the sun, the moon, the planets, and the earth, and when darkness was enveloped in darkness, then the Mother, the Formless One, Mahā-Kālī, the Great Power, was one with Mahā-Kāla [this is the masculine form of the same name], the Absolute.

"Śyāmā-Kālī has a somewhat tender aspect and is worshiped in the Hindu households. She is the Dispenser of boons and the Dispeller of fear. People worship Rakṣā-Kālī, the Protectress, in times of epidemic, famine, earthquake, drought, and flood. Śmaśāna-Kālī is the embodiment of the power of destruction. She resides in the cremation ground, surrounded by corpses, jackals, and terrible female spirits. From Her mouth flows a stream of blood, from Her neck hangs a garland of human heads, and around Her waist is a girdle made of human hands.

"After the destruction of the universe, at the end of a great cycle, the Divine Mother garners the seeds for the next creation. She is like the elderly mistress of the house, who has a hotch-potch-pot in which she keeps different articles for the household use. (*All laugh.*) Oh, yes! Housewives have pots like that, where they keep sea-foam, blue pills, small bundles of seeds of cucumber, pumpkin, and gourd, and so on. They take them out when they want them. In the same way, after the destruction of the universe, my Divine Mother, the Embodiment of Brahman, gathers together the seeds for the next creation. After the creation the Primal Power dwells in the universe itself. She brings forth this phenomenal world and then pervades it. In the Vedas creation is likened to the spider and its web. The spider brings the seb out of itself and then remains in it. God is the container of the universe and also what is contained in it.

"Is Kālī, my Divine Mother, of a black complexion? She appears black because She is viewed from a distance; but when intimately known She is no longer so. The sky appears blue at a distance; but look at air close by and you will find that it has no color. The water of the ocean looks blue at a distance, but when you go near and take it in your hand, you find that it is colorless."

Śrī Rāmakrishna, filled with love for the Goddess, then sang to her two songs of the Bengali devotee and yogī Rāmprasād, after which he resumed his talk.

"The Divine Mother is always sportive and playful. This universe is Her play. She is self-willed and must always have her own way. She is full of bliss. She gives freedom to one out of a hundred thousand."

A BRĀHMO DEVOTEE: "But, sir, if She likes She can give freedom to all. Why, then, has She kept us bound to the world?"

ŚRĪ RĀMAKRISHNA: "That is Her will. She wants to continue playing with Her created beings. In a game of hide-and-seek the running about soon stops if in the beginning all the players touch the 'granny.' If all touch her, then how can the game go on? That displeases her. Her pleasure is in continuing the game.

"It is as if the Divine Mother said to the human mind in confidence, with a sign from Her eye, 'Go and enjoy the world.' How can one blame the mind? The mind can disentangle itself from worldliness if, through her grace, She makes it turn toward Herself."

Singing again the songs of Rāmprasād, Śrī Rāmakrishna interrupted his discourse, but then continued. "Bondage is of the mind, and freedom is also of the mind. A man is free if he constantly thinks: 'I am a free soul. How can I be bound, whether I live in the world or in the forest? I am a child of God, the King of Kings. Who can bind me?' If bitten by a snake, a man may get rid of its venom by saying emphatically, 'There is no poison in me.' In the same way, by repeating with grit and determination, 'I am not bound, I am free,' one really becomes so—one really becomes free.

"Once someone gave me a book of the Christians. I asked him to read it to me. It talked about nothing but sin. (*To Keshab Chandra Sen:*) Sin is the only thing one hears of at your Brāhmo Samāj, too. The wretch who constantly says, 'I am bound, I am bound,' only succeeds in being bound. He who says day and night, 'I am a sinner, I am a sinner,' really becomes a sinner.

"One should have such burning faith in God that one can say: 'What? I have repeated the name of God, and can sin still cling to me? How can I be a sinner any more? How can I be in bondage any more?'

"If a man repeats the name of God, his body, mind, and everything become pure. Why should one talk about sin and hell, and such things? Say but once, 'O Lord, I have undoubtedly done wicked things, but I won't repeat them.' And have faith in his name."

Śrī Rāmakrishna sang:

> If only I can pass away repeating Durgā's name;
> How canst Thou then, O Blessed One,
> Withhold from me deliverance,
> Wretched though I may be? . . .

Then he said: "To my Divine Mother I prayed only for pure love, I offered flowers at Her Lotus Feet and prayed to Her: 'Mother, here is Thy virtue, here is Thy vice. Take them both and grant me only pure love for Thee. Here is Thy knowledge, here is Thy ignorance. Take them both and grant me only pure love for Thee. Here is Thy purity, here is Thy impurity. Take them both, Mother, and grant me only pure love for Thee. Here is Thy dharma, here is Thy adharma. Take them both, Mother, and grant me only pure love for Thee."

In Tantra the theistic attitude practically obliterates the abstract ideal of the Formless Brahman (*nirguṇa brahman*) in favor of Brahman-in-the-Guṇas (*saguṇa brahman*)—the Lord (*īśvara*), the personal God; and the latter is represented by the Tāntrics preferably in the female aspect, since in this the nature of Māyā-Śakti is most immediately affirmed. The Tāntric development supported the return to power in popular Hinduism of the figure of the Mother Goddess of the innumerable names—Devī, Durgā, Kālī, Pārvatī, Umā, Satī, Padmā, Candī,

Tripura-sundarī, etc.—whose cult, rooted in the Neolithic past, had been overshadowed for a period of about a thousand years by the male divinities of the patriarchal Āryan pantheon. The Goddess began to reassert herself in the period of the later Upaniṣads. She is today the chief divinity again. All the consorts of the various gods are her manifestations, and, as the śakti or "power" of their husbands, represent the energy that has brought the latter into manifestation. Moreover, as Mahāmāyā, the Goddess personifies the World Illusion, within the bounds and thralldom of which exist all forms whatsoever, whether gross or subtle, earthly or angelic, even those of the highest gods. She is the primary embodiment of the transcendent principle, and as such the mother of all names and forms. "God Himself," states Rāmakrishna, "is Mahāmāyā, who deludes the world with Her illusion and conjures up the magic of creation, preservation, and destruction. She has spread this veil of ignorance before our eyes. We can go into the inner chamber only when She lets us pass through the door." It is entirely possible that in this reinstatement of the Goddess, both in the popular cults and in the deep philosophy of the Tantra, we have another sign of the resurgence of the religiosity of the non-Āryan, pre-Āryan, matriarchal tradition of Dravidian times.

NOTE

[1] Brahman as Being (*sat*), Consciousness (*cit*), and Bliss (*ānanda*).

Yoga: Immortality and Freedom

MIRCEA ELIADE

Mircea Eliade had a distinguished career as professor at the Divinity School of the University of Chicago and has been highly regarded as one of the foremost interpreters of religion in the modern period. His formative work in the field was done in India, where he went as a young man to study in Calcutta. In the following selection, a direct result of that early period of field study in India, Eliade clearly presents the theory of the Chakras and their relation to the concept of Kundalini, central to Tantric practices. Eliade's work, based on thorough scientific principles, is an exposition of the reality of the classical theory of Tantra and Kundalini, rather than a new interpretation of it such as can be found in the earlier selection by Chaudhuri.

THE *CAKRAS*

According to Hindu tradition, there are seven important *cakras*, which some authorities refer to the six plexuses and the *sutura frontalis.*

1. The *mūlādhāra* (*mūla* = root) is situated at the base of the spinal column, between the anal orifice and the genital organs (sacrococcygeal plexus). It has the form of a red lotus with four petals, on which are inscribed in gold the letters *v*, *ṣ*, *ś*, and *s*. In the middle of the lotus is a yellow square, emblem of the element earth (*pṛthivī*); at the center of the square is a triangle with its apex downward, symbol of the *yoni*, and called Kāmarūpa; at the center of the triangle is the *svayambhū-liṇga* (the *liṇga* existing by itself), its head as brilliant as a jewel. Coiled eight times (like a serpent) around it, as brilliant as lightning, sleeps Kuṇḍalinī, blocking the opening of the *liṇga* with her mouth (or her head). In this way Kuṇḍalinī obstructs the *brahmadvāra* (the "door of Brahman") and access to the *suṣumṇā*. The *mūlādhāra cakra* is related to the cohesive power of matter, to inertia, the birth of sound, the sense of smell, the *apāna* breath, the gods Indra, Brahmā, Ḍākinī, Śakti, etc.
2. The *svādhiṣṭhāna cakra,* also called *jalamaṇḍala* (because its *tattva* is *jala* = water) and *meḍhrādhāra* (*meḍhra* = penis), is situated at the base of the male genital organ (sacral plexus). Lotus with six vermilion petals inscribed with the letters, *b, bh, m, r, l.* In the middle of the lotus, a white half-moon, mystically related to Varuṇa. In the middle of the moon, a *bīja-mantra,* at the center of which is Viṣṇu flanked by the goddess Cākinī. The *svādhiṣṭhāna cakra* is related to the element water, the color white, the *prāṇa* breath, the sense of taste, the hand, etc.
3. The *maṇipūra* (*maṇi* = jewels; *pūra* = city) or *nābhiṣṭhāna* (*nābhi* = umbilicus), situated in the lumbar region at the level of the navel (epigastric plexus). Blue lotus with ten petals and the letters *ḍ, ḍh, ṇ, t, th, d, dh, n, p.*

ph. In the middle of the lotus, a red triangle and on the triangle the god Mahārudra, seated on a bull, with Lākinī Śakti (blue in color) beside him. This *cakra* is related to the element fire, the sun, *rajas* (menstrual fluid), the *samāna* breath, the sense of sight, etc.

4. The *anāhata* (*anāhahata śabd* is the sound produced without contact between two objects; i.e., a mystical sound); region of the heart, seat of the *prāṇa* and of the *jīvātman*. Color, red. Lotus with twelve golden petals (letters *k, kḥ, g, gḥ*, etc.). In the middle, two interlaced triangles forming a Solomon's seal, in the center of which is another golden triangle enclosing a shining *liṇga*. Above the two triangles is Īśvara with the Kākinī Śakti (red in color). The *anāhata cakra* is related to the element air, the sense of touch, the phallus, the motor force, the blood system, etc.
5. The *viśuddha cakra* (the *cakra* of purity); region of the throat (laryngeal and pharyngeal plexus, at the junction of the spinal column and the medulla oblongata), seat of the *udāna* breath and of the *bindu*. Lotus with sixteen petals of smoky purple (letters *a, ā, i, u, ū*, etc.). Within the lotus a blue area, in the center of which is a white circle containing an elephant. On the elephant rests the *bīja-mantra h* (*Haṇg*), supporting Sadāśiva, half silver, half golden, for the god is represented under his androgynous aspect (*ardhanāriśvara*). Seated on a bull, he holds in his many hands a multitude of objects and emblems proper to him (*vajra*, trident, bell, etc.). One half of his body constitutes the Sadā Gāurī, with ten arms and five faces (with three eyes each). The *viśuddha cakra* is related to the color white, the ether (*ākāśa*), sound, the skin, etc.
6. The *ājñā* (= order, command) *cakra*, situated between the eyebrows (cavernous plexus). White lotus with two petals bearing the letters *h* and *kṣ*. Seat of the cognitive faculties: *buddhi, ahaṃkāra, manas*, and the *indryas* (= the senses) in their "subtle" modality. In the lotus is a white triangle, apex downward (symbol of the *yoni*); in the middle of the triangle a white *liṇga*, called *itara* (the "other"). Here is the seat of Paramaśiva. The *bīja-mantra* is OṂ. The tutelary goddess is Hākinī; she has six faces and six arms and is seated on a white lotus.
7. The *sahasrāra cakra:* at the top of the head, represented under the form of a thousand-petaled lotus (*sahasrā* = thousand), head down. Also called *brahmasthāna, brāhmarandhra, nirvāṇa cakra*, etc. The petals bear all the possible articulations of the Sanskrit alphabet, which has fifty letters (50 × 20). In the middle of the lotus is the full moon, enclosing a triangle. It is here that the final union (*unmanī*) of Śiva and Śakti, the final goal of tantric *sādhana*, is realized, and here the *kuṇḍalinī* ends its journey after traversing the six *cakras*. We should note that the *sahasrāra* no longer belongs to the plane of the body, that it already designates the plane of transcendence—that this explains why writers usually speak of the doctrine of the "six" *cakras*.

There are other *cakras*, of less importance. Thus, between the *mūlādhāra* and the *svādhiṣṭhāna* is the *yoniṣṭhāna*; this is the meeting place of Śiva and Śakti, a place of bliss, also called (like the *mūlādhāra*) *kāmarūpa*. It is the source of desire and, on the carnal level, an anticipation of the union Śiva-Śakti, which is accomplished in the *sahasrāra*. Near the *ājñā cakra* are the *manas cakra* and the *soma cakra*, related to the intellective functions and to certain yogic experiences. Near the *ājñā cakra* again is the *kārana-rūpa*, seat of the seven "causal forms," which are held to produce and constitute the "subtle" and the "physical" bodies. Finally, other texts refer to a number of *ādhāras* (= supports, receptacles), situated between the *cakras* or identified with them.

The Buddhist tantras speak of only four *cakras*, situated respectively in the umbilical, cardiac, and laryngeal regions and the cerebral plexus; this last *cakra*, the most important, is called *uṣṇīṣa-kamala* (lotus of the head) and corresponds to the *sahasrāra* of the Hindus. The three lower *cakras* are the sites of the three *kāyas* (bodies): *nirmāṇa-kāya* in the umbilical *cakra*, *dharma-kāya* in the *cakra* of the heart, *sambhoga-kāya* in the *cakra* of the throat. But there are anomalies and contradictions in regard to the number and locations of these *cakras*. As in the Hindu traditions, the *cakras* are associated with *mudrās* and goddesses: Locanā, Māmakī, Pāṇḍarā, and Tārā. In connection with the four *cakras*, the *Hevajra-tantra* gives a long list of quaternities: four *tattvas*, four *aṇgas*, four moments, four noble truths *(ārya-sattva)*, etc.

Viewed in projection, the *cakras* constitute a *maṇḍala* whose center is marked by the *brāhmarandhra*. It is in this "center" that the rupture of plane occurs, that the paradoxical act of transcendence—passing beyond *saṃsāra*, "emerging from time"—is accomplished. The symbolism of the *maṇḍala* is also a constituent element of Indian temples and sacred edifices in general: viewed in projection, a temple is a *maṇḍala*. We may, then, say that any ritual ambulation is equivalent to an approach to the center and that entrance into a temple repeats the initiatory entrance into a *maṇḍala* or the passage of the *kuṇḍalinī* through the *cakras*. Then again, the body is transformed both into a microcosm (with Mount Meru as center) and into a pantheon, each region and each "subtle" organ having its tutelary divinity, its *mantra*, its mystical letter, etc. Not only does the disciple identify himself with the cosmos; he also rediscovers the genesis and destruction of the universe in his own body. As we shall see, tantric *sādhana* comprises two stages: (1) cosmicization of the human being and (2) transcendence of the cosmos—that is, its destruction through the unification of opposites ("sun"—"moon," etc.). The pre-eminent sign of this transcendence is found in the final act of the *kuṇḍalinī's* ascent—its union with Śiva, at the summit of the skull, in the *sahasrāra*.

KUṆḌALINĪ

Some aspects of the *kuṇḍalinī* have already been mentioned; it is described at once under the form of a snake, of a goddess, and of an "energy." The *Haṭhayogapradīpikā* (III, 9) presents it in the following terms: "Kuṭilāṇgī ('crooked bodied'), Kuṇḍalinī, Bhujangī ('a she-serpent'), Śakti, Īśvarī, Kuṇḍalī, Arundhatī—all these words are synonymous. As a door is opened with a key, so the Yogī opens the door of *mukti* (deliverance) by opening Kuṇḍalinī by means of Haṭha Yoga." When the sleeping goddess is awakened through the grace of the *guru*, all the *cakras* are quickly traversed. Identified with Śabdabrahman, with *OṂ*,

Kuṇḍalinī possesses all the attributes of all gods and all goddesses. Under the form of a snake, it dwells in the midpoint of the body (*dehamadhyayā*) of all creatures. Under the form of Paraśakti, *kuṇḍalinī* manifests itself at the base of the trunk (*ādhāra*); under the form of Paradevatā, it dwells in the middle of the knot at the center of the *ādhāra,* whence the *nāḍīs* issue. Kuṇḍalinī moves in the *suṣumṇā* by the force aroused in the inner sense (*manas*) by the *prāṇa,* is "drawn upward through the *suṣumṇā* as a needle draws a thread." Kuṇḍalinī is awakened by *āsanas* and *kumbhakas*; "then the Prāṇa becomes absorbed in Śūnya" (the Void).

The awakening of the *kuṇḍalinī* arouses an intense heat, and its progress through the *cakras* is manifested by the lower part of the body becoming as inert and cold as a corpse, while the part through which the *kuṇḍalinī* passes is burning hot. The Buddhist tantras even more strongly emphasize the fiery nature of the *kuṇḍalinī.* According to the Buddhists, the Śakti (also called Caṇḍālī, Ḍombī, Yoginī, Nairāmani, etc.) lies sleeping in the *nirmāṇa-kaya* (umbilical region); the adept wakes her by producing the *bodhicitta* in that region, and her awakening is revealed by the sensation of a "great fire." The *Hevajra-tantra* says that "the Caṇḍālī burns in the navel," and that when everything is burned, the moon (situated in the forehead) lets fall drops of nectar. A poem of Guñjarīpāda's presents the phenomenon in veiled terms: "The lotus and the thunder meet together in the middle and through their union Caṇḍālī is ablaze; that blazing fire is in contact with the house of the Ḍombī—I take the moon and pour water. Neither scorching heat nor smoke is found, but it enters the sky through the peak of Mount Meru." We shall return to the signification of this symbolic vocabulary. But let us now note the following fact: according to the Buddhists, the secret force lies asleep in the umbilical region; through yogic practice, it is kindled and, like a fire, reaches the two higher *cakras* (*dharma* and *sambhoga*), enters them, reaches the *uṣṇīṣa-kamala* (corresponding to the *sahasrāra* of the Hindus), and, having "burned" everything in its path, returns to the *nirmāṇa-kaya.* Other texts explain that this heat is obtained by the "transmutation" of sexual energy.

The relations between the *kuṇḍalinī* (or the Caṇḍālī) and fire, the production of inner fire by the ascent of the *kuṇḍalinī* through the *cakras,* are facts requiring emphasis at this point. The reader will remember that, according to some myths, the goddess Śakti was created by the fiery energies of all the gods. Then, too, it is well known that the production of "inner heat" is a very old "magical" technique, which reached its fullest development in shamanism. We may note here and now that tantrism adheres to this universal magical tradition, although the spiritual content of its principal experience—the "fire" kindled by the ascent of the *kuṇḍalinī*—is on quite another plane than that of magic or shamanism. In any case, we shall have occasion, again in connection with the production of "inner heat," to examine the symbiosis tantrism-shamanism attested in certain Himālayan practices.

It is impossible to set about waking the *kuṇḍalinī* without the spiritual preparation implied by all the disciplines we have just discussed. But its actual awakening and its journey through the *cakras* are brought about by a technique whose essential element is arresting respiration (*kumbhaka*) by a special bodily position (*āsana, mudrā*). One of the most frequently used methods of arresting the breath is that prescribed by the *khecarīmudrā:* obstructing the cavum by turning the tongue back and inserting the tip of it into the throat.[1] The abundant salival secretion thus produced is interpreted as celestial ambrosia (*amṛta*) and the flesh of the tongue itself as the "flesh of the cow" that the yogin "eats." This symbolical interpretation of a "physiological situation" is not without interest; it is an attempt to express the fact that the yogin already participates in "transcendence": he transgresses the strictest of Hindu prohibitions (eating cow flesh)—that is, he is no longer conditioned, is no longer in the world; hence he tastes the celestial ambrosia.

This is but one aspect of the *khecarīmudrā;* we shall see that the obstruction of the cavum by the tongue and the arrest of breathing that follows are accompanied by a sexual practice. Before presenting it, we shall add that the awakening of the *kuṇḍalinī* brought about in this way is only the beginning of the exercise; the yogin further endeavors both to keep the *kuṇḍalinī* in the median "duct" (*suṣumṇā*) and to make it rise through the *cakras* to the top of the head. Now, according to the tantric authors themselves, this effort is rarely successful. We quote the following enigmatic passage from the *Tārākhaṇḍa* (it also occurs in the *Tantrasāra*): "Drinking, drinking, again drinking, drinking fall down upon earth; and getting up and again drinking, there is no re-birth." This is an experience realized during total arrest of respiration. The commentary interprets: "During the first stage of Shaṭchakra Sâdhana [= penetration of the *cakras*], the Sâdhaka [disciple] cannot supress his breath for a sufficiently long time at a stretch to enable him to practice concentration and meditation in each centre of Power [*cakra*]. He cannot therefore, detain Kuṇḍalinî within the Sushumnâ longer than his power of Kumbhaka [arresting respiration] permits. He must, consequently, come down upon earth—that is the Mûlâdhâra Chakra—which is the center of the element, earth, after having drunk of the heavenly ambrosia [which can be understood in various ways, and especially as the copious salivation provoked by obstruction of the cavum]. The Sâdhaka must practice this again and again, and by constant practice the cause of re-birth . . . is removed."

To hasten the ascent of the *kuṇḍalinī*, some tantric schools combined corporal positions (*mudrā*) with sexual practices. The underlying idea was the necessity of achieving simultaneous "immobility" of breath, thought, and semen. The *Gorakṣa Saṃhitā* (61–71) states that during the *khecarīmudrā* the *bindu* (= sperm) "does not fall" even if one is embraced by a woman. And, "while the *bindu* remains in the body, there is no fear of Death. Even if the *bindu* has reached the fire (is ejaculated) it straightway returns, arrested . . . by the *yonimudrā*." The *Haṭhayogapradīpikā* (III, 82) recommends the *vajrolīmudrā* for the same purpose. Milk and an "obedient woman" are required. The yogin "should draw back up again with the *meḍhra* (the *bindu* discharged in *maithuna*)," and the same act of reabsorption must also be performed by the woman. "Having drawn up his own discharged *bindu* [the Yogi] can preserve (it). . . . By the loss of *bindu* (comes) death, from its retention, life." All these texts insist upon the interdependence between the breath, psychomental experience, and the *semen virile*. "So long as the air (breath) moves, *bindu* moves; (and) it becomes stationary (when the air) ceases to move. The Yogī should, therefore, control the air and obtain immovability. As long as *prāṇa* remains in the body, life (*jīva*) does not depart."

We see, then, that the *bindu* is dependent upon the breath and is in some sort homologized to it; for the departure of the one as of the other is equivalent to death. The *bindu* is also homologized to the *citta,* and thus there is finally homology of the three planes on which either "movement" or "immobility" are exercised: *prāṇa, bindu, citta.* As the *kuṇḍalinī* is weakened by a violent stoppage—whether of respiration or of seminal emission—it is important to know the structure and ramifications of the homology among the three planes. The more so as the texts do not always make it clear to which plane of reference they apply.

NOTES

[1] To succeed, the frenum of the tongue must be cut beforehand.

Buddhist Texts Through the Ages

EDWARD CONZE

Edward Conze was the foremost interpreter of Buddhism to the West in the post World War II period. Among his many publications, this volume of excerpts from Buddhist texts is one of the most useful short introductions to the primary sources of the Buddhist tradition. The examples of Tantric texts that follow introduce the reader to the theological aspects of Tantra and its practice in Kundalini Yoga, as well as to theories of the mandala and the use of mantra in meditation. In these theologies and practices, the physical body itself becomes "divinized" through the "magical" means of the meditation technique. These are striking features of Tantra theory and practice that distinguish it from the yoga practices of other Hindu and Buddhist traditions.

THE ATTAINMENT OF THE REALIZATION OF WISDOM AND MEANS *by Anangavajra*

Salutation to Vajrasattva!

Having made salutation again and again to him who consists of Wisdom and Means, of him I now speak, in whom there arises the threefold body of Buddhahood: the matchless Dharma-Body, pure in essence and untrammelled by the whole veil of unreal phenomena, the Sambhoga-Body which is the basis for the spreading of the Good Law, the Nirmana-Body which is effected in diverse forms. Of him I Anangavajra shall now briefly speak for the benefit of all beings who are ignorant of his true nature.

The wise explain phenomenal existence, that deluder of simple minds, as the result of a false construction, the essence of which is the notion of existence. Thence there arises continually and under varying forms a whole mass of defilements, so hard to bear, and a vast accumulation of karma. For those whose minds thus cling to falsehood these two produce perpetual suffering of diverse kinds, of which the chief are death and birth. So long as the notion of existence remains the fixation of men who tarry in the prison of phenomenal life, what good can they do themselves or others in such deprivation of wisdom? And so the wise who desire to bring joy to the triple world and remove their own mental confusion must abandon altogether the fixation of this notion of existence. But having abandoned the fixation of existence, the wise man should not conceive a state of non-existence, for if indeed he should discriminate between these two, his conceits are not destroyed. Far preferable is a conception of existence, but never the conceit of non-existence; a burning lamp may be extinguished, but if it is already extinguished, then what course should one pursue? So long as one conceives of existence, one remains aware of existence (and there is always the possibility of realizing its true nature), but there can be no realization of the end of space which has here no beginning, and nothing can be realized as mere non-existence, for it would be deprived of all means. It would be useless to oneself and others, for it would be as non-existent as flowers that grow in the sky. Thus those who desire the fruit of Buddhahood should renounce the notion of existence because it is deluding like a magical display, but they should also renounce the notion of non-existence, for it is non-existent. O Wise Ones, do ye now hearken, for in so far as one renounces both extremes, the state in which one abides is neither Samsara nor Nirvana, for one has renounced these two.

The non-substantiality of things which is realized by reflection and by discriminating between the act of knowing and what is known, is called the essence of Wisdom.

Because one is passionately devoted to all beings who have failed to extricate themselves from a whole flood of suffering, this passionate devotion of which their suffering is the cause is known as Compassion. In that one thereby brings a man to the desired end by a combination of appropriate measures; it is also called the Means.

The mingling of both, which is like that of water and milk, is known as Wisdom-Means in a union free of duality. It is the essence of Dharma, to which nothing may be added and from which nothing may be withdrawn. It is free from the two notions of subject and object, free from being and non-being, from characterizing and characteristics; it is pure and immaculate in its own nature. Neither duality nor non-duality, calm and tranquil it consists in all things, motionless and unflurried; such is Wisdom-Means which may be known intuitively. It is this that is called the supreme and wondrous abode of all Buddhas, the Dharma-sphere, the divine cause of the perfection of bliss. It is Nirvana Indeterminate (apratiṣṭhitanirvāṇa) and is frequented by the Buddhas of the Past, Present and Future; it is the blissful stage of self-consecration (svādhiṣṭhāna), the beatitute of the Perfection of Wisdom. The three Buddha-bodies, the three Buddhist vehicles, mantras in their innumerable thousands, mudras and mandala-circles, phenomenal existence and that which transcends it, all arise from the same source; gods and asuras and men, disembodied spirits and whatever else exists, all spring from her and return here to their cessation. It abides always in all things like a wish-granting gem; it is the final stage of Enjoyment and Release. It is here that the Blessed Ones met in times past and so became Buddhas, and it is here that those intent on the good of the world become Buddhas now and will always do so in future. It is called the Great Bliss, for it consists of bliss unending; it is the Supreme One, the Universal Good, the producer of Perfect Enlightenment. The great sages define this truth, which is the supreme bliss of self and others, as the union of limitless Compassion which is intent alone on the destruction of all the world's suffering, and of perfect Wisdom which is free from all attachment and is an accumulation of knowledge which may not be reckoned, so great is its diversity.

II

And now in order that the jewel of this truth may be obtained, I Anangavajra, my mind penetrated with compassion, will discourse a little concerning the Means, but briefly, for it is just as has been taught by the Buddhas of old. It should benefit all those who are bewildered by the delusion of phenomenal existence which is so hard to destroy. But even Buddhas may not speak on this matter, saying that this is thus, for since it must be realized intuitively, it cannot be contained in words. And so it has been revealed as a process by means of the practices of many sutras and mantras by those Buddhas of the Past, Present and Future who have brought joy to the world. But this process itself can in no wise express the actual knowledge of tradition and so on, and since there is no connection between the sound and the true meaning, it merely shows the characteristics of a particular treatise. So a wise man must resort to a good master, for without him the truth cannot be found even in millions of ages. And if truth is not found, the final goal can never be reached, just as without a seed a plant will not grow even in the best and clearest of fields. So when in the course of life one comes upon masters with this truth, those who are teachers of Wisdom-Means, firm in line of succession, wondrous as a wish-granting gem and established in the path that is free from querulous thought, one should honour them to the best of one's ability if final perfection of the self is really one's aim. It is by means of their splendour that the bliss of infinite enlightenment is gained which is the highest goal possible for all beings in this triple world of moving and motionless things. So good men who desire their own perfection always pay with their whole being full honour to their master, who is the bestower of infinite rewards. They abandon envy and malignancy, and pride and self-conceit, their determination set on enlightenment and the concept of weariness renounced, and thus they always honour their guru, master of the world, who bestows sucess in all things. They are zealous in conduct and unwavering in mind and without thought for their own affairs they bow with head to his feet three times with great devotion. Thereby they gain by the grace of their guru and without any obstruction that truth supreme which is taught by all the Buddhas. It is eternal, resplendent and pure, the abode of the conquerors, the divine substance in all things and the source of all things. Just as a sun-stone shines brightly from the proximity of the sun-light which dispels the enclosing darkness, even so does the jewel of a pupil's mind, freed from the murkiness of impurity, light up from the proximity of a world-teacher who is bright with the fire of the practice of truth. As soon as he becomes thus enlightened, ablaze with the jewel of truth, this happy pupil is quickly approached by the Buddhas of the ten directions whose thoughts are pervaded with compassion, and by them he is firmly consecrated as a perfect Buddha. From the consecration of these Buddhas he becomes himself the equal of all Buddhas.

Thus those who abandon the cloak of pride and envy and deceit, and firm in their resolve serve with unequalled faith their master who embodies the traditional succession, they certainly shall gain the jewel of truth, which is the abode of all Buddha-qualities, for they have won supreme enlightenment which is the possession of all the Blessed Ones.

III

Now the rite of consecration, the basis of the triple world, shall be expounded for the benefit of all practisers who aim to reach the stage of Vajrasattva. When a wise man is consecrated in the tradition of the way of mantras, all Buddhas become manifest to him in the mandala which is their abode. So the thoughtful man who fears the loss of this sacramental experience does not let go of the Lord of the realms of endless worlds, when once he has attained the process of self-consecration. By perfected Buddhas it is taught that in the mantra-way there exists in its absolute aspect that sacramental experience of Vajrasattva and the other divine beings which is so hard to withstand. And so a son of the Conquerors for the sake of this consecration strives with all his might, and with the observation of all due ceremony approaches his vajraguru, that ocean of all good qualities. Having found a yogini with wondrous eyes and endowed with youth and beauty, he should deck her with fair raiment, and garlands and the scent of sandal-wood, and commit her to his master, and together with her he should honour and praise him with all his might, using scents and garlands and milk and other kinds of offerings. Then placing in all faith his knees upon the ground, he should beseech his master with this hymn of praise, stretching forth his hands in supplication:

"Hail to thee, womb of the Void, who art free of all conceits, omniscient one, thou mass of knowledge, knowledge personified, all hail to thee!

"Thou, teacher of the pure essence of truth which makes an end of worldly knowledge, O Vajrasattva, born of the nonsubstantiality of all things, hail to thee!

"From you, O Lord, there ever rise into existence Buddhas and Bodhisattvas, who possess as their good qualities the great perfections, O Thought of Enlightenment, hail to thee!

"From you, O Lord of the world, come the Three Jewels and the Great Way and the whole triple world with its moving and motionless things. All hail to thee!

"Thou art wondrous as a wish-granting gem for producing the welfare of the world; O Buddha-Son, most blessed, who dost command all the blessed ones, all hail to thee!

"It is by your grace, O ocean of good qualities, that I have known the truth, and now, O Omniscient One, may thou favour me with the vajra-consecration.

"Do thou favour me with the secret of all Buddhas, even as it has been taught by the Supreme Vajra of Thought, embodiment of the Vajra of the Dharma.

"If once I leave your lotus-feet, I have no other way to go. Therefore do thou, Lord, conqueror of the foe which is samsara, have compassion."

Then the worthy vajra-guru, filled with sympathy and intent on good, makes manifest his compassion and calls the pupil into the mandala. It is strewn with the five kinds of delectable things resplendent with a canopy spread above; it resounds with bells and cymbals, and is pleasant with flowers and incense and garlands and heavenly perfumes; it is the most wondrous resort of Vajrasattva and other divine beings and is prepared for union with the yogini. Then joining with the yogini the most worthy master places the Thought of Enlightenment in the lotus-vessel, which is the abode of the Buddhas. Next he consecrates the pupil, now joined with the yogini, with chowries and umbrellas held high and with propitious songs and verses. Having thus bestowed upon him the consecration, that excellent gem, he should give him the fivefold sacrament, delightful, divine, essentially pure. It consists of the precious jewel, with camphor, red sandal-wood and vajra-water, and as fifth component the empowering mantra. "This is your sacrament, Beloved, prescribed by former Buddhas. Do thou protect it always, and hearken now to the vow you must keep. Never must you take life, and never abandon the Three Jewels, and never abandon your master. This is the vow you transgress at your peril."

Then he should give leave of departure to that pupil, who having been thus consecrated with the Thought of Enlightenment, has become freed from his evil ways and is now the foremost son of the Buddhas. "O best of beings, until you enter final enlightenment, cause the excellent wheel of the Law to revolve everywhere throughout all the quarters. He who is the essence of Wisdom-Means is said to be like a wish-granting gem, for he is free of suffering and free of attachment, so serve now the cause of living beings!"

Having thus received the consecration and his leave to depart, joyful and content, he should pronounce pleasant words, which cause joy to the world. "Today my birth has proved fruitful, and fruitful is now my life. Today I am born in the Buddha-family, and I am now a son of the Buddhas. Thou, O Lord, hast rescued me from the ocean of ages, vast in its terror, and disturbed with the waves which are our births, from the mud of molestations, which is so hard to cross. I know that it is by your grace that I have emerged; I know that I am free from desire, and that all past influences are cast away." Then falling with devotion at his master's feet, his eyes wide-open with joy, he adds: "And may thou please inform me whatever thing thou desirest most." Then spontaneously his compassionate master should accept the offer, for it will tend to the destruction of the pupil's acquisitiveness and will be for his good. Thus the pupil, having gained the prize he sought, should present his gift with praise and worship, and once again should supplicate his master: "Now upon me has descended your favour, O foremost of Buddhas, and by this favour of yours I achieve the highest enlightenment, such as is your possession. And having gained highest Buddhahood, that stage which is praised by all pre-eminent ones, there I shall establish all beings who wander in the triple world."

Having understood in one's heart a pupil's aspiration, one should grant him consecration in conformity with the methods prescribed, and then in one's own devotion to a discipline so vast and profound, one should give him, also verbally, the jewel of consecration. And he, now a true yogin, who has received the great consecration of the Thought of Enlightenment, which is the matchless enjoyment of perfection of the Bearer of the Vajra, the acquisition of Lakshmi, having received his leave of departure and with his mind set on gaining victory over that evil foe which is the triple world, should raise his thought to enlightenment and pursue the great and immaculate way.

Note on the Mandala

The Mandala is a circle of symbolic forms, which is either just mentally produced for special purposes of meditation, or actually marked out on the ground for the purpose of special ritual. Its function is always the same, namely as a means towards the reintegration of the practitioner.

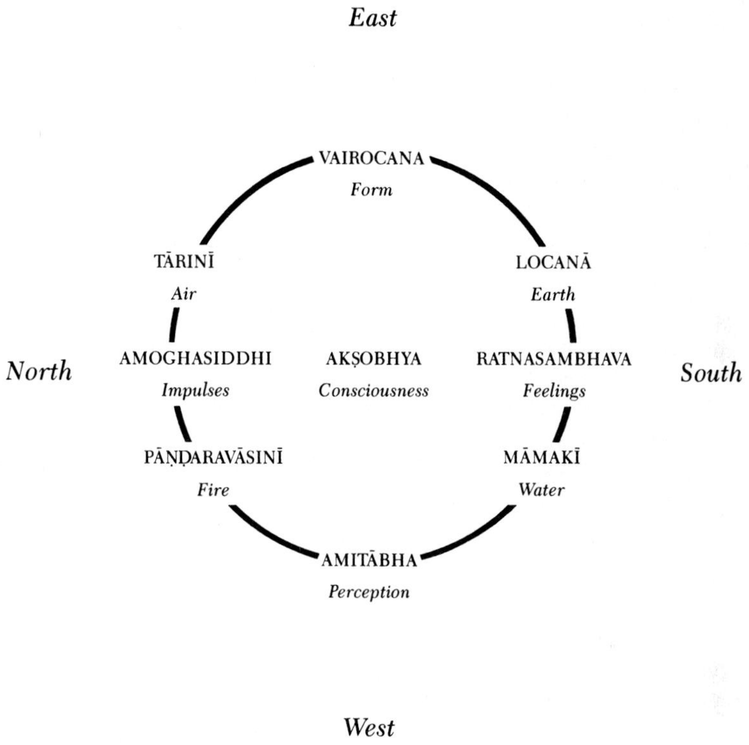

The Five Buddhas and the Four Goddesses and some of the Elements of the Universe they symbolize

There is a vast variety of different sets of symbolic forms, but they all follow the same fundamental pattern: that is to say, one symbol at the centre, which represents absolute truth itself, and other symbols arranged at the various points of the compass, which represent manifested aspects of this same truth. The simplest set of such symbolic forms consists of the group of the Five Buddhas or Tathagatas, who in the extract that now follows are conceived of as emanating from their respective seed-syllables (for they are each associated with a particular sound, the correct intoning of which will conjure them forth), and taking up their positions to the centre, east, south, north and west. It will also be seen that they are each associated with one of the five skandhas, with one of the five fundamental evils (wrath, passion, envy, malignity and delusion), with one of the five Wisdoms, with a particular taste, with a particular season, with a particular time of the day and with a particular substance used for sacramental purposes.

They serve therefore as symbols not only of the final perfection that is the goal of the aspirant, but also as symbols of those very evils which bind him to existence and also of his own fivefold personality and of the fivefold sacrament employed in the rite. The set of five symbolizes then both the condition of nirvana and the condition of samsara, as well as the means towards the realization of their essential identity, which is the aim of all this endeavour. The condition of perfect identity is represented by a sixth Buddha, Vajrasattva, who here appears in the head-dress of Akshobhya, the central Buddha.

These six Buddhas are associated with the various watches of the day and the different seasons, for they comprehend not only all space, but also all time.

At the intermediate points of the compass are manifested four goddesses, who represent the four material elements, earth, fire, water and air, the basic constituents of material things. This set is completed with a fifth goddess at the centre, who represents space, but is also identical in essence with Vajrasattva, for at the centre, which is the absolute, the Perfection of Wisdom itself, no distinctions can be made whatsoever.

It will be noted that she is identified with the Sanskrit vowel-series, while the Buddhas between them make up the whole Sanskrit consonantal series. The mandala therefore symbolizes all sound in addition to all the things already mentioned above, for the right employment of intoned sound has now become a very important part of Buddhist practice at this stage.

This short note should serve as a sufficient guide to the extract that follows.

THE FIVEFOLD MANIFESTATION

Having made obeisance to Vajrasattva, unevolved and supreme, I shall expound in brief for the welfare of my pupils the fivefold manifestation.

With the mantra: OM ĀḤ HŪṂ one ensures protection for the site, for oneself and for the performance. Then upon this site which has been prepared with perfumes and so on one should worship the Five Buddhas and the Five Yoginis in the centre of the four-sided mandala.

There in the centre one envisages the syllable PAṂ of many colours. This turns into a fair eight-petalled lotus of many colours with the red syllable RAṂ at its centre. This becomes a solar disk, upon which is a dark-blue HŪṂ whence arises Akshobhya with one face and two arms, in the crossed-legged-posture and making the "earth-touching" gesture. His body exhibits the 32 major and the 80 minor marks of perfection, for it is the repository of the whole host of excellent qualities, the ten powers, fearlessness and the rest. It is without apertures, flesh or bone, for it is neither true nor false like pure light reflected in a mirror. He is black in colour because he is permeated with great compassion and his symbol is a black vajra which embodies the five constituents of the pure absolute. He has no hair on head or body and he is clad in religious garb. His head is marked with Vajrasattva, for he is himself the essence of Vajrasattva, where there is no distinction between Void and Compassion. He also embodies cause and effect, of which the characteristic is the Void in that it comprehends all possible forms. He is the Dharma-Body because he is the embodiment of the Buddhas who are unconditioned; he is the Sambhoga-Body because he is a pure reflection; he is the Nirmana-Body because he represents the true nature of constructive consciousness; he is the Self-Existing Body (svābhāvika) because this is the single flavour of those three bodies. So it is said:

> "Dharma is mind unconditioned;
> Sambhoga is characterized by reciprocal enjoyment;
> Nirmana is that which is variously created;
> Svabhavika is that which is innate in everything."

As he is undefiled by discriminating thought and so on, he is of the Vajra-family, and the Vajra-family is undefiled by worldlings. His external and internal bases of symbolization are: wrath, vajra-water, midday, pungency, revealed knowledge, space, sound, the CA-series of consonants. His mantra is: OṂ ĀḤ VAJRADHṚK HŪṂ.

Vajrasattva, born of the syllable HŪṂ, is pure white with two arms and one face; he holds a vajra and a bell. He is the true nature of mind. He represents astringency, Autumn, the consonantal series YA RA LA VA and the second half of the night up to dawn. He is another name of the absolute.

Then on the eastern petal on a lunar disk appears Vairocana, born of the syllable OṂ; he is white in colour with a white disk as his symbol and makes the gesture indicating enlightenment. He embodies the skandha of form and the nature of delusion; he is symbolized by dung; he is of the Tathagata-family; he consists in the Mirror-like Knowledge, he represents Winter, sweetness, the KA-series of consonants and the morning watch. His mantra is: OṂ ĀḤ JINAJIK HŪṂ.

Then on the southern petal on a solar disk appears Ratnasambhava, born of the syllable TRĀṂ, yellow in colour with a gem as his symbol and making the gesture of giving. He embodies the skandha of feeling and the nature of malignity (paiśunya); he is symbolized by blood; he is of the Gem-family; he consists in the Knowledge of Sameness; he represents Spring, saltiness, the TA-series of consonants and the third watch. His mantra is: OṂ ĀḤ RATNADHṚK HŪṂ.

Then on the western petal on a solar disk appears Amitabha, born of the syllable HRĪḤ, red in colour with a lotus as his symbol and in the pose of contemplation. He embodies the skandha of perception and the nature of passion; he is symbolized by seed; he is of the Lotus-family; he consists in Perceptual Knowledge; he represents Summer, bitterness, the TA-series of consonants and the evening watch. His mantra is: OṂ ĀḤ ĀROLIK HŪṂ.

Then on the northern petal on a solar disk appears Amoghasiddhi, born of the syllable KHAṂ, dark-green in colour with a crossed vajra as his symbol and making the gesture of protection. He embodies the skandha of impulses and the nature of envy (īrṣyā); he is symbolized by flesh; he is of the Karma-family; he consists in the Knowledge of Needful Activity; he represents the season of the rains, the hot taste, the PA-series of consonants and the first watch of the night. His mantra is: OṂ ĀḤ PRAJNĀDHṚK HŪṂ. These are all seated in the cross-legged posture with two arms and one face, wearing the Ushuisha but with no hair on head or body; they are clad in religious garb and exhibit the 32 major and 80 minor marks of physical perfection, for they are repositories of the whole host of excellent qualities, the ten powers, fearlessness and the rest. Theirs are Sambhoga-Bodies, free of apertures, flesh and bones, pure light and naught else, like reflections in a mirror and free from such concepts as truth and falsehood. But these four, Vairocana, Ratnasambhava, Amitabha and Amoghasiddhi, who embody the constituents of Form, Feeling, Perception and Impulses, possess indistinguishably the Dharma-Body which is the unconditioned essence of Buddhas, and the Body of Constructive Consciousness, for they all comprise the Self-Existing Body which is the single flavour of the other three bodies. Thus in order to show that they are all mere consciousness they are signed on the head with the sign of Akshobhya, and in order to show that consciousness, which is non-substantial, is the essence of Void and Compassion, Akshobhya is signed with Vajrasattva. Thereby it is asserted that the world consists of cause and effect, which is just the single flavour of Samsara and Nirvana. So it is said:

"When thought is realized as Void and Compassion indistinguishable, that indeed is the teaching of the Buddha, of the Law, of the Assembly. As sweetness is the nature of sugar and heat of fire, so voidness is the nature of all elements." And so it is said: "Thorough knowledge of Samsara is Nirvana."

On the south-eastern petal on a lunar disk appears Locana, born of a white syllable LOṂ, white in colour and with a disk as her symbol. She represents the element earth, belongs to the Tathagata-family and is affected by delusion. Her mantra is: OṂ ĀḤ LOṂ HŪṂ SVĀHĀ.

To the south-west on a lunar disk appears Mamaki, born of a black syllable MĀṂ, black in colour and with a black vajra as her symbol. She represents the element water, belongs to the Vajra-family and is affected by wrath. Her mantra is: OṂ ĀḤ MĀṂ HŪṂ SVĀHĀ.

To the north-west on a lunar disk appears Panduravasini, born of the syllable PĀṂ, red in colour and with a lotus as her symbol. She represents the element fire, belongs to the Lotus-family and is affected by passion. Her mantra is: OṂ ĀḤ PĀṂ HŪṂ SVĀHĀ.

To the north-east on a lunar disk appears Tarini, born of the syllable TĀṂ which is gold and dark-green in colour, herself dark-green and with a dark-coloured lotus as her symbol. She represents the element air, belongs to the Karma-family and is affected by Envy.

These four are sixteen years old, endowed with uncommon loveliness and youth as though they were beauty herself. Like the Buddhas they are possessed of the essence of the four Buddha-bodies; they are ravishing to the mind, the repository of the qualities of all the Buddhas and of the very nature of the Five Tathagatas.

In their midst is Nāyikā, the essence of ĀLI (the vowel series); she possesses the true nature of Vajrasattva and is Queen of the Vajra-realm. She is known as the Lady, as Suchness, as Void, as Perfection of Wisdom, as Limit of Reality, as Absence of Self.

Unit IX

The Descent to Heaven: The Tibetan Book of the Dead

Birth, Marriage, Sickness, and Death

GIUSEPPE TUCCI

Giuseppe Tucci is one of the leading Tibetologists of our period. He was an expert in Tibetan language and explaining special characteristics of Tibetan religion, such as the Mandala which he interpreted in the Indian context as well. The following selection presents Tucci's insights into the popular religion of Tibet through The Tibetan Book of the Dead, *which explains in a priestly style the fate of the life-stream entity. The concepts behind the text are the common beliefs of traditional Tibetan society. This selection also introduces the reader to other related beliefs and religious practices.*

Family life followed the same pattern everywhere, varied only by those moments of joy, sadness or worry known to all of us. Birth, marriage, sickness and death were accompanied by their special customs, differing a little from place to place and according to the wealth of the family involved.

Childbirth was attended by none of the care and concern known in western countries. Birth was regarded as a completely natural function, and the physical strength necessary to the mother was not weakened by worries about hygiene.

Prayers and talismans would help to make the birth safe and easy, according to age-old custom, and the intervention of the exorcist or the village lama would frequently be requested, but doctors and midwives were unknown. The woman was left to herself, and only in rich families would she stay in bed, or be looked after, for a week or so. In the country she would resume her normal activities after two or three days. Neither would the new-born child receive the care and protection given in our society. It might be washed once in a while, and since fuel was scarce and expensive the water would as likely as not be cold, or else be slightly warmed by the mother, who would take a mouthful at a time and sprinkle it over the baby's body. Much faith was placed in the talismans and chantings of the lamas summoned to welcome the new-born child, but infant mortality was very high. Accurate information is difficult to come by, as there were no registrar's offices and no statistics, but it is probably no exaggeration to say that, up to the tenth year of life, it was over fifty per cent.

Male children were the favourites in Tibet, but owing to the high status enjoyed by women, the birth of a girl was in the end as joyful an occasion as the birth of a boy. Three or four days after the birth would come the celebration. Friends and relations would visit the family, bringing gifts of food and the usual silk or gauze scarf, a necessary part of any ceremony. But a few days later a much more important event occurred, the casting of the baby's horoscope. There was usually an astrologer, or *tsipa,* in every village. If by any chance there was none, the nearest one would be sent for. The hour, date and month of the birth were noted, so that the horoscope could be carefully cast, and the favourable and unfavourable moments in the life of the child accurately foretold, according to the influences of the Elements. Sometimes, especially in families of the poorer classes, the day when the child was born would give it its first name: for example, a boy born on Friday would be called Pasang, the Tibetan word for Friday. Other more personal names with an auspicious meaning might then be taken, such as Töndrup, "he whose aims shall be achieved" or "he who has fulfilled the wishes of his parents," a name having the double advantage of evoking that of the Buddha Śākyamuni who was called Sarvārthasiddha, its Sanskrit equivalent. These names were changed if later the boy entered a monastic order. Then a religious name would be given him, probably chosen from those of the great teachers of the sect he joined.

The giving of a name and the casting of the horoscope were celebrated with much solemnity, not only because of the Tibetan's natural enjoyment of festivities, but also in order to create a happy and therefore auspicious atmosphere at the moment when the new-born child was assuming his proper place and asserting his own identity in the world.

Marriage was of course another event to be celebrated with solemnity. It must be noted that Tibetan women enjoyed considerable freedom and, in practice if not in theory, lived on the same level as the men. They were never segregated or expected to be bashful in the way law or custom decreed in other parts of the East. They were also fortunate in possessing marked practical ability. Many women helped their husbands in business affairs or even acted on their own account, and were more likely to increase the family fortunes than to squander them, because of this business sense. It was not unheard of for a civil servant to find some excuse to stay at home and direct his own affairs, sending his wife off to the office to do his work in his absence. I have myself frequently met the wives of local governors who were temporarily carrying out their husbands' duties, and I have always admired the strictly ordered way in which they ran their own houses as well as the way in which they conducted their business affairs. In the family, the mother had great authority, being responsible for the upbringing of the children and the household expenditure. Even when the sons of the family brought home wives, the mother would remain the central figure.

Much has been written on the subject of polyandry, or the practice of a woman taking more than one husband by marrying not only the man she has been betrothed to but also his brothers. This custom did exist in every part of Tibet, but it was not as common as is generally believed. Many explanations for it have been given; perhaps it arose because more females than males died in infancy, or many women as well as men entered monastic orders, or because it was desirable to keep the family and the property undivided, especially among the richer families. But monogamous marriages were the rule in all classes of society. The children of a polyandrous union were considered to be the issue of the eldest brother, to whom the mother was first married, and his brothers, even if they were in fact the fathers of some of the children, would be called uncle by all. In such a union it was generally arranged so that only one of the brothers was at home at any one time, the others being absent on business or for other reasons. In some parts of Tibet, when the wife retired with one of the men, his boots were left outside the door.

Because of the position of women and the flexibility of family life, young couples were always acquainted before marriage, although the consent of the parents was a necessary formality. Girls being free to go out and meet young men in the course of the day's work or at a celebration, the choice of a partner rested with the young people and depended on mutual attraction. Sometimes the girl was already a mother. In the case of rich families, however, marriage became more of a contract as the financial interests of the parents were involved, as well as the feelings of the couple, and the desire to save a heritage in peril could override the preferences of the children.

The marriage ceremony had none of the complexity found in India. The lamas participated, but only because their presence was necessary on all occasions in life. The wedding itself was perhaps the most secular event of all, accompanied by no religious rites. Certain preliminaries were necessary. First of all, it had to be established that there was no close kinship between the two families. Marriage between cousins was frowned on, causing a scandal when it did occur, but a man could marry his brother's widow, or even his father's, provided, of course, that she was not his mother. Many of the rules which so severely govern marriage in India were completely absent in Tibet. The horoscopes of the couple were studied to ensure that they were not incompatible: in the case of any serious clashes the marriage would not take place. There were, however, certain difficulties that could be smoothed over by consulting astrological manuals. The astrologer was all-important during this preliminary phase, and when the first decisions had been taken, the intermediary was called in. He was either a member of the family or a professional go-between, and his duty was to obtain the permission of the bride's family, to negotiate her dowry and to establish the kind of presents, jewels and clothes that the bridegroom would give on the day of the wedding. When agreement had been reached, the bride's family would accept the chhang which the intermediary had brought with him to offer at this moment. This meeting did no more than confirm an already existing agreement but it gave an air of solemnity to the proceedings. The intermediary's job was sometimes deliberately made to appear more difficult than it was in fact, as this treating between the two families, the long discussions, the banquets and the exchanging of chhang were a welcome break in the monotony of everyday life and no one was anxious to hasten things unduly. Sometimes the discussions were spread over a number of days, difficulties and obstacles being artificially created to give an air of drama. Chhang was of importance throughout the negotiations and had an almost religious significance at the wedding ceremony. Words which could bring to mind others, similar but with a different, unlucky meaning, were carefully avoided at this time, but objects with auspicious names were freely exchanged, so that a joyful harmonious atmosphere was preserved. It was seen that no sterile woman approached during the ceremonies as her presence could be unlucky and mean that the couple would be childless.

The wedding day was chosen by the astrologer and the eve spent in banqueting and drinking chhang, friends of the two families and the intermediary all participating. The bride's mother would put the finishing touches to her daughter's dress and the necessary rites were performed to ensure that the phuklha would not desert the bride's family and follow her to her new dwelling. In addition to the advice and warnings given to the bride by her mother, in rich families long sermons on the duties of married life were often delivered by some eloquent lama.

The marriage procession started off at dawn, led by a man dressed in white, mounted on a white horse and carrying the *si pa ho* to ward off evil. This was a square sheet of paper bearing a picture of a tortoise with its limbs clasping a circle. In the circle, symbols would be drawn to show the course of time and astrological combinations causing good or evil events. The position of the planets at the bride's birth was also shown, so as to identify her place in the cosmic order and portend the favourable or unfavourable events in her life. Then came the bride, also on horseback, dressed magnificently in many coloured garments and wearing her finest trinkets, but with her face covered by a white scarf as a sign of the modesty every girl of good family was supposed to show at her wedding. It was customary for her to weep as she left her family, while the procession set off, with everyone shouting "*Trashi-delek*," "May it be auspicious and well." The procession stopped outside the bridegroom's house. Now came the most dramatic moment of all. The house was locked and barred and a long dialogue ensued during which questions and answers were exchanged in order to establish a state of complete mutual trust so that the uncertainty of the waiting period might be replaced by one of calm and happiness. Finally the door was opened and the bride admitted, to be received by her parents-in-law and to take her place on the chair reserved for her. The contract was sealed by the exchange of scarves and copious servings of chhang and celebrations were soon in full swing.

The bride had not come to live in the house of the bridegroom for good, however. She remained there for three to seven days after the ceremony, and then returned to her parents' home. There was no fixed time for her to take up permanent residence with her husband, this depending on the age of the couple and local custom.

Considering the absence of a binding civil or religious marriage ceremony and the freedom of Tibetan women, it is hardly surprising that divorce was not uncommon, taking place by mutual consent. Long discussions were necessary to resolve the most important problem, the division of the property. The woman, if no blame was attached to her, would ask for the return of her dowry, and was entitled to claim some of the husband's possessions if she took the children of the marriage with her, as was her right. The village buzzed momentarily with the gossip, but it was soon forgotten since such events were not unexpected.

A man is a man the world over, and the Tibetan did not take it kindly if his wife committed adultery, for all the independence she might have. On discovering that his wife was unfaithful, the husband had the right to cut off the tip of her nose. The victim would try to heal the wound by putting a black covering over it.

It was, therefore, customary to say, when something arousing scorn took place, "It's a case for cutting noses." Wife-murder was not infrequent; but in this case the law took over and the culprit had to pay blood-money.

The Tibetan approached illness with a mixture of religious convictions and a belief in magic together with the most advanced medical ideas from India, China and Iran—and even Galen had been heard of—all countries where research into medicine and pharmacology had been pursued in a constructive fashion. In India there is a substantial body of writings on medicine, which reached Tibet together with sacred literature and aroused widespread interest. Various manuals, which also included treatments and theories from China, helped to disseminate medical knowledge.

But medicines, drugs and surgery were effective only up to a point. If the right prayers were not said and divine intervention was refused, then no medicine would bring about a cure. At the end of every book on medicine, as its conclusion, there was a section on the "mantras," or phrases to be recited by the sick man himself or by the lama, to invoke divine power to enter the patient and the drugs he was taking. The medical manuals state that cures can be effected by the use of medicine, dieting, courses of treatment, bleeding and exorcisms and religious ceremonies, thus affirming that the psychological element was an essential one for the Tibetan. No clear demarcation was made between the body and the soul; the two were one, so that the soul, helped by the right religious and magic utterances, was able to act on the body and re-establish the right balance of humours, the disturbance of which had caused the illness in the first place. My friend, Lopsang Tenpa, the president of the Mentsikhang Medical College in Lhasa, once told me that in his opinion western medicine, which he knew a little about from his travels in India, was of course incomplete, as it was built entirely on human knowledge and discoveries which were frequently misunderstood. "But our own," he continued with great certainty, "has the advantage of being revealed to us. The Great Physician, the Doctor who cures body and soul at once, the Buddha, who is a manifestation of cosmic consciousness and absolute wisdom, has shown it to man."

The famous medical school of Chakpori in Lhasa was therefore primarily a temple, with images ranged on the altars around the "God of the colour of lapis-lazuli," the chief of eight divinities called the *Menlha*. These were the supreme doctors, since they taught men how to cure sin, the most powerful disharmony causing physical infirmity.

At Mentsikhang there was an extensive array of herbs and drugs of all kinds, roots, powders and plants, some from India and China as well as from Tibet itself. Medicines were usually made up in the form of pills, which were appropriately blessed before being distributed to those in need of them, with instructions not only on the dose to be taken, but on the times of day they were to be swallowed and the prayers that were to be said with them. Medicine did not isolate man from the universe but he was considered to be so bound up with the elements that any treatment which did not take these into account would be not only ineffective but positively harmful. Cosmic harmony had to be preserved, as the most important factor in good health.

Most of the herbs used as medicine were gathered on a mountain to the north of Lhasa, near the monastery of Sera. This mountain was the Garden of Aesculapius to Tibetan doctors and great care was taken in the choice of the plants, which could be beneficial or harmful, more or less effective, depending on the season, or whether they grew on sunlit or northern slopes. Not all medicines were extracted from herbs: sometimes metals, particularly gold, silver, zinc, iron and mercury, or stones, were used, as was the custom in India. Other remedies were based on the belief in strange analogies between certain substances and parts of the body, as in the folk-medicine of any other country. So the gall of oxen was used to treat weaknesses of the eyes, the spleen for abscesses, the tongue of a dog was often used to heal wounds and the liver of a dog was a remedy for leprosy.

The basis of Tibetan medicine was the Indian theory of Humours, which in turn springs from the ethical concept that physical imbalance is associated with spiritual imbalance. In other words, it is the work of karma, which controls the course of human life. Sickness, which originates in sin, is born with karma, and as soon as sin begins to trouble the human conscience, tainting man's original innocence and serenity, evil gives rise to illness. Karma operates under three guises, as greed, wrath and torpor. These three defects have their parallels in the three humours, wind, bile and phlegm. The theory and practice of medicine were based on this fundamental scheme, and by means of endless classifications and ramifications the whole system of interconnected causes, signs, symptoms and cures of the various kinds of sickness could be shown.

These three humours permeate the whole body, but they govern, above all, the brain, the abdomen and the bowels respectively. When one of the humours prevails over the others, illness results. Wind influences the bones, ears, skin, flesh and arteries; bile governs the blood, sweat, eyes, liver and intestines; and phlegm has its effect on the chyle, flesh, fat, marrow and semen, the nose, tongue, lungs, spleen, liver, kidneys, stomach and bladder.

Surgery was rarely resorted to, and then only in the case of wounds, abscesses and the like. Even here it was necessary to establish the cause of the disability. Cauterization was the treatment for malfunctioning of wind, blood-letting and cold compresses for disorders of the bile, cauterization and hot compresses for disorders of phlegm. Surgical instruments were few; the lance, the cautery, the *mebum*—a vessel in which paper was burnt and which was then applied to the part of the body to be bled—and the "sucking-horn," which drew blood off by suction.

The time comes when no medicine is effective, meaning that the patient is the victim of his karma, and the sum of his sins is overtaking him, or else that demons have gained possession of him, amongst them the planets, always believed to be hostile. Some diseases are, in fact, caused exclusively by them, for instance, epilepsy and apoplexy. The *lu* and the *sadak*, spirits from beneath the earth, are also easily offended and most vindictive. A slight upset, even involuntary, can call down extreme wrath on man. Then medicine is no use and the doctor becomes an exorcist.

The reading of any book whose subject matter seemed particularly apt, or of sacred texts in general, helped in such cases, for they represented the Buddha's "verbal plane," embodied in sounds. The mere recitation of these writings had a strange but positive magical power, as of light dispelling darkness. Monks from the temple or the village lama were called to the sick man's bedside to read or murmur the recommended books, the comprehension of which was immaterial. Even the lama did not always fully understand the text he was reading. The act of reciting the appropriate words was, in itself, effective through the magical power of the sound produced. In this way the divine presence was evoked and healing would result. Another method was to buy from the butcher a goat or a sheep about to be slaughtered. The animal was then set free after a red ribbon had been tied round its neck. As long as the ribbon lasted, the animal was safe from recapture, and the saving of life in this way was also the saving of the sick man.

The moment of death was a decisive one, governing the future destiny of the victim. In case there was no immediate attainment of nirvana there was a limit of forty-nine days between the end of one life and the beginning of another, and the dying man's fate depended on his clear awareness in the moments preceding his end. Acts are important according to Buddhism, but only conscious acts, as consciousness gives responsibility to the action, and this responsibility will bring the action to fruition. An involuntary action can bring no positive result. Buddhism acknowledges no soul, but in Tibet certain old beliefs persisted and it was thought that a certain spiritual entity analogous with the soul existed, so that Buddhism there was occasionally tinged with these superstitions among some people of lower doctrinal experience. The continuance of our personality, which is destined to dissolve at death, is entrusted to the *nampurshé*, or consciousness. This preserves within itself the karmic remains of past existences, modified according to the kind of life led, which have determined the conditions of our present life. This consciousness, enriched by its new experiences, orders our future life at the moment of death, for it then has the power to project itself into a new set of components of the physical personality, such as every living being possesses. Whether these are real or illusory is a question for the Buddhist metaphysicians.

This new future depends on us, on our karma. So, on his deathbed, it was necessary to draw the dying man's attention, or rather, that of his consciousness, which contains his personality, to a consideration of what was about to befall him. There were existences far worse than that of a human being in store for sinners: rebirth as an animal, descent into various hells, or becoming one of the *yidak*, wretched beings who perpetually wander through space, tormented by thirst and hunger.

Terrible visions of hell, where the damned suffer indescribable torments, had been brought from India. It was thought that these wretched beings believed they suffered, rather than that they suffered in fact, for everything is merely a product of the imagination, though to the uninitiated it has the value of reality. The Tibetan, with his tendency to the macabre, drew an even grimmer picture of hot and cold hells and frightful tortures, which are dwelt on at length in a hair-raising literature, the *dêlo*. This is a series of accounts given by those who, on the brink of death, caught a glimpse of life beyond the tomb, but then returned to tell of the terrifying things they saw.

An idea prevalent in Indian religious thinking is found also in Tibet. The luminous awareness of what we are and what awaits us acts, at the moment of departure, as a factor for our salvation since, as we see in the *Song of the Blessed One*, "the dying man attains the same level as his thoughts occupied in his last moment." To this end, a lama specialized in these rites would read an appropriate book, whispering the words into the ear of the dying man or of the corpse. It mattered little if the man was already dead or not, as his consciousness would remain hovering about his body for several days. The book is called the *Bardo-thödröl*, "the book whose mere recitation, when heard by the dead man during the period of his intermediate existence between the life he has left and the one he is to enter in forty-nine days' time, will lead him to salvation." None of this applied, of course, to the saintly who, at the moment of death, were absorbed, in full consciousness, into the wonderful clear light which is the Absolute itself made manifest, or who vanished into the sky in the splendour of a rainbow.

For more ordinary beings the intermediate state with its attendant dangers began. Apparitions were perceived by the dead man's consciousness, the most exalted being flashes of contrasting light, which dazzled or invited him, a tumultuous roaring, shapes which frightened or allured him. Then came seven days during which the five primary forces were revealed in the form of the five supreme Buddhas, the fivefold source of everything that exists. The five Buddhas each appear with a "Mother," coupled with whom they become manifest in the universe with all its variety. They were arranged in the form of a cross, representing infinite space, divided by two intersecting axes to mark the five directions of divine experience: the centre, east, south, west and north. The secret of their mysterious works was revealed and the dead man's consciousness was made aware of their true meaning. The forces thus represented are present in all of us and go to make up our personality of which they form the underlying pattern; they are therefore also the means of salvation, when our gnosis, on understanding their nature, absorbs them. This is the knowledge that annihilates, bringing us back from the apparent to the real, a return to our origin.

When this salvation through recognition did not take place, a progressive decline set in; visions of deities, which the dead man should think of as springing from his own body, would appear. He must win the conviction that all images, including the divine, are the creation of his own mind, emerging from and returning to himself. The man who was not saved by such recognition would then see six lights which corresponded to six kinds of existence, those of gods, of men, of demons, of animals, of ghosts (*yidak*) and of beings in hell. These lights are pleasing and attractive; and he who surrendered to their persuasion, unmindful of what they represented, would be caught up in that existence whose light most forcefully struck or attracted him. This happened because desire would take over in this welter of images and volition, unless the purifying of gnosis intervened; what we are is the effect of the ideas in which we believe.

The fearsome gods would appear and had to be disposed of in the same way. When recognition was absent, such visions would be regarded as the god of death, and death would be believed to be a reality, and the dead man caught up in the succeeding phases of the karmic process, since all events are caused by our own will. Whatever can death be, let alone the god of death, in a philosophy which reduces everything to a nebulous set of images? How can there be anyone that dies, or causes to die, from the point of view of this doctrine, according to which everything dissolves into the colourless, motionless light of cosmic consciousness?

After the thirteenth day, the dead man's consciousness, which up to now had stayed in the vicinity of the corpse, harried by the wind of karma, and occupying a slender body in a state of unrest, began to wander about in profound grief. After trying nine times in vain to re-enter the recently abandoned body, it vainly attempted to enter another. Frightening visions disturbed it and chased it from place to place. Liberation must be found in the knowledge that these visions too are illusory. Divine mercy had only to be invoked for them to flee away, provided that the mind could remain in a state of complete tranquility during the ordeal.

Many creatures would be saved, but many others would be led by their karma to embark on a new incarnation. In this case new visions of rebirth now appeared before the consciousness and corresponded to the four continents into which the world is divided according to Indo-Tibetan cosmology. So a man had to be able to distinguish the signs of Jambudvīpa, the continent blessed with regular appearances of the Buddha, from those of the other worlds where the teachings of the Buddha have not spread. He could, alternatively, allow himself to be drawn towards the heavens, the habitation of the gods, but would have to flee the deceitful images of the world of the demons, of animals, the hungry ghosts, or hell. And he would have to be alert and wary, as karma would call up frightening visions to tempt him to take refuge in caves or lotus flowers, where he might expect to find escape from the turmoil surrounding him and the demons pursuing him. These places of shelter would only be the entrances to unhappy forms of existence, during which he would have to expiate his past misdeeds.

But not all was lost even in this sad situation since, by composing one's mind and meditating on the god Tamdrin, there were still two courses open: one was to reach a paradise of the Buddha, the other, the choice of the womb. The first goal was attained by concentrating the will towards this miraculous rebirth with complete and undivided faith. The second course was embarked on after the continent had been chosen, and it was not impossible to be born into a good family, so as to live to the advantage of one's fellow-creatures, if this was desired with real sincerity and determination. No further effort was to be made, and a desire for one particular womb or a feeling of aversion for another were avoided, as such feelings could reflect badly on the coming existence and would favour the development of karma. Once the wish had been formed, the mind was left to rest in the thought of the three Sacred Gems, the Buddha, the Law, and the Monastic Community.

Such then was the way of salvation, except in the case of saints. The consciousness left the body through a tiny hole at the point where the cranial bones meet on the top of the head, the lama pulling out a hair to provide the way of escape. Since it was believed that the consciousness wandered in the vicinity for several days, the corpse was left in the house for a period of three to seven days, or even longer. The dead man was dressed in his best clothes and placed in a sitting position, his hands crossed on his lap. Only the bodies of highly esteemed lamas were washed, and then this was done with water made fragrant with herbs. The dead man was not left in the room where he died but was taken to another room where the picture of his personal deity (*yi dam*) was hung above him. He could not be touched for three days. If a member of the family were to touch his feet, for example, the deceased would run the risk of going to hell. A lama would remain near him for several days, reading aloud from a sacred book and reciting the prayers for the dead: the rich would send for seven or fourteen monks from the nearest monastery to perform this rite. Those who were very wealthy would make a generous gift to the monastery, so that a funeral service could be held with the participation of all the monks, and alms would be given to the poor. Meanwhile the astrologer would establish the most appropriate time for the burial, after consulting the dead man's horoscope and each day a funeral meal would be eaten by the members of the family, friends, and the officiating lama. Some of the food was put in front of the corpse or his image. This food offering was placed before the image of the dead man for forty-nine days, the whole of the period of the intermediate existence. The image itself consisted of a piece of paper with the Sanskrit syllable *nri* ("man") and the word *tshelêdêpa* (dead) written on it, followed by the name of the deceased. Every seven days the piece of paper was burnt and another put in its place, and from the colour of the flames the lama was able to infer what new form of existence the dead man was progressing towards. The ashes were carefully collected and mixed with water to make little conical stūpas, like the tshatsha mentioned previously, and put on the household altar or in a place set aside for the purpose. On the appointed day, the corpse was taken to the cemetery. A further banquet brought the family together again and the last prayers were said, while the dead man was warned not to come near the house any more nor to bring misfortune to other members of the family. In some places a white scarf was tied to the corpse's feet, the other end being held by the lama. Then the body was carried out of the house by the *Ragyapas*, whose job it was to see to the last funeral rites.

There was no general method of disposing of the body, the custom varying from place to place and with the climate. In places where fuel was scarce one common method was to take the body to the top of a mountain reserved for the purpose. It was then left to the vultures or beasts of prey, after the Ragyapas had cut it up to make it easier for the animals to consume. They even stripped off the flesh with their large knives, and threw it to the crows and vultures, or broke the bones in pieces. A few days later the area was visited again and the remaining bones were crushed to a powder so that nothing of the body was left. In places where fuel was plentiful, the corpse was cremated; in other parts it was thrown into a river. This is an old custom and sometimes, where the body was disposed of on a mountain-top, an effigy was made of the dead man, which was clothed and thrown into the river in place of the actual corpse.

In the case of venerated lamas and also the Dalai Lamas and Panchen Lamas, the remains were placed in the hollow interior of chhörtens.

The lama would assist the members of the dead man's family to purify the house, death being a contamination, calling down supernatural, harmful forces. There was also a danger that the dead man himself might return to his earthly home to take vengeance for some offence, but the lamas performed the necessary rites to restore sanctity to the house and put everything to rights.

The *Bardo* of the Moments of Death

W. Y. EVANS-WENTZ

W. Y. Evans-Wentz was an English scholar, educated at Oxford, where he also taught. He dedicated an important part of his life's work to Indian religious subjects. His translation of The Tibetan Book of the Dead, *published in collaboration with a Tibetan–scholar Lama, has entered world literature as an original classic often compared to* The Egyptian Book of the Dead, *which concerned itself with the voyage of the "soul" into the underworld. Comparisons are sometimes invidious, but C. G. Jung in his introduction to the text had the following to say about the two texts: "Unlike* The Egyptian Book of the Dead, *which always prompts one to say too much or too little,* The Bardo Thodol *offers one an intelligible philosophy addressed to human beings rather than to gods. . . ." Reading the selection from* The Bardo, *the reader becomes familiar with one of the most important Eastern religious texts.*

[Instructions on the symptoms of death, or the first stage of the Chikhai Bardo*: the primary clear light seen at the moment of death]*

The first, the setting-face-to-face with the Clear Light, during the Intermediate State of the Moments of Death, is:

Here [some there may be] who have listened much [to religious instructions] yet not recognized; and [some] who, though recognizing, are, nevertheless, weak in familiarity. But all classes of individuals who have received the practical teachings [called] *Guides* will, if this be applied to them, be set face to face with the fundamental Clear Light; and, without any Intermediate State, they will obtain the Unborn *Dharma-Kāya*, by the Great Perpendicular Path.[1]

The manner of application is:

It is best if the *guru* from whom the deceased received guiding instructions can be had; but if the *guru* cannot be obtained, then a brother of the Faith; or if the latter is also unobtainable, then a learned man of the same Faith; or, should all these be unobtainable, then a person who can read correctly and distinctly ought to read this many times over. Thereby [the deceased] will be put in mind of what he had [previously] heard of the setting-face-to-face and will at once come to recognize that Fundamental Light and undoubtedly obtain Liberation.

As regards the time for the application [of these instructions]:

When the expiration hath ceased, the vital-force will have sunk into the nerve-centre of Wisdom[2] and the Knower[3] will be experiencing the Clear Light of the natural condition.[4] Then, the vital-force,[5] being thrown backwards and flying downwards through the right and left nerves,[6] the Intermediate State momentarily dawns.

The above [directions] should be applied before [the vital-force hath] rushed into the left nerve [after first having traversed the navel nerve-centre].

The time [ordinarily necessary for this motion of the vital-force] is as long as the inspiration is still present, or about the time required for eating a meal.[7]

Then the manner of the application [of the instructions] is:

When the breathing is about to cease, it is best if the Transference hath been applied efficiently; if [the application] hath been inefficient, then [address the deceased] thus:

O nobly-born (so and so by name), the time hath now come for thee to seek the Path [in reality]. Thy breathing is about to cease. Thy *guru* hath set thee face to face before with the Clear Light; and now thou art about to experience it in its Reality in the *Bardo* state, wherein all things are like the void and cloudless sky, and the naked, spotless intellect is like unto a transparent vacuum without circumference or centre. At this moment, know thou thyself; and abide in that state. I, too, at this time, am setting thee face to face.

Having read this, repeat it many times in the ear of the person dying, even before the expiration hath ceased, so as to impress it on the mind [of the dying one].

If the expiration is about to cease, turn the dying one over on the right side, which posture is called the "Lying Posture of a Lion." The throbbing of the arteries [on the right and left side of the throat] is to be pressed.

If the person dying be disposed to sleep, or if the sleeping state advances, that should be arrested, and the arteries pressed gently but firmly.[8] Thereby the vital-force will not be able to return from the median-nerve[9] and will be sure to pass out through the Brahmanic aperture.[10] Now the real setting-face-to-face is to be applied.

At this moment, the first [glimpsing] of the *Bardo* of the Clear Light of Reality, which is the Infallible Mind of the *Dharma-Kāya*, is experienced by all sentient beings.

The interval between the cessation of the expiration and the cessation of the inspiration is the time during which the vital-force remaineth in the median-nerve.[11]

The common people call this the state wherein the consciousness-principle[12] hath fainted away. The duration of this state is uncertain. [It dependeth] upon the constitution, good or bad, and [the state of] the nerves and vital-force. In those who have had even a little practical experience of the firm, tranquil state of *dhyāna,* and in those who have sound nerves, this state continueth for a long time.[13]

In the setting-face-to-face, the repetition [of the above address to the deceased] is to be persisted in until a yellowish liquid beginneth to appear from the various apertures of the bodily organs [of the deceased].

In those who have led an evil life, and in those of unsound nerves, the above state endureth only so long as would take to snap a finger. Again, in some, it endureth as long as the time taken for the eating of a meal.

In various *Tantras* it is said that this state of swoon endureth for about three and one-half days. Most other [religious treatises] say for four days; and that this setting-face-to-face with the Clear Light ought to be persevered in [during the whole time].

The manner of applying [these directions] is:

If [when dying] one be by one's own self capable [of diagnosing the symptoms of death], use [of the knowledge] should have been made ere this.[14] If [the dying person be] unable to do so, then either the *guru*, or a *shiṣḥya*, or a brother in the Faith with whom the one [dying] was very intimate, should be kept at hand, who will vividly impress upon the one [dying] the symptoms [of death] as they appear in due order [repeatedly saying, at first] thus:[15]

Now the symptoms of earth sinking into water are come.[16]

When all the symptoms [of death] are about to be completed, then enjoin upon [the one dying] this resolution, speaking in a low tone of voice in the ear:

O nobly-born (or, if it be a priest, O Venerable Sir), let not thy mind be distracted.

If it be a brother [in the Faith], or some other person, then call him by name, and [say] thus:

O nobly-born, that which is called death being come to thee now, resolve thus: "O this now is the hour of death. By taking advantage of this death, I will so act, for the good of all sentient beings, peopling the illimitable expanse of the heavens, as to obtain the Perfect Buddhahood, by resolving on love and compassion towards [them, and by directing my entire effort to] the Sole Perfection."

Shaping the thoughts thus, especially at this time when the *Dharma-Kāya* of Clear Light [in the state] after death can be realized for the benefit of all sentient beings, know that thou art in that state; [and resolve] that thou wilt obtain the best boon of the State of the Great Symbol,[17] in which thou art, [as follows]:

"Even if I cannot realize it, yet will I know this *Bardo*, and, mastering the Great Body of Union in *Bardo*, will appear in whatever [shape] will benefit [all beings] whomsoever:[18] I will serve all sentient beings, infinite in number as are the limits of the sky."

Keeping thyself unseparated from this resolution, thou shouldst try to remember whatever devotional practices thou wert accustomed to perform during thy lifetime.[19]

In saying this, the reader shall put his lips close to the ear, and shall repeat it distinctly, clearly impressing it upon the dying person so as to prevent his mind from wandering even for a moment.

After the expiration hath completely ceased, press the nerve of sleep firmly; and, a *lāma*, or a person higher or more learned than thyself, impress in these words, thus:

Reverend Sir, now that thou art experiencing the Fundamental Clear Light, try to abide in that state which now thou art experiencing.

And also in the case of any other person the reader shall set him face-to-face thus:

O nobly-born (so-and-so), listen. Now thou art experiencing the Radiance of the Clear Light of Pure Reality. Recognize it. O nobly-born, thy present intellect,[20] in real nature void, not formed into anything as regards characteristics or colour, naturally void, is the very Reality, the All-Good.[21]

Thine own intellect, which is now voidness, yet not to be regarded as of the voidness of nothingness, but as being the intellect itself, unobstructed, shining, thrilling, and blissful, is the very consciousness,[22] the All-good Buddha.[23]

Thine own consciousness, not formed into anything, in reality void, and the intellect, shining and blissful,—these two,—are inseparable. The union of them is the *Dharma-Kāya* state of Perfect Enlightenment.[24]

Thine own consciousness, shining, void, and inseparable from the Great Body of Radiance, hath no birth, nor death, and is the Immutable Light—Buddha Amitābha.[25]

Knowing this is sufficient. Recognizing the voidness of thine own intellect to be Buddhahood, and looking upon it as being thine own consciousness, is to keep thyself in the [state of the] divine mind[26] of the Buddha.[27]

Repeat this distinctly and clearly three or [even] seven times. That will recall to the mind [of the dying one] the former [i.e. when living] setting-face-to-face by the *guru*. Secondly, it will cause the naked consciousness to be recognized as the Clear Light; and, thirdly, recognizing one's own self [thus], one becometh permanently united with the *Dharma-Kāya* and Liberation will be certain.

[Instructions concerning the second stage of the Chikhai Bardo: *the secondary clear light seen immediately after death]*

Thus the primary Clear Light is recognized and Liberation attained. But if it be feared that the primary Clear Light hath not been recognized, then [it can certainly be assumed] there is dawning [upon the deceased] that called the secondary Clear Light, which dawneth in somewhat more than a mealtime period after that the expiration hath ceased.[29]

According to one's good or bad *karma*, the vital-force floweth down into either the right or left nerve and goeth out through any of the apertures [of the body]. Then cometh a lucid condition of the mind.[30]

To say that the state [of the primary Clear Light] endureth for a meal-time period [would depend upon] the good or bad condition of the nerves and also whether there hath been previous practice or not [in the setting-face-to-face].

When the consciousness-principle getteth outside [the body, it sayeth to itself], "Am I dead, or am I not dead?" It cannot determine. It seeth its relatives and connexions as it had been used to seeing them before. It even heareth the wailings. The terrifying *karmic* illusions have not yet dawned. Nor have the frightful apparitions or experiences caused by the Lords of Death[31] yet come.

During this interval, the directions are to be applied [by the *lāma* or reader]:

There are those [devotees] of the perfected stage and of the visualizing stage. If it be one who was in the perfected stage, then call him thrice by name and repeat over and over again the above instructions of setting-face-to-face with the Clear Light. If it be one who was in the visualizing stage, then read out to him the introductory descriptions and the text of the Meditation on his tutelary deity,[32] and then say,

O thou of noble-birth, meditate upon thine own tutelary deity.—[Here the deity's name is to be mentioned by the reader.[33]] Do not be distracted. Earnestly concentrate thy mind upon thy tutelary deity. Meditate upon him as if he were the reflection of the moon in water, apparent yet inexistent [in itself]. Meditate upon him as if he were a being with a physical body.

So saying, [the reader will] impress it.

If [the deceased be] of the common folk, say,

Meditate upon the Great Compassionate Lord.[34]

By thus being set-face-to-face even those who would not be expected to recognize the *Bardo* [unaided] are undoubtedly certain to recognize it.

Persons who while living had been set face to face [with the Reality] by a *guru*, yet who have not made themselves familiar with it, will not be able to recognize the *Bardo* clearly by themselves. Either a *guru* or a brother in the Faith will have to impress vividly such persons.[35]

There may be even those who have made themselves familiar with the teachings, yet who, because of the violence of the disease causing death, may be mentally unable to withstand illusions. For such, also, this instruction is absolutely necessary.

Again [there are those] who, although previously familiar with the teachings, have become liable to pass into the miserable states of existence, owing to breach of vows or failure to perform essential obligations honestly. To them, this [instruction] is indispensable.

If the first stage of the *Bardo* hath been taken by the forelock, that is best. But if not, by application of this distinct recalling [to the deceased], while in the second stage of the *Bardo*, his intellect is awakened and attaineth liberation.

While on the second stage of the *Bardo*, one's body is of the nature of that called the shining illusory-body.[36]

Not knowing whether [he be] dead or not, [a state of] lucidity cometh [to the deceased.].[37] If the instructions be successfully applied to the deceased while he is in that state, then, by the meeting of the Mother-Reality and the Offspring-Reality,[38] *karma* controlleth not.[39] Like the sun's rays, for example, dispelling the darkness, the Clear Light on the Path dispelleth the power of *karma*.

That which is called the second stage of the *Bardo* dawneth upon the thought-body.[40] The Knower hovereth within those places to which its activities had been limited. If at this time this special teaching be applied efficiently, then the purpose will be fulfilled; for the *karmic* illusions will not have come yet, and, therefore, he [the deceased] cannot be turned hither and thither [from his aim of achieving Enlightenment].

NOTES

1 Text: *Yar-gyi-sang-thal-chen-po*: the "Great Straight Upward Path." One of the Doctrines peculiar to Northern Buddhism is that spiritual emancipation, even Buddhahood, may be won instantaneously, without entering upon the *Bardo* Plane and without further suffering on the age-long pathway of normal evolution which traverses the various worlds of *sangsāric* existence. The doctrine underlies the whole of the *Bardo Thödol*. Faith is the first step on the Secret Pathway. Then comes Illumination; and, with it, Certainty; and, when the Goal is won, Emancipation. But here again success implies very unusual proficiency in *yoga*, as well as much accumulated merit, or good *karma*, on the part of the devotee. If the disciple can be made to see and to grasp the Truth as soon as the *guru* reveals it, that is to say, if he has the power to die consciously, and at the supreme moment of quitting the body can recognize the Clear Light which will dawn upon him then, and become one with it, all *sangsāric* bonds of illusion are broken asunder immediately: the Dreamer is awakened into Reality simultaneously with the mightly achievement of recognition.

2 Here, as elsewhere in our text, "nerve-centre" refers to a psychic nerve-centre. The psychic nerve-centre of Wisdom is located in the heart.

3 Text: *Shespa* (pron. *Shepa*): "Mind," "Knower," i.e., the mind in its knowing, or cognizing, functions.

4 Text: *Sprosbral* (pron. *Todal*): "devoid of formative activity;" i.e., the mind in its natural, or primal, state. The mind in its unnatural state, that is to say, when incarnate in a human body, is, because of the driving force of the five senses, continuously in thought-formation activity. Its natural, or discarnate, state is a state of quiescence, comparable to its condition in the highest of *dhyāna* (or deep meditation) when still united to a human body. The conscious recognition of the Clear Light induces an ecstatic condition of consciousness such as saints and mystics of the West have called Illumination.

5 Text: *rlung* (pron. *lung*): "vital-air," or "vital-force," or psychic-force."

6 Text: *risa-gyas-gyon* (pron. *tsa-yay-yōn*): "right and left [psychic] nerves;" Skt. *Pingāla-nādī* (right [psychic] nerve) and *Idā-nādī* (left [psychic] nerve).

7 When this text first took form the reckoning of time was, apparently, yet primitive, mechanical time-keeping appliances being unknown. A similar condition still prevails in many parts of Tibet, where the period of a meal-time is frequently mentioned in old religious books—a period of from twenty minutes to half an hour in duration.

8 The dying person should die fully awake and keenly conscious of the process of death; hence the pressing of the arteries.

9 Skt. of text: *dhutih* (pron. *dutī*), meaning "median-nerve," but lit. "trijunction." V. S. Apte's *Sanscrit-English Dictionary* gives *dhūti* as the only similar word, defined as "shaking" or "moving," which, if applied to our text, may refer to the vibratory motion of the psychic force traversing the median-nerve as its channel."—Lama Kazi Dawa-Samdup.

"*Dutī* may also mean 'throwing away,' or 'throwing out,' with reference to the outgoing of the consciousness in the process of death."—Sj. Atal Bihari Ghosh.

10 If non-distracted, and alertly conscious, at this psychological moment, the dying person will realize, through the power conferred by the reading of the *Thödol*, the importance of holding the vital-force in the median-nerve till it passes out thence through the Aperture of Brāhma.

11 After the expiration has ceased, the vital-force (lit. "inner-breath") is thought to remain in the median-nerve so long as the heart continues to throb.

[12]Text: *rnam-shes* (pron. *man-she*): Skt. *vijñāna* or, preferably, *chaitanya:* "conscious-principle" or "object-knowing principle."

[13]Sometimes it may continue for seven days, but usually only for four or five days. The consciousness-principle, however, save in certain conditions of trance, such as a *yogī*, for example, can induce, is not necessarily resident in the body all the while; normally it quits the body at the moment called death, holding a subtle magnetic-like relationship with the body until the state referred to in the text comes to an end. Only for adepts in *yoga* would the departure of the consciousness-principle be accomplished without break in the continuity of the stream of consciousness, that is to say, without the swoon state referred to.

The death process is the reverse of the birth process, birth being the incarnating, death the discarnating of the consciousness-principle; but, in both alike, there is a passing from one state of consciousness into another. And, just as a babe must wake up in this world and learn by experience the nature of this world, so, likewise, a person at death must wake up in the *Bardo* world and become familiar with its own peculiar conditions. The *Bardo* body, formed of matter is an invisible or ethereal-like state, is an exact duplicate of the human body, from which it is separated in the process of death. Retained in the *Bardo* body are the consciousness-principle and the psychic nerve-system (the counterpart, for the psychic or *Bardo* body, of the physical nerve-system of the human body).

[14]The full meaning implied is that not only should the person about to die diagnose the symptoms of death as they come, one by one, but that he should also, if able, recognize the Clear Light without being set face to face with it by some second person.

[15]Compare the following instructions, from *Ars Moriendi* (fifteenth century), Comper's ed. "When any of likelihood shall die [i.e., is likely to die], then it is most necessary to have a special friend, the which will heartily help and pray for him, and therewith counsel the sick for the weal [i.e., health] of his soul."

[16]The three chief symptoms of death (which the text merely suggests by naming the first of them, it being taken for granted that the reader officiating will know the others and name them as they occur), with their symbolical counterparts, are as follows: (1) a bodily sensation of pressure, "earth sinking into water;" (2) a bodily sensation of clammy coldness as though the body were immersed in water, which gradually merges into that of feverish heat, "water sinking into fire;" (3) a feeling as though the body were being blown to atoms, "fire sinking into air." Each symptom is accompanied by visible external changes in the body, such as loss of control over facial muscles, loss of hearing, loss of sight, the breath coming in gasps just before the loss of consciousness, whereby *lāmas* trained in the science of death detect, one by one, the interdependent psychic phenomena culminating in the release of the *Bardo* body from its human-plane envelope. The translator held that the science of death, as expounded in this treatise, has been arrived at through the actual experiencing of death on the part of learned *lāmas*, who, when dying, have explained to their pupils the very process of death itself, in analytical and elaborate detail.

[17]In this state, realization of the Ultimate Truth is possible, providing sufficient advance on the Path has been made by the deceased before death. Otherwise, he cannot benefit now, and must wander on into lower and lower conditions of the *Bardo*, as determined by *karma*, until rebirth.

[18]The Tibetan of the text is here unusually concise. Literally rendered it is, 'will appear in whatever will subdue [for beneficial ends] whomsoever.' To subdue in this sense any sentient being of the human world, a form which will appeal religiously to that being is assumed. Thus, to appeal to a Shaivite devotee, the form of Shiva is assumed; to a Buddhist, the form of the Buddha Shakya Muni; to a Christian, the form of Jesus; to a Moslem, the form of the Prophet; and so on for other religious devotees; and for all manners and conditions of mankind a form appropriate to the occasion—for example, for subduing children, parents, and vice versa; for *shiṣḥyas, gurus,* and vice versa; for common people, kings or rulers; and for kings, ministers of state.

[19]Compare the following, from *The Book of the Craft of Dying*, in *Bodleian MS. 423* (*circa* fifteenth century): "Also, if he that shall die have long time and space to be-think himself, and be not taken with hasty death, then may be read afore him, of them that be about him, devout histories and devout prayers, in the which he most delighted in when he was in heal [i.e. health]."

[20]Text: *Shes-rig* (pron. *She-rig*) is the intellect, the knowing or cognizing faculty.

[21]Text: *Chös-nyid Kün-tu-bzang-po* (pron.*Chö-nyid Küntu-zang-po*), Skt. *Dharma-Dhātu Samanta-Bhadra,* the embodiment of the *Dharma-Kāya*, the first state of Buddhahood. Our Block-Print text, in error here, gives for the All-Good (*Kuntu-Zang-po*, meaning "All-Good Father") *Kuntu-Zang-mo,* which means "All-Good Mother." According to the Great Perfectionist School, the Father is that which appears, or phenomena, the Mother is that which is conscious of the phenomena. Again, Bliss is the Father, and the Voidness perceiving it, the Mother; the Radiance is the Father, and the Voidness perceiving it, the Mother; and, as in our text here, the intellect is the Father, the Voidness the Mother. The repetition of "void" is to emphasize the importance of knowing the intellect to be in reality void (or of the nature of voidness), i.e., of the unborn, uncreated, unshaped Primordial.

[22]Text: *Rig-pa,* meaning "consciousness" as distinct from the knowing faculty by which it cognizes or knows itself to be. Ordinarily, *rig-pa* and *shes-rig* are synonymous; but in an abstruse philosophical treatise, as herein, *rig-pa* refers to the consciousness in its purest and most spiritual (i.e., supramundane) aspect, and *shes-rig* to the consciousness in that grosser aspect, not purely spiritual, whereby cognizance of phenomena is present.

In this part of the *Bardo Thödol* the psychological analysis of consciousness or mind is particularly abstruse. Wherever the text contains the word *rig-pa* we have rendered it as "consciousness," and the word *shes-rig* as "intellect," or else, to suit the context, *rig-pa* as "consciousness" and *shes-rig* as "consciousness of phenomena," which is "intellect."

[23]Text: *Kun-tu-bzang-po*: Skt. *Samanta* ("All" or "Universal" or "Complete," *Bhadra* ("Good" or "Beneficent"). In this state, the experiencer and the thing experienced are inseparably one and the same, as, for example, the yellowness of gold cannot be separated from gold, nor saltness from salt. For the normal human intellect this transcendental state is beyond comprehension.

[24]From the union of the two states of mind, or consciousness, implied by the two terms *rig-pa* and *shes-rig*, and symbolized by the All-Good Father and the All-Good Mother, is born the state of the *Dharma-Kāya*, the state of Perfect Enlightenment, Buddhahood. The *Dharma-Kāya* ("Body of Truth") symbolizes the purest and the highest state of being, a state of supramundane consciousness, devoid of all mental limitations or obscurations which arise from the contact of the primordial consciousness with matter.

[25]As the Buddha-Samanta-Bhadra state is the state of the All-Good, so the Buddha-Amitābha state is the state of the Boundless Light; and, as the text implies, both are, in the last analysis, the same state, merely regarded from two viewpoints. In the first, is emphasized the mind of the All-Good, in the second, the enlightening *Bodhi* power, symbolized as Buddha Amitābha (the personification of the Wisdom faculty), Source of Life and Light.

[26]Text: *dgongs-pa* (pron. *gong-pa*): "thoughts" or "mind," and, being in the honorific form, "divine mind."

[27]Realization of the Non-*Sangsāra*, which is the Voidness, the Unbecome, the Unborn, the Unmade, the Unformed, implies Buddhahood, Perfect Enlightenment—the state of the Divine Mind of the Buddha. Compare the following passage, from *The Diamond* [or Immutable] *Sūtra*, with its Chinese commentary: "Every form or quality of phenomena is transient and illusive. When the mind realizes that the phenomena of life are not real phenomena, the Lord Buddha may then be clearly perceived."—(*Chinese Annotation*: "The spiritual Buddha must be realized within the mind, otherwise there can be no true perception of the Lord Buddha.")

[28]If, when dying, one be familiar with this state, in virtue of previous spiritual (or *yogic*) training in the human world, and have power to win Buddhahood at this all-determining moment, the Wheel of Rebirth is stopped, and Liberation instantaneously achieved. But such spiritual efficiency is so very rare that the normal mental condition of the person dying is unequal to the supreme feat of holding on to the state in which the Clear Light shines; and there follows a progressive descent into lower and lower states of the *Bardo* existence, and then rebirth. The simile of a needle balanced and set rolling on a thread is used by the *lāmas* to elucidate this condition. So long as the needle retains its balance, it remains on the thread. Eventually, however, the law of gravitation affects it, and it falls. In the realm of the Clear Light, similarly, the mentality of a person dying momentarily enjoys a condition of balance, or perfect equilibrium, and of oneness. Owing to unfamiliarity with such a state, which is an ecstatic state of non-ego, of subliminal consciousness, the consciousness-principle of the average human being lacks the power to function in it; *karmic* propensities becloud the consciousness principle with thoughts of personality, of individualized being, of dualism, and, losing equilibrium, the consciousness-principle falls away from the Clear Light. It is ideation of ego, of self, which prevents the realization of *Nirvāṇa* (which is the "blowing out of the flame of selfish longing"); and so the Wheel of Life continues to turn.

[29]Immediately after the passing of the vital-force into the median-nerve, the person dying experiences the Clear Light in its primitive purity, the *Dharma-Kāya* unobscured; and, if unable to hold fast to that experience, next experiences the secondary Clear Light, having fallen to a lower state of the *Bardo*, wherein the *Dharma-Kāya* is dimmed by *karmic* obscurations.

[30]Text: *shes-pa*, rendered here as "mind." The translator has added the following comment: "The vital-force, passing from the navel psychic-nerve centre, and the principle of consciousness, passing from the brain psychic-nerve centre, unite in the heart psychic-nerve centre, and in departing thence from the body, normally through the Aperture of Brāhma, produce in the dying person a state of ecstasy of the greatest intensity. The succeeding stage is less intense. In the first, or primary, stage, is experienced the Primary Clear Light, in the second stage, the Secondary Clear Light. A ball set bounding reaches its greatest height at the first bound; the second bound is lower, and each succeeding bound is still lower until the ball comes to rest. Similarly is it with the consciousness-principle at the death of a human body. Its first spiritual bound, directly upon quitting the earth-plane body, is the highest; the next is lower. Finally, the force of *karma* having spent itself in the after-death state, the consciousness-principle comes to rest, a womb is entered, and then comes rebirth in this world."

[31]Text: *Gshin-rje* (pron. *Shin-je*): "Lord of Death;" but the plural form is allowable and preferable here.

[32]Cf. the following, from *The Craft to Know Well to Die*, chap. IV, Comper's ed. (p. 73): "And after he [the person dying] ought to require the apostles, the martyrs, the confessors and the virgins, and in special all the saints that he most loved ever."

[33]The favourite deity of the deceased is the tutelary (Tib.*yi-dam*), usually one of the Buddhas or Bodhisattvas, of whom Chenrazee is the most popular.

[34]Text: *Jo-vo-thugs-rje-chen-po* (pron.*Jo-wo-thu-ji-chen-po*): "Great Compassionate Lord," synonymous with Tib. *Spyan-ras-gzigs* (pron. *Chen-rā-zi*): Skt. *Avalokiteshvara.*

[35]A person may have heard a detailed description of the art of swimming and yet never have tried to swim. Suddenly thrown into water he finds himself unable to swim. So with those who have been taught the theory of how to act in the time of death and have not applied, through *yogic* practices, the theory: they cannot maintain unbroken continuity of consciousness; they grow bewildered at the changed conditions; and fail to progress or to take advantage of the opportunity offered by death, unless upheld and directed by a living *guru*. Even with all that a *guru* can do, they ordinarily, because of bad *karma*, fail to recognize the *Bardo* as such.

[36]Text: *dag-pahi-sgyu-lus* (pron. *tag-pay-gyu-lü*): "pure (or shining) illusory body." Skt. *māyā-rūpa*. This is the ethereal counterpart of the physical body of the earth-plane, the "astral-body" of Theosophy.

[37]With the departure of the consciousness-principle from the human body there comes a psychic thrill which gives way to a state of lucidity.

[38]Text: *Chös-nyid-ma-bu*: Skt. *Dharma Mātri Putra*: "Mother and Offspring Reality (or Truth)." The Offspring-Truth is that realized in this world through practising deep meditation (Skt. *dhyāna*). The Mother-Truth is the Primal or Fundamental Truth, experienced only after death whilst the Knower is in the *Bardo* state of equilibrium, ere *karmic* propensities have erupted into activity. What a photograph is compared to the object photographed, the Offspring-Reality is to the Mother-Reality.

[39]Lit., "*karma* is unable to turn the mouth or head," the figure implied being that of a rider controlling a horse with a bridle and bit. In the *Tantra of the Great Liberation*, there is this similar passage: "The man blinded by the darkness of ignorance, the fool caught in the meshes of his actions, and the illiterate man, by listening to this Great Tantra, are released from the bonds of *karma*."

[40]Text: *yid-kyi-lüs* (pron. *yid-kyi-lü*), "mental-body," "desire-body," or "thought-body."

I Am Recognized

HEINRICH HARRER

Heinrich Harrer traveled to Tibet in the period prior to the Communist takeover of the country. He lived there for a considerable period of time and wrote a much acclaimed travel book about his adventures in what was then a rarely visited region of the world. In this selection, Harrer tells about his experiences revisiting Tibet in 1982 and how despite the fact that much had changed, he was able to rediscover familiar landmarks and also renew friendships with Tibetans he had not seen for thirty years. The selection puts in context the tremendous changes that are taking place in Tibetan culture and the great sense of loss these changes provoked in a European devoted to the Tibetan people and their way of life.

It was the spring of 1982. Thirty years later. I was back in Lhasa, and alone at last. I had risen early and could go wherever I wished. Without a chaperon—what bliss! My time was precious and I wanted to miss as little as possible. First I strolled to Lingkhor, an eight-kilometre (five-mile) pilgrimage road around Lhasa; today it survives in sections only. I recalled our arrival in Lhasa over thirty years ago, when we were paying our courtesy calls. That time we had also gone to see the monk-minister on Lingkhor, and I now remembered his words: "In our ancient writings there is a prophecy that a great power from the north will make war on Tibet, destroy religion, and make itself master of the whole world. . . ."

Lingkhor no longer exists in its old form—running through gardens full of flowers and past picturesque corners. I was surprised to see that on an asphalted road, with buses and lorries circulating, pilgrims were prostrating themselves—until I realized that I was on Lingkhor already. I saw a few stonemasons chiselling figures of deities, and some incense-burners were alight in the road around a place of offerings canopied by prayer-flags and guarded by an old monk. I strolled on along the transformed Lingkhor and discovered a still delightful spot by a tributary of the Kyichu. I knew the spot well from the past, and I knew that the 'Blue Buddha' was here reflected in the water's surface. Several times on this trip I asked myself if I was really in the country which for seven years had been my home—but at this spot, for the first time, I felt that, outwardly at least, nothing had changed. The faithful passing this freshly repainted rock relief touch the sacred rock with their foreheads, their backs and their hands—just as they had done in the past.

I stopped for a long time, until scarcely anyone was in sight. Lhasa was far away, long and weary shadows were already slipping down the rocks, and the little stream was rushing along melodiously. I discovered some left-overs of the New Year—old tins with the green sprouts of barley. I gazed up at the Blue Buddha, his left hand gripping a thunderbolt, a symbol of immutability and indestructibility. Numerous other incarnations from the lamaist pantheon surround the central divine figure, among them, enthroned, the eleven-headed Chenrezi, the "god of mercy," whose incarnation is the Dalai Lama. In the dusk the many prayer-flags were weaving among the trees, and I did not find it difficult to believe that the spot was still inhabited by the spirits which guard the religion and the gods. Unlike the old days, there were hardly any beggars; only ducks beating their wings, as they had always done, whenever one of the few pilgrims fed them at this last romantic spot in Lhasa.

As I returned to that city an almost spectral silence hung over houses and streets. A few nomads were approaching; they asked me for a picture of the Dalai Lama and told me they had been on the move for five months; they would stay a week in Lhasa. I went on until I came to Barkhor, the inner ring-road enclosing the Tsuglagkhang, the most sacred of Tibetan temples. In the old days the city's entire life was concentrated on Barkhor, where most of the shops were. I had then written in my diary:

> Barkhor has its heyday at the New Year. Here all religious ceremonies and processions begin and end. In the evenings, especially on holy days, the faithful move along Barkhor in huge numbers; they mumble their prayers, and many of them measure out the distance by flinging their bodies to the ground. Yet this inner ring-road also has a less pious aspect: pretty young women there display their colourful costumes, their turquoise and coral jewellery, they flirt with the nobles, and the local beauties of easier virtue also find there what they are after. The centre of commerce, of social life and of gossip—that is Barkhor.

The Mani and lama singers no longer exist. They used to sit on the ground in the old days, a *thangka* would hang on the wall, illustrating the life of some saint, and in a sing-song voice they would relate to their listeners the most wondrous tales, all the while turning their prayer-wheels.

This time, too, Barkhor was swarming with people whose features radiated contentment. Many of the older women recognized me, started to cry, and asked if I could tell them anything about the Dalai Lama, or if I had a picture of him for them. I asked them how many Tibetans were still *nangpa* after thirty years of communism. *Nangpa* means "within" and it denotes people within the Buddhist faith. "About 100 per cent," was their brief answer. "What's become of Po-la, the one with the beard, surely there were two of you?" They meant Peter Aufschnaiter. They offered me pots and ritual vessels of copper, brass and bronze, old and beautiful ones, which they wanted to sell for very little money, for 50 to 100 *yüan,* about the same number of Swiss francs. Dozens of young Tibetans, who could not have known me from the past, soon crowded around us. They were astonished and amused to find a stranger speaking their Lhasa dialect.

In reply to a question from me our Chinese guide had told me: "Of course you can buy those things." But everyone knew that it was forbidden, and I felt fairly sure the pots would be taken away from us before our departure. A pity, since both sides would have benefited from the transaction. And the example of other countries has shown what a blessing it was that Europeans purchased or carried off artistic treasures which would otherwise have been destroyed. Thus a few museums at least can testify today to the culture of the Tibetan people.

Here on Barkhor, at the foot of the Potala, one could just about visualize how enchanting this city was in the past. The Tibetans invariably used only stone and timber to erect their beautiful buildings. Just imagine: those blocks of granite were joined together without any cement, only with clay, and the gigantic Potala, constructed in this manner, has withstood all earthquakes.

But now the "modern age" has arrived. Following the Chinese invasion, everything was transformed except the small inner kernel of the city. All round, as far as the eye can roam, there is now a sea of hideous tin roofs. I stood on the roof of the Potala and was blinded by the ugly tin; I had to shut my eyes. The whole atmosphere of the city was gone. I talked about it to Wangdü, who is presently responsible for the maintenance of the Potala. I reminded him of the many times we had talked about building a new Lhasa with a great canal, with fresh water between Norbulingka and the Tölung valley. Aufschnaiter and I had already drawn up plans and made drawings. All the banks were to be covered with flowers, like the hanging-gardens of Babylon, and trees were to be planted in front of the palace. Instead, there is now this sad wasteland of tin roofs. Wangdü remembered our plans perfectly well; he assured me he would use his influence to have all the hideousness removed and instead have traditional timber-and-stone houses erected at the foot of the Potala.

The vast part of Shugtri-Lingka is also irretrievably lost. It started at the dense cluster of houses in the little village of Shö, below the Potala, where the Dalai Lama's state printing-press and stabling were located, as well as the prison, and extended all the way to the Kyichu. At the centre of the park stood a stone throne, Shugtri, which for certain rare ceremonies served the Dalai Lama as a seat. Here I used to walk with the Nepalese ambassador, twice a year, in order—as mentioned before—to pick asparagus. Now this too is a sea of cheap hutments and tinroofed houses.

The Potala, the emblem of Lhasa, has survived everything—the centuries, several earthquakes and, worst of all, the destructive fury of the Red Guards. Maybe its spacious internal courtyard, where the fantastic black-hatted dancers used to perform, will again be the scene of similar events for the benefit of tourists. But never again will spectators sit at different levels of the palace, robed in precious brocade and brilliant shimmering silk: on the top floor the Dalai Lama with his three attendants; on the floor below him the solemn figure of the Regent; next the ministers, the parents of the Lama-king and the rest of his relations; finally among them, in those days, myself—staring respectfully at Tibet's high dignitaries.

Instead of brocade and silk, future spectators will wear blue-and-green uniforms, and identical caps on their heads. My unforgettable impressions of the various colourful hats, the splendid variety of garments and furs now belong to the past—only in museums will people be able to admire that pomp. There is no doubt that intelligent tourists can do something to preserve these ancient cultural values. One would need to support the Tibetans tactfully and sensitively, assuring them that it is of the utmost importance that ancient customs and rites are preserved before—as might so easily happen—it is too late. We Europeans have also frequently been late—sometimes, unfortunately, too late—in realizing the importance of cultivating and safeguarding our cultural heritage.

Buddhism: Mantrayāna and Vajrayāna

FOKKE SIERKSMA

The following selection by Fokke Sierksma places the Tantric material in a different context from that of such writers as Eliade, Zimmer and Campbell. Sierksma views Tantra as the development of an antinomian religious movement of special appeal to those elements in Hindu-Buddhist society that were "disenfranchised" by the dominant social groups. This selection offers an insight into the way religious movements develop in response to social and psychological pressures. This view does not invalidate the religious or mystical claims of Tantra, but rather helps to explain the constituency that was prepared to accept its novel viewpoint.

Some three or four centuries after its inception between 200 and 100 B.C., magical formulas penetrated into the Mahāyāna, which could serve to command good or evil forces. If mysticism is the most international form of spiritualised religion, magic is the most international form of religion directed towards material ends. Spells and magical means of killing an enemy at a distance, finding a treasure, making a woman fall in love, or identifying witches and rendering them harmless, are to be found in practically all primitive or complex cultures. Those who would defend the principle that Buddhism always remained the same, may point out traces of magic even in the earliest texts, but that does not alter the fact that in the third phase of this religion the stream of spells becomes a deluge. Magic became an integral part of Buddhism. In the fourth century A.D. Hsi Ts'o-chih wrote so enthusiastically about one particular Chinese Mahāyāna monastery, where fasting and study was not shirked, and where no magic was performed, that one can only conclude he had discovered an exception to the general rule.

India would not be India, if it had not been attempted to make something of this development also and to raise it to a higher level. The magic spells—*mantras*—were not only used for profane ends such as the death of an enemy or the love of a woman, they were also used in mystic, magic rituals, in which the adept visualised a god in the yogic manner, and then identified himself with him, in order to obtain by the quick way that liberation which in the Mahāyāna could only be procured through moral purification and philosophical knowledge. The text of these rituals is found in the *tantras*, whose magic character forms a contrast with the *sūtras* of the first two phases of Buddhism. This was the origin of Tantrism or Mantrayāna. Thus the vehicle of magic spells includes both the secret formulas for compassing deserving and undeserving profane ends, and the liturgical and yogic texts leading to speedy liberation. For that matter, these two aims were by no means mutually exclusive. That they should be pursued together was extremely characteristic of the great pressure obviously exerted by the people, and remained typical of the third phase right into Tibet. One might say of the Hīnayāna élite, who sought their own salavation and took it for granted that they should be maintained by the laity in the mean time, that they wanted to have their cake and eat it. In the Mantrayāna this could be applied to the people, who wanted speedy and complete spiritual liberation while wishing to obtain material advantages by the same magical mystical way.

This was not yet the end of the concessions. Around the 7th century A.D. the cult of Śiva, already discussed in the first part of this chapter, became a redoubtable foe of Buddhism. The adherents of this god attempted to obtain their desire by violent means, as an Indian philosopher puts it. The manner in which they professed their faith is a clear protest against their society, which they confronted in a way calculated to shock: they went naked, carrying a club as a symbol of sexuality and aggressivity, jabbered nonsense, made obscene gestures in the presence of women, drank wine from human skulls, and meditated in cemeteries. Caste was not acknowledged by them, from which one may conclude that their action was a continual protest on the part of outcasts and other victims of India's social organisation. If in the beginning of this chapter Śiva was called an explosion of autochthonous religion, here he must also be termed an outburst of social rebellion. In both cases he is the rude and untutored representative of the natural appetites. The fact that he is also the great yogin, shows that outcasts too are people with their own ideas of human dignity.

Śiva had a female partner, who became increasingly important in this period owing to the influence of Śaktism. The followers of this sect were of opinion, that God was a woman and expressed this by contemplation of the female genital organ, either in effigy or in actuality. This short-cut identifying mind and nature was most probably also a form of social protest, and at the same time a re-activation of autochthonous religion in stirring times of social and political unrest. The sect entered into partnership with Śaivism, bringing Śiva's female companion, the Devī, into high honour. We even find indications that there were differences of opinion as to whether Śiva or his wife was the more important. The later philosophical and mystic Śaivism resolves the matter thus: the Devī is the active *śakti*—*i.e.*, creative energy—of the spiritual Creator Śiva who rests in himself, and thus she forms the link between the God and his world and worshippers. The less philosophical and more drastic followers of the god would on occasion solve the problem by copulating with women in cemeteries. The woman was then the śakti of the man. Outside cemeteries also, sacral *coitus* and śakti played a more and more important part in Śaivism as it was popular.

As Śiva was the great, divine yogin and many yogins were to be counted among his followers, the practice of yoga was also influenced by the sexualising tendencies in religious life. But the yogin did not require a corporeal woman, for since ancient times yoga was just as self-sufficient as early Buddhism, which has sometimes been termed pure yoga. "What need have I of any outer woman? I have an inner woman within myself." The name of this inner woman is Kuṇḍalinī, and the yogin who practices this form of yoga imagines her as a coiled snake situated in the lowermost "lotus-centre" of his body, about at the level of the genitals. The uppermost centre is located in the brain, and there Śiva rests in himself. Between these two there are five other lotus-centres. After the physical and mental training indispensable to all yoga, the yogin must now concentrate all his psychical and mental forces on doing what a fakir does with a real snake: "arousing" the Kuṇḍalinī-snake and inducing it to stretch itself upward. He leads Kuṇḍalinī upwards in such a way that she penetrates the various lotus-centres—in authentic yoga each centre signifies a new and higher state of consciousness—until in the brain-centre she unites herself with her lord and consort Śiva, and the yogin attains the highest enlightenment. A psychological commentary upon this matter would require a separate volume, so that here there is only place for those few remarks strictly necessary to our purpose. A goddess who raises herself up as a snake and pierces lotus flowers, who at the beginning and the end of her "course" envelops the sweet liṅga of Śiva with her mouth and drinks nectar, is an evident symbol of bisexuality. As sexual symbols are not only expressions of sexuality, but not infrequently also express structures and trends of human nature, bisexuality equals autarky, certainly so in this instance, where a man consciously strives to unite nature and mind—phallus and brain, his virility and the woman in himself—within his own body and mind. This will towards autarky is also made evident in the fact, that the yogin's semen or the power of his semen is considered to rise up and descend again transformed into the nectar of Śiva's union. The clearest proof is found in the power obtained by the yogin, to destroy and re-create the world. For as he leads Kuṇḍalinī upwards, he simultaneously demolishes the visible world step by step, so that at the union of Śiva and his consort it vanishes completely. Then as Kuṇḍalinī is led back centre by centre—*i.e.*, when the yogin systematically returns to normal consciousness—the world is built up again. If the yogin were to become angry, writes a commentator, he would be able to destroy all the worlds. The difference with mysticism, including Buddhist mysticism, is evident: the mystic comes to non-being or blows out like a flame, the Śaiva yogin seeks power, the power of God himself. Not seldom, he imagines himself to be God. It is significant for our subject that this God in apprehension beholds, in the various states of consciousness he creates, both peaceable and terrifying gods, and that the latter are the "lords of dissolution," those who preside over the destruction of the world. In this yoga both sexuality and aggressivity serve the lust of power, assuming itself capable of destroying the world and building it anew, like the great god Śiva himself. This means that aggressivity dominates over sexuality in an autism, which recognises mysticism as a danger and conquers it. Yoga and magic often go together. When the Buddha met a yogin who proudly declared that after many years of effort he could walk over the river, the reply was: what a pity to spend so much effort; a little further on they will take you across for a penny. This yoga stands between mysticism and magic, as the latter was described in connection with the Mantrayāna.

The people, themselves fond of employing magic, held the yogins in high honour. Greedy for marvels, their imagination added numerous miracles to the yogins' actual achievements, and so 84 yogins became the legendary 84 Siddhas or miracle-workers. It is significant, that both Śaivism and Tantric Buddhism knew the 84 Siddhas, and that a number of names were the same in both lists. This shows that for wandering yogins and their admirers and followers of that milieu, the line of demarcation between Buddhism and Śaivism was growing faint at that time. Other facts too point in the direction of a syncretism in which particularly Buddhism was the loser. The boundaries between the practices of wandering yogins and the yogic "communal" rituals in these circles are often not clear either. A tantra translated by Sir John Woodroffe gives a good impression of the spirit and practice of a community of the Śaiva milieu. The symbolism is not only sexual, but also sensual. The tantra breathes an aggressively sectarian spirit. Wine, meat, fish, parched corn and coition have a central role. Wine is very important for the union of Śiva and his partner, and the Praise of Wine, which is sung, contains the obvious and thereby rather dubious argument that it is not the quantity of wine, but its effect upon the drinker that counts. Almost comical is the rule with regard to sexual continence: in the daytime one must abstain, but at night one is free. Intolerance is very great. Whoever practises a different religion is sure of death and hell, and the fierce and terrifying gods are dancing in delight at the thought of the flesh and bones of scoffing unbelievers whom they will devour. The background of this barely sublimated sexuality and aggressivity is evidently of a social nature. A protest against the caste system is clearly audible. No wonder these unfortunate outcasts seek compensation for their sense of inferiority. The members of this church are elevated far above society and its morals. Secrets and secrecy shelter the vulnerability of this proud consciousness of self.

In connection with our subject, two matters claim attention: the Terrifying Gods and the social background of such religious milieus. The terrifying aspect of the god Śiva is the night-side of his aspect of fertility and salvation, and together these form the numinous unity of contrast which characterises every great god, though the stress laid on his demonic activity is probably due to the contempt and enmity the Aryan rulers felt at first for the native god. In Kuṇḍalinī-yoga the lords of dissolution are symbols of the power of the yogin who can destroy the world. Though far more personal than in the case of Śiva himself, this aggressivity is yet sublimated to a very great degree, and is entirely consonant with the system of yoga engaged in. In the sect described above, whose members practised no more than a weak infusion of yoga, the terrifying deities are the signs of an understandable, nevertheless petty-minded hatred. Within their own circle they practised a scantly spiritualised sexuality, externally they ventilated a furious aggressivity. There is no question of any relation between these two forces in human life. One can only think of a hypothesis in child and adolescent psychology, according to which a child can only love certain people if it first hates others. It has to hate these, because otherwise it would simultaneously love and hate its parents and others, and would itself succumb in the conflict.

In social respect we have here the milieu of what Max Weber called "plebeian religions," which are to be found in various complex cultures. They are the religions of the "underdogs," who cannot assert themselves and who take their revenge by regarding themselves as the true élite in their religion, "far better than the rest." Among themselves they are most fraternal, linked by the same fate, and towards the outside world they are extremely aggressive, in this instance attacking the castes and the monks! Naturally they do not recognise rank and class, any more than the system of values of the existing social order, and this is formulated as being above good and evil. To these characteristics, studied by

Weber and set in a wider context by Mühlmann we may add, in a word, that all these religions are dominated by the psychology of resentment. One finds the same thing with non-white peoples under white colonial rule: he who feels inferior wants to eat his cake and have it, wants, for instance, to remain a Papuan and to become white, to remain outcast and to have a religion of a philosophical tint. Much would-be philosophising in the tantras should, I think, be seen in this light. The true proportions are easily seen, when one hears that the tantras, which announce themselves as composed for great minds, are written in an abominable style.

It is not yet certain how things went exactly, but everything points to it that this social protest, manifested in a secret religion with a strong sexual element, also impressed a not inconsiderable part of the Mantrayāna Buddhists. Though fiercely opposed by other milieus of the Mantrayāna, it managed to gain a wide acceptance, and so the Mantrayāna was converted into the Vajrayāna. Tantric Buddhism, then, is a general term for this phase of Buddhism as expressed in the tantras, Mantrayāna indicates more particularly the magical, and Vajrayāna the magico-sexual trend. It is extremely improbable that—as Govinda and others think—the sexually directed magic and the sexual mysticism of the tantras should have originated in Buddhist circles. Already in prehistoric times India, like many other cultures, was acquainted with the zestful and frank adoration of sexual symbols and mother goddesses whose sexual aspect predominated. The Aryan rulers could not in the end prevent this autochthonous religion from reasserting itself, as it did in the religions of Śiva and Viṣṇu. The social underdogs seized upon these indigenous elements to oppose the existing society and to shock it. For evident reasons, the disposition to shock was coupled with a desire for secrecy, so that in Śaivism and Śaktism sex and the sexual act, in themselves capable of filling man with awe and leading him to the numinous, not infrequently acquired a dubious character. The close connection between the two religions also points to a Hindu origin of śaktism in the milieu concerned, while the resistance of Mantrayāna circles to the Vajrayāna rounds off the argument. Sexual conceptions and practices were of indigenous origin, and penetrated into Śaivism and also into that sector of Buddhism where the faithful were sensitive to such a form of rebellion and protest.

The popularity of Śaktism must have compelled the Buddhist leaders to adopt these conceptions, if they did not want to lose their followers to their great competitor, Śaivism. Thus one can read in many texts of the Vajrayāna that buddhahood lies in the vagina. In some places the frontiers between Buddhism and Hinduism must have been very vague indeed at this time, between the seventh and eighth century A.D. The common opposition to the higher classes, to the brahmans and to the Mahāyāna monks with their ritualistic and theological paraphernalia will certainly have carried more weight than religious subtilities. Meanwhile Buddhism, essentially asocial, seems always to have retained something of this character, and to have had a special attraction for those who were asocial by necessity of choice. Only Buddhism produced Zen, profound as it is eccentric, and that comes from China, where "mad monks" were a familiar phenomenon. The Vajrayāna in India did not only draw its adherents from the lowest classes, many Vajrayāna teachers were of good family, but had come down through addiction to wine and women. Perhaps this asocial character—even more asocial, for anti-hindu, than Śaivism—was one of the reasons why Buddhism did not disappear in the sexualistic dark, where all cats are gray. Another factor may have been, that particularly social underdogs fight out group rivalry in the ideological field, because they simply have no other battle-ground and ideology is their only sign of status. However this may be, the Buddhist faithful and, or their yogic leaders managed to keep their identity and in spite of all Hinduistic intrusions they have given the Vajrayāna a clear Buddhist stamp.

Naturally, there was a great resemblance between these Buddhist milieus and the Hinduist, particularly the Śaivan circles. Not an interminable, century-long wrestling after salvation in the endless chain of reincarnation (and social misery!), but liberation here and now by the so-called short path was craved. It was not only dislike of theologians, priests and monks, but also their own religious "haste" and illiteracy, which caused these people to despise the way of knowledge and insight and give vent to the magical desire for action in the rituals. Tantrism is ritualism. The urgent, intolerable longing after liberation found an outlet in the sexuality of Śaktism. The Mahāyāna theory, that the buddha-spark glows latent in every human being, formed a useful starting-point for the tantric thesis, that salvation and liberation are found in the body. Much stress was laid on *coitus reservatus*, coition without ejaculation, but in the tantric twilight one need not have a "dirty mind" to suppose that there were two possibilities, and that many of the faithful simply did not have the time and the capacity for strict practice of yoga. Sexually also these people adhered to the magical theory of like by like, the body was to be redeemed by the body. Saṁsāra = Nirvāṇa!—thus the Buddhist Tantrists. For much the greater part of the faithful, the goal of even seriously practised yoga was not the *unio mystica* but immortality, and magical power in this life was the highest good. Usually yoga was practised more as an attempt, the good intention counting more than the result. When the Hevajra tantra enjoins "Try it one fortnight with zeal," one can hardly take this seriously, though Snellgrove does. Even an Occidental can see that a fortnight's yoga is too much like the quick courses for this, that and the other that are advertised nowadays. Most results are founded on auto-suggestion. The yoga of the tantric laymen should be regarded in the same light as their conviction that they are the spiritual élite, that the badly written tantras are literature for highly developed minds, etc. Their jealous secrecy alone betrays that they have stuck half-way between actual inferiority and imagined superiority, half-way between natural sexuality and sublimated mysticism, in ritualism and magic. This is not ridiculous or contemptible. Humanly speaking, it is saddening, almost tragic, and completely understandable, like that "aping the white man" of primitives who have not the possibility and the means truly to follow him. Scientifically, however, it is a phenomenon to be observed and stated. Very convincing are in this respect the following words from a list of tantric code-terms: passion, power, corpse, naked, ejaculation, semen, coition, brain-pan, food, faeces, urine, blood.

Buddhism remained true to its anti-hinduist tradition though, and impressed its own mark upon the new religious forms. Although the Vajrayāna is such a heterogeneous collection of spells, magic, ritualism, yoga, speculation, sexual symbolism and practice that it almost defies definition, they succeeded by holding fast to a few basic concepts. When the rituals were written down, this process was already going on. Indeed, in the tantras one finds gross magic beside Buddhist passages. Later on monasteries and universities produced many commentaries in which this tantrism, which could be taken either way, was interpreted entirely in the spiritual and mystic sense. Although magical elements seldom disappeared entirely, so that it is often to be defined as a mysticism of *illuminati*, Buddhism attained in many of these commentaries to the very heights of religious life. For comparison, the theological and mystical commentaries of Christianity upon the Israelitic eroticism of the Song of Songs may be mentioned, even if the case is not quite the same.

In the first place, the concept of buddhahood was maintained and the idea, that every human being is to become a buddha. Even if this is ultimately a matter of form, it yet remains evidence of deliberate continuity, which made it possible that also the Hinayāna and the Mahāyāna writings were afterwards profoundly studied in Tibet. Then they held to Voidness as the final mystery and the foundation of all that is. Although not an original Buddhist concept, it could fairly often come to sound like the Buddha's anātmatā: the doctrine, that there is no self and no immortal soul. How the cool Buddhist analysis could sometimes revert to keen aggressivity, may be shown in passing by the words of a marginal figure: the spirit which creates the illusory world in which we move, must be killed like a rat. The concept of Voidness was particularly fitted to bring order into the chaos of buddhas, bodhisattvas, gods, rituals and symbols, because thus this chaos was brought *sub specie aeternitatis* and could be recognised as essentially void—a point of view which can also be shared in the twentieth century. Tucci very aptly uses the word gnosis for the mysticism and pseudo-mysticism of the Vajrayāna. There are striking resemblances with gnosticism, as also with kabbalism, so that a comparative study might yield interesting results. The symbol of Voidness became the *vajra*, from which this form of Buddhism derives its name. Originally the thunderbolt of the Vedic god Indra, it also appeared in Hinduism as the weapon of Śiva. The Vajrayāna did not simply take over this symbol, but interpreted it anew as the pure, indestructible diamond of Voidness, so that Vajrayāna is best translated as the diamond vehicle. If enlightenment was considered as the mystic union of the male and female principle, the vajra symbolised man and penis, the lotus (in the ritual also a bell) woman and vagina.

Another difference is the Buddhist reinterpretation of the union with the śakti, who in Hinduism symbolised the active energy which created the phenomenal world, while the male god was regarded as the passive spiritual principle, resting in itself. Now whether this mystic union was imagined as a normal copulation, or as a yogic *coitus reservatus*, or as the symbol of a purely spiritual mystic union, in Buddhism the śakti was not active, but passive, and did not symbolise energy, but insight, the supreme wisdom, embraced by the male as a symbol of his desire for insight and good works. Zimmer already drew attention to this reinterpretation, and Snellgrove and Govinda have made it quite a matter of principle. Justly so indeed, for it cannot be denied that the Buddhists usually regarded the union of god and goddess, of man and śakti, as the union of the active impulse towards salvation, manifested in good works and ritual actions, with the passive, quiescent knowledge or wisdom, of *upāya* with *prajñā*. But they were certainly not consistent in this respect. Bhattacharyya speaks here, for instance, of "Lord Mind and Lady Vacuity," which agrees with the reversal of the active and passive roles, but not with the interpretation of the male as compassion and good works. Besides, in Tibet we shall meet with interpretations that are closer to Śaivism than to Buddhism. Altogether the concept of śakti with its original background does not seem to have entirely disappeared. Not to trespass against the Buddhist reinterpretation of the union of compassion and wisdom, however, the śakti will henceforth be called *mudrā* (= seal), when a woman of flesh and blood is meant, and *yum* (= mother) when referring to a goddess appearing in a vision, in imagination or as a work of art.

In meditation and contemplation the Buddhist also differed clearly from the Śaivite. The latter felt to the last a difference and a distance between himself and his god. In the Buddhist liturgies, the *sādhanas*, the worshipper identifies himself with his god so completely, that one may really speak of identity. Here an essential piece of Buddhism proves to have been preserved. The complete indifference of the Buddha towards the gods, who are illusions like all living beings, has been slightly intensified under the influence of the philosophy of Voidness and the ideas concerning the creative and destructive power of the yogin, and become a unique kind of atheism: the gods exist, in so far as man does not control them. The followers of the Vajrayāna "conceived of a god in knowledge of his non-existence," as it is formulated in the Hevajra tantra.

Typical of the Buddhism of those who adhered to the diamond vehicle was also the creation of new gods. The Hevajra already referred to is simply the hypostasis of an invocation, upon the principle of *nomina numina:* he vajra. But the scale was turned by their god Yamāntaka, who conquered the Hindu god of death Yama, and took his place. Nothing affords better proof of their withdrawal from and hatred of Hinduism, including Śaivism, than the liquidation and complete negation of no one less than the god of death, known and feared by all Hindus. Just as the visualising of a god in knowledge of his non-existence determines the Buddhist character of the Vajrayāna from the religious aspect, so that character is sharply underlined in a social respect by Yamāntaka's liquidation of Yama, who for centuries had ruled as king of the dead, in the Vedic religion, in Hinduism, in Hinayāna and Mahāyāna, with the particular function of judging the dead and inexorably, but justly, determining their next rebirth.

Finally, mention must be made of the work of the theologians. They not only interpreted the sexual and other material symbols and practices mystically in a gnostic fashion—for instance, coition is mystic union, killing an enemy is killing the illusion of the self, one's mother or sister as mudrā means different levels of truth—in their great and flourishing universities they also introduced a systematic order into the continually increasing pantheon of buddhas, bodhisattvas and gods who do not exist, but are real in man as long as he has not transcended himself and thereby the gods. At the head of the five families, to which the gods had been assigned, stood Vajradhara or Vajrasattva, who *is* Voidness. The heads of the five separate families, which correspond to the five components of the human body, are the buddhas Vairocana, Akṣobhya, Ratnasambhava, Amitābha, and Amoghasiddhi. In view of the subject of this book, Akṣobhya calls for particular attention, for his family includes all the terrifying deities (with one exception), and his bodhisattva is Vajrapāṇi. His family is that of the vajra, its colour is blue, the sin assigned to this family and which can be conquered by it is wrath, and the corresponding component of the body is consciousness. These associations will be considered later, for the moment we may point out that, entirely in agreement with the spirit of ancient Buddhism, they are rather of a psychological than of a theological nature. It is at once evident, that *e.g.*, the correspondence between components of the body and gods admits of a theory of religious projection, and a mystic way whereby insight can be gained with regard to the non-entity of both body and gods, and the inexpressible Voidness realised.

As it is not possible to give a complete sketch of the Vajrayāna, it will be best to close the chapter with a concrete example, which may convey more than a too summary and too abstract exposé. In Snellgrove's translation of the Hevajra tantra, much praised by specialists, there is a passage which sounds emotional. Here the author of this most important tantra seems to let himself go for a moment, and as people often say more in such a sally than in collected argument, this passage may perhaps be more typical than others. The god Hevajra himself speaks, in part, as follows: "Without bodily form how should there be bliss? Of bliss one could not speak. The world is pervaded by bliss, which pervades and is itself pervaded. Just as the perfume of a flower depends upon the flower, and without the flower becomes impossible, likewise without form and so on, bliss would not be perceived. I am existence, I am non-existence, I am the Enlightened One for I am enlightened concerning what things are. But me they do not know, those fools, afflicted by indolence. I dwell in the Sukhāvatī, in the vagina of the Vajra Woman, in that place which is symbolized by the syllable E, in that casket of buddha-gems. I am the teacher, and I am the doctrine, I am the disciple endowed with good qualities. I am the goal, and I am the trainer. I am in the world and supramundane. My nature is that of Innate Joy and I come at the end of the Joy that is Perfect and at the beginning of the Joy of Cessation. So be assured, my son, it is like a lamp in darkness. I am the Master with the thirty-two marks, the Lord with the eighty characteristics and I dwell in the Sukhāvatī, in the vagina of the female in the name of semen. Without it—the semen—there is no bliss and again without bliss it—the semen—cannot be. Since they are ineffective one without the other, bliss is found in union with the divinity. So the Enlightened One is neither existence nor non-existence; he has a form with arms and faces and yet in highest bliss is formless. So the whole world is the Innate, for the Innate is its essence. Its essence too is nirvāṇa when the mind is in a purified state. The divine form consists of just something born, for it is a repository of arms and faces and colours, and moreover arises by the normal influence of past actions. With the very poison, a little of which would kill any other being, a man who understands poison would dispell another poison. Just as a man who suffers with flatulence is given beans to eat, so that wind may overcome wind in the way of a homeopathic cure, so existence is purified by existence in the countering of discursive thought by its own kind. Just as water entered in the ear is drawn out again by water, so also the notion of existing things is purified by appearances. Just as those who have been burned by fire must suffer again by fire, so those who have been burned by the fire of passion must suffer the fire of passion. Those things by which men of evil conduct are bound, others turn into means and gain thereby release from the bonds of existence. By passion the world is bound, by passion too it is released, but by heretical buddhists this practice of reversals is not known."

This mystical-seeming pantheism of the body, which for a moment appears a pantheism of the semen, makes the impression of a medley of sexuality—of a rather drastic kind in a religious context—, philosophical speculation and comparisons which are clear, but not always delicate. The aggressive sally against the fools who understand nothing of this religious homoeopathy, is characteristic, also in that it is directed against heretical buddhists. Heretics are of course always the others. It is rather piquant, that the later commentators always use the term "fools" for those who take the tantras literally, whilst here it is those who do not do so, who are called fools. For though a mystical interpretation is certainly possible—one need only think of the commentaries on the Song of Songs—there can be little doubt that here it is the redemption of the body by the body that is preached, and in a high strain. Without the body there is no bliss and, to be exact, this bliss depends on the semen. The world is pervaded by bliss and the god himself is this world. The nature and the essence of the god is the innate joy, which comes at the end of the joy that is perfect, and at the beginning of the joy of cessation. The reader who might find the essential nature of this god obscure, will easily understand the sequence 1) joy, 2) perfect joy, 3) innate joy and 4) joy of cessation, if he considers that it is an analysis of cohabitation as experienced by the male. The commentators interpret these four phases as four yogic conditions of consciousness. They have a perfect right to do so, all the more as they do it subtly and consistently. But it is clear that tantrism is a very different thing in its origin and in its mystical re-interpretation, and that the tantras, which themselves already make the impression of having been worked over, do indeed offer an ambiguous choice. To speak with Snellgrove: it was indeed the razor's edge.

If we turn off the flood-lights of the commentaries, then Hevajra stands before us in this tantra as a tight-rope dancer who slips more than once, and that on the carnal side of his rope. To take an instance, there is his description of a fair mudrā, whose charms include a sweet breath and whose *pudenda* are fragrant as lotus flowers, who is calm and resolute and delightful, and his comment on this description: "By vulgar men, indeed, she would be classed as first-rank." The fact that the thought can arise, how vulgar men would judge of the beautiful woman destined for the ritual, is somewhat curious. Many a man will prefer simply to be vulgar in this respect. In the part called the Manifestation of Hevajra, he sinks into the bliss of union with his beloved, whereupon the other goddesses of his court begin calling on him to arise. One of them cries: "Embrace me in the union of great bliss, and abandon the condition of Voidness." Even if we leave a very wide margin for the secret language of mysticism, no one familiar with it will believe in this polygamous mysticism, which is a mere projection of the ritual celebrated by a yogin who is provided with a number of so-called tantric assistants. As this tantra also contains directions for a magical rite to gain the favours of a woman, we may with good reason suppose this yoga to carry an uneasy conscience. What is propagated here, is actually a copulating form of religion, either factual or imaginary. The only condition is purification of the spirit. In view of the very sensebound symbolism and magic, it is to be supposed that in most cases this purity was rather a matter of intention than of fact. The Hevajra tantra is too much impregnated with rankling resentment, sensuousness and sensuality, to give credit to a high level of sublimation. In such cases a secret language is not a sign of profundity, but of ambiguousness.

That there can be no question of mysticism, not even of spurious mysticism or authentic yoga, becomes apparent when Hevajra awakens from his sweet slumber in the lap of Nairātmyā. For then he plants his feet firmly on the ground, and threatens gods and titans in a terrible way, quite without provocation. Mysticism seeks permanent liberation from the self, Hevajra uses sex and aggressivity for the unbounded inflation of his self, in line with the I, I, I, in the quotation given above. This aggressivity is a mixture of the quick anger often manifested by the yogins, so easily offended because their appetite for power is never sate, and the rankling resentment felt by social underdogs, as the tantric Buddhists originally were. Some fortnight's practice of yoga by the faithful will probably have caused both factors to coalesce. It must have been satisfying for outcasts to be able to say: "One is oneself the Destroyer, the Creator, the King, the Lord." Psychologically, this aggressive urge to power is incompatible with sexuality and love, in and outside mysticism. Division and union, destruction and synthesis, maintaining and losing the self, magic and mysticism are unlinked and irreconcilable opposites here. While the Buddha taught, that affective ties with the phenomenal world must be cut, to become free from the illusion of the self, the voluptuous enjoyment of power and sex alternate here, making the ties with the world as thick as cables in magic of mystic pretension, and inflating the ego till it becomes the Lord of Creation himself. So when Hevajra says he is black and terrible, but that peace rules in his heart, one simply does not believe him. One would like to know whether the mystic commentators refer to this, and if so what they said, but most of their work is still unpublished and untranslated. Govinda says the true Buddhist does not seek power, and then Hevajra is a deplorable Buddhist. If Buddhism maintained its identity in essential points, this was not the case in all points. Śiva, the great destroyer and re-creator, had more influence on Hevajra and his followers in point of mysticism than the Buddha. It was necessary to go into this matter, for it was figures like this Hevajra who brought Buddhism to Tibet, figures like one of the 84 Siddhas, who "makes butter with his śakti," and who flamed into anger like their Terrifying Gods. If the reader, unaccustomed to such material, should be confused by the pluriform content of this chapter, he will at any rate sympathise with the farmers and herdsmen of Tibet, who also had to work through this strange, bizarre world. To put it quite simply, the Tibetans must have been pretty much taken aback, when yogins of the Hevajra type came to preach a new religion to them, a religion in which complicated theories justified men's vagrancy in company with women, and explained as representative of a higher order the commerce of the sexes, which the Tibetans had regarded as belonging to the natural human and animal world (several sexual words in their language refer to humans and animals both, as seems perfectly natural to any herdsman or farmer). Tibet did not easily give in. And afterwards, when the Vajrayāna had gained currency there, it was found that the Tibetans had resolved many dubious matters into their component parts by realistic analysis. However many complications the process of acculturation caused, one's general impression is, that the cool and realistic outlook in Tibet at any rate considerably desexualised the Vajrayāna. Tibet was inclined to call a spade a spade, and to set limits. Much was to happen, however, before this was attained.

Unit X

From Darkness to Light: The Mystery Religion of Ancient Greece

The Bacchae

EURIPIDES

Euripides (480–406 B.C.E.) wrote Prometheus Bound, The Trojan Women *and* The Bacchae. *Reprinted here is the last section of* The Bacchae *in which Pentheus plots the disruption of Dionysus's rituals and Dionysus plots his revenge on Pentheus. This selection offers a good glimpse of the Bacchic orgies themselves as well as the attitudes of many of the characters toward the mysterious rites.*

DIONYSUS: Women! this man is in our net; he goes
To find his just doom 'mid the Bacchanals.
Dionysus, to thy work! thou'rt not far off;
Vengeance is ours. Bereave him first of sense;
Yet be his frenzy slight. In his right mind
He never had put on a woman's dress;
But now, thus shaken in his mind, he'll wear it.
A laughing-stock I'll make him to all Thebes,
Led in a woman's dress through the wide city,
For those fierce threats in which he was so great.
But I must go, and Pentheus—in the garb
Which wearing, even by his own mother's hand
Slain, he goes down to Hades. Know he must
Dionysus, son of Zeus, among the gods
Mightiest, yet mildest to the sons of men.

Strophe

CHORUS: O when, through the long night,
With fleet foot glancing white,
Shall I go dancing in my revelry,
My neck cast back, and bare
Unto the dewy air,
Like sportive fawn in the green meadow's glee?
Lo, in her fear she springs
Over th' encircling rings,
Over the well-woven nets far off and fast;
While swift along her track
The huntsman cheers his pack,
With panting toil, and fiery storm-wind haste.
Where down the river-bank spreads the wide meadow,
Rejoices she in the untrod solitude.
Couches at length beneath the silent shadow
Of the old hospitable wood.

What is wisest? what is fairest,
Of god's boons to man the rarest?
With the conscious conquering hand
Above the foeman's head to stand.
What is fairest still is dearest.

Antistrophe

Slow come, but come at length,
In their majestic strength,
Faithful and true, the avenging deities:
And chastening human folly,
And the mad pride unholy,
Of those who to the gods bow not their knees.
For hidden still and mute,
As glides their printless foot,
The impious on their winding path they hound.
For it is ill to know,
And it is ill to do,
Beyond the law's inexorable bound.
'Tis but light cost in his own power sublime
To array the godhead, whoso'er he be;
And law is old, even as the oldest time,
Nature's own unrepealed decree.

What is wisest? what is fairest,
Of god's boons to man the rarest?
With the conscious conquering hand
Above the foeman's head to stand.
What is fairest still is rarest.

Epode

Who hath 'scaped the turbulent sea,
And reached the haven, happy he!
Happy he whose toils are o'er,
In the race of wealth and power!
This one here, and that one there,
Passes by, and everywhere
Still expectant thousands over
Thousand hopes are seen to hover.
Some to mortals end in bliss;
Some have already fled away:
Happiness alone is his
That happy is to-day.

DIONYSUS: Thou art mad to see that which thou shouldst not see.
And covetous of that thou shouldst not covet.

Pentheus! I say, come forth! Appear before me,
Clothed in the Bacchic Maenads' womanly dress;
Spy on thy mother and her holy crew,
Come like in form to one of Cadmus' daughters.

(*Enter* PENTHEUS.)

PENTHEUS: Ha! now indeed two suns I seem to see,
A double Thebes, two seven-gated cities;
Thou, as a bull, seemest to go before me,
And horns have grown upon thine head. Art thou
A beast indeed? Thou seem'st a very bull.
DIONYSUS: The god is with us; unpropitious once,
But now at truce: now seest thou what thou shouldst see?
PENTHEUS: What see I? Is not that the step of Ino?
And is not Agave there, my mother?
DIONYSUS: Methinks 'tis even they whom thou behold'st;
But lo! This tress hath strayed out of its place,
Not as I braided it, beneath thy bonnet.
PENTHEUS: Tossing it this way now, now tossing that,
In Bacchic glee, I have shaken it from its place.
DIONYSUS: But we, whose charge it is to watch o'er thee,
Will braid it up again. Lift up thy head.
PENTHEUS: Braid as thou wilt, we yield ourselves to thee.
DIONYSUS: Thy zone is loosened, and thy robe's long folds
Droop outward, nor conceal thine ankles now.
PENTHEUS: Around my right foot so it seems, yet sure
Around the other it sits close and well.
DIONYSUS: Wilt thou not hold me for thy best of freinds,
Thus strangely seeing the coy Bacchanals?
PENTHEUS: The thyrsus—in my right hand shall I hold it?
Or thus am I more like a Bacchanal?
DIONYSUS: In thy right hand, and with thy right foot raise it.
I praise the change of mind now come o'er thee.
PENTHEUS: Could I not now bear up upon my shoulders
Cithaeron's crag, with all the Bacchanals?
DIONYSUS: Thou couldst if 'twere thy will. In thy right mind
Erewhile thou wast not; now thou art as thou shouldst be.
PENTHEUS: Shall I take levers, pluck it up with my hands,
Or thrust mine arm or shoulder 'neath its base?
DIONYSUS: Destroy thou not the dwellings of the nymphs,
The seats where Pan sits piping in his joy.
PENTHEUS: Well hast thou said; by force we conquer not
These women. I'll go hide in yonder ash.
DIONYSUS: Within a fatal ambush wilt thou hide thee,
Stealing, a treacherous spy, upon the Maenads.

PENTHEUS: And now I seem to see them there like birds
Couching on their soft beds amid the fern.

DIONYSUS: Art thou not therefore set as watchman o'er them?
Thou'lt seize them—if they do not seize thee first.

PENTHEUS: Lead me triumphant through the land of Thebes!
I, only I, have dared a deed like this.

DIONYSUS: Thou art the city's champion, thou alone.
Therefore a strife thou wot'st not of awaits thee.
Follow me! thy preserver goes before thee;
Another takes thee hence.

PENTHEUS: Mean'st thou my mother?

DIONYSUS: Aloft shalt thou be borne.

PENTHEUS: O the soft carriage!

DIONYSUS: In thy mother's hands.

PENTHEUS: Wilt make me thus luxurious?

DIONYSUS: Strange luxury, indeed!

PENTHEUS: 'Tis my desert.

(*Exit* PENTHEUS.)

DIONYSUS: Thou art awful!—awful! Doomed to awful end!
Thy glory shall soar up to the high heavens!
Stretch forth thine hand, Agave!—ye her kin,
Daughters of Cadmus! To a terrible grave
Lead I this youth! Myself shall win the prize—
Bromius and I; the event will show the rest.

(*Exit* DIONYSUS.)

Strophe

CHORUS: Ho! fleet dogs and furious, to the mountains, ho!
Where their mystic revels Cadmus' daughters keep.
Rouse them, goad them out,
'Gainst him, in woman's mimic garb concealed,
Gazer on the Maenads in their dark rites unrevealed.
First his mother shall behold him on his watch below,
From the tall tree's trunk or from the wild scaur steep;
Fiercely will she shout—
"Who the spy upon the Maenads on the rocks that roam
To the mountain, to the mountain, Bacchanals, has come?"
Who hath borne him?
He is not of woman's blood—
The lioness!
Or the Libyan Gorgon's brood?

Come, vengeance, come, display thee!
With thy bright sword array thee!

The bloody sentence wreak
On the dissevered neck
Of him who god, law, justice hath not known,
Echion's earth-born son.

Antistrophe

He, with thought unrighteous and unholy pride,
'Gainst Bacchus and his mother, their orgies' mystic mirth
Still holds his frantic strife,
And sets him up against the god, deeming it light
To vanquish the invincible of might.
Hold thou fast the pious mind; so, only so, shall glide
In peace with gods above, in peace with men on earth,
Thy smooth painless life.
I admire not, envy not, who would be otherwise:
Mine be still the glory, mine be still the prize,
By night and day
To live of the immortal gods in awe;
Who fears them not
Is but the outcast of all law.

Come, vengeance, come display thee!
With thy bright sword array thee!
The bloody sentence wreak
On the dissevered neck
Of him who god, law, justice has not known,
Echion's earth-born son.

Epode

Appear! appear!
Or as the stately steer!
Or many-headed dragon be!
Or the fire-breathing lion, terrible to see.
Come, Bacchus, come 'gainst the hunter of the Bacchanals,
Even now, now as he falls
Upon the Maenads' fatal herd beneath,
With smiling brow,
Around him throw
The inexorable net of death.

(*Enter a* MESSENGER.)

MESSENGER: O house most prosperous once throughout all Hellas!
House of the old Sidonian!—in this land
Who sowed the dragon's serpent's earth-born harvest—
How I deplore thee! I a slave, for still
Grieve for their master's sorrows faithful slaves.

CHORUS: What's this? Aught new about the Bacchanals?
MESSENGER: Pentheus hath perished, old Echion's son.
CHORUS: King Bromius, thou art indeed a mighty god!
MESSENGER: What sayst thou? How is this? Rejoicest thou,
O woman, in my master's awful fate?
CHORUS: Light chants the stranger her barbarous strains;
I cower not in fear for the menace of chains.
MESSENGER: All Thebes thus void of courage deemest thou?
CHORUS: O Dionysus! Dionysus! Thebes
Hath o'er me now no power.
MESSENGER: 'Tis pardonable, yet it is not well,
Woman, in other's miseries to rejoice.
CHORUS: Tell me, then, by what fate died the unjust—
The man, the dark contriver of injustice?
MESSENGER: Therapnae having left the Theban city,
And passed along Asopus' winding shore,
We 'gan to climb Cithaeron's upward steep—
Pentheus and I (I waited on my lord),
And he that led us on our quest, the stranger—
And first we crept along a grassy glade,
With silent footsteps, and with silent tongues,
Slow moving, as to see, not being seen.
There was a rock-walled glen, watered by a streamlet,
And shadowed o'er with pines; the Maenads there
Sate, all their hands busy with pleasant toil;
And some the leafy thyrsus, that its ivy
Had dropped away, were garlanding anew;
Like fillies some, unharnessed from the yoke;
Chanted alternate all the Bacchic hymn.
Ill-fated Pentheus, as he scarce could see
That womanly troop, spake thus: "Where we stand, stranger,
We see not well the unseemly Maenad dance:
But mounting on a bank, or a tall tree,
Clearly shall I behold their deeds of shame."
A wonder then I saw that stranger do.
He seized an ash-tree's high heaven-reaching stem,
And dragged it down, dragged, dragged to the low earth;
And like a bow it bent. As a curved wheel
Becomes a circle in the turner's lathe,
The stranger thus that mountain tree bent down
To the earth, a deed of more than mortal strength.
Then seating Pentheus on those ash-tree boughs,
Upward he let it rise, steadily, gently
Through his hands, careful lest it shake him off;
And slowly rose it upright to its height,
Bearing my master seated on its ridge.

There was he seen, rather than saw the Maenads,
More visible he could not be, seated aloft.
The stranger from our view had vanished quite.
Then from the heavens a voice, as it should seem
Dionysus, shouted loud, "Behold! I bring,
O maidens, him that you and me, our rites,
Our orgies laughed to scorn; now take your vengeance."
And as he spake, a light of holy fire
Stood up, and blazed from earth straight up to heaven.
Silent the air, silent the verdant grove
Held its still leaves; no sound of living thing.
They, as their ears just caught the half-heard voice,
Stood up erect, and rolled their wondering eyes.
Again he shouted. But when Cadmus' daughters
Heard manifest the god's awakening voice,
Forth rushed they, fleeter than the wingéd dove,
Their nimble feet quick coursing up and down.
Agave first, his mother, then her kin,
The Maenads, down the torrent's bed, in the grove,
From crag to crag they leaped, mad with the god.
And first with heavy stones they hurled at him,
Climbing a rock in front; the branches some
Of the ash-tree darted; some like javelins
Sent their sharp thyrsi through the sounding air,
Pentheus their mark: but yet they struck him not;
His height still baffled all their eager wrath.
There sat the wretch, helpless in his despair.
The oaken boughs, by lightning as struck off,
Roots torn from the earth, but with no iron wedge,
They hurled, but their wild labours all were vain.
Agave spake, "Come all, and stand around,
And grasp the tree, ye Maenads; soon we will seize
The beast that rides thereon. He will ne'er betray
The mysteries of our god." A thousand hands
Were on the ash, and tore it from the earth:
And he that sat aloft, down, headlong, down
Fell to the ground, with thousand piteous shrieks,
Pentheus, for well he knew his end was near.
His mother first began the sacrifice,
And fell on him. His bonnet from his hair
He threw, that she might know and so not slay him,
The sad Agave. And he said, her cheek
Fondling, "I am thy child, thine own, my mother!
Pentheus, whom in Echion's house you bare.
Have mercy on me, mother! For his sins,
Whatever be his sins, kill not thy son."

She, foaming at the mouth, her rolling eyeballs
Whirling around, in her unreasoning reason,
By Bacchus all possessed, knew, heeded not.
She caught him in her arms, seized his right hand,
And, with her feet set on his shrinking side,
Tore out the shoulder—not with her own strength:
The god made easy that too cruel deed.
And Ino laboured on the other side,
Rending the flesh: Autonoe, all the rest,
Pressed fiercely on, and there was one wild din—
He groaning deep, while he had breath to groan,
They shouting triumph; and one bore an arm,
One a still-sandalled foot; and both his sides
Lay open, rent. Each in her bloody hand
Tossed wildly to and fro lost Pentheus' limbs.
The trunk lay far aloof, 'neath the rough rocks
Part, part amid the forest's thick-strewn leaves,
Not easy to be found. The wretched head,
Which the mad mother, seizing in her hands,
Had on a thyrsus fixed, she bore aloft
All o'er Cithaeron, as a mountain lion's,
Leading her sisters in their Maenad dance.
And she comes vaunting her ill-fated chase
Unto these walls, invoking Bacchus still,
Her fellow-hunter, partner in her prey,
Her triumph—triumph soon to end in tears!
I fled the sight of that dark tragedy,
Hastening, ere yet Agave reached the palace.
Oh! to be reverent, to adore the gods,
This is the noblest, wisest course of man,
Taking dread warning from this dire event.

(*Exit* MESSENGER.)

CHORUS: Dance and sing
In Bacchic ring,
Shout, shout the fate, the fate of gloom,
Of Pentheus, from the dragon born;
He the woman's garb hath worn,
Following the bull, the harbinger, that led him to his doom.
O ye Theban Bacchanals!
Attune ye now the hymn victorious,
The hymn all glorious,
To the tear, and to the groan!
O game of glory!
To bathe the hands besprent and gory,
In the blood of her own son.

But I behold Agave, Pentheus' mother
Nearing the palace with distorted eyes
Hail we the ovation of the Evian god.

(*Enter* AGAVE, *mad, bearing Pentheus' head.*)

AGAVE: O ye Asian Bacchanals!
CHORUS: Who is she on us who calls?
AGAVE: From the mountains, lo! we bear
To the palace gate
Our new-slain quarry fair.
CHORUS: I see, I see! and on thy joy I wait.
AGAVE: Without a net, without a snare,
The lion's cub, I took him there
CHORUS: In the wilderness, or where?
AGAVE: Cithaeron—
CHORUS: Of Cithaeron what?
AGAVE: Gave him to slaughter.
CHORUS: O blest Agave!
AGAVE: In thy song extol me.
CHORUS: Who struck him first?
AGAVE: Mine, mine, the glorious lot.
CHORUS: Who else?
AGAVE: Of Cadmus—
CHORUS: What of Cadmus' daughter?
AGAVE: With me, with me, did all the race
Hound the prey.
CHORUS: O fortunate chase!
AGAVE: The banquet share with me!
CHORUS: Alas! what shall our banquet be?
AGAVE: How delicate the kid and young!
The thin locks have but newly sprung
Over his forehead fair.
CHORUS: 'Tis beauteous as the tame beasts' cherished hair.
AGAVE: Bacchus, hunter known to fame!
Did he not our Maenads bring
On the track of this proud game?
A mighty hunter is our king!
Praise me! praise me!
CHORUS: Praise I not thee?
AGAVE: Soon with the Thebans all, the hymn of praise
Pentheus my son will to his mother raise:
For she the lion prey hath won,
A noble deed and nobly done.
CHORUS: Dost thou rejoice?
AGAVE: Ay, with exulting voice
My great, great deed I elevate,
Glorious as great.

CHORUS: Sad woman, to the citizens of Thebes
Now show the conquered prey thou bearest hither.
AGAVE: Ye that within the high-towered Theban city
Dwell, come and gaze ye all upon our prey,
The mighty beast by Cadmus' daughter ta'en;
Nor with Thessalian sharp-pointed javelins,
Nor nets, but with the white and delicate palms
Of our own hands. Go ye, and make your boast,
Trusting to the spear-maker's useless craft:
We with these hands have ta'en our prey, and rent
The mangled limbs of this grim beast asunder.
Where is mine aged sire? Let him draw near!
And where is my son Pentheus? Let him mount
On the broad stairs that rise before our house;
And on the triglyph nail this lion's head,
That I have brought him from our splendid chase.

(*Enter* CADMUS *and attendants, with Pentheus' body.*)

CADMUS: Follow me, follow, bearing your sad burden,
My servants—Pentheus' body—to our house;
The body that with long and weary search
I found at length in lone Cithaeron's glens;
Thus torn, not lying in one place, but wide
Scattered amid the dark and tangled thicket.
Already, as I entered in the city
With old Teiresias, from the Bacchanals,
I heard the fearful doings of my daughter.
And back returning to the mountain, bear
My son, thus by the furious Maenads slain.
Her who Actaeon bore to Aristaeus,
Autonoe, I saw, and Ino with her
Still in the thicket goaded with wild madness.
And some one said that on her dancing feet
Agave had come hither—true he spoke;
I see her now—O most unblessed sight!
AGAVE: Father, 'tis thy peculiar peerless boast
Of womanhood the noblest t' have begot—
Me—me the noblest of that noble kin.
For I the shuttle and the distaff left
For mightier deeds—wild beasts with mine own hands
To capture. Lo! I bear within mine arms
These glorious trophies, to be hung on high
Upon thy house: receive them, O my father!
Call thy friends to the banquet feast! Blest thou!
Most blest, through us who have wrought such splendid deeds.
CADMUS: Measureless grief! Eye may not gaze on it,
The slaughter wrought by those most wretched hands.

Oh! what a sacrifice before the gods!
All Thebes, and us, thou callest to the feast.
Justly—too justly, hath King Bromius
Destroyed us, fatal kindred to our house.

AGAVE: Oh! how morose is man in his old age,
And sullen in his mien. Oh! were my son
More like his mother, mighty in his hunting,
When he goes forth among the youth of Thebes
Wild beasts to chase! But he is great alone,
In warring on the gods. We two, my sire,
Must counsel him against his evil wisdom.
Where is he? Who will call him here before us
That he may see me in my happiness?

CADMUS: Woe! woe! When ye have sense of what ye have done,
With what deep sorrow, sorrow ye! To th' end,
Oh! could ye be, only as now ye are,
Nor happy were ye deemed, nor miserable.

AGAVE: What is not well? For sorrow what the cause?

CADMUS: First lift thine eyes up to the air around.

AGAVE: Behold! Why thus commandest me to gaze?

CADMUS: Is all the same? Appears there not a change?

AGAVE: 'Tis brighter, more translucent than before.

CADMUS: Is there the same elation in thy soul?

AGAVE: I know not what thou mean'st; but I become
Conscious—my mind is settling down.

CADMUS: Canst thou attend, and plainly answer me?

AGAVE: I have forgotten, father, all I said.

CADMUS: Unto whose bed wert thou in wedlock given?

AGAVE: Echion's, him they call the Dragon-born.

CADMUS: Who was the son to thy husband thou didst bear?

AGAVE: Pentheus, in commerce 'twixt his sire and me.

CADMUS: And whose the head thou holdest in thy hands?

AGAVE: A lion's; thus my fellow-hunters said.

CADMUS: Look at it straight: to look on't is no toil.

AGAVE: What see I? Ha! what's this within my hands?

CADMUS: Look on't again, again: thou wilt know too well.

AGAVE: I see the direst woe that eye may see.

CADMUS: The semblance of a lion bears it now?

AGAVE: No: wretch, wretch that I am; 'tis Pentheus' head!

CADMUS: Even ere yet recognized thou might'st have mourned him.

AGAVE: Who murdered him? How came he in my hands?

CADMUS: Sad truth! Untimely dost thou ever come!

AGAVE: Speak; for my heart leaps with a boding throb.

CADMUS: 'Twas thou didst slay him, thou and thine own sisters.

AGAVE: Where died he? In his palace? In what place?

CADMUS: There where the dogs Actaeon tore in pieces.

AGAVE: Why to Cithaeron went the ill-fated man?
CADMUS: To mock the god, to mock the orgies there.
AGAVE: But how and wherefore had we thither gone?
CASMUS: In madness!—the whole city maddened with thee.
AGAVE: Dionysus hath destroyed us! Late I learn it.
CADMUS: Mocked with dread mockery; no god ye held him.
AGAVE: Father! Where's the dear body of my son?
CADMUS: I bear it here, not found without much toil.
AGAVE: Are all the limbs together, sound and whole?
And Pentheus, shared he in my desperate fury?
CADMUS: Like thee he was, he worshiped not the god.
All, therefore, are enwrapt in one dread doom.
You, he, in whom hath perished all our house,
And I who, childless of male offspring, see
This single fruit—O miserable!—of thy womb
Thus shamefully, thus lamentably dead—
Thy son, to whom our house looked up, the stay
Of all our palace he, my daughter's son,
The awe of the whole city. None would dare
Insult the old man when thy fearful face
He saw, well knowing he would pay the penalty.
Unhonoured now, I am driven from out mine home;
Cadmus the great, who all the race of Thebes
Sowed in the earth, and reaped that harvest fair.
O best beloved of men, thou art now no more,
Yet still art dearest of my children thou!
No more, this grey beard fonding with thine hand,
Wilt call me thine own grandsire, thou sweet child,
And fold me round and say, "Who doth not honour thee?
Old man, who troubles or afflicts thine heart?
Tell me, that I may 'venge thy wrong, my father!"
Now wretchedest of men am I. Thou pitiable—
More pitiable thy mother—sad thy kin.
O if there be who scorneth the great gods,
Gaze on this death, and know that there are gods.
CHORUS: Cadmus, I grieve for thee. Thy daughter's son
Hath his just doom—just, but most piteous.
AGAVE: Father, thou seest how all is changed with me:
I am no more the Maenad dancing blithe,
I am but the feeble, fond, and desolate mother.
I know, I see—ah, knowledge best unknown!
Sight best unseen!—I see, I know my son,
Mine only son!—alas! no more my son.
O beauteous limbs, that in my womb I bare!
O head, that on my lap wast wont to sleep!
O lips, that from my bosom's swelling fount

Drained the delicious and soft-oozing milk!
O hands, whose first use was to fondle me!
O feet, that were so light to run to me!
O gracious form, that men wondering beheld!
O haughty brow, before which Thebes bowed down!
O majesty! O strength! by mine own hands—
By mine own murderous, sacrilegious hands—
Torn, rent asunder, scattered, cast abroad!
O thou hard god! was there no other way
To visit us? Oh! if the son must die,
Must it be by the hand of his own mother?
If the impious mother must atone her sin,
Must it be but by murdering her own son?

Mysteries and Asceticism

WALTER BURKERT

Walter Burkert is a renowned scholar of the classics. In this selection, Burkert discusses the general background of Greek mystery religions and then turns his attention to the particular mysteries of Eleusis, Bacchus, Orpheus, and Pythagoras. Burkert emphasizes the individual aspect of the mysteries.

MYSTERY SANCTUARIES

General Considerations

Greek religion, bound to the polis, is public religion to an extreme degree. Sacrificial processions and communal meals, loud prayers and vows, temples visible from afar with splendid votive displays—this is the image of *eusebeia,* this guarantees the integration of the individual into the community. Whoever refuses to take part incurs suspicions of *asebeia.* Yet at the same time there were always secret cults, accessible only through some special, individual initiation, the mysteries. In Greek to initiate is *myein* or else *telein,* the initiate is called *mystes,* and the whole proceedings *mysteria,* while *telesterion* is the special building where initiations take place. The ceremony can also be called *telete,* but this word is also used for religious celebrations generally. *Orgia* too is a word for ritual which is used especially for mysteries: to celebrate in exaltation, having been transformed to a higher status by initiation, is *orgiazein.* Most famous and best known, were the mysteries of Eleusis; for the Athenians the Eleusinian festival was quite simply *ta mysteria*. But this evidently was just the most prominent exemplar of a widespread class of similar institutions.

Secrecy was radical, though it remained an open question whether in mysteries the sacred was forbidden, *aporrheton*, or unspeakable, *arrheton* in an absolute sense. The image which epitomizes the mysteries is the basket closed with a lid, the *cista mystica*: only the initiate knows what this *kiste* conceals; the snake which curls around the *kiste* or protrudes from under the lid points to unspeakable terror. Pagan authors never went beyond circumspect allusions, and the Christian writers who strove to tear off the veil of secrecy were seldom able to produce more than vague insinuations. It is by happy coincidence that one Gnostic writer has revealed a few essential details about Eleusis.

The scholar will attempt to draw tangents, as it were, around the hidden centre, making use of the totality of those allusions. First, there is the aspect of initiation as such. *Initia* is the Latin equivalent for *mysteria*. Secret societies are known from many civilizations; they all have their initiations, whereby the degree of solidarity achieved is in direct relation to the hardships of access. It is possible that mysteries arose from puberty initiations. In Eleusis, with the exception of the "child from the hearth," only adults are initiated, and at an earlier stage access was probably limited to Athenian citizens. Yet Greek mysteries only exist in the true sense if and insofar as initiation is open to both sexes and also to non-citizens. Second, there is the agrarian aspect. Demeter and Dionysus are gods of important mysteries; the drinking of the barley potion or the drinking of wine are central ceremonies. Yet to derive mysteries from agrarian magic is at best a conjecture about prehistory. For the Athenians as we know them, mysteries and corn stand side by side as the two gifts of Demeter; but, on the other hand, the wine festival of the *Anthesteria* or the seed magic of the *Thesmophoria*, the agrarian celebrations of *Proerosia* or *Kalamaia*, are not mysteries. It may rather be asked, even without the prospect of a certain answer, whether at the basis of mysteries there were prehistoric drug rituals, some festival of immortality which, through the expansion of consciousness, seemed to guarantee some psychedelic Beyond. A third and undeniable aspect of the mysteries is the sexual aspect: genital symbols, exposures, and occasionally veritable orgies, in the common sense, are attested. Puberty initiation, agrarian magic, and sexuality may unite in the great experience of life overcoming death. Finally, there is the aspect of myth: mysteries are accompanied by tales—some of which may be secret, *hieroi logoi*—mostly telling of suffering gods. The *mystai* in turn do suffer something in the initiation. Yet the assertion that the *mystes* himself suffers the fate of the god who would thus himself be the first *mystes* does not hold true generally. Suffering easily goes together with the initiation aspect. Deadly terror provoked and dispelled in ritual can be experienced and interpreted as anticipation and overcoming of death. The concept of rebirth admittedly appears only in late Hellenism. In the background there appears once more the sacred deed in general, the encounter with death in sacrifice as such. Precisely for this reason mysteries do not constitute a separate religion outside the public one; they represent a special opportunity for dealing with gods within the multifarious framework of polytheistic polis religion. In Crete, we are told, the very rituals which were absolutely secret in Samothrace or Eleusis were performed in public.

That for the *mystes* death will lose its terror, that he gains the guarantee of a blessed life in another world, is not expressly stated in all of the mysteries we know about, but this promise stands very much to the fore in many of them. Here the different aspects are seen to fuse with one another: the certainty of life attained by intoxication and sexual arousal goes together with insight into the cycle of nature. At the same time the special status attained through initiation is claimed to be valid even beyond death: the orgiastic festival of the *mystai* continues to be held in the afterlife. Yet if the chance of initiation has been let slip in this life, it is impossible to make up for the omission after death. Impressive mythical images bring home this impossibility: Oknos, hesitation personified, is an old man who sits in Hades plaiting a cord which his ass immediately eats away; the uninitiated are carrying water in sieves up to a leaking vessel, aimlessly and endlessly.

Secret societies and initiations are doubtless very ancient institutions. A Neolithic basis for the mysteries may be assumed. Both the Demeter and Dionysos mysteries show specific relations to the ancient Anatolian Mother Goddess. And yet that which is older than the developed polis system could also lead beyond it. The discovery of the individual is the great event that is seen to occur in Greece in the seventh and sixth centuries. Personalities such as Archilochus, Alcaeus and Sappho are the first to exhibit their self-conscious ego in literature. The capability of individual decision and the search for private fulfilment in life are not absent from religion either. Alongside participation in the polis festivals as fixed by the calendar there emerges the interest in something special, chosen by oneself, and hence in additional initiations and mysteries. At the same time individual death, which is built into the system of communal life as an unquestionable fact, becomes a personal problem more than before; thus promises of help extending over and beyond one's own death are sure to find attentive ears. Some dynamic movement seems to enter the static system of religion. After 600 various mysteries are seen to arise or come into the foreground. Over and above the ancient type of clan and family initiations, special sanctuaries such as Samothrace and Eleusis gain increasing influence in a society becoming more and more mobile. At the same time, movements which are not bound to established sanctuaries and their ancestral customs are seen to press even more radical claims; this is true of the *Bakchika* and *Orphika*. The autonomy of the individual attains its peak when collective ritual is left behind and rules are set for a life on one's own responsibility.

Eleusis

Their secrecy notwithstanding, the mysteries of Eleusis are more extensively documented than any other single Greek cult. This is true whether one considers the archaeological evidence of the sanctuary buildings, the epigraphical evidence of sacred laws and other relevant inscriptions, the prosopography of the priests, the iconography of the gods and heroes, or the many reflections in literature, both poetry and philosophy. From the earliest testimony, the Eleusinian section of the Homeric *Hymn to Demeter*, to the proscription of the cult by Theodosius and the destruction of the sanctuary by the Goths about 400 A.D., we survey a period of a thousand years. During this time the cult drew men and women from all of Greece and later from the whole of the Roman Empire and, as is affirmed over and over again, brought them happiness and comfort. According to Diodorus, it was the great age and the untouchable purity of the cult that constituted its special fame. The unique position of Athens in Greek literature and philosophy made this fame spread everywhere.

The secret which was open to thousands every year was, of course, profaned repeatedly. Yet the sources available to us, both iconography and texts, keep to the rule that only allusions are admitted. We are told that Demeter found and met with her daughter in Eleusis; this is the mythical disguise of what happened at the mysteries. Only Christian writers strove to violate the rules. Clement of Alexandria gives the password, *synthema,* of the Eleusinian *mystai,* and a Gnostic writer records a proclamation of the hierophant and names what was shown as the high point of the celebration: an ear of corn cut in silence. There is no reason to doubt his testimony.

The celebration of the mysteries was in the hands of two families, the Eumolpidai, who provide the hierophant and the Kerykes who provide the torch bearer, *dadouchos,* and the sacred herald, *hierokeryx.* In addition there is a priestess of Demeter who lives permanently in the sanctuary. The first *telesterion* proper, the hall for initiations, was built in the time of Pisistratus on the site of a temple-like structure from the time of Solon. The cult on the site, however, can be traced back into the Geometric Age and indeed to Mycenaean predecessors.

The initiation, *myesis,* was an act of individual choice. Most but not all Athenians were initiated. Women, slaves, and foreigners were admitted. The first part of the initiation could take place at various times, either in Eleusis or at the affiliated sanctuary, the Eleusinion above the Agora of Athens. The first act was the sacrifice of a young pig. Each *mystes* had to bring his piglet. According to one description the *mystes* took a bath in the sea together with his piglet. He gives the animal in his stead to its death. Myth associated the death of the pig with Persephone sinking into the earth, just as at the *Thesmophoria* festival.

There follows a purification ceremony for which the Homeric *Hymn* has Demeter herself set the example. Without speaking a word she sits down on a stool which is covered by a ram fleece, and she veils her head. Thus reliefs show Heracles at his initiation, veiled and sitting on a ram fleece, while either a winnowing fan is held over him or a torch is brought up close to him from beneath. In ancient interpretation this would be purification by air and by fire; for the blindfolded *mystes* these must be disquieting, threatening experiences. On the reliefs there follows the encounter with Demeter, Kore, and the *kiste.* This probably points to the festival proper: "As long as you have not reached the *Anaktoron,* you are not initiated."

The *synthema* gives information on successive stages of the initiation rites, yet in veiled terms such as one initiate would use to another to let him know that he has fulfilled all that is prescribed: "I fasted, I drank from the *kykeon,* I took out of the *kiste,* worked, placed back in the basket (*kalathos*) and from the basket into the *kiste.*" Clement himself apparently was unable to give further details, but intimated that this must be something obscene. This has led scholars to make guesses about genital symbols contained in *kalathos* and *kiste*. Yet there is an allusion in Theophrastus to the tools of working, of grinding corn, that early men "consigned to secrecy and encountered as something sacred," evidently in Demeter mysteries. This indicates that mortar and pestle were hidden in the basket, the instruments, in fact, for preparing the *kykeon*. This is a barley drink, a kind of barley-groat broth seasoned with pennyroyal.

There was also some tradition of the mysteries by word of mouth, explanations probably in the form of myths of which we know nothing. Aristotle states, however, that the important thing was not to learn anything but to suffer or experience (*pathein*) and to be brought into the appropriate state of mind through the proceedings.

The *Mysteria* proper are a major festival which has its fixed place in the calendar, in the autumn month of *Boedromion*. The main public event is the great procession from Athens to Eleusis along the Sacred Way, a distance of over thirty kilometers. This took place on the 19th of *Boedromion*. Prior to this, on the 14th day of the month, the "sacred things" had been brought from Eleusis to the Eleusinion in Athens by the *epheboi*. The hierophant opened the festal period with the pronouncement, *prorrhesis*, that those "who are not of pure hands or speak an incomprehensible tongue" should keep away. According to the general interpreatation this excluded murderers and barbarians. It is characteristic of an archaic tradition that no mention is made of purity of heart. On the 16th of the month the *mystai* went together to the sea in the bay of Phaleron in order to purify themselves by bathing. On the 18th they remained at home, probably observing a fast.

The procession which sets off towards Eleusis on the 19th, escorting the sacred things which priestesses carry in closed *kistai*, is pervaded by a mood of dancing, indeed almost ecstasy. The rhythmic shout *Iakch' o Iakche*, resounds again and again and articulates the movement of the crowd. In this shout one can discern the name of a divine being, Iakchos, a *daimon* of Demeter, as it was said, or rather an epithet of Dionysus, as many believed. Bundles of branches called *bakchoi* were swung to the rhythm. In 480 after the Persians had conquered the mainland, a Greek witnessed a cloud of dust, as if from 30,000 men, out of which the Iakchos shout resounded, rising from Eleusis and moving towards Salamis where the Greek army was stationed: the festival prevented by the war was miraculously celebrating itself, as it were, and from it came strength and victory to the Athenians.

When the procession reached the boundary between Athens and Eleusis where there were some small streams, a piece of grotesque buffoonery called *gephyrismoi* was enacted on one of the bridges: masked figures made fun of the passing *mystai* with mockery and obscene gestures. Thus in myth Iambe or Baubo had cheered up Demeter. As soon as the stars became visible the *mystai* broke their fast; the appearance of the stars signals the beginning of the night which is reckoned as belonging to the following day, the 20th of the month. Meanwhile the procession had arrived at the sanctuary. The temples of Artemis and Poseidon, sacrificial altars, and a "fountain of beautiful dances," *Kallichoron*, could all still be visited freely, but behind them lay the gateway to the precinct which, on pain of death, no one but the initiates might enter.

The gates were open to the *mystai*. We know that immediately beyond the entrance there is a grotto, though this is not a great marvel of nature and hence could scarcely have been the starting-point of the cult on the site as a whole. It was dedicated to Pluto, Lord of the Underworld, whom the *mystai* thus approached. The celebration proper took place in the *Telesterion*, a quite unique kind of building. Whereas the appeal of a normal Greek temple lies primarily in its exterior façade with the dark interior simply housing the cult image, the *Telesterion* is built to hold several thousand people at a time, watching as the hierophant showed the sacred things. Two classes of celebrants were distinguished, the *mystai* who took part for the first time, and the *epoptai*, watchers, who were present for at least the second time. They saw what the *mystai* did not yet see; perhaps the latter had to veil themselves at certain phases of the celebration. Each *mystes* had his *mystagogos* who escorted him into the sanctuary. In the centre was the *Anaktoron*, a rectangular, oblong, stone construction with a door at the end of one of its longer sides; there the throne of the hierophant was placed. He alone might pass through the door into the interior of this building. The *Anaktoron* remained unmoved throughout the various phases of the construction of the temple and initiation halls; a piece of natural unhewn rock was left exposed inside. There was no true entrance to the nether world, no chasm, no possibility of acting out a journey into the underworld. The great fire under which the hierophant would officiate obviously burned on top of the *Anaktoron*. Accordingly the roof of the *Telesterion* had a kind of skylight, *opaion*, as an outlet for the smoke. Thus the *Anaktoron* properly belongs in the class of altars with pit chambers, and it may also be compared to the *megaron* of Lykosoura: Even at Eleusis the mysteries were probably celebrated in the open air around a fire, before the building activities of the Solonian and Pisistratean epoch.

Darkness shrouded the crowd thronged in the hall of mysteries as the priests proceeded to officiate by torchlight. Dreadful, terrifying things were shown until finally a great light shone forth "when the Anaktoron was opened" and the hierophant "appeared from out of the *Anaktoron* in the radiant nights of the mysteries." We do not know the true course of events and have difficulty in co-ordinating the various allusions. Was there a sacred marriage of hierophant and priestess? In myth, Demeter places the son of the king of Eleusis into the fire on the hearth, so that the horrified mother is led to believe that the child is being burned, whereas the goddess is actually bestowing immortality on the child. In ritual, one child, *pais*, is always initiated from the hearth, a role which was regarded as a great distinction. A badly damaged relief shows Demeter sitting enthroned while beside her two figures hold out torches towards a cowering child. Apotheosis by fire seems to be indicated; what is present is destruction. In ritual animals were probably killed and burned, with certain manipulations of the sacrificial remains.

Yet it was not terror, but the assurance of blessing that had to prevail. The blessings of the mysteries are expressed in three ways. The *mystes* sees Kore, who is called up by the hierophant by strokes of a gong; as the underworld opens up, terror gives way to the joy of reunion. Then the hierophant announces a divine birth: "The Mistress has given birth to a sacred boy, Brimo the Brimos." Finally, he displays the ear of corn cut in silence. The question as to who the boy was and who his mother was seems to have been answered in different ways: either Iakchos-Dionysos, son of Persephone, or Plutos, Wealth, son of Demeter. Wealth proper is the produce of the corn harvest that banishes poverty and hunger. Vase paintings from the fourth century show a boy with a horn of plenty between the Eleusinian goddesses, surrounded in one case by sprouting corn ears—Plutos personified. The child can easily be identified with the ear of corn, yet cutting or mowing has a further association, that of castration. Growing, thriving, blossoming is brought to a halt with the cutting of the harvester's sickle; and yet in the ear of corn cut down there lies the force for new life.

Further festal activities surround the celebration of the mysteries, dances, and a great bull sacrifice which the *epheboi* execute in the court of the sanctuary and which surely provides a rich meal, as in Andania. Finally, two special vessels, *plemochoai,* are filled and poured out, one towards the west, the other towards the east, while people, looking up to the sky, shout "rain!" and, looking down to the earth, "conceive!" In Greek this is a magical rhyme, *hye—kye.* Perhaps there were dances across the Rharian field where according to myth the first corn grew.

What, in the tribal mysteries and in the case of the gods of Samothrace, is perhaps implied but not explicitly stated becomes the true and universal claim of Eleusis: the mysteries, taking from death its terror, are the guarantee of a better fate in the afterworld. The Homeric *Hymn to Demeter,* Pindar, and Sophocles all leave no doubt about this. The words of the *Hymn* are "blessed is he who has seen this among earthly men; but he who is unitiated in the sacred rites and who has no portion, never has the same lot once dead down in the murky dark." Pindar says "Blessed is he who has seen this and thus goes beneath the earth; he knows the end of life, he knows the beginning given by Zeus," and Sophocles: "Thrice blessed are those mortals who have seen these rites and thus enter into Hades: for them alone there is life, for the others all is misery." In the prose of Isocrates, this becomes the statement that the *mystai* "have more pleasing hopes for the end of life and for all eternity." Simply but emphatically the same message is repeated on the funeral inscription of a hierophant of the Imperial Age: he had shown to the *mystai* "that death is not an evil but something good."

The formula thrice blessed must have been part of the Eleusinian liturgy. Whence it could draw its force of conviction remains a mystery to us. If there was a doctrine or myth of Eleusis on which this faith was explicitly based it has been lost. The images of a blessed afterworld that appear in literature, the symposium of the *hosioi* and a gentle sun shining in the underworld, are not specifically Eleusinian, but elaborations in narrative and poetry of quite different levels and without any official authority. Alongside there was the tradition, which Athenians assiduously spread, that Demeter had given corn to mankind and thus founded civilization at Eleusis. The pictures of Triptolemos setting out in his winged chariot to bestow Demeter's gift on the entire world begin in the middle of the sixth century. The Athenians were bold enough to demand officially that first fruit offerings from the whole world be sent to Eleusis. Thus the importance of Eleusis seems to be transposed to this world, but it is not exhausted in it. There are conjectures about shocking events behind the secret, notably orgies and drugs. Yet precisely the analogy of the Indo-Iranian drug ritual, the Soma/Haoma festival shows that a ritual can persist when the original drug has long been forgotten and replaced by harmless substances. Perhaps the night of the mysteries was not so very different from an Orthodox Easter festival or a Western Christmas. It is remarkable that the concept of immortality is never mentioned in connection with Eleusis. Death remains a reality, even if it is not an absolute end, but at the same time a new beginning. There is another kind of life, and this, at all events, is good. Attention has been drawn to the saying from St. John's Gospel that a grain of wheat must die if it is to bring forth fruit. For "from the dead comes nourishment and growth and seeds." The ear of corn cut and shown by the hierophant can be understood in this way. Euripides has one character in his play *Hypsipyle* comment on the death of a child in these words: "One buries children, one gains new children, one dies oneself; and this men take heavily, carrying earth to earth. But it is necessary to harvest life like a fruit-bearing ear of corn, and that the one be, the other not." This may be seen as a deeper level of worldly piety than that attained by vows and sacrifice in normal, self-interested *eusebeia.*

Bacchic Mysteries

The cult of Dionysos is very ancient in Greece, and yet it is seen to be in a process of continual change. It is no coincidence that outburst and revolution belong to the very essence of this god. Revolutionary innovations can be discerned from the middle of the seventh century. Archilochus, who boasts how he can strike up the *dithyrambos* for Lord Dionysos, is made in legend the founder of Dionysiac phallic processions. About 600, burlesque scenes set in an atmosphere of Dionysian revelry spread like wildfire in Corinthian vase painting, Fat Dancers, whose mummery suggests a grotesque nakedness, are shown dancing, drinking wine, and playing all sorts of tricks. According to tradition it was Arion who at just this time invented the *dithyrambos* in Corinth. We know that in Corinth the family clan of the Bacchiadai, who traced their ancestry back to Dionysos himself, was overthrown in the same period by the tyrant Kypselos who was succeeded by his son Periandros. Accordingly, a new and popular form fitted to the milieu of craftsmen, seems to have taken the place of the old, gentilicial Dionysos cult. Almost simultaneously the tyrant Cleisthenes of Sikyon developed the cult of Dionysos at the expense of a traditional cult of Adrastos. The Athenian innovations in the age of the tyrants followed shortly afterwards—the Great Dionysia with the *dithyrambos*, the *tragodoi* of Thespis and the *satyroi* of Pratinas of Phleius, that is, tragedy and the satyr play. Around 530/20 the iconography of the Dionysiac *thiasos* with satyrs and maenads achieved its fixed, canonical form, while dithyramb and tragedy became part of high literature. Behind these innovations there is clearly an impulse directed against the nobility, which comes from the lower classes of craftsmen and peasants from whom the tyrants drew their support.

Societies of raving women, maenads, and *thyiades*, are no doubt also very ancient, even if direct evidence is available only from later periods. They break out of the confines of their women's quarters and make their way to the mountain. Characteristically the social roles are fixed, as is the calendar: the women of a given city rave at a given time, at the annual festival of *Agrionia* or *Lenaia*, which often gives its name to the month—there is also a month *Thyios* in some calendars—or else they rave every second year at the trieteric festival. True ecstasy, though, remains incalculable: "the *narthex* bearers are many, the *bakchoi* are few."

Dionysos is the god of the exceptional. As the individual gains in independence, the Dionysos cult becomes a vehicle for the separation of private groups from the polis. Alongside public Dionysiac festivals there emerge private Dionysos mysteries. These are esoteric, they take place at night; access is through an individual initiation, *telete*. As a symbolic Beyond, closed and mysterious, the Bacchic grotto or cave appears. The role of the sexes becomes less important: there are male as well as female *mystai*. In contrast to the mysteries of Demeter and the Great Gods, these mysteries are no longer bound to a fixed sanctuary with priesthoods linked to resident families; they make their appearance wherever adherents can be found. This presupposes a new social phenomenon of wandering priests who lay claim to a tradition of *orgia* transmitted in private succession.

The oldest testimony to *bakchoi* and *mystai* is in Heraclitus; Herodotus' description of the fate of the hellenized Scythian King Skyles refers to the middle of the fifth century: in the Greek city of Olbia, Skyles had himself initiated (*telein*) "to Dionysos Bakcheios" at his own wish, even though a divine sign should have warned him against it "as he was on the point of taking upon himself the initiation (*telete*)." He "completed the initiation" and proceeded to rave through town with the *thiasos* of the god. Scythians observed him doing this, and it cost him his throne and his life. Here Bacchic initiations are neither a spontaneous outburst nor a public festival; admission rests on personal application, there is a preparatory period, a tradition of sacred rites, and finally the integration into the group of the initiates. A man, a foreigner no less, is admitted for initiation.

Herodotus, who with this story directs some scarcely concealed criticism against a cult he knows, expressly refers to Miletus as the mother city of Olbia. The same cult of Dionysos Baccheios appears in a third-century inscription from Miletus. Both men and women, we learn, are initiated, but the initiations should be undertaken separately for each sex by priests and priestesses respectively. Omophagy, the eating of raw flesh, which in myth appears as the gruesome high point of Dionysiac frenzy, is mentioned. *Oreibasia*, the procession to the mountain, is also attested in Miletus. The polis asserts its precedence in that no one is allowed to make sacrifice before the polis.

Euripides in his *Bacchae* is dramatizing the old myth which has all the women of the city spontaneously overcome by the god. And yet he also has men going to the mountain, and the leader of the *thiasos* is a man—in reality, as the spectators know, the god Dionysos in person. This man boasts of having received the *orgia* from the god himself. His duty is to show them and to pass them on. This process is secret; nothing of it may be divulged to someone who does not present himself to the *bakcheia*. Even the advantage to be gained, which attracts the initiates, remains secret. The celebrations are nocturnal. Here, therefore, the mythical uprising of the women overlaps with the practice of secret celebrations indifferent to sex and resting on personal initiation. To these Bacchic mysteries belongs a blessing rivalling Eleusis: "O blessed he who knows the initiations of the gods."

For Plato, finally, Dionysos is master of the telestic madness which is to be distinguished from prophetic, musical, and erotic-philosophical madness. The god acts through purifications (*katharmoi*) and initiations (*teletai*), bringing release from "illness and grievous affliction" which manifest themselves in a family on account of an ancient wrath. One must surrender to the madness and allow oneself to be seized by the god in order to become free and well, not only for the present but for all the future.

Dionysiac initiation is fulfilled in raving, *baccheia*. The initiate is turned into a *bacchos*. This state of frenzy is blessedness, compellingly expressed in the entrance song of Euripides' *Bacchae*: earth is transformed into a paradise with milk, wine, and honey springing from the ground; maenads offer their breasts to a fawn. Yet at the very centre of this paradise there is murderous savagery when the frenzied ones become irresistible hunters of animal and man striving towards the climax of dismemberment, the "delight of eating raw flesh." An atavistic spring of vital energy breaks through the crust of refined urban culture. Man, humbled and intimidated by normal everyday life, can free himself in the orgies from all that is oppressive and develop his true self. Raving becomes divine revelation, a centre of meaning in the midst of a world that is increasingly profane and rational.

True ecstasy has its own laws and sources, even if dance and rhythmic music can promote it to a special degree; this is evident in the play of Euripides. Nevertheless, there are two very specific stimulants that belong to Dionysos, which cannot have been missing even in the secret celebrations: alcohol and sexual excitement, the drinking of wine and phallos symbolism. Two complexes from the Hellenistic age reveal details about the further development of Bacchic mysteries: the infamous *Bacchanalia* that were suppressed in Italy by Rome in 186 B.C. with extreme brutality, and the magnificent frescoes of the *Villa dei Misteri* in Pompeii that date from the time of Caesar. Whereas in regard to the *Bacchanalia* it is asserted that the initiation consisted *inter alia* in suffering a homosexual act, in the frescoes a large, erect phallos in a winnowing basket is depicted next to the god; a woman is present to unveil it; blows with a rod are to be suffered too. The forms of Bacchic initiation probably varied a great deal from group to group, and from period to period, with the extent of these variations stretching from outdoor picnics to an existential turning-point in life, from sublime symbolism to downright orgies. It is possible that old forms of puberty initiation were still preserved in sexual initiation; not virgins, but only women could be *bacchai*, and married couples could be initiated together. Usually the purer forms of religion have better chances of longevity than orgies in the modern sense.

Liberation from former distress and from the pressures of everyday life, an encounter with the divine through an experience of the force and meaning of life, are present in Dionysiac initiation. But hopes for the future, for death and the afterlife were no less a part of the secret advantage promised to those who knew. This is shown above all by the gold leaf from Hipponion which came to light only in 1969.

Bacchic Hopes for an Afterlife

Gold leaves from tombs in southern Italy, Thessaly, and Crete, with hexameter texts that give instruction to the dead about the path to be followed in the other world, have long been known and much discussed. The oldest and most extensive text comes from Hipponion-Vibo Valentia and is dated with certainty to about 400 B.C. by its archaeological context. The beginning is barely intelligible and probably corrupt; then it runs:

> In the house of Hades there is a spring to the right, by it stands a white cypress; here the souls, descending, are cooled. Do not approach this spring! Further you will find cool water flowing from the lake of recollection. Guardians stand over it who will ask you in their sensible mind why you are wandering through the darkness of corruptible Hades. Answer: I am a son of the earth and of the starry sky; but I am desiccated with thirst and am perishing: therefore give me quickly cool water flowing from the lake of recollection. And then the subjects of the Chthonian King (?) will have pity and will give you to drink from the lake of recollection . . . And indeed you are going a long, sacred way which also other *mystai* and *bacchoi* gloriously walk.

This last statement is only in the text from Hipponion, while the main part, the scene before the guardians at the lake, is attested, with minor variations, in several other specimens.

Mystai and *bacchoi* walk a sacred way, the goal of which is eternal bliss, indeed the Island of the Blessed; Pindar calls this "the path of Zeus." "You will rule with the other heroes," says another text. Knowledge and certainty of this is gained through initiation. "Blessed are they all by the part they have in the initiations that release from affliction," is said again in Pindar, in a mourning song which points to the lot of the pious. To the sacred way which the *mystai* walk there corresponds in this world the path to the mountain, the *oreibasia*: afterlife is repetition of the mysteries. Accordingly, in the *Frogs* Aristophanes has the chorus of *mystai* in the underworld continue to celebrate the Iakchos procession on the Sacred Way to Eleusis. In parallel to this we find Bacchic mysteries too aiming at a blessed state in the afterworld.

The scenery of the underworld drawn in these texts is both impressive and enigmatic: the white cypress at the dangerous spring, the guardians at the cool water, the password with which the initiate claims cosmic status: "son of heaven and earth." The guardians at the water, and the question and answer have striking parallels in the Egyptian *Book of the Dead.* To recollection some forgetting must correspond. Plato's myth tells of the "plain of Lethe" where the souls drink from the river Ameles, Indifference, before reincarnation. Whether reincarnation is also presupposed by the gold-leaf texts or whether, in a Homeric manner, a gloomy Hades of unconsciousness is threatening those who do not know, must remain an open question. Recollection at any rate guarantees the better lot. "Awake recollection of the sacred initiation in the *mystai*," the goddess of Memory is implored in a late Orphic hymn; in Plato, forgetting what the soul has seen at the high point of its initiation makes it fall into the depths. Recollection, memory is valued highest among Pythagoreans.

Herodotus reports that Egyptians do not bury their dead in woollen cloths, but in linen, and he adds: "This corresponds to the so-called *Orphica* and *Bacchica* which are in fact Egyptian and Pythagorean. The Bacchic cult prescription which has to do with death and burial is thus at the same time placed under the name of Orpheus. Herodotus is writing in southern Italy about 430 B.C.; he therefore is quite close to the text of Hipponion in time and place, a text which like Herodotus speaks of Bacchic expectations in a funerary context; even the references to Egyptians and Pythagoreans fit remarkably well with what emerges from the gold leaves.

From a grave precinct at Cumae in southern Italy comes an inscription that has long been known: "It is not permitted that someone should lie here who has not been made to celebrate as a *bacchos*." The special graveyard for the *bebachcheumenoi* openly corresponds to their exceptional lot in the other world. The initiation is not reserved for one sex: both the inscription and the Hipponion text use the general, masculine form, but in the tomb of Hipponion a woman was buried.

One detail of Bacchic ritual relating to the underworld appears in literature: "Those who were initiated into the *Bacchica* were garlanded with poplar, because this plant is chthonic, and so is Dionysos the son of Persephone." The poplar, it is said, grows on the Acheron, and it was a wreath of poplar twigs that Heracles put on after he had conquered Cerberus. Thus myth and ritual give expression to the bond with the nether world and the conquest of death at the same time; this is explicitly connected with the chthonic Dionysos, son of Persephone.

Thus gold leaves, the inscription, and the literary texts fit together: by the fifth century at the latest there are Bacchic mysteries which promise blessedness in the afterlife. Implied is the concept of *baccheia* that designates ecstasy in the Dionysiac *orgia*, in which reality, including the fact of death, seems to dissolve. Our knowledge of the accompanying rites, myths, and doctrines, remains, of course, very fragmentary.

Bacchic hopes for an afterlife are hinted at in a veiled form by funerary gifts, as found in graves from the fourth century onwards. At Derveni not far from Thessaloniki the ashes of a noble Macedonian were buried about 330 B.C. in a magnificent gilt bronze krater, which is richly decorated with Dionysiac scenes; at another burial at the same place an Orphic book, part of which is preserved, was burnt along with the corpse. More modest yet more abundant and continuous is the evidence which comes from the funeral vases of southern Italy. Again and again the dead man and his grave are placed in a Dionysiac atmosphere; in one case Orpheus is clearly portrayed as mediator of other worldly blessedness. Later Greek-Punic grave steles from Lilybaeum show the heroicized dead surrounded by the emblems of the Dionysiac *orgia-tympanon, kymbala, krotala,* ivy leaf, and both *kiste* and *kalathos,* the true badge of mysteries. Plutarch, too, was a Dionysiac *mystes* and could still find a source of comfort in this at the death of a child. In all this we can discern, however faintly, an incalculable religious current that is not to be underestimated. It would, however, be an inadmissible generalization to claim that all bacchic *teletai* were concerned exclusively or even primarily with the afterlife. There is no evidence at all for a unified organization; the *teletai* remained as multiform as their god.

There is a second group of gold leaves which were found exclusively in two extraordinary grave mounds in Thurioi in southern Italy; they date to the fourth century B.C. In one of them several people, who had probably died by lightning, had been buried successively; death by heavenly fire seemed to announce an extraordinary status in the afterworld. In the texts from this grave the dead man proclaims: "I come pure from the pure, queen of the chthonic, Eukles, Eubuleus and you other immortal gods; for I am proud to be myself also of your blessed race." Two texts added: "I have paid penance for works not just," while the third adds some especially imaginative verses: "I flew away from the circle of heavy grief and pains. I stepped with swift feet on the longed-for wreath. I dived beneath the lap of the mistress, the chthonic queen." There follows a blessing with the unprecedented claim: "Happy and blessed one, god will you be instead of a mortal;" then an enigmatic conclusion: "Kid I fell in the milk." A similar text comes from the larger grave mound in which a single man was buried, to whom ritual worship was paid: "Be glad that you have suffered what you never suffered before. A god you have become from a man. Kid you fell in the milk." The proclamation of beatitude, the suffering, and the code word of the kid most probably refer to some initiation ritual; Persephone, queen of the dead, plays a special role in these mysteries. The rest remains obscure: does the kid who fell in the milk point to a proverb or an initiation sacrifice? Does the lap of the goddess hint at a ritual of rebirth? With the promise of apotheosis these texts go beyond everything else that is known from Greek mysteries of the Classical Age. The order of what is set down and apportioned, Themis and Moira, seems to dissolve. Here indeed mysteries infringe upon the system of traditional Greek religion. Yet this seems to have remained an isolated case for a long time.

The Hymn to Demeter

GEORGE DEFOREST LORD

In this selection, George deForest Lord, distinguished Bodman Professor of English at Yale University, translates the Hymn to Demeter. The Hymn was put into writing about 600 B.C.E. *It tells the story of the abduction and rape of Persephone and thus of the origin of Eleusis as a sacred place. Eleusis was the location of the reunion of Demeter and Persephone, mother and daughter. This poem is the best existing evidence for the nature of the mysteries at Eleusis.*

I sing now of fair-haired Demeter, reverend goddess,
of her and her trim-ankled daughter whom Hades
carried away, a gift from Zeus, the Thunderer, the all-seeing
as she was playing far from her mother,
lady of the golden blade and shining fruit,
with the ripe daughters of Ocean, gathering flowers
through the soft meadow—roses, crocuses, fair violets,
irises also and hyacinths and the narcissus Earth
put forth by the will of Zeus to please the Host of Many,
a snare for the blooming girl and a radiant wonder.

It was a marvel for immortals or mortal men to see:
from its root sprang a hundred blooms, and its fragrance
made heaven and the whole earth and the salt swell
of the sea laugh for joy. Astounded, the girl stretched forth
both hands to seize the lovely plaything;
but earth with its wide paths yawned there, in the plain of Nysa,
and the Host of Many, son of Kronos, with his immortal horses,
sprang upon her, he who has many names.
He seized her unwilling and lamenting and bore her off
on his golden car, and she called on her father,
the glorious son of Kronos, on high.
Neither god nor mortal heard her voice, nor did
the rich-fruited olive trees. Only gentle Hecate,
Persaeus' shining-haired daughter, heard her
from her cave, and Lord Helios, Hyperion's
splendid son, as she called.

But her father sat aloof and apart from the gods
in his prayer-thronged temple receiving
sweet offerings from mortal men.
So he, that son of Kronos, of many names, host

and ruler of many, bore her off by Zeus's leave,
his own brother's child, all unwilling, on his immortal car.

As long as she still saw earth and starry heaven
and the flowing sea, where fishes shoal, and the rays of the sun,
as long as she still hoped to see her mother
and the race of immortals, hope calmed her great mind
despite her grief. . . .
Mountain heights and the sea's depths
echoed her immortal voice, and her queenly mother
heard her.

A pang pierced her heart, and she ripped the fine veil
from her shining hair, cast off her dark cloak,
and flew like a bird over firm land and yielding sea, searching.
But no god or mortal wished to speak the truth;
no birds of omen came as true messengers.
Nine days the lady Deo roamed the earth
with flaming torches in her hands, so grieved
she never tasted ambrosia or drank sweet nectar
or sprinkled water on her body.
The tenth dawn, Hecate met her, holding a torch,
and spoke to her aloud and told the news:

"Lady Demeter, bringer of seasons and bright gifts,
what god or mortal seized Persephone
and pierced your loving heart with sorrow?
I heard a voice but whose I could not tell,
but let me tell you briefly what I know."

So, then, spoke Hecate. The daughter of fair-haired Rhea
answered not a word, but sped away with her,
holding bright torches. They came to Helios,
watcher of gods and men. Standing before his horses
Demeter, shining goddess, asked of him:

"Helios, pay regard to me, goddess as I am,
if ever I did anything to cheer your heart and spirit.
Through the fruitless air I heard the piercing cries of
the lovely daughter I bore, sweet scion of my body,
as if seized by force, but I saw nothing.
But you, surveying the whole earth and sea,
casting your rays through the bright air,
Tell me truly if you have seen my dear child anywhere.
What god or mortal has seized her against
her will and mine, and so made off?"

So she asked. And the son of Hyperion answered,
"Queen Demeter, daughter of fair-haired Rhea,

you shall know, for I greatly reverence and pity you
in your grief for your slim-ankled child.
No other immortal is to blame but Zeus,
the cloud-gatherer, who gave her to Hades,
his own brother, to be his blooming wife.
Hades took her crying aloud with his steeds
down to the misty darkness. But, Goddess,
cease your lamentations; you mustn't be
Relentless in your anger. Aïdoneus, Host of Many,
is no unfitting husband among immortals for your child,
being your own brother from the same seed.
As for honor, he has a third share in
that triple division made in the beginning,
and is lord over all those he dwells among."

So he spoke, and called his horses; at his word
they whirled the swift chariot off like
long-winged birds.

But grief more terrible and wild
gripped Demeter's heart; so enraged was she
at dark-clouded Kronides
that she shunned the gods' gatherings
and high Olympus, and went among the towns and fields
of men and their rich labors, long hiding her form.
No man or low-girt woman
knew who she was until she reached
the house of wise Celeus, lord of Eleusis,
a town rich in sacrifices. In deep sorrow
she sat by the road near the Maiden Well,
where local women drew water, a shady place
above which an olive bush grew. She looked like
an ancient woman past childbearing and the gifts
of wreath-loving Aphrodite, the kind that nurses
the children of kings who deal justice,
or like the housekeepers in their echoing halls.
There the daughters of Celeus, Eleusis' son,
saw her as they came to draw water easily
and bear it in bronze pitchers to their dear
father's house. There were four of them, like goddesses
in the flower of girlhood: Callidice,
Cleisidice, lovely Demo, and Callithöe,
the eldest. They knew her not—
it is hard for mortals to recognize gods—
but standing near her they spoke winged words:
"Old woman, whence and who are you of folk born long ago?
Why have you left the city and not approached a house?

In their shady halls are women just your age,
and others younger, who would welcome you."

. . . "I am the honored Demeter, the greatest help
of gods and men. But now let the people build me
a great temple and beneath it an altar below the steep walls
on a rising hill above Kallichoron.
I myself shall introduce rites so that later
you may propitiate me by their performance."

At these words the goddess changed her size and form
and sloughed off old age, and a lovely scent came
from her robes, and from her body and her golden hair
radiance, like lightning, filled the house.
And so she left the palace.

Metaneira staggered and was speechless
for a while and failed to lift her late-born son from the ground.
But his sisters heard him wailing pitifully and sprang
down from their couches, and one took him
in her bosom, another revived the fire, while
a third sped on tender feet to rouse her mother
from the fragrant chamber. So they gathered
round the squirming child and washed him
and squeezed him lovingly. He was not comforted
because much less skillful nurses held him now.

All night, quaking with fear, they tried
to appease the glorious goddess, but as dawn appeared
they told great Celeus all that had happened,
as fair-crowned Demeter had ordered.
So Celeus summoned the multitudes to meet
and told them to build a fine temple for
rich-haired Demeter and an altar on a hillock.
They obeyed his commands swiftly,
and the temple grew by divine decree.
As for the child, he grew like an immortal.

Now when they had finished building
and ceased their work, they all went home.
But fair-haired Demeter stayed apart from the blessed gods,
wasted with yearning for her buxom daughter.
Then she brought on mankind a cruel year
over all-nourishing earth: the seed didn't sprout,
for rich-haired Demeter suppressed it;
in the fields the oxen drew their curved ploughs in vain,
and in vain did white barley-seed fall to earth.
So, with famine she would have destroyed

the whole race of men and robbed Olympians
of their due gifts and sacrifices,
had not Zeus perceived this and pondered in his heart.

First, he sent Iris of the golden wings
to summon fair-haired Demeter, the beautiful.
Gladly she obeyed dark-clouded Kronides
and sped through intervening space.
She reached Eleusis rich in sacrifice,
a city fragrant and strong. She found
veiled Demeter in her temple and spoke winged words:

"Demeter, Father Zeus, whose wisdom never fails,
bids you join the company of immortals; come, then,
and let not his bidding pass unheeded."
Thus Iris implored. Demeter was unmoved.

Then the Father sent all the blessed Olympians,
and, one by one, they came, hailing her
and offering lovely gifts and whatever rights
she might choose among the deathless gods.
Yet no one could move her in her strong wrath.
Adamant, she rejected their pleas:
Never, she swore, would she set foot on Olympus
nor let fruit spring from the ground, not until
she saw her lovely daughter with her own eyes.

When he heard this, all-seeing Zeus
the Thunderer, sent Hermes of the golden wand
to Erebus to win over Hades with soft words
that he might lead chaste Persephone up from the misty gloom
to join the gods, so her mother might see her
and give up her wrath.

Promptly Hermes left the house of Olympus
and sped down to the depths of earth.
He found Lord Hades at home, seated on a couch,
his shy, reluctant bride beside him,
yearning for her mother. But she was far off, brooding
on her plan, apart from the gods.
The strong slayer of Argos drew near and said:
"Dark-haired Hades, king of the departed,
Zeus bids me bring noble Persephone
from Erebus so her mother can see her
and so give up her anger at the gods.
For now she plots an awful deed, to kill off
the feeble tribes of men by hiding the seed
beneath the earth, extinguishing the honors

due the undying gods. Her anger is terrible.
She shuns the gods and sequesters herself
in her fragrant temple in Eleusis' rocky hold."

So he said. And Aïdoneus, ruler of the dead,
smiled a grim smile and yielded to King Zeus.
He urged Persephone the wise to go, saying:
"Go now, Persephone, to your dark-robed mother,
go with a gentle spirit, be not cast down.
I shall be no unfitting husband for you
among immortals. Father Zeus is my brother.
While you are here you shall rule all that lives and moves.
Any who defraud you or fail to appease your power
by due rites and rich gifts shall suffer forever."

At this Persephone sprang up, rejoicing.
But, surreptitiously, he gave her a seed
of pomegranate to eat, taking care that she
stay not forever with grave Demeter
of the dark robe.

Then Aïdoneus, Host of Many, harnessed
his deathless horses to the golden chariot.
Persephone mounted and Argeïphontes took reins and whip
and sped from the hall. Quickly they covered their course:
neither sea nor rivers nor grassy glens nor
peaks of mountains checked the undying steeds,
and they clove the deep air as they went.

When Demeter saw them she rushed
like a Maenad down mountain thickets,
while on her side Persephone leapt from the chariot
and ran to her mother and hugged her
passionately. But while Demeter held her child
her heart misgave her, fearing some trick,
so she stopped caressing her and asked right off:
"Surely, child, you touched no food while you were
below? Tell me all. If you have not,
you shall return from loathsome Hades to live
with me and your father, dark-clouded Zeus,
son of Kronos, honored by the deathless ones,
but if you have tasted anything you must go back
down to the secret places of the earth,
to dwell there a third of the revolving year.
Even so, for the rest, you shall be with me
and the other deathless gods. When the earth blooms
with fragrant flowers of all kinds, from your dark kingdom
you shall arise again as a marvel to gods and men.

Tell me now how he took you away to his realm
of gloomy darkness. What trick did the strong
Host of Many beguile you with?"

Lovely Persephone replies,
"Mother, here is the plain truth.
When Hermes came with the message from Zeus,
my father, and the other sons of Heaven,
bidding me come back from Erebus so
you might see me and give up your anger
with the gods, I leaped for joy;
but secretly he put a pomegranate seed
into my mouth and made me taste it against my will.

"I'll tell you also how he snatched me away
by the deep plot of my father, son of Kronos,
and carried me down to the depths of the earth.

"We were all playing in a lovely meadow,
Leucippe and Phaeno and Electra and Ianthe,
Melita also and Iache with Rhodea and Callirhoë,
and Melobosis and Tyche and Ocyrhoë,
fair as a blossom, Chryseis, Ianeira, Acaste,
Rhodope and Pluto and charming Calypso.
Styx was there and Urania and fair Calaxaura
with warlike Pallas and Artemis, who loves arrows.
We were playing and gathering sweet blossoms
in our hands—soft crocuses mixed with iris
and hyacinths and roses and lilies, marvelous
to see, and the narcissus wide earth made grow
yellow as crocus. That one I plucked, ravished
with delight; but the earth split open.
The strong Lord, Host of Many,
sprang forth and bore me off unwilling in his car of gold
beneath the earth. I cried piercingly.
This is all true, much as it grieves me to tell the tale."

So, with one heart, they cheered each other
with many embraces, and their grief changed to joy.
Then bright-coifed Hecate came and embraced
holy Demeter's daughter again and again;
from that time Hecate was her companion and mentor.

And Zeus, who knows it all, sent rich-haired Rhea
as messenger to bring dark-cloaked Demeter
to join the family of gods, and he promised
to grant her whatever rights she chose among the
immortal gods and agreed that their daughter

should descend to the dark gloom
for a third of the revolving year; the rest
she should spend with her mother and the other immortals.

So he decreed, and the goddess complied, rushing
from the heights of Olympus. She came
to the Rharian Plain, once fertile corn-land,
now dry, bare and leafless,
since by Demeter's design the grass was hid.
But as spring waxed it was soon to be alive
with long-eared wheat, its furrows filled
with the mowers' work soon to be bound
in sheaves. There she first landed from
the fruitless ether, and the goddesses rejoiced
when they saw each other. Then
bright-haired Rhea said to Demeter,
"Come, my daughter. Far-seeing Zeus
the Thunderer calls you to join the family of gods
and has sworn to grant whatever rights you choose
among them and has agreed on allotting
the time your daughter will spend in chthonic gloom
and the time she shall spend with you
and the other deathless gods.
Come, daughter, obey; don't be adamant
in your anger. Increase for men life-giving harvests."

So spoke Rhea, and crowned Demeter
did not refuse.

Promptly she made fruit spring from the rich land
so the whole wide earth was laden with leaves and flowers.
Then she went to the kings who deal justice,
Triptolemos, Diocles the horseman, strong Eumolpus
and Celeus, leader of the people, and taught them
how to conduct her rites and showed them all her
mysteries as well—awful mysteries that none
may pry into or utter or transgress in any way,
for deep awe of the gods seals their lips.
Blessed is he among men on earth who has seen
these mysteries, but those uninitiated,
who have no part in them, do not prosper;
once dead they waste away down in the dark gloom.

When the bright goddess had taught them all,
they went to the gathering of the other immortals
on Olympus. And there they dwell beside Zeus,
who loves to thunder, awful and reverend goddesses.

Happy indeed is the man they freely love,
for they send Ploutos as a guest in his house,
Ploutos who gives mortals wealth.

And now, queen of the sweet Eleusinian
land and sea-girt Paros and rocky Antron,
Lady, giver of good gifts, bringer of seasons,
queen Deo, be gracious, you and your daughter
Persephone, and for my song grant me
heartwarming substance.

The Teachings of Pythagoras

OVID

In mathematics, music, astrology, and every other science Pythagoras sought the mystical unity of all things in the world. In this selection from Ovid, *Pythagoras presents what might well be the insight or mystic knowledge (gnosis) of one who had passed through the initiation rituals of Eleusis. Pythagoras's vision of the world is a rational explanation of the mystical enlightenment.*

There was a man here, Samian born, but he
Had fled from Samos, for he hated tyrants
And chose, instead, an exile's lot. His thought
Reached far aloft, to the great gods in Heaven,
And his imagination looked on visions
Beyond his mortal sight. All things he studied
With watchful eager mind, and he brought home
What he had learned and sat among the people
Teaching them what was worthy, and they listened
In silence, wondering at the revelations
How the great world began, the primal cause,
The nature of things, what God is, whence the snows
Come down, where lightning breaks from, whether wind

Or Jove speaks in the thunder from the clouds,
The cause of earthquakes, by what law the stars
Wheel in their courses, all the secrets hidden
From man's imperfect knowledge. He was first
To say that animal food should not be eaten,
And learnèd as he was, men did not always
Believe him when he preached "Forbear, O mortals,
To spoil your bodies with such impious food!
There is corn for you, apples, whose weight bears down
The bending branches; there are grapes that swell
On the green vines, and pleasant herbs, and greens
Made mellow and soft with cooking; there is milk
And clover-honey. Earth is generous
With her provision, and her sustenance
Is very kind; she offers, for your tables,
Food that requires no bloodshed and no slaughter.
Meat is for beasts to feed on, yet not all
Are carnivores, for horses, sheep, and cattle
Subsist on grass, but those whose disposition
Is fierce and cruel, tigers, raging lions,
And bears and wolves delight in bloody feasting.
Oh, what a wicked thing it is for flesh
To be the tomb of flesh, for the body's craving
To fatten on the body of another,
For one live creature to continue living
Through one live creature's death. In all the richness
That Earth, the best of mothers, tenders to us,
Does nothing please except to chew and mangle
The flesh of slaughtered animals? The Cyclops
Could do no worse! Must you destroy another
To satiate your greedy-gutted cravings?
There was a time, the Golden Age, we call it,
Happy in fruits and herbs, when no men tainted
Their lips with blood, and birds went flying safely
Through the air, and in the fields the rabbits wandered
Unfrightened, and on little fish was ever
Hooked by its own credulity: all things
Were free from treachery and fear and cunning,
And all was peaceful. But some innovator,
A good-for-nothing, whoever he was, decided,
In envy, that what lions ate was better,
Stuffed meat into his belly like a furnace,
And paved the way for crime. It may have been
That steel was warmed and dyed with blood through killing
Dangerous beasts, and that could be forgiven

On grounds of self-defense; to kill wild beasts
Is lawful, but they never should be eaten.

One crime leads to another: first the swine
Were slaughtered, since they rooted up the seeds
And spoiled the season's crop; then goats were punished
On vengeful altars for nibbling at the grape-vines.
These both deserved their fate, but the poor sheep,
What had they ever done, born for man's service,
But bring us milk, so sweet to drink, and clothe us
With their soft wool, who give us more while living
Than ever they could in death? And what had oxen,
Incapable of fraud or trick or cunning,
Simple and harmless, born to a life of labor,
What had they ever done? None but an ingrate,
Unworthy of the gift of grain, could ever
Take off the weight of the yoke, and with the axe
Strike at the neck that bore it, kill his fellow
Who helped him break the soil and raise the harvest.
It is bad enough to do these things; we make
The gods our partners in the abomination,
Saying they love the blood of bulls in Heaven.
So there he stands, the victim at the altars,
Without a blemish, perfect (and his beauty
Proves his own doom), in sacrificial garlands,
Horns tipped with gold, and hears the priest intoning:
Not knowing what he means, watches the barley
Sprinkled between his horns, the very barley
He helped make grow, and then is struck
And with his blood he stains the knife whose flashing
He may have seen reflected in clear water.
Then they tear out his entrails, peer, examine,
Search for the will of Heaven, seeking omens.
And then, so great man's appetite for food
Forbidden, then, O human race, you feed,
You feast, upon your kill. Do not do this,
I pray you, but remember: when you taste
The flesh of slaughtered cattle, you are eating
Your fellow-workers.
"Now, since the god inspires me,
I follow where he leads, to open Delphi,
The very heavens, bring you revelation
Of mysteries, great matters never traced
By any mind before, and matters lost
Or hidden and forgotten, these I sing.

There is no greater wonder than to range
The starry heights, to leave the earth's dull regions,
To ride the clouds, to stand on Atlas' shoulders,
And see, far off, far down, the little figures
Wandering here and there, devoid of reason,
Anxious, in fear of death, and so advise them,
And so make fate an open book.
"O mortals,
Dumb in cold fear of death, why do you tremble
At Stygian rivers, shadows, empty names,
The lying stock of poets, and the terrors
Of a false world? I tell you that your bodies
Can never suffer evil, whether fire
Consumes them, or the waste of time. Our souls
Are deathless; always, when they leave our bodies,
They find new dwelling-places. I myself,
I well remember, in the Trojan War
Was Panthous' son, Euphorbus, and my breast
Once knew the heavy spear of Menelaus.
Not long ago, in Argos, Abas' city,
In Juno's temple, I saw the shield I carried
On my left arm. All things are always changing,
But nothing dies. The spirit comes and goes,
Is housed wherever it wills, shifts residence
From beasts to men, from men to beasts, but always
It keeps on living. As the pliant wax
Is stamped with new designs, and is no longer
What once it was, but changes form, and still
Is pliant wax, so do I teach that spirit
Is evermore the same, though passing always
To ever-changing bodies. So I warn you,
Lest appetite murder brotherhood, I warn you
By all the priesthood in me, do not exile
What may be kindred souls by evil slaughter.
Blood should not nourish blood.
"Full sail, I voyage
Over the boundless ocean, and I tell you
Nothing is permanent in all the world.
All things are fluent; every image forms,
Wandering through change. Time is itself a river
In constant movement, and the hours flow by
Like water, wave on wave, pursued, pursuing,
Forever fugitive, forever new.
That which has been, is not; that which was not,
Begins to be; motion and moment always

In process of renewal. Look, the night,
Worn out, aims toward the brightness, and sun's glory
Succeeds the dark. The color of the sky
Is different at midnight, when tired things
Lie all at rest, from what it is at morning
When Lucifer rides his snowy horse, before
Aurora paints the sky for Phoebus' coming.
The shield of the god reddens at early morning,
Reddens at evening, but is white at noonday
In purer air, farther from earth's contagion.
And the Moon-goddess changes in the nighttime,
Lesser today than yesterday, if waning,
Greater tomorrow than today, when crescent.

Our bodies also change. What we have been,
What we now are, we shall not be tomorrow.
There was a time when we were only seed,
Only the hope of men, housed in the womb,
Where Nature shaped us, brought us forth, exposed us
To the void air, and there in light we lay,
Feeble and infant, and were quadrupeds
Before too long, and after a little wobbled
And pulled ourselves upright, holding a chair,
The side of the crib, and strength grew into us,
And swiftness; youth and middle age went swiftly
Down the long hill toward age, and all our vigor
Came to decline, so Milon, the old wrestler,
Weeps when he sees his arms whose bulging muscles
Were once like Hercules', and Helen weeps
To see her wrinkles in the looking glass:
Could this old woman ever have been ravished,
Taken twice over? Time devours all things
With envious Age, together. The slow gnawing
Consumes all things, and very, very slowly.

Not even the so-called elements are constant.
Listen, and I will tell you of their changes.
There are four of them, and two, the earth and water,
Are heavy, and their own weight bears them downward,
And two, the air and fire (and fire is purer
Even than air) are light, rise upward
If nothing holds them down. These elements
Are separate in space, yet all things come
From them and into them, and they can change
Into each other. Earth can be dissolved
To flowing water, water can thin to air,

And air can thin to fire, and fire can thicken
To air again, and air condense to water,
And water be compressed to solid earth.
Nothing remains the same: the great renewer,
Nature, makes form from form, and, oh, believe me
That nothing ever dies. What we call birth
Is the beginning of a difference,
No more than that, and death is only ceasing
Of what had been before. The parts may vary,
Shifting from here to there, hither and yon,
And back again, but the great sum is constant.

The earth has something animal about it,
Living almost, with many lungs to breathe through,
Sending out flames, but the passages of breathing
Are changeable; some caverns may be closed
And new ones open whence the fire can issue.
Deep caves compress the violent winds, which drive
Rock against rock, imprisoning the matter
That holds the seeds of flame, and this bursts blazing
Ignited by the friction, and the caves
Cool when the winds are spent. The tars and pitches,
The yellow sulphur with invisible burning,
Are no eternal fuel, so volcanoes,
Starved of their nourishment, devour no longer,
Abandon fire, as they have been abandoned.

Far to the north, somewhere around Pallene,
The story goes, there is a lake where men
Who plunge nine times into the chilly waters
Come out with downy feathers over their bodies.
This I do not believe, nor that the women
Of Scythia sprinkle their bodies with magic juices
For the same purpose and effect.
"However,
There are stranger things that have been tried and tested
And these we must believe. You have seen dead bodies,
Rotten from time or heat, breed smaller creatures.
Bury the carcasses of slaughtered bullocks,
Chosen for sacrifice (all men know this),
And from the putrid entrails will come flying
The flower-culling bees, whose actions prove
Their parenthood, for they are fond of meadows,
Are fond of toil, and work with hopeful spirit.
The horse, being warlike, after he is buried
Produces hornets. Cut a sea-crab's claws,
Bury the rest of the body, and a scorpion

Comes from the ground. And worms that weave cocoons
White on the leaves of the trees, as country people,
Know well, turn into moths with death's-head marking.
The mud holds seeds that generate green frogs,
Legless at first, but the legs grow, to swim with,
And take long jumps with, later. And a bear-cub,
New-born, is not a bear at all, but only
A lump, hardly alive, whose mother gives it
A licking into shape, herself as model.
The larvae of the honey-bearing bees,
Safe in hexagonal waxen cells, are nothing
But wormlike bodies; feet and wings come later.
Who would believe that from an egg would come
Such different wonders as Juno's bird, the peacock,
Jove's eagle, Venus' dove, and all the fliers?
Some people think that when the human spine
Has rotted in the narrow tomb, the marrow
Is changed into a serpent.
"All these things
Have their beginning in some other creature,
But there is one bird which renews itself
Out of itself. The Assyrians call it the phoenix.
It does not live on seeds nor the green grasses,
But on the gum of frankincense and juices
Of cardamon. It lives five centuries,
As you may know, and then it builds itself
A nest in the highest branches of a palm-tree,
Using its talons and clean beak to cover
This nest with cassia and spikes of spikenard,
And cinnamon and yellow myrrh, and there
It dies among the fragrance, and from the body
A tiny phoenix springs to birth, whose years
Will be as long. The fledgling, gaining strength
To carry burdens, lifts the heavy nest,
His cradle and the old one's tomb, and bears it
Through the thin air to the city of the Sun
And lays it as an offering at the doors
Of the Sun-god's holy temple.
"Wonders, wonders!
The same hyena can be male or female,
To take or give the seed of life, at pleasure,
And the chameleon, a little creature
Whose food is wind and air, takes on the color
Of anything it rests on. India, conquered,
Gave Bacchus, tendril-crowned, the tawny lynxes
Whose urine, when it met the air, was hardened

Becoming stone; so coral also hardens
At the first touch of air, while under water
It sways, a pliant weed.
"The day will end,
The Sun-god plunge tired horses in the ocean
Before I have the time I need to tell you
All of the things that take new forms. We see
The eras change, nations grow strong, or weaken,
Like Troy, magnificent in men and riches,
For ten years lavish with her blood, and now
Displaying only ruins and for wealth
The old ancestral tombs. Sparta, Mycenae,
Athens, and Thebes, all flourished once, and now
What are they more than names? I hear that Rome
Is rising, out of Trojan blood, established
On strong and deep foundations, where the Tiber
Comes from the Apennines. Rome's form is changing
Growing to greatness, and she will be, some day,
Head of the boundless world; so we are told
By oracles and seers. I can remember
When Troy was tottering ruinward, a prophet,
Helenus, son of Priam, told Aeneas
In consolation for his doubts and weeping
'O son of Venus, if you bear in mind
My prophecies, Troy shall not wholly perish
While you are living: fire and sword will give you
Safe passage through them; you will carry on
Troy's relics, till a land, more friendly to you
Than your own native soil, will give asylum.
I see the destined city for the Trojans
And their sons' sons, none greater in all the ages,
Past, present, or to come. Through long, long eras
Her famous men will bring her power, but one,
Sprung from Iulus' blood, will make her empress
Of the whole world, and after earth has used him
The heavens will enjoy him, Heaven will be
His destination.' What Helenus told Aeneas,
I have told you, I remember, and I am happy
That for our kin new walls, at last, are rising,
That the Greek victory was to such good purpose.

We must not wander far and wide, forgetting
The goal of our discourse. Remember this:
The heavens and all below them, earth and her creatures,
All change, and we, part of creation, also
Must suffer change. We are not bodies only,

But wingèd spirits, with the power to enter
Animal forms, house in the bodies of cattle.
Therefore, we should respect those dwelling-places
Which may have given shelter to the spirit
Of fathers, brothers, cousins, human beings
At least, and we should never do them damage,
Not stuff ourselves like the cannibal Thyestes.
An evil habit, impious preparation,
Wicked as human bloodshed, to draw the knife
Across the throat of the calf, and hear its anguish
Cry to deaf ears! And who could slay
The little goat whose cry is like a baby's,
Or eat a bird he has himself just fed?
One might as well do murder; he is only
The shortest step away. Let the bull plow
And let him owe his death to length of days;
Let the sheep give you armor for rough weather,
The she-goats bring full udders to the milking.
Have done with nets and traps and snares and springes,
Bird-lime and forest-beaters, lines and fish-hooks.
Kill, if you must, the beasts that do you harm,
But, even so, let killing be enough;
Let appetite refrain from flesh, take only
A gentler nourishment."

Unit XI

Where There Was No Path: Arthurian Legends and the Western Way

De Excidio et Conquestu Britanniae

GILDAS

Gildas was a Welshman who, like many others at the time, fled Britain and settled in Brittany on the northern coast of France. This selection from De Excidio et Conquestu Britanniae *("The Cutting-Up and Conquest of Britain"), written about 540, is the oldest surviving reference to Arthur. Gildas chronicles the invasions of the Anglo-Saxons, whom from this selection it is clear that Gildas hates.*

FROM CHAPTER 25

And thus some of the wretched remnants [of the Britons], caught in the mountains, were slaughtered in large numbers; others, weakened by hunger, came forward and surrendered to their enemies [the Saxons] to be slaves forever, if for all that they were not immediately cut to pieces, which was the greatest kindness that remained to them; others went to lands across the sea with loud lamentation, just as if in the manner of the chief oarsman intoning in this way under the billows of the sails: "Thou hast made us like sheep for slaughter, and has scattered us among the nations" [Psalms 44:11]; others, entrusting life always with suspicious mind to the mountainous country, overhanging hills, steep fortified places, densest forests, and sea caverns, stood firm on their native soil, although in a state of fear. Then, some time intervening, when these most cruel plunderers had gone back home, under the leader Ambrosius Aurelianus, a moderate man, who by chance alone of the Roman nation had survived in the shock of so great a calamity—his parents, undoubtedly of royal rank, having perished in the same disaster, his progeny today having very much degenerated from the excellence of their ancestors—[the remnants of the Britons] gained strength and challenging the conquerors to battle, by God's favor the victory fell to them.

FROM CHAPTER 26

Since that period, at one time our countrymen, at another the enemy, were victorious . . . up to the year of the besieging of Mount Badon, when almost the last but not the least slaughter of these hangdogs took place, and which, as I know, begins the forty-fourth year (one month having passed already), which is also the year of my birth.

Historia Brittonum

NENNIUS

In this selection from Historia Brittonum, *Nennius, an Anglo-Saxon, describes the customs and legends of the Britons before and during the Anglo-Saxon conquest. Nennius wrote in the year 800 after the conversion of Britain to Christianity and thus makes Arthur a Christian.*

FROM CHAPTER 56

At that time the Saxons grew powerful in great numbers and increased in Britain. But now that Hengist was dead, his son Octha crossed over from the left side of Britain to the kingdom of the Kentishmen, and from him are sprung the kings of the Kentishmen. Then Arthur fought against them [the Saxons] in those days together with the kings of Britain, but he was himself the leader of battles. The first battle was at the mouth of the river which is called Glein; the second, third, fourth, and fifth on another river, which is called Dubglas and is in the region of Linnuis; the sixth battle on a river which is called Bassas. The seventh was the battle in the wood of Celidon, that is Cat Coit Celidon. The eighth was the battle at the castle Guinnion, in which Arthur carried the image of Saint Mary, the perpetual Virgin, on his shoulders and the pagans were put to flight on that day, and there was great slaughter of them by the virtue of our Lord, Jesus Christ, and by the virtue of Saint Mary the Virgin, His Mother. The ninth battle was fought in the city of the Legion. The tenth battle he fought on the shore of the river, which is called Tribruit. The eleventh was the battle waged on the mountain, which is called Agned. The twelfth was the battle at Mount Badon, in which on one day nine hundred and sixty men fell to the ground during one onset of Arthur; and no one overthrew them save himself alone, and in all the battles he emerged the victor.

FROM CHAPTER 73

There is another marvelous thing in the region which is called Buelt. There is at that place a pile of stones and one stone placed over and above this heap with the footprint of a dog on it. When he was hunting the boar Troynt, Cabal, who was the dog of Arthur the soldier, impressed his footprint on the stone, and Arthur afterwards gathered together a pile of stones under the one on which was the footprint of his dog, and it is called Carn Cabal. And men come and carry the stone away in their hands for the space of a day and a night, and on the next day it is found back on its pile.

There is another marvel in the region which is called Ercing. There is found at that place a tomb near a fountain, which is called Licat Anir, and the name of the man who is buried in the sepulchral mound was thus designated Anir. He was the son of Arthur the soldier, and he was the one who killed him in the same place and buried him. And men come to measure the sepulchral mound, sometimes six, sometimes nine, sometimes twelve, sometimes fifteen feet in length. Whatever way you measure it in alternation, the second time you will not find its measurement the same, and I have tested it myself.

Historia Regum Britanniae

GEOFFREY OF MONMOUTH

Geoffrey of Monmouth (1100–1154) was a Welsh clerk from a minor religious order, who spent most of his life in Oxford before returning to Wales in 1151 as Bishop of St. Assaph. As is obvious from this selection, Geoffrey was a Welshman, a Celt, and fiercely nationalistic. In his book Historia Regum Britanniae *(The History of the Kings of Britain), Geoffrey attempts to trace the history of the Briton people to the royal family of Troy and provides the first fully developed account of Arthur. The book was written in 1136 and translated by Robert Wace into French in 1155, which brought the story of Arthur to France. This selection presents many of the familiar characters of romance such as Uther Pendragon, Merlin, Kay, and Bedevere.*

BOOK I, CHAPTER 1

Epistle Dedicatory to Robert, Earl of Gloucester

Oftentimes in turning over in mine own mind the many themes that might be subject-matter of a book, my thoughts would fall upon the plan of writing a history of the Kings of Britain, and in my musings thereupon meseemed it a marvel that, beyond such mention as Gildas and Bede have made of them in their luminous tractate, nought could I find as concerning the kings that had dwelt in Britain before the Incarnation of Christ, nor nought even as concerning Arthur and the many others that did succeed him after the Incarnation, albeit that their deeds be worthy of praise everlasting and be as pleasantly rehearsed from memory by word of mouth in the traditions of many peoples as though they had been written down. Now, whilst I was thus thinking upon such matters, Walter, Archdeacon of Oxford, a man learned not only to the art of eloquence, but in the histories of foreign lands, offered me a certain most ancient book in the British language that did set forth the doings of them all in due succession and order from Brute, the first King of the Britons, onward to Cadwallader, the son of Cadwallo, all told in stories of exceeding beauty. At his request, therefore, albeit that never have I gathered gay flowers of speech in other men's little gardens, and am content with mine own rustic manner of speech and mine own writing-reeds, have I been at the pains to translate this volume into the Latin tongue. For had I besprinkled my page with high-flown phrases, I should only have engendered a weariness in my readers by compelling them to spend more time over the meaning of the words than upon understanding the drift of my story.

Unto this my little work, therefore, do thou, Robert, Earl of Gloucester, show favour in such wise that it may be so corrected by thy guidance and counsel as that it may be held to have sprung, not from the poor little fountain of Geoffrey of Monmouth, but rather from thine own deep sea of knowledge, and to savour of thy salt. Let it be held to be thine own offspring, as thou thyself art offspring of the illustrious Henry, King of the English. Let it be thine, as one that hath been nurtured in the liberal arts by philosophy, and called unto the command of our armies by thine own inborn prowess of knighthood; thine, whom in these our days Britain haileth with heart-felt affection as though in thee she had been vouchsafed a second Henry.

CHAPTER 19

After this victory Uther marched unto the city of Dumbarton and made ordinance for settling that province, as well as for restoring peace everywhere. He also went round all the nations of the Scots, and made that rebellious people lay aside their savage ways, for such justice did he execute throughout the lands as never another of his predecessors had ever done before him. In his days did misdoers tremble, for they were dealt punishment without mercy. At last, when he had stablished his peace in the parts of the North, he went to London and bade that Octa and Eosa should be kept in prison there. And when the Easter festival drew nigh, he bade the barons of the realm assemble in that city that he might celebrate so high holiday with honour by assuming the crown thereon. All obeyed accordingly, and repairing thither from the several cities, assembled together on the eve of the festival. The King, accordingly, celebrated the ceremony as he had proposed, and made merry along with his barons, all of whom did make great cheer for that the King had received them in such joyful wise. For all the nobles that were there had come with their wives and daughters as was meet on so glad a festival. Among the rest, Gorlois, Duke of Cornwall, was there, with his wife Igerna, that in beauty did surpass all the other dames of the whole of Britain. And when the King espied her amidst the others, he did suddenly wax so fain of her love that, paying no heed unto none of the others, he turned all his attention only upon her. Only unto her did he send dainty tit-bits from his own dish; only unto her did he send the golden cups with messages through his familiars. Many a time did he smile upon her and spake merrily unto her withal. But when her husband did perceive all this, straightway he waxed wroth and retired from the court without leave taken. Nor was any that might recall him thither, for that he feared to lose the one thing that he loved better than all other. Uther, waxing wroth hereat, commanded him to return and appear in his court that he might take lawful satisfaction for the affront he had put upon him. And when Gorlois was not minded to obey the summons, the King was enraged beyond all measure and sware with an oath that he would ravage his demesnes so he hastened not to make him satisfaction. Forthwith, the quarrel betwixt the two abiding unsettled, the King gathered a mighty army together and went his way into the province of Cornwall and set fire to the cities and castles therein. But Gorlois, not daring to meet him in the field for that he had not so many armed men, chose rather to garrison his own strong places until such time as he obtained the succour he had besought from Ireland. And, for that he was more troubled upon his wife's account than upon his own, he placed her in the Castle of Tintagel on the seacoast, as holding it to be the safer refuge. Howbeit, he himself betook him into the Castle of Dimilioc, being afeared that in case disaster should befall him both might be caught in one trap. And when message of this was brought unto the King, he went unto the castle wherein Gorlois had ensconced him, and beleaguered him and cut off all access unto him. At length, at the end of a week, mindful of his love for Igerna, he spake unto one of his familiars named Ulfin of Rescraddeck: "I am consumed of love for Igerna, nor can I have no joy, nor do I look to escape peril of my body save I may have possession of her. Do thou therefore give me counsel in what wise I may fulfil my desire, for, and I do not, of mine inward sorrow shall I die." Unto whom Ulfin: "And who shall give thee any counsel that may avail, seeing that there is no force that may prevail whereby to come unto her in the Castle of Tintagel? For it is situate on the sea, and is on every side encompassed thereby, nor none other entrance is there save such as a narrow rock doth furnish, the which three armed knights could hold against thee, albeit thou wert standing there with the whole realm of Britain beside thee. But, and if Merlin the prophet would take the matter in hand, I do verily believe that by his counsel thou mightest compass thy heart's desire."

The King, therefore, believing him, bade Merlin be called, for he, too, had come unto the leaguer. Merlin came forthwith accordingly, and when he stood in presence of the King, was bidden give counsel how the King's desire might be fulfilled. When he found how sore tribulation of mind the King was suffering, he was moved at beholding the effect of a love so exceeding great, and saith he: "The fulfilment of thy desire doth demand the practice of arts new and unheard of in this thy day. Yet know I how to give thee the semblance of Gorlois by my leechcrafts in such sort as that thou shalt seem in all things to be his very self. If, therefore, thou art minded to obey me, I will make thee like unto him utterly, and Ulfin will I make like unto Jordan of Tintagel his familiar. I also will take upon me another figure and will be with ye as a third, and in such wise we may go safely unto the castle and have access unto Igerna." The King obeyed accordingly, and gave heed strictly unto that which Merlin enjoined him. At last, committing the siege into charge of his familiars, he did entrust himself unto the arts and medicaments of Merlin, and was transformed into the semblance of Gorlois. Ulfin was changed into Jordan, and Merlin into Brithael in such sort as that none could have told the one from the other. They then went their way toward Tintagel, and at dusk hour arrived at the castle. The porter, weening that the Duke had arrived, swiftly unmade the doors, and the three were admitted. For what other than Gorlois could it be, seeing that in all things it seemed as if Gorlois himself were there? So the King lay that night with Igerna and enjoyed the love for which he had yearned, for as he had beguiled her by the false likeness he had taken upon him, so he beguiled her also by the feigned discourses wherewith he did full artfully entertain her. For he told her he had issued forth of the besieged city for naught save to see to the safety of her dear self and the castle wherein she lay, in such sort that she believed him every word, and had no thought to deny him in aught he might desire. And upon that same night was the most renowned Arthur conceived, that was not only famous in after years, but was well worthy of all the fame he did achieve by his surpassing prowess.

CHAPTER 20

In the meantime, when the beleaguering army found that the King was not amongst them, they did unadvisedly make endeavour to breach the walls and challenge the besieged Duke to battle. Who, himself also acting unadvisedly, did straightway sally forth with his comrades in arms, weening that his handful of men were strong enow to make head against so huge a host of armed warriors. But when they met face to face in battle, Gorlois was amongst the first that were slain, and all his companies were scattered. The castle, moreover, that they had besieged was taken, and the treasure that was found therein divided, albeit not by fair casting of lots, for whatsoever his luck or hardihood might throw in his way did each man greedily clutch in his claws for his own. But by the time that this outrageous plundering had at last come to an end messengers had come unto Igerna to tell her of the Duke's death and the issue of the siege. But when they beheld the King in the likeness of the Duke sitting beside her, they blushed scarlet, and stared in amazement at finding that he whom they had just left dead at the leaguer had thus arrived hither safe and sound, for little they knew what the medicaments of Merlin had accomplished. The King therefore, smiling at the tidings, and embracing the countess, spake saying: "Not slain, verily, am I, for lo, here thou seest me alive, yet, natheless, sore it irketh me of the destruction of my castle and the slaughter of my comrades, for that which next is to dread is lest the King should overtake us here and make us prisoners in this castle. First of all, therefore, will I go meet him and make my peace with him, lest a worst thing befall us." Issuing forth accordingly, he made his way unto his own army, and putting off the semblance of Gorlois again became Uther Pendragon. And when he understood how everything had fallen out, albeit that he was sore grieved at the death of Gorlois, yet could he not but be glad that Igerna was released from the bond of matrimony. Returning, therefore, to Tintagel, he took the castle, and not the castle only, but Igerna also therein, and on this wise fulfilled he his desire. Thereafter were they linked together in no little mutual love, and two children were born unto them, a son and a daughter, whereof the son was named Arthur and the daughter Anna.

BOOK IX, CHAPTER 1

After the death of Uther Pendragon, the barons of Britain did come together from the divers provinces unto the city of Silchester, and did bear on hand Dubric, Archbishop of Caerleon, that he should crown as king, Arthur, the late King's son. For sore was need upon them; seeing that when the Saxons heard of Uther's death they had invited their fellow-countrymen from Germany, and under their Duke Colgrin were bent upon exterminating the Britons. They had, moreover, entirely subdued all that part of the island which stretcheth from the river Humber, as far as the sea of Caithness. Dubric, therefore, sorrowing over the calamities of the country, assembled the other prelates, and did invest Arthur with the crown of the realm. At that time Arthur was a youth of fifteen years, of a courage and generosity beyond compare, whereunto his inborn goodness did lend such grace as that he was beloved of well-nigh all the peoples in the land. After he had been invested with the ensigns of royalty, he abided by his ancient wont, and was so prodigal of his bounties as that he began to run short of wherewithal to distribute amongst the huge multitude of knights that made repair unto him. But he that hath within him a bountiful nature along with prowess, albeit that he be lacking for a time, natheless in no wise shall poverty be his bane for ever. Wherefore did Arthur, for that in him did valour keep company with largesse, make resolve to harry the Saxons, to the end that with their treasure he might make rich the retainers that were of his own household. And herein was he monished of his own lawful right, seeing that of right ought he to hold the sovereignty of the whole island in virtue of his claim hereditary. Assembling, therefore, all the youth that were of his allegiance, he made first for York. And when Colgrin was ware of this, he got together his Saxons, Scots, and Picts, and came with a mighty multitude to meet him nigh the river Douglas, where, by the time the battle came to an end, the more part of both armies had been put to the sword. Natheless, Arthur won the day, and after pursing Colgrin's flight as far as York, did beleaguer him within that city. Thereupon, Baldulf, hearing of his brother's flight, made for the besieged city with six thousand men to relieve him. For, at the time his brother had fought the battle, he himself was upon the seacoast awaiting the arrival of Duke Cheldric, who was just coming from Germany to their assistance. And when he had come within ten miles of the city, he was resolved to make a night march and fall upon them by surprise. Howbeit, Arthur was ware of his purpose, and bade Cador, Duke of Cornwall, go meet him that same night with six hundred horse and three thousand foot. He, choosing a position on the road whereby the enemy were bound to march, surprised them by an assault on the sudden, and cutting up and slaying the Saxons, drave Baldulf off in flight. Baldulf, distressed beyond measure that he could convey no succour to his brother, took counsel with himself in what wise he might have speech of him, for he weened that so he might get at him, they might together devise some shift for the safety of them both. Failing all other means of access unto him, he shaved off his hair and his beard, and did upon him the habit of a jongleur with a ghittern, and walking to and fro within the camp, made show as had he been a minstrel singing unto the tunes that he thrummed the while upon his ghittern. And, for that none suspected him, by little and little he drew nigh unto the walls of the city, ever keeping up the disguise he had taken upon him. At last he was found out by some of the besieged, who thereupon drew him up with cords over the wall into the city and brought him unto his brother, who, overjoyed at the sight of him, greeted him with kisses and embraces. At last, after talking over every kind of shift, when they had fallen utterly into despair of ever issuing forth, the messengers they had sent into Germany returned, bringing with them unto Scotland six hundred ships full of stout warriors under Duke Cheldric; and when Arthur's counsellors heard tell of their coming, they advised him to hold the leaguer no longer, for that sore hazard would it be to do battle with so mighty a multitude of enemies as had now arrived.

CHAPTER 2

Arthur, therefore, in obedience to the counsel of his retainers, retired him into the city of London. Hither he summoned all the clergy and chief men of his allegiance and bade them declare their counsel as to what were best and safest for him to do against this inroad of the Paynim. At last, by common consent of them all, messengers are sent unto King Hoel in Brittany with tidings of the calamitous estate of Britain. For Hoel was sister's son unto Arthur, born unto Budicius, King of the Bretons. Where, so soon as he heard of the invasion wherewith his uncle was threatened he bade fit out his fleet, and mustering fifteen thousand men-at-arms, made for Southampton with the first fair wind. Arthur received him with all honour due, and the twain embraced the one the other over and over again.

CHAPTER 3

A few days later they set forth for the city of Kaerliudcoit, then besieged by the Paynim already mentioned, the which city lieth upon a hill betwixt two rivers in the province of Lindsey, and is otherwise called Lincoln. Accordingly, when they had come thither with their whole host, they did battle with the Saxons and routed them with no common slaughter, for upon that day fell six thousand of them, some part drowned in the rivers and some part smitten of deadly weapons. The residue, in dismay, forsook the siege and fled, but Arthur stinted not in pursuit until they had reached the forest of Caledon, wherein they assembled again after the fight and did their best to make a stand against him. When the battle began, they wrought sore havoc amongst the Britons, defending themselves like men, and avoiding the arrows of the Britons in the shelter afforded by the trees. When Arthur espied this he bade the trees about that part of the forest be felled, and the trunks set in a compass around them in such wise as that all ways of issuing forth were shut against them, for he was minded to beleaguer them therein until they should be starven to death of hunger. This done, he bade his companies patrol the forest, and abode in that same place three days. Whereupon the Saxons, lacking all victual and famishing to death, besought leave to issue forth upon convenant that they would leave all their gold and silver behind them so they might return unto Germany with nought but their ships only. They promised further to give them tribute from Germany and to leave hostages for the payment thereof. Arthur, taking counsel thereupon, agreed unto their petition, retaining all their treasure and the hostages for payment of the tribute, and granting only unto them bare permission to depart. Natheless, whilst that they were ploughing the seas as they returned homeward, it repented them of the covenant they had made, and tacking about, they returned into Britain, making the shore at Totnes. Taking possession of the country, they devastated the land as far as the Severn sea, slaying the husbandmen with deadly wounds. Marching forth from thence they made for the country about Bath and besieged that city. When word of this was brought unto the King, astonied beyond measure at their wicked daring, he bade judgment be done upon their hostages and hanged them out of hand, and, abandoning the expedition whereby he intended to repress the Picts and Scots, hurried away to disperse the leaguer. Howbeit, that which did most sorely grieve him in this strait was that he was compelled to leave his nephew Hoel behind him lying sick in the city of Dumbarton. When at last he arrived in the province of Somerset, and beheld the leaguer nigh at hand, he spake in these words: "For that these Saxons, of most impious and hateful name, have disdained to keep faith with me, I, keeping my faith unto my God, will endeavour me this day to revenge upon them the blood of my countrymen. To arms, therefore, ye warriors, to arms, and fall upon yonder traitors like men, for, of a certainty, by Christ's succour, we cannot fail of victory!"

CHAPTER 4

When he had thus spoken, the holy Dubric, Archbishop of Caerleon, went up on to the top of a certain mount and cried out with a loud voice:

"Ye men that be known from these others by your Christian profession, take heed ye bear in mind the piety ye owe unto your country and unto your fellow-countrymen, whose slaughter by the treachery of the Paynim shall be unto ye a disgrace everlasting save ye press hardily forward to defend them. Fight ye therefore for your country, and if it be that death overtake ye, suffer it willingly for your country's sake, for death itself is victory and a healing unto the soul, inasmuch as he that shall have died for his brethren doth offer himself a living sacrifice unto God, nor is it doubtful that herein he doth follow in the footsteps of Christ who disdained not to lay down His own soul for His brethren. Whosoever, therefore, amongst ye shall be slain in this battle, unto him shall that death be as full penance and absolution of all his sins, if so be he receive it willingly on this wise."

Forthwith, thus cheered by the benison of the blessed man, each one hastened to arm him to do his bidding, and Arthur himself doing upon him a habergeon worthy of a king so noble, did set upon his head a helm of gold graven with the semblance of a dragon. Upon his shoulders, moreover, did he bear the shield that was named Pridwen, wherein, upon the inner side, was painted the image of holy Mary, Mother of God, that many a time and oft did call her back unto his memory. Girt was he also with Caliburn, best of swords, that was forged within the Isle of Avallon; and the lance that did grace his right hand was called by the name Ron, a tall lance and a stout, full meet to do slaughter withal. Then, stationing his companies, he made hardy assault upon the Saxons that after their wont were ranked wedge-wise in battalions. Natheless, all day long did they stand their ground manfully maugre the Britons that did deliver assault upon assault against them. At last, just verging upon sundown, the Saxons occupied a hill close by that might serve them for a camp, for, secure in their numbers, the hill alone seemed all the camp they needed. But when the morrow's sun brought back the day, Arthur with his army clomb up to the top of the hill, albeit that in the ascent he lost many of his men. For the Saxons, dashing down from the height, had the better advantage in dealing their wounds, whilst they could also run far more swiftly down the hill than he could struggle up. Howbeit, putting forth their utmost strength, the Britons did at last reach the top, and forthwith close with the enemy hand to hand. The Saxons, fronting them with their broad chests, strive with all their endeavour to stand their ground. And when much of the day had been spent on this wise, Arthur waxed wroth at the stubbornness of their resistance, and the slowness of his own advance, and drawing forth Caliburn, his sword, crieth aloud the name of Holy Mary, and thrusteth him forward with a swift onset into the thickest press of the enemy's ranks. Whomsoever he touched, calling upon God, he slew at a single blow, nor did he once slacken in his onslaught until that he had slain four hundred and seventy men singlehanded with his sword Caliburn. This when the Britons beheld, they followed him up in close rank dealing slaughter on every side. Colgrin and Baldulf his brother fell amongst the first, and many thousands fell besides. Howbeit, as soon as Cheldric saw the jeopardy of his fellows, he turned to flee away.

CHAPTER 5

The King having won the victory, bade Cador, Duke of Cornwall, pursue the enemy, while he himself hastened his march into Scotland, for word had thence been brought him that the Scots and Picts were besieging Hoel in the city of Dumbarton, wherein, as I have said, he was lying afflicted of grievous sickness, and sore need it was he should come swiftly to his succour lest he should be taken by the barbarians along with the city. The Duke of Cornwall, accordingly, accompanied by ten thousand men, was not minded, in the first place, to pursue the fleeing Saxons, deeming it better to make all speed to get hold of their ships and thus forbid their embarking therein. As soon as he had taken possession of the ships, he manned them with his best soldiers, who could be trusted to take heed that no Paynim came aboard, in case they should flee unto them to escape. Then he made best haste to obey Arthur's orders by following up the enemy and slaying all he overtook without mercy. Whereupon they, who but just now had fallen upon the Britons with the all fury of a thunderbolt, straightway sneak off, faint of heart, some into the depths of the forest, others into the mountains and caves, anywhither so only they may live yet a little longer. At last, when they found all shelter failing, they march their shattered companies into the Isle of Thanet. Thither the Duke of Cornwall follows hard upon their heels, smiting them down without mercy as was his wont; nor did he stay his hand until after Cheldric had been slain. He compelled them to give hostages for the surrender of the whole residue.

CHAPTER 6

Having thus established peace, he marched towards Dumbarton, which Arthur had already delivered from the oppression of the barbarians. He next led his army into Moray, where the Scots and Picts were beleaguered, for after they had thrice been defeated in battle by Arthur and his nephew they had fled into that province. When they had reached Loch Lomond, they occupied the islands that be therein, thinking to find safe refuge; for this lake doth contain sixty islands and receiveth sixty rivers, albeit that but a single stream doth flow from thence unto the sea. Upon these islands are sixty rocks plain to be seen, whereof each one doth bear an eyrie of eagles that there congregating year by year do notify any prodigy that is to come to pass in the kingdom by uttering a shrill scream all together in concert. Unto these islands accordingly the enemy had fled in order to avail them of the protection of the lake. But small profit reaped they thereby, for Arthur collected a fleet and went round about the inlets of the rivers for fifteen days together, and did so beleaguer them as that they were famished to death of hunger by thousands. And whilst that he was serving them out on this wise arrived Gillamaur, King of Ireland, with a mighty host of barbarians in a fleet, to bring succour unto the wretched islanders. Whereupon Arthur left off the leaguer and began to turn his arms against the Irish, whom he forced to return unto their own country, cut to pieces without mercy. When he had won the victory, he again gave all his thoughts to doing away utterly the race of the Scots and Picts, and yielded him to treating them with a cruelty beyond compare. Not a single one that he could lay hands on did he spare, insomuch as that at last all the bishops of the miserable country assembled together with all the clergy of their obedience, and came unto him barefoot, bearing relics of the saints and the sacred objects of the church, imploring the King's mercy for the safety of their people. As soon as they came into his presence, they prayed him on their bended knees to have pity on the down-trodden folk, for that he has visited them with pains and penalties enow, nor was any need to cut off the scanty few that still survived to the last man. Some petty portion of the country he might allot unto them whereon they might be allowed to bear the yoke of perpetual bondage, for this were they willing to do. And when they had besought the King on this wise, he was moved unto tears for very pity, and, agreeing unto the petition of the holy men, granted them his pardon.

CHAPTER 12

When the high festival of Whitsuntide began to draw nigh, Arthur, filled with exceeding great joy at having achieved so great success, was fain to hold high court, and to set the crown of the kingdom upon his head, to convene the Kings and Dukes that were his vassals to the festival so that he might the more worshipfully celebrate the same, and renew his peace more firmly amongst his barons. Howbeit, when he made known his desire unto his familiars, he, by their counsel, made choice of Caerleon wherein to fulfil his design. For, situate in a passing pleasant position on the river Usk in Glamorgan, not far from the Severn sea, and abounding in wealth above all other cities, it was the place most meet for so high a solemnity. For on the one side thereof flowed the noble river aforesaid whereby the Kings and Princes that should come from oversea might be borne thither in their ships; and on the other side, girdled about with meadows and woods, passing fair was the magnificence of the kingly palaces thereof with the gilded verges of the roofs that imitated Rome. Howbeit, the chiefest glories thereof where the two churches, one raised in honour of the Martyr Julius, that was right fair graced by a convent of virgins that had dedicated them unto God, and the second, founded in the name of the blessed Aaron, his companion, the main pillars whereof were a brotherhood of canons regular, and this was the cathedral church of the third Metropolitan See of Britain. It had, moreover, a school of two hundred philosophers learned in astronomy and in the other arts, that did diligently observe the courses of the stars, and did by true inferences foretell the prodigies which at that time were about to befall unto King Arthur. Such was the city, famed for such abundance of things delightsome, that was now busking her for the festival that had been proclaimed. Messengers were sent forth into the divers kingdoms, and all that owed allegiance throughout the Gauls and the neighbour islands were invited unto the court. Came accordingly Angusel, King of Albany, that is now called Scotland; Urian, King of them of Moray; Cadwallo Lawirh, King of the Venedotians, that now be called the North Welsh; Stater, King of the Demeti, that is, of the South Welsh; Cador, King of Cornwell; the Archbishops of the three Metropolitan Sees, to wit, of London and York, and Dubric of Caerleon. He, Primate of Britain and Legate of the Apostolic See, was of so meritorious a piety that he could make whole by his prayers any that lay oppressed of any malady. Came also the Earls of noble cities; Morvid, Earl of Gloucester; Mauron of Worcester; Anaraut of Salisbury; Arthgal of Cargueir, that is now called Warwick; Jugein from Leicester; Cursalem from Caichester; Kimmarc, Duke of Canterbury; Galluc of Salisbury; Urbgennius from Bath; Jonathal of Doreset; Boso of Rhydychen, that is Oxford. Besides the Earls came champions of no lesser dignity, Donaut map Papo; Cheneus map

Coil; Peredur map Eridur; Grifuz map Nogoid; Regin map Claud; Eddelein map Cledauc; Kincar map Bangan; Kimmarc; Gorbonian map Goit; Clofaut; Run map Neton; Kimbelin map Trunat; Cathleus map Catel; Kinlith map Neton, and many another beside, the names whereof be too long to tell. From the neighbour islands came likewise Gillamaur, King of Ireland; Malvasius, King of Iceland; Doldavius, King of Gothland; Gunvasius, King of the Orkneys; Loth, King of Norway; Aschil, King of the Danes. From the parts oversea came also Holdin, King of Flanders; Leodegar, Earl of Boulogne; Bedevere the Butler, Duke of Normandy; Borel of Maine; Kay the Seneschal, Duke of Anjou; Guitard of Poitou; the Twelve Peers of the Gauls whom Guerin of Chartres brought with him; Hoel, Duke of Brittany, with the Barons of his allegiance, who marched along with such magnificence of equipment in trappings and mules and horses as may not easily be told. Besides all these, not a single Prince of any price on this side Spain remained at home and came not upon the proclamation. And no marvel, for Arthur's bounty was of common report throughout the whole wide world, and all men for his sake were fain to come.

CHAPTER 13

When all at last were assembled in the city on the high day of the festival, the archbishops were conducted unto the palace to crown the King with the royal diadem. Dubric, therefore, upon whom the charge fell, for that the court was held within his diocese, was ready to celebrate the service. As soon as the King had been invested with the ensigns of kingship, he was led in right comely wise to the church of the Metropolitan See, two archbishops supporting him, the one upon his right hand side the other upon his left. Four Kings, moreover, to wit, those of Scotland, Cornwall, and North and South Wales, went before him, bearing before him, as was their right, four golden swords. A company of clerics in holy orders of every degree went chanting music marvellous sweet in front. Of the other party, the archbishops and pontiffs led the Queen, crowned with laurel and wearing her own ensigns, unto the church of the virgins dedicate. The four Queens, moreover, of the four Kings already mentioned, did bear before her according to wont and custom four white doves, and the ladies that were present did follow after her rejoicing greatly. At last, when the procession was over, so manifold was the music of the organs and so many were the hymns that were chanted in both churches, that the knights who were there scarce knew which church they should enter first for the exceeding sweetness of the harmonies in both. First into the one and then into the other they flocked in crowds, nor, had the whole day been given up to the celebration, would any have felt a moment's weariness thereof. And when the divine services had been celebrated in both churches, the King and Queen put off their crowns, and doing on lighter robes of state, went to meat, he to his palace with the men, she to another palace with the women. For the Britons did observe the ancient custom of the Trojans, and were wont to celebrate their high festival days, the men with the men and the women with the women severally. And when all were set at table according as the rank of each did demand, Kay the Seneschal, in a doublet furred of ermines, and a thousand youths of full high degree in his company, all likewise clad in ermines, did serve the meats along with him. Of the other part, as many in doublets furred of vair did follow Bedevere the Butler, and along with him did serve the drinks from the divers ewers into the manifold-fashioned cups. In the palace of the Queen no less did numberless pages, clad in divers brave liveries, offer their service each after his office, the which were I to go about to describe I might draw out my history into an endless prolixity. For at that time was Britain exalted unto so high a pitch of dignity as that it did surpass all other kingdoms in plenty of riches, in luxury of adornment, and in the courteous wit of them that dwelt therein. Whatsoever knight in the land was of renown for his prowess did wear his clothes and his arms all of one same colour. And the dames, no less witty, would apparel them in like manner in a single colour, nor would they deign have the love of none save he had thrice approved him in the wars. Wherefore at that time did dames wax chaste and knights the nobler for their love.

King Arthur and His Knights

SIR THOMAS MALORY

Sir Thomas Malory, who died in the year 1472, was the last medieval romance writer. Malory was a knight who spent most of his life in prison for petty crimes and vandalism. In King Arthur and His Knights, *Malory attempts to bring together and synthesize all the legends surrounding Arthur. It is a magnificent work. These selections recount the birth, the rise to power, and the death of Arthur. The first selection, "Merlin," covers much of the same material covered by Geoffrey of Monmouth. The second selection, "The Day of Destiny," recounts the theme that Arthur will come again.*

MERLIN

It befell in the days of Uther Pendragon, when he was king of all England and so reigned, that there was a mighty duke in Cornwall that held war against him long time, and the duke was called the duke of Tintagel. And so by means King Uther sent for this duke, charging him to bring his wife with him, for she was called a fair lady and a passing wise, and her name was called Igraine.

So when the duke and his wife were come unto the king, by the means of great lords they were accorded both. The king liked and loved this lady well, and he made them great cheer out of measure, and desired to have lain by her. But she was a passing good woman and would not assent unto the king. And then she told the duke her husband, and said,

"I suppose that we were sent for that I should be dishonoured. Wherefore, husband, I counsel you that we depart from hence suddenly, that we may ride all night unto our own castle."

And in like wise as she said so they departed, that neither the king nor none of his council were ware of their departing. Also soon as King Uther knew of their departing so suddenly, he was wonderly wroth; then he called to him his privy council and told them of the sudden departing of the duke and his wife. Then they advised the king to send for the duke and his wife by a great charge:

"And if he will not come at your summons, then may ye do your best; then have ye cause to make mighty war upon him."

So that was done, and the messengers had their answers; and that was this, shortly, that neither he nor his wife would not come at him. Then was the king wonderly wroth; and then the king sent him plain word again and bade him be ready and stuff him and garnish him, for within forty days he would fetch him out of the biggest castle that he hath.

When the duke had this warning, anon he went and furnished and garnished two strong castles of his, of the which the one hight Tintagel, and the other castle hight Terrabil. So his wife Dame Igraine he put in the castle of Tintagel, and himself he put in the castle of Terrabil, the which had many issues and posterns out. Then in all haste came Uther with a great host and laid a siege about the castle of Terrabil, and there he pight many pavilions. And there was great war made on both parties and much people slain.

Then for pure anger and for great love of fair Igraine the king Uther fell sick. So came to the King Uther Sir Ulfius, a noble knight, and asked the king why he was sick.

"I shall tell thee," said the king, "I am sick for anger and for love of fair Igraine, that I may not be whole."

"Well, my lord," said Sir Ulfius, "I shall seek Merlin and he shall do you remedy, that your heart shall be pleased."

So Ulfius departed and by adventure he met Merlin in a beggar's array, and there Merlin asked Ulfius whom he sought, and he said he had little ado to tell him.

"Well," said Merlin, "I know whom thou seekest, for thou seekest Merlin; therefore seek no farther, for I am he. And if King Uther will well reward me and be sworn unto me to fulfil my desire, that shall be his honour and profit more than mine, for I shall cause him to have all his desire."

"All this will I undertake," said Ulfius, "that there shall be nothing reasonable but thou shalt have thy desire."

"Well," said Merlin, "he shall have his intent and desire. And therefore," said Merlin, "ride on your way, for I will not be long behind."

Then Ulfius was glad and rode on more than a pace till that he came to King Uther Pendragon and told him he had met with Merlin.

"Where is he?" said the king.

"Sir," said Ulfius, "he will not dwell long."

Therewithal Ulfius was ware where Merlin stood at the porch of the pavilion's door, and then Merlin was bound to come to the king. When King Uther saw him he said he was welcome.

"Sir," said Merlin, "I know all your heart every deal. So ye will be sworn unto me, as ye be a true king anointed, to fulfil my desire, ye shall have your desire."

Then the king was sworn upon the four Evangelists.

"Sir," said Merlin, "this is my desire: the first night that ye shall lie by Igraine ye shall get a child on her; and when that is born, that it shall be delivered to me for to nourish thereas I will have it; for it shall be your worship and the child's avail as mickle as the child is worth."

"I will well," said the king, "as thou wilt have it."

"Now make you ready," said Merlin. "This night ye shall lie with Igraine in the castle of Tintagel. And ye shall be like the duke her husband, Ulfius shall be like Sir Brastias, a knight of the duke's, and I will be like a knight that hight Sir Jordanus, a knight of the duke's. But wait ye make not many questions with her nor her men, but say ye are diseased, and so hie you to bed and rise not on the morn till I come to you, for the castle of Tintagel is but ten mile hence."

So this was done as they devised. But the duke of Tintagel espied how the king rode from the siege of Terrabil. And therefore that night he issued out of the castle at a postern for to have distressed the king's host, and so through his own issue the duke himself was slain or ever the king came at the castle of Tintagel. So after the death of the duke King Uther lay with Igraine, more than three hours after his death, and begat on her that night Arthur; and or day came, Merlin came to the king and bade him make him ready, and so he kissed the lady Igraine and departed in all haste. But when the lady heard tell of the duke her husband, and by all record he was dead or ever King Uther came to her, then she marvelled who that might be that lay with her in likeness of her lord. So she mourned privily and held her peace.

Then all the barons by one assent prayed the king of accord betwixt the lady Igraine and him. The king gave them leave, for fain would he have been accorded with her; so the king put all the trust in Ulfius to entreat between them. So by the entreaty at the last the king and she met together.

"Now will we do well," said Ulfius. "Our king is a lusty knight and wifeless, and my lady Igraine is a passing fair lady; it were great joy unto us all an it might please the king to make her his queen."

Unto that they all well accorded and moved it to the king. And anon, like a lusty knight, he assented thereto with good will, and so in all haste they were married in a morning with great mirth and joy.

And King Lot of Lothian and of Orkney then wedded Margawse that was Gawain's mother, and King Nentres of the land of Garlot wedded Elaine. All this was done at the request of King Uther. And the third sister, Morgan le Fay, was put to school in a nunnery, and there she learned so much that she was a great clerk of necromancy. And after she was wedded to King Uriens of the land of Gore, that was Sir Ywain le Blanchemain's father.

Then Queen Igraine waxed daily greater and greater. So it befell after within half a year, as King Uther lay by his queen, he asked her, by the faith she owed to him, whose was the child within her body. Then was she sore abashed to give answer.

"Dismay you not," said the king, "but tell me the truth, and I shall love you the better, by the faith of my body!"

"Sir," said she, "I shall tell you the truth. The same night that my lord was dead, the hour of his death, as his knights record, there came into my castle of Tintagel a man like my lord in speech and in countenance, and two knights with him in likeness of his two knights Brastias and Jordanus, and so I went unto bed with him as I ought to do with my lord; and the same night, as I shall answer unto God, this child was begotten upon me."

"That is truth," said the king, "as ye say, for it was I myself that came in the likeness. And therefore dismay you not, for I am father of the child;" and there he told her all the cause how it was by Merlin's counsel. Then the queen made great joy when she knew who was the father of her child.

Soon came Merlin unto the king and said,

"Sir, ye must purvey you for the nourishing of your child."

"As thou wilt," said the king, "be it."

"Well," said Merlin, "I know a lord of yours in this land that is a passing true man and a faithful, and he shall have the nourishing of your child; and his name is Sir Ector, and he is a lord of fair livelihood in many parts in England and Wales. And this lord, Sir Ector, let him be sent for for to come and speak with you, and desire him yourself, as he loveth you, that he will put his own child to nourishing to another woman and that his wife nourish yours. And when the child is born let it be delivered to me at yonder privy postern unchristened."

So like as Merlin devised it was done. And when Sir Ector was come he made fiaunce to the king for to nourish the child like as the king desired; and there the king granted Sir Ector great rewards. Then when the lady was delivered the king commanded two knights and two ladies to take the child bound in a cloth of gold, "and that ye deliver him to what poor man ye meet at the postern gate of the castle." So the child was delivered unto Merlin, and so he bare it forth unto Sir Ector, and made an holy man to christen him, and named him Arthur. And so Sir Ector's wife nourished him with her own pap.

Then within two years King Uther fell sick of a great malady. And in the meanwhile his enemies usurped upon him and did a great battle upon his men and slew many of his people.

"Sir," said Merlin, "ye may not lie so as ye do, for ye must to the field, though ye ride on an horse-litter. For ye shall never have the better of your enemies but if your person be there, and then shall ye have the victory."

So it was done as Merlin had devised, and they carried the king forth in an horse-litter with a great host towards his enemies, and at St. Albans there met with the king a great host of the North. And that day Sir Ulfius and Sir Brastias did great deeds of arms, and King Uther's men overcame the Northern battle and slew many people and put the remnant to flight; and then the king returned unto London and made great joy of his victory.

And then he fell passing sore sick, so that three days and three nights he was speechless; wherefore all the barons made great sorrow and asked Merlin what counsel were best.

"There is none other remedy," said Merlin, "but God will have His will. But look ye all barons be before King Uther to-morn, and God and I shall make him to speak."

So on the morn all the barons with Merlin came tofore the king. Then Merlin said aloud unto King Uther,

"Sir, shall your son Arthur be king after your days of this realm with all the appurtenance?"

Then Uther Pendragon turned him and said in hearing of them all,

"I give him God's blessing and mine, and bid him pray for my soul, and righteously and worshipfully that he claim the crown upon forfeiture of my blessing;" and therewith he yielded up the ghost. And then was he interred as longed to a king, wherefore the queen, fair Igraine, made great sorrow, and all the barons.

Then stood the realm in great jeopardy long while, for every lord that was mighty of men made him strong, and many weened to have been king. Then Merlin went to the Archbishop of Canterbury and counselled him for to send for all the lords of the realm and all the gentlemen of arms, that they should to London come by Christmas upon pain of cursing, and for this cause, that Jesu, that was born on that night, that He would of His great mercy show some miracle, as He was come to be King of mankind, for to show some miracle who should be rightwise king of this realm. So the Archbishop, by the advice of Merlin, sent for all the lords and gentlemen of arms that they should come by Christmas even unto London; and many of them made them clean of their life, that their prayer might be the more acceptable unto God.

So in the greatest church of London, whether it were Paul's or not the French book maketh no mention, all the estates were long or day in the church for to pray. And when matins and the first mass was done there was seen in the churchyard, against the high altar, a great stone four square, like unto a marble stone, and in midst thereof was like an anvil of steel a foot on high, and therein stuck a fair sword naked by the point, and letters there were written in gold about the sword that said thus: "WHOSO PULLETH OUT THIS SWORD OF THIS STONE AND ANVIL IS RIGHTWISE KING BORN OF ALL ENGLAND." Then the people marvelled and told it to the Archbishop.

"I command," said the Archbishop, "that ye keep you within your church and pray unto God still; that no man touch the sword till the high mass be all done."

So when all masses were done all the lords went to behold the stone and the sword. And when they saw the scripture some essayed, such as would have been king, but none might stir the sword nor move it.

"He is not here," said the Archbishop, "that shall achieve the sword, but doubt not God will make him known. But this is my counsel," said the Archbishop, "that we let purvey ten knights, men of good fame, and they to keep this sword."

So it was ordained, and then there was made a cry that every man should essay that would for to win the sword. And upon New Year's Day the barons let make a jousts and a tournament, that all knights that would joust or tourney there might play. And all this was ordained for to keep the lords together and the commons, for the Archbishop trusted that God would make him known that should win the sword.

So upon New Year's Day, when the service was done, the barons rode unto the field, some to joust and some to tourney. And so it happed that Sir Ector, that had great livelihood about London, rode unto the jousts, and with him rode Sir Kay, his son, and young Arthur that was his nourished brother; and Sir Kay was made knight at All Hallowmass afore. So as they rode to the jousts-ward Sir Kay had lost his sword, for he had left it at his father's lodging, and so he prayed young Arthur for to ride for his sword.

"I will well," said Arthur, and rode fast after the sword.

And when he came home the lady and all were out to see the jousting. Then was Arthur wroth, and said to himself, "I will ride to the churchyard and take the sword with me that sticketh in the stone, for my brother Sir Kay shall not be without a sword this day." So when he came to the churchyard Sir Arthur alight and tied his horse to the stile, and so he went to the tent and found no knights there, for they were at the jousting. And so he handled the sword by the handles, and lightly and fiercely pulled it out of the stone, and took his horse and rode his way until he came to his brother Sir Kay and delivered him the sword.

And as soon as Sir Kay saw the sword he wist well it was the sword of the stone, and so he rode to his father Sir Ector and said,

"Sir, lo here is the sword of the stone, wherefore I must be king of this land."

When Sir Ector beheld the sword he returned again and came to the church, and there they alight all three and went into the church, and anon he made Sir Kay to swear upon a book how he came to that sword.

"Sir," said Sir Kay, "by my brother Arthur, for he brought it to me."

"How gat ye this sword?" said Sir Ector to Arthur.

"Sir, I will tell you. When I came home for my brother's sword I found nobody at home to deliver me his sword, and so I thought my brother Sir Kay should not be swordless, and so I came hither eagerly and pulled it out of the stone without any pain."

"Found ye any knights about this sword?" said Sir Ector.

"Nay," said Arthur.

"Now," said Sir Ector to Arthur, "I understand ye must be king of this land."

"Wherefore I?" said Arthur, "and for what cause?"

"Sir," said Ector, "for God will have it so, for there should never man have drawn out this sword but he that shall be rightwise king of this land. Now let me see whether ye can put the sword thereas it was and pull it out again."

"That is no mastery," said Arthur, and so he put it in the stone. Therewithal Sir Ector essayed to pull out the sword and failed.

"Now essay," said Sir Ector unto Sir Kay. And anon he pulled at the sword with all his might, but it would not be.

"Now shall ye essay," said Sir Ector to Arthur.

"I will well," said Arthur, and pulled it out easily.

And therewithal Sir Ector kneeled down to the earth and Sir Kay.

"Alas!" said Arthur, "my own dear father and brother, why kneel ye to me?"

"Nay, nay, my lord Arthur, it is not so. I was never your father nor of your blood, but I wot well ye are of an higher blood than I weened ye were," and then

Sir Ector told him all, how he was betaken him for to nourish him and by whose commandment, and by Merlin's deliverance.

Than Arthur made great dole when he understood that Sir Ector was not his father.

"Sir," said Ector unto Arthur, "will ye be my good and gracious lord when ye are king?"

"Else were I to blame," said Arthur, "for ye are the man in the world that I am most beholding to, and my good lady and mother your wife that as well as her own hath fostered me and kept. And if ever it be God's will that I be king as ye say, ye shall desire of me what I may do, and I shall not fail you. God forbid I should fail you."

"Sir," said Sir Ector, "I will ask no more of you but that ye will make my son, your foster brother Sir Kay, seneschal of all your lands."

"That shall be done," said Arthur, "and more, by the faith of my body, that never man shall have that office but he while he and I live."

Therewithal they went unto the Archbishop and told him how the sword was achieved and by whom. And on Twelfth-day all the barons came thither and to essay to take the sword who that would essay, but there afore them all there might none take it out but Arthur. Wherefore there were many lords wroth, and said it was great shame unto them all and the realm to be overgoverned with a boy of no high blood born. And so they fell out at that time, that it was put off till Candlemas, and then all the barons should meet there again; but always the ten knights were ordained to watch the sword day and night, and so they set a pavilion over the stone and the sword, and five always watched.

So at Candlemas many more great lords came hither for to have won the sword, but there might none prevail. And right as Arthur did at Christmas he did at Candlemas, and pulled out the sword easily, whereof the barons were sore aggrieved and put it off in delay till the high feast of Easter. And as Arthur sped afore so did he at Easter. Yet there were some of the great lords had indignation that Arthur should be king, and put it off in a delay till the feast of Pentecost. Then the Archbishop of Canterbury, by Merlin's providence, let purvey then of the best knights that they might get, and such knights as Uther Pendragon loved best and most trusted in his days, and such knights were put about Arthur as Sir Baudwin of Britain, Sir Kaynes, Sir Ulfius, Sir Brastias; all these with many other were always about Arthur day and night till the feast of Pentecost.

And at the feast of Pentecost all manner of men essayed to pull at the sword that would essay, but none might prevail but Arthur, and he pulled it out afore all the lords and commons that were there. Wherefore all the commons cried at once,

"We will have Arthur unto our king! We will put him no more in delay, for we all see that it is God's will that he shall be our king, and who that holdeth against it we will slay him!"

And therewithal they kneeled at once, both rich and poor, and cried Arthur mercy because they had delayed him so long. And Arthur forgave them, and took the sword between both his hands and offered it upon the altar where the Archbishop was, and so was he made knight of the best man that was there.

And so anon was the coronation made, and there was he sworn unto his lords and the commons for to be a true king, to stand with true justice from thenceforth the days of this life. Also then he made all lords that held of the crown to come in and to do service as they ought to. And many complaints were made unto Sir Arthur of great wrongs that were done since the death of King Uther, of many lands that were bereaved lords, knights, ladies, and gentlemen; wherefore King Arthur made the lands to be given again unto them that ought them.

When this was done, that the king had stablished all the countries about London, then he let make Sir Kay seneschal of England; and Sir Baudwin of Britain was made constable, and Sir Ulfius was made chamberlain, and Sir Brastias was made warden to wait upon the North from Trent forwards, for it was that time the most party the king's enemies. But within few years after Arthur won all the North, Scotland and all that were under their obeissance, also Wales. A part of it held against Arthur, but he overcame them all as he did the remnant, through the noble prowess of himself and his knights of the Round Table.

THE DAY OF DESTINY

As Sir Mordred was ruler of all England, he let make letters as though that they had come from beyond the sea, and the letters specified that King Arthur was slain in battle with Sir Lancelot. Wherefore Sir Mordred made a parliament, and called the lords together, and there he made them to choose him king. And so was he crowned at Canterbury, and held a feast there fifteen days.

And afterward he drew him unto Winchester, and there he took Queen Guinevere, and said plainly that he would wed her (which was his uncle's wife and his father's wife). And so he made ready for the feast, and a day prefixed that they should be wedded; wherefore Queen Guinevere was passing heavy, but spake fair, and agreed to Sir Mordred's will.

And anon she desired of Sir Mordred to go to London to buy all manner things that longed to the bridal. And because of her fair speech Sir Mordred trusted her and gave her leave; and so when she came to London she took the Tower of London and suddenly in all haste possible she stuffed it with all manner of victual, and well garnished it with men, and so kept it.

And when Sir Mordred wist this he was passing wroth out of measure. And short tale to make, he laid a mighty siege about the Tower and made many assaults, and threw engines unto them, and shot great guns. But all might not prevail, for Queen Guinevere would never, for fair speech neither for foul, never to trust unto Sir Mordred to come in his hands again.

Then came the Bishop of Canterbury, which was a noble clerk and an holy man, and thus he said unto Sir Mordred:

"Sir, what will ye do? Will you first displease God and sithen shame yourself and all knighthood? For is not King Arthur your uncle, and no farther but your mother's brother, and upon her he himself begat you, upon his own sister? Therefore how may you wed your own father's wife? And therefore, sir," said the Bishop, "leave this opinion, other else I shall curse you with book, bell and candle."

"Do thou thy worst," said Sir Mordred, "and I defy thee!"

"Sir," said the Bishop, "and wit you well I shall not fear me to do that me ought to do. And also ye noise that my lord Arthur is slain, and that is not so, and therefore ye will make a foul work in this land!"

"Peace, thou false priest!" said Sir Mordred, "for an thou chafe me any more, I shall strike off thy head."

So the Bishop departed, and did the cursing in the most orgulust wise that might be done. And then Sir Mordred sought the Bishop of Canterbury for to have slain him. Then the Bishop fled, and took part of his goods with him, and went nigh unto Glastonbury. And there he was a priest-hermit in a chapel, and lived in poverty and in holy prayers; for well he understood that mischievous war was at hand.

Then Sir Mordred sought upon Queen Guinevere by letters and sonds, and by fair means and foul means, to have her to come out of the Tower of London; but all this availed nought, for she answered him shortly, openly and privily, that she had liefer slay herself than be married with him.

Then came there word unto Sir Mordred that King Arthur had araised the siege from Sir Lancelot and was coming homeward with a great host to be avenged upon Sir Mordred; wherefore Sir Mordred made write writs unto all the barony of this land, and much people drew unto him. For then was the common voice among them that with King Arthur was never other life but war and strife, and with Sir Mordred was great joy and bliss. Thus was King Arthur depraved and evil said of; and many there were that King Arthur had brought up of nought, and given them lands, that might not then say him a good word.

Lo ye Englishmen, see ye not what a mischief here was? For he that was the most king and noblest knight of the world, and most loved the fellowship of noble knights, and by him they all were upholden, and yet might not these Englishmen hold them content with him. Lo thus was the old custom and the usages of this land, and men say that we of this land have not yet lost that custom. Alas! this is a great default of us Englishmen, for there may no thing us please no term.

And so fared the people at that time: they were better pleased with Sir Mordred than they were with the noble King Arthur, and much people drew unto Sir Mordred and said they would abide with him for better and for worse. And so Sir Mordred drew with a great host to Dover, for there he heard say that King Arthur would arrive, and so he thought to beat his own father from his own lands. And the most party of all England held with Sir Mordred, for the people were so newfangle.

And so as Sir Mordred was at Dover with his host, so came King Arthur with a great navy of ships and galleys and carracks, and there was Sir Mordred ready awaiting upon his landing, to let his own father to land upon the land that he was king over.

And anon King Arthur drew him with his host down by the seaside westward, toward Salisbury. And there was a day assigned betwixt King Arthur and Sir Mordred, that they should meet upon a down beside Salisbury, and not far from the seaside. And this day was assigned on Monday after Trinity Sunday, whereof King Arthur was passing glad that he might be avenged upon Sir Mordred.

Then Sir Mordred araised much people about London, for they of Kent, Sussex and Surrey, Essex, Suffolk and Norfolk held the most party with Sir Mordred. And many a full noble knight drew unto him and also to the king; but they that loved Sir Lancelot drew unto Sir Mordred.

So upon Trinity Sunday at night King Arthur dreamed a wonderful dream, and in his dream him seemed that he saw upon a chafflet a chair, and the chair was fast to a wheel, and thereupon sat King Arthur in the richest cloth of gold that might be made. And the king thought there was under him, far from him, an hideous deep black water, and therein was all manner of serpents and worms and wild beasts, foul and horrible. And suddenly the king thought that the wheel turned up-so-down, and he fell among the serpents, and every beast took him by a limb. And then the king cried as he lay in his bed, "Help! help!"

And then knights, squires and yeomen awaked the king, and then he was so amazed that he wist not where he was. And then so he awaked until it was nigh day, and then he fell on slumbering again, not sleeping nor thoroughly waking. So the king seemed verily that there came Sir Gawain unto him with a number of fair ladies with him. So when King Arthur saw him he said,

"Welcome, my sister's son, I weened ye had been dead. And now I see thee on live, much am I beholden unto Almighty Jesu. Ah, fair nephew, what been these ladies that hither be come with you?"

"Sir," said Sir Gawain, "all these be ladies for whom I have foughten for, when I was man living. And all these are those that I did battle for in righteous quarrels, and God hath given them that grace at their great prayer, because I did battle for them for their right, that they should bring me hither unto you. Thus much hath given me leave God for to warn you of your death: for an ye fight as to-morn with Sir Mordred, as ye both have assigned, doubt ye not ye shall be slain, and the most party of your people on both parties. And for the great grace and goodness that Almighty Jesu hath unto you, and for pity of you and many more other good men there shall be slain, God hath sent me to you of His especial grace to give you warning that in no wise ye do battle as to-morn, but that ye take a treatise for a month-day. And proffer you largely, so that to-morn ye put in a delay. For within a month shall come Sir Lancelot with all his noble knights, and rescue you worshipfully, and slay Sir Mordred and all that ever will hold with him."

Then Sir Gawain and all the ladies vanished, and anon the king called upon his knights, squires, and yeomen, and charged them mightly to fetch his noble lords and wise bishops unto him. And when they were come the king told them of his avision: that Sir Gawain had told him and warned him that an he fought on the morn he should be slain. Then the king commanded Sir Lucan the Butler and his brother Sir Bedivere the Bold, with two bishops with them, and charged them in any wise to take a treatise for a month-day with Sir Mordred:

"And spare not, proffer him lands and goods as much as you think reasonable."

So then they departed and came to Sir Mordred where he had a grim host of an hundred thousand. And there they entreated Sir Mordred long time, and at the last Sir Mordred was agreed for to have Cornwall and Kent by King Arthur's days; and after that all England, after the days of King Arthur. Then were they condescended that King Arthur and Sir Mordred should meet betwixt both their hosts, and every each of them should bring fourteen persons. And so they came with this word unto Arthur. Then said he,

"I am glad that this is done," and so he went into the field.

And when King Arthur should depart he warned all his host that an they see any sword drawn, "look ye come on fiercely and slay that traitor, Sir Mordred, for I in no wise trust him." In like wise Sir Mordred warned his host that "an ye see any manner of sword drawn look that ye come on fiercely and so slay all that ever before you standeth, for in no wise I will not trust for this treatise." And in the same wise said Sir Mordred unto his host: "for I know well my father will be avenged upon me."

And so they met as their pointment was, and were agreed and accorded thoroughly. And wine was fette, and they drank together. Right so came out an adder of a little heathbush, and it stang a knight in the foot. And so when the knight felt him so stung, he looked down and saw the adder; and anon he drew his sword to slay the adder, and thought none other harm. And when the host on both parties saw that sword drawn, then they blew beams, trumpets, and horns, and shouted grimly, and so both hosts dressed them together. And King Arthur took his horse and said, "Alas, this unhappy day!" And so rode to his party, and Sir Mordred in like wise.

And never since was there seen a more dolefuller battle in no Christian land, for there was but rushing and riding, foining and striking, and many a grim word was there spoken of either to other, and many a deadly stroke. But ever King Arthur rode throughout the battle of Sir Mordred many times and did full nobly, as a noble king shold do, and at all times he fainted never. And Sir Mordred did his devour that day and put himself in great peril.

And thus they fought all the long day, and never stinted till the noble knights were laid to the cold earth. And ever they fought still till it was near night, and by then was there an hundred thousand laid dead upon the earth. Then was King Arthur wood wroth out of measure, when he saw his people so slain from him.

And so he looked about him and could see no mo of all his host, and good knights left no mo on live but two knights: the tone was Sir Lucan de Butler and his brother, Sir Bedivere; and yet they were full sore wounded.

"Jesu mercy!" said the king, "where are all my noble knights become? Alas, that ever I should see this doleful day! For now," said King Arthur, "I am come to mine end. But would to God," said he, "that I wist now where were that traitor Sir Mordred that hath caused all this mischief."

Then King Arthur looked about and was ware where stood Sir Mordred leaning upon his sword among a great heap of dead men.

"Now, give me my spear," said King Arthur unto Sir Lucan, "for yonder I have espied the traitor that all this woe hath wrought."

"Sir, let him be," said Sir Lucan, "for he is unhappy. And if ye pass this unhappy day ye shall be right well revenged. And, good lord, remember ye of your night's dream and what the spirit of Sir Gawain told you to-night, and yet God of His great goodness hath preserved you hitherto. And for God's sake, my lord, leave off this, for, blessed be God, ye have won the field: for yet we been here three on live, and with Sir Mordred is not one of live. And therefore, if ye leave off now, this wicked day of Destiny is past!"

"Now tide me death, tide me life," said the king, "now I see him yonder alone, he shall never escape mine hands! For at a better avail shall I never have him."

"God speed you well!" said Sir Bedivere.

Then the king gat his spear in both his hands, and ran toward sir Mordred, crying and saying,

"Traitor, now is thy death-day come!"

And when Sir Mordred saw King Arthur he ran until him with his sword drawn in his hand, and there King Arthur smote Sir Mordred under the shield with a foin of his spear throughout the body more than a fathom. And when Sir Mordred felt that he had his death wound he trust himself with the might that he had up to the burr of King Arthur's spear, and right so he smote his father, King Arthur, with his sword holding in both his hands, upon the side of the head, that the sword pierced the helmet and the tay of the brain. And therewith Mordred dashed down stark dead to the earth.

And noble King Arthur fell in a swough to the earth, and there he swooned oftentimes, and Sir Lucan and Sir Bedivere oftentimes hove him up. And so weakly betwixt them they led him to a little chapel not far from the sea, and when the king was there, him thought him reasonably eased.

Then heard they people cry in the field.

"Now go thou, Sir Lucan," said the king, "and do me to wit what betokens that noise in the field."

So Sir Lucan departed, for he was grievously wounded in many places; and so as he rode he saw and harkened by the moonlight how that pillers and robbers were come into the field to pille and to rob many a full noble knight of brooches and bees and of many a good ring and many a rich jewel. And who that were not dead all out, there they slew them for their harness and their riches.

When Sir Lucan understood his work he came to the king as soon as he might, and told him all what he had heard and seen.

"Therefore by my rede," said Sir Lucan, "it is best that we bring you to some town."

"I would it were so," said the king, "but I may not stand, my head works so . . . Ah, Sir Lancelot!" said King Arthur, "this day have I sore missed thee! And alas, that ever I was against thee! For now have I my death, whereof Sir Gawain me warned in my dream."

Then Sir Lucan took up the king the tone party and Sir Bedivere the other party, and in the lifting up the king swooned, and in the lifting Sir Lucan fell in a swoon, that part of his guts fell out of his body; and therewith the noble knight his heart brast. And when the king awoke he behold Sir Lucan, how he lay foaming at the mouth and part of his guts lay at his feet.

"Alas," said the king, "this is to me a full heavy sight, to see this noble duke so die for my sake, for he would have holpen me that had more need of help than I! Alas, that he would not complain him, for his heart was so set to help me. Now Jesu have mercy upon his soul!"

Then Sir Bedivere wept for the death of his brother.

"Now leave this mourning and weeping, gentle knight," said the king, "for all this will not avail me. For wit thou well an I might live myself, the death of Sir Lucan would grieve me evermore. But my time passeth on fast," said the king. "Therefore," said King Arthur unto Sir Bedivere, "take thou here Excalibur, my good sword, and go with it to yonder water's side; and when thou comest there, I charge thee throw my sword in that water, and come again and tell me what thou seest there."

"My lord," said Sir Bedivere, "your commandment shall be done, and lightly bring you word again."

So Sir Bedivere departed. And by the way he beheld that noble sword, and the pomell and the haft was all precious stones. And then he said to himself, "If I throw this rich sword in the water, thereof shall never come good, but harm and loss." And then Sir Bedivere hid Excalibur under a tree, and so soon as he might he came again unto the king and said he had been at the water and thrown the sword into the water.

"What saw thou there?" said the king.

"Sir," he said, "I saw nothing but waves and winds."

"That is untruly said of thee," said the king. "And therefore go thou lightly again, and do my commandment as thou art to me lief and dear: spare not but throw it in."

Then Sir Bedivere returned again and took the sword in his hand; and yet him thought sin and shame to throw away that noble sword. And so eft he hid the sword and returned again and told the king that he had been at the water and done his commandment.

"What sawest thou there?" said the king.

"Sir," he said, "I saw nothing but waters wap and waves wan."

"Ah, traitor unto me and untrue," said King Arthur, "now hast thou betrayed me twice! Who would ween that thou who has been to me so lief and dear, and also named so noble a knight, that thou would betray me for the riches of this sword? But now go again lightly; for thy long tarrying putteth me in great jeopardy of my life, for I have taken cold. And but if thou do now as I bid thee, if ever I may see thee, I shall slay thee mine own hands, for thou wouldest for my rich sword see me dead."

Then Sir Bedivere departed and went to the sword and lightly took it up, and so he went unto the water's side. And there he bound the girdle about the hilt, and threw the sword as far into the water as he might. And there came an arm and an hand above the water, and took it and cleight it, and shook it thrice and brandished, and then vanished with the sword into the water.

So Sir Bedivere came again to the king and told him what he saw.

"Alas!" said the king, "help me hence, for I dread me I have tarried over long."

Then Sir Bedivere took the king upon his back and so went with him to the water's side. And when they were there, even fast by the bank hoved a little barge with many fair ladies in it, and among them all was a queen, and all they had black hoods. And all they wept and shrieked when they saw King Arthur.

"Now put me into that barge," said the king.

And so he did softly, and there received him three ladies with great mourning. And so they set him down, and in one of their laps King Arthur laid his head. And then the queen said,

"Ah, my dear brother! Why have you tarried so long from me? Alas, this wound on your head hath caught overmuch cold!"

And anon they rowed fromward the land, and Sir Bedivere beheld all those ladies go fromward him. Then Sir Bedivere cried and said,

"Ah, my lord Arthur, what shall become of me, now ye go from me and leave me here alone among mine enemies?

"Comfort thyself," said the king, "and do as well as thou mayst, for in me is no trust for to trust in. For I must into the vale of Avalon to heal me of my grievous wound. And if thou hear nevermore of me, pray for my soul!"

But ever the queen and ladies wept and shrieked, that it was pity to hear. And as soon as Sir Bedivere had lost sight of the barge he wept and wailed, and so took the forest and went all that night.

And in the morning he was ware, betwixt two holts hoar, of a chapel and an hermitage. Then was Sir Bedivere fain, and thither he went, and when he came into the chapel he saw where lay an hermit grovelling on all fours, fast thereby a tomb was new graven. When the hermit saw Sir Bedivere he knew him well, for he was but little tofore Bishop of Canterbury, that Sir Mordred fleamed.

"Sir," said Sir Bedivere, "what man is there here interred that you pray so fast for?"

"Fair son," said the hermit, "I wot not verily but by deeming. But this same night, at midnight, here came a number of ladies and brought here a dead corse and prayed me to inter him. And here they offered an hundred tapers, and gave me a thousand besants."

"Alas," said Sir Bedivere, "that was my lord King Arthur, which lieth here graven in this chapel."

Then Sir Bedivere swooned, and when he awoke he prayed the hermit that he might abide with him still, there to live with fasting and prayers:

"For from hence will I never go," said Sir Bedivere, "by my will, but all the days of my life here to pray for my lord Arthur."

"Sir, ye are welcome to me," said the hermit, "for I know you better than ye ween that I do: for ye are Sir Bedivere the Bold, and the full noble duke Sir Lucan de Butler was your brother."

Then Sir Bedivere told the hermit all as you have heard tofore, and so he beleft with the hermit that was beforehand Bishop of Canterbury. And there Sir Bedivere put upon him poor clothes, and served the hermit full lowly in fasting and in prayers.

Thus of Arthur I find no more written in books that been authorised, neither more of the very certainty of his death heard I never read, but thus was he led away in a ship wherein were three queens; that one was King Arthur's sister, Queen Morgan le Fay, the tother was the Queen of North Galis, and the third was the Queen of the Waste Lands.

Now more of the death of King Arthur could I never find, but that these ladies brought him to his grave, and such one was interred there which the hermit bare witness that sometime Bishop of Canterbury. But yet the hermit knew not in certain that he was verily the body of King Arthur; for this tale Sir Bedivere, a knight of the Table Round, made it to be written.

Yet some men say in many parts of England that King Arthur is not dead, but had by the will of our Lord Jesu into another place; and men say that he shall come again, and he shall win the Holy Cross. Yet I will not say that it shall be so, but rather I would say: here in this world he changed his life. And many men say that there is written upon the tomb this:

HIC IACET ARTHURUS REX QUONDAM REXQUE FUTURUS

And thus leave I here Sir Bedivere with the hermit that dwelled that time in a chapel beside Glastonbury, and there was his hermitage. And so they lived in prayers and fastings and great abstinence.

And when Queen Guinevere understood that King Arthur was dead and all the noble knights, Sir Mordred and all the remnant, then she stole away with five ladies with her, and so she went to Amesbury. And there she let make herself a nun, and weared white clothes and black, and great penance she took upon her, as ever did sinful woman in this land. And never creature could make her merry, but ever she lived in fasting, prayers and alms-deeds, that all manner of people marvelled how virtuously she was changed.

The Development of Arthurian Romance

ROGER SHERMAN LOOMIS

Roger Sherman Loomis taught at Columbia University and is a leading Arthurian scholar. In these two essays, he discusses the ways in which early accounts of Arthur and Celtic myths spread to France and became transformed into Arthurian romance. Loomis's concern is with the prototypes for the major characters of the Arthurian romances.

THE *MABINOGION*

In 1849 Lady Charlotte Guest, the literary wife of a steel magnate, completed the publication of twelve Welsh tales in prose, together with a translation which became a minor English classic and inspired two of Tennyson's *Idylls of the King*, as well as Peacock's rollicking burlesque, the *Misfortunes of Elphin*. The title which Lady Guest gave to the collection, the *Mabinogion* (pronounced Mabinóg-yon), does not mean, as she supposed, "tales for children," but "tales of a hero's birth, infancy, and youth." There is, however, a wide variety in the nature of the narratives, and only three conform to the correct definition. To complicate matters still further, the first four tales are called the Four Branches of the Mabinogi, though only one of these, relating the birth and boyhood of Gwri of the Golden Hair, seems to deserve the title.

The Four Branches were probably composed by a single author about 1060, and represent a blending of various strands of Celtic myth which had, to begin with, little or no connection with each other. Matthew Arnold recognized in his *Lectures on Celtic Literature* both the charm and the true nature of the material: "Who is the mystic Arawn, the king of Annwn, who changed semblance for a year with Pwyll, prince of Dyved, and reigned in his place? These are no medieval personages; they belong to an older mythological world." And again: "The very first thing that strikes one in reading the *Mabinogion* is how evidently the medieval storyteller is pillaging an antiquity of which he does not fully possess the secret; he is like a peasant building his house on the site of Halicarnassus or Ephesus." The researches of the last hundred years have fully confirmed Arnold's impression and have demonstrated that in the Four Branches, along with myths of British origin, there are similar elements directly borrowed from Ireland.

Very few scholars have realized the great importance of the Four Branches of the Mabinogi for the study of Arthurian romance. Since Arthur himself is absent and none of the prominent knights and ladies of his court is easily recognizable, the tales until lately have been generally ignored. But take the very first episode—the compact of friendship beween Arawn, the supernatural huntsman, and Pwyll; Pwyll's lying with Arawn's wife in Arawn's shape; his fidelity and chastity under temptation. While, on the one hand, there is a clear affinity to the Irish story of Manannan and Fiachna's wife, there is an even more remarkable relationship to the experiences of Gawain at the Green Knight's castle in the English poetic masterpiece. Not that this Welsh text of the eleventh century was the literary source, even at several removes, of the fourteenth-century English poem; but both were indebted, apparently, to a well-known Welsh tradition of the temptation of a hero by the wife of the huntsman king, Arawn. In other supernatural figures of the Four Branches, we have the prototypes of the enchantress Vivien, who beguiled Merlin, and of the Maimed King of the Grail romances, Wagner's Amfortas. Originally, of course, these mythical personages had no relationship to the tales circulating about Arthur, but by the twelfth century they had been attracted into his orbit.

The earliest surviving story in which the heroic Arthur is surrounded by figures drawn from this primitive world and by others assembled from folklore and history is included in the *Mabinogion* under the title *Kulhwch and Olwen*. Probably copied down about 1100, it might be called a *mabinogi* (though it is not one of the Four Branches), for it recounts the fortunes of a prince from birth to marriage, and frequently refers to him as a *mab*; that is, a boy or youth. On the other hand, it foreshadows distinctly the typical romance of two centuries later: the hero is a near kinsman of Arthur; love is the impelling force; quests form the staple of his adventures; and his reward is the hand of a beautiful bride. Arthur, at last elevated to the rank of "sovereign prince of this island," presides over a host of warriors, and when Kulhwch names them we are not surprised to meet again our previous acquaintances: Kei, Bedwir, Lluch (spelled Llwch), Mabon son of Modron, and Manawidan son of Llyr.

Kulhwch and Olwen differs markedly from the Four Branches in the comparative paucity of strictly mythological elements. The main plot is known to folklorists as the "Giant's Daughter," and it is not hard to discern a likeness in outline to a famous example of the type: the Greek romance of Jason, Medea, and the Golden Fleece. Prince Kulhwch is put under a taboo by his stepmother never to marry anyone except Olwen, daughter of the Giant Ysbaddaden. Filled with longing, although he has never seen her, he seeks the help of Arthur and all his host, naming each one. Never was a more heterogeneous company. Together with those already mentioned there were Gildas the saint, Taliesin the bard, and several "helpful companions," like those who joined the Argonauts. There was Ear son of Hearer, who though buried seven fathoms below ground could hear an ant rising in the morning fifty miles away. There was the Tracker, who could track down the swine which had been carried off seven years before he was born. Accompanied by a troop of Arthur's men, Kulhwch entered Ysbaddaden's hall. "Where are those rascal servants and those ruffians of mine?" said the giant. "Raise up the forks under my eyelids that I may see my future son-in-law." But only after Kulhwch has flung a poisoned spear through his eyeball will Ysbaddaden state the conditions under which he will consent to Kulhwch's becoming his son-in-law; namely, thirty-nine tasks which must be performed in order to provide a suitable wedding feast and to enable Ysbaddaden to appear properly shaved and groomed for the occasion. With the aid of Arthur and his men, Kulhwch carries out the tasks, one by one, and obtains scissors, a comb, and a razor from between the ears of the savage boar, Twrch Trwyth. He then presents himself before the giant; Cadw of Pictland shaves him, cutting off at the same time the flesh and the ears, and Goreu chops off his head. "That night Kulhwch slept with Olwen, and she was his only wife." Thus ends this curious medley of the grotesque, the gruesome, and the romantic.

Fitted into the main plot of the "Giant's Daughter" are many traditional stories. We have already met Twrch Trwyth as the boar Troit in Nennius's list of *mirabilia*; and Arthur's hunting of the beast from its lair in Ireland, across St George's Channel, over a devious course through South Wales, and down into Cornwall is represented as one of the most perilous of the tasks assigned to Kulhwch, and it is treated with extraordinary geographical precision. Similarly, the raid of Arthur on Annwn to procure a magic cauldron, which was the subject of the cryptic poem treated in the first chapter, is converted into another of the tasks imposed by Ysbaddaden. Here the strange glamour has been dispelled, and we have only a prosaic account of an attack on the King of Ireland's steward to obtain a cauldron to boil meat for Kulhwch's wedding banquet.

But the general impression is one of ebullient vigour and wide-ranging fancy. And there are passages of poetic charm such as the description of Olwen, whose hair was yellower than the flower of the broom, whose skin whiter than the foam of the wave, and whose eye was fairer than that of the thrice-mewed falcon. And look at Kulhwch as he canters off to Arthur's court. "A gold-hilted sword was at his thigh, the blade of which was of gold, and he carried a gold-chased buckler with the colour of heaven's lightning in it, and the boss was of ivory. Before him were two brindled, white-breasted greyhounds, having strong collars of rubies about their necks, reaching from the shoulder to the ear. The one that was on the left side bounded across to the right side, and the one on the right to the left, and like two sea-swallows sported around him."

Although the tale of Kulhwch could not have been read outside Wales, Cornwall, and Brittany, and was not the source of any French romance, one should not overlook the fact that Chrétien de Troyes, who wrote the first French Arthurian romances, used two of the situations already employed by his Welsh predecessor. One has only to read the account of Kulhwch's arrival and reception at Arthur's court, and then turn to Chrétien's story of Perceval's arrival and reception at the same court, in order to perceive that here is no accidental resemblance, that here we have two versions of the same Welsh tradition. Likewise the dialogue between Kulhwch and a herdsman has its counterpart in Chrétien's *Ivain*.

The most sophisticated effort in Lady Guest's collection is the *Dram of Rhonabwy*. Composed at least a hundred years later than *Kulhwch*, it was designed to create the phantasmagoric effect of a dream, and also to serve as a memory test for a reciter, as the last sentence proclaims: "No one, neither bard nor story-teller, knows the *Dream* without a book because of the number of colours that were on the horses and all the different kinds of rare colours both on the arms and the panoply, and on the precious mantles and the magic stones."

The opening scene offers a deliberate contrast to this opulence and splendour. One stormy night Rhonabwy, a man of Powys, took refuge in a filthy, smoky cow-barn and went to sleep on an ox-hide. He dreamed that he was riding with two companions towards a ford of the Severn where it passed from Wales into England, and was pursued and overtaken by a youth clad in green and yellow and mounted on a green-and-yellow horse. Rhonabwy learned that his pursuer had deliberately provoked the battle of Camlann, in which, as we remember, Arthur and Medraut (Modred) fell. But, with a dreamlike indifference to chronology, when Rhonabwy reached the ford, he saw Arthur still alive, apparently of huge size, seated on an island in the river, attended by a bishop and a white-skinned youth costumed in black. Arthur smiled grimly when Rhonabwy and his companions approached, because such "little fellows" as these were now the defenders of Britain. Troop after troop of horsemen, each troop uniformly garbed in a single colour, rode up, and we learn with astonishment that they were mustering for the battle of Badon, to take place at midday.

The army now set out along the valley of the Severn, and was joined by a Scandinavian troop led by King Mark and a Danish troop under Edern. Before the walls of Caer Baddon, here equated with Bath, the host dismounted. Kai came dashing into the midst, he and his horse clad in mail as white as the water-lily, with rivets as red as blood.

There follows a scene of dreamlike irrelevance. Arthur started playing a game with Owain (Malory's Uwaine) son of Urien on a board of silver with gold pieces. Messengers, each minutely described, arrived one after another to complain that Arthur's squires were wounding and killing Owain's ravens, and though Owain protested Arthur refused each time to restrain the squires. At last, in a fury, Owain commanded that his standard be raised, whereupon the ravens recovered their magical powers, swooped on the squires, and tore them to pieces. It was now Arthur's turn to beg Owain to intervene, but in vain. Only when Arthur crushed the golden pieces on the board did Owain cause his standard to be lowered and peace returned.

The impending battle of Baddon did not take place, for Arthur's prospective foe, Osla, the Octa of Nennius, sent horsemen to ask for a truce. A council was held, the truce was agreed to, Kai rose to urge all who chose to follow Arthur to be with him that night in Cornwall. With the consequent commotion Rhonabwy awoke.

Quite different in character from *Kulhwch* and the *Dream of Rhonabwy* is a group of three Arthurian romances, *Geraint,* the *Lady of the Fountain,* and *Peredur.* In each the sequence of incidents corresponds more or less closely to that in one of three poems by Chrétien de Troyes, who wrote in the latter half of the twelfth century; in each, mythological and primitive elements are much less conspicuous; and the geography of the *Lady of the Fountain* and *Peredur* evinces none of that precise knowledge of Wales and its borders as do the two tales previously discussed. There has been a long and heated controversy about whether their resemblance to Crétien's famous romances is due to rather hazy and inaccurate reminiscences or to the indebtedness of both to a common French or Anglo-Norman source. On grounds of antecedent probability one might guess that the Welsh tales are re-tellings of Chrétien's celebrated stories. But a detailed examination of the resemblances and differences does not support this conjecture. *Geraint* and the *Lady of the Fountain* seem rather to be based on the lost work of a single author who wrote in French about the middle of the twelfth century two extraordinarily realistic and cleverly constructed novelettes dealing with love and marriage. Geraint, as readers of Tennyson's *Idylls* will remember, won his beautiful bride Enid by his victory over the Knight of the Sparrowhawk, was moved by a false suspicion of her fidelity to put her through a series of severe tests, and at the same time demonstrated his prowess anew; and so, his faith in her and her pride in him restored, they lived happily thereafter. The *Lady of the Fountain* is cast in a similar mould. Owain son of Urien won his beautiful bride by his victory over the Knight of the Fountain, but, after a brief honeymoon, lost her love by his absorption in knightly sports. Only after a period of estrangement did he win his lady's forgiveness by his prowess, and live with her thereafter at Arthur's court till the end of her days. Though in each of these romances there are still supernatural features—the dwarf king in one, the storm-making spring in the other—the natural motivation and the concern with practical problems of conduct show that the unknown author lived in a very different world from that inhabited by the author of *Kulhwch.* It should not be forgotten that *Geraint* and the *Lady of the Fountain* are not translations from a written text but rather re-tellings based on memory, and the conclusions especially have suffered as a result.

The *Mabinogion,* then, consists of four different strata in the development of Arthurian fiction: 1) the Four Branches, made up largely of mythical material, Welsh and Irish, as yet unattached to Arthur; 2) *Kulhwch,* in which we find a slight amount of this same material, together with a preponderance of widespread folktale motifs, the whole being attracted into the orbit of Arthur; 3) the *Dream of Rhonabwy,* a highly successful experiment in combining very archaic traditions about Arthur and Owain with recent pseudo-historical additions, to give the vivid impression of a dream; 4) three romances which, though incorporating much material of ultimate Welsh and Irish origin, have their immediate sources in French or Anglo-Norman compositions of the twelfth century. Still the continuity of the tradition may be discerned sometimes dimly, sometimes clearly, in its variegated manifestations. And Arthur, who had vanished from this earthly scene in the early sixth century, an obscure figure, remembered only by his fellow Britons, has become for their descendants, seven centuries later, not merely a glorious king, but an emperor, and for the French and Ango-Normans the centre of the most adventurous circle of knights that the world had ever seen.

THE INTERMEDIARIES

A perusal of the previous chapter leads inevitably to the conclusion that some powerful forces had been at work between the eleventh and the thirteenth centuries to transform Arthur from an insular into an international figure, from the subject of a purely Welsh and Cornish legend into a king whose literary fame rivalled that of Charlemagne, even in France itself. It is the function of this chapter to determine what these forces were.

Though the problem has been confused by plausible guesses and unwarranted claims, the evidence points clearly to two main agents in the establishment of Arthur's prestige and the popularization outside Celtic lands of what is called the Matter of Britain. The first was the wide-ranging activity of professional Breton story-tellers, *conteurs*, who, speaking French, were welcomed as entertainers wherever that language was understood. The second was the sensation produced by the *History of the Kings of Britain* of Geoffrey of Monmouth, first, in its Latin form, among the learned, and then, in French translation, among the courtly classes.

It will be remembered that in the course of the Anglo-Saxon occupation of what was to be England, thousands of Britons fled from their island home southwards across the sea, and there in what is now called Brittany their descendants multiplied and flourished. Through contact with their French-speaking neighbours many of them became bilingual. They still kept in touch, though, with their cousins, the Welsh and Cornish, across the Channel, and as the Arthurian legend developed and expanded in Wales into a multitude of such prose tales as *Kulhwch,* so the Bretons took them up. A class of wandering minstrels, with histrionic talents, found that this novel material captivated barons and their ladies, not only in Brittany but wherever French was understood. More and more they adapted the fantastic tales to French tastes, manners, and standards of rationality, costumed their characters according to the latest mode, and introduced all the pageantry of chivalry. Their audiences, somewhat bored by a monotonous diet of epics dealing with the quarrels and wars of Charlemagne and his paladins, were fascinated by the new and various tales of love and marvel and adventure, and were easily persuaded to accept the Breton image of Arthur as the nonpareil of kings.

The Norman Conquest opened up new territory. After 1066 William awarded many fiefs all over England to Breton knights who had helped him on the field of Senlac; and, of course, in these transplanted Breton households, as well as in Anglo-Norman halls, the Breton entertainers were welcome. It was thus by way of Brittany that Arthur returned, as it were, to reconquer the land of his historic exploits and become a hero for the Anglo-Normans. And it was as a result of the enormous popularity enjoyed by the Breton tales told in French that we find French romances of Arthur and his knights rendered into Welsh, as we saw in the last chapter; namely, *Geraint,* the *Lady of the Fountain,* and *Peredur.*

There is a rival theory which at first glance seems very plausible. After 1066 there were many direct contacts between the Anglo-Normans and the Welsh, especially in South Wales, and relations were sometimes friendly. What more natural and even inevitable than that tales resembling *Kulhwch* should be transmitted directly across the border to England, without any Breton intermediaries? But so far no one has produced contemporary testimony to direct transmission of Arthurian stories from the Welsh to the Anglo-Normans; there is no close affinity between *Kulhwch* and the Four Branches of the Mabinogi, on the one hand, and Anglo-Normans and Middle English romances, on the other; and these last, most of them derived from French sources, betray in their nomenclature traces of Breton transmission.

To be sure, a certain story-teller named Bleheris or Breri (Welsh Bleddri) enjoyed a prodigious reputation, both in Britain and in France, for his tales of Gawain, the Round Table, and the Grail, and it seems that, having become expert in French, he excelled the Bretons at their own game. But it is unlikely that he ever committed his tales to parchment, and it is impossible to find, even in the French poems which cite his authority, any strain of pure Welsh tradition. The solitary instance of Bleheris does not disprove the preponderant share of the Breton minstrels in the dissemination of Arthurian matter both on the Continent and in England. This oral tradition continued well into the thirteenth century, long after French writers had been busy drawing on this reservoir of story for their own experiments in prose and verse. In brief, the Breton *conteurs* were the direct inspirers, the immediate progenitors, of French Arthurian romance. Once grasp this fact, and you have the answer to many puzzles.

Geoffrey of Monmouth, was the second great intermediary between the Celtic peoples and the non-Celts in spreading the renown of Arthur. The latest authorities believe that he was born of Breton parents who had settled at Monmouth, and that he obtained an excellent education, probably at Oxford, which, though not yet the seat of a university, was already attracting scholars. He was certainly living there between 1129 and 1151, and, as the title *magister* suggests, may have been himself a sort of classics don. Ordained a priest in 1152, he was promptly elevated to the bishopric of St Asaph in North Wales. In 1153 he witnessed at Westminster a charter assuring to Henry of Anjou the succession to the English throne, and in 1155 he died. His literary patrons and presumably his friends were two successive bishops of Lincoln and the powerful illegitimate son of Henry I, Robert Earl of Gloucester. Evidently Geoffrey was on familiar terms with some of the highest ecclesiastical and political figures of his day.

Though Geoffrey was destined to wear the mitre, he never displayed any religious feeling nor any deep concern about the Church. He was ambitious, as his flattering dedications prove; and he was quite unscrupulous, for the *History of the Kings of Britain*, which he claimed to have translated from an ancient book imported from Brittany, was one of the world's most brazen and successful frauds.

Shortly before completing it, he indulged in another piece of mystification, a little book in Latin containing alleged prophecies of Merlin. For centuries previous a poet named Myrddin had been famous in Wales as a soothsayer, and his reputation had spread to Anglo-Norman circles. About 1134 Geoffrey set out to satisfy curiosity about the mage, and boldly composed a set of predictions, not only covering in cryptic style the period up to his own time but also extending into a future filled with lurid catastrophes. Merlin's prediction about Arthur shows that Geoffrey had at least sketched in his mind a glorious career of conquest for the British hero. "The Boar of Cornwall shall bring succour [to the Britons] and shall trample the necks [of the Saxons] under his feet. The islands of the ocean shall be subdued to his power, and he shall possess the forests of Gaul. The house of Romulus shall dread his fierceness, and his end shall be doubtful. In the mouths of peoples he shall be celebrated, and his deeds shall be food to the tellers of tales." Highly significant are the cautious remark about the possibility of Arthur's survival and the reference to story-tellers who found in him a subject which provided them with their daily bread.

When, about 1136, Geoffrey completed and published the *History* he included the *Prophecies of Merlin* as the seventh book. In the preface to the larger work he again alluded to the deeds of Arthur and other kings of Britain as proclaimed by many peoples from memory. He then cites as his source an old book in the British language, dealing with these kings, from Brutus, the first, to Cadwalader, the last. If there was any such ancient tome Geoffrey could not have taken much from it—at most, the names of apocryphal early British rulers, such as Leir; but the story of King Leir and his daughters is Geoffrey's own clever remodelling of the Buddhist parable of a man and his three friends which reached the West in the eleventh century as part of *Barlaam and Josaphat.* Needless to say, the *History* contains a considerable amount of veracious chronicling when the material is drawn from Julius Caesar, Bede, Henry of Huntingdon, etc. But even when Geoffrey had reliable material he used it with cynical ingenuity to create fiction. When he composed his elaborate description of Arthur's second coronation at Caerleon-on-Usk, and found that Welsh Arthurian tradition, as known to him, did not supply enough names of guests for so august an occasion, he picked at random names from ancient Welsh genealogies. And when he could find no record of the kings who succeeded Arthur he arbitrarily took the names from Gildas's invective against several kings reigning simultaneously in Britain, and strung them along seriatim.

It is interesting to observe what the wily "historian" did with Nennius. From that source Geoffrey lifted the meagre notion that the first conqueror and colonizer of Albion was a Trojan named Brutus, and he then proceeded to stretch it out into a prose *Aeneid,* with Brutus as the hero. The story of Vortigern's tower and the clairvoyant boy also came from Nennius. Geoffrey arbitrarily identified the boy with the youthful Merlin. The name of Uther Pendragon is known to genuine Welsh tradition, but it was probably the misinterpretation of a passage in Nennius which led Geoffrey to make him the father of Arthur. The battles with the Saxons derive from the same source, but their number has been reduced from twelve to three, and these have been treated with much imaginary detail and with much attention to the geography of the campaign. For the supreme victory of Badon, arbitrarily equated with Bath, Arthur armed himself with his shield Pridwen, the sword Caliburn, and the spear Ron—names derived more or less directly from Wales. Apparently to avoid monotony, Geoffrey inserted in the midst of this campaigning an account from Nennius of the marvels of Loch Lomond and of a Welsh whirlpool.

Three of the most important elements in Arthur's career Geoffrey derived from the oral tradition of the *conteurs,* though he may well have met them in written form. For the begetting of the future conqueror by Uther on Gorlois's wife in Tintagel castle he found ready to his hand a local Cornish legend. As for the great climax, Arthur's victory over the Emperor Lucius Hiberus and the legions of Rome, and as for Gawain's (Walwanus's) part in it, there is evidence that an old Welsh tradition of a war between Arthur and his vassal Lluch, of which there are other traces, came to Geoffrey in a Breton-French form. Quite arbitrarily he exalted Lluch to the imperial throne in order to make Arthur's triumph the more splendid. The ironic reversal of fortune, the treachery of Modred, Arthur's nephew, the forced marriage of Guenevere, the pursuit of Modred into Cornwall, and the fatal battle on the River Camel—all this, we may safely believe, represents the tradition which had grown up about the battle of Camlann, first recorded, as we know, in the *Annals of Wales*. To the statement that even the renowned King Arthur was mortally wounded, Geoffrey added the contradictory belief of the Bretons that he was borne away to the isle of Avalon for the healing of his wounds. Evidently on the burning issue of Arthur's survival the Oxford scholar could be quoted on either side. However ardently his Breton friends desired Arthur's messianic return, his Anglo-Norman patrons would find it embarrassing.

Literary and traditional sources provided the great bulk of Geoffrey's matter, but he drew some scenes from life. There can be little doubt that his description of the coronation ceremony at Caerleon and the sports which followed was a clever adaptation of the festivities at the crowning of King Stephen. And most striking is this additon: "The women . . . esteemed no one worthy of their love but such as had given proof of their valour in three several battles. . . . Thus the ladies were made chaste and the knights more noble because of their love. . . . The knights engaged in a game on horseback in imitation of a battle; the ladies, looking on from the top of the walls, excited in them wild flames of love as is customary in such sports." Here, then, is the first reference in England to a tournament, the first reference to ladies "whose bright eyes rain influence and judge the prize," and, even more momentous, the earliest assertion in medieval literature that love between the sexes was an ennobling and refining force. Not that this faker of history was the inventor of what we may call romantic love, any more than he invented the tournament, but evidently he believed in making Arthur's court a mirror of the latest fashions in sport and sentiment.

The more one studies the *History of the Kings of Britain* and the methods of its composition, the more one is astonished at the author's impudence, and the more one is impressed with his cleverness, his art. Written in a polished but not ornate style, displaying sufficient harmony with learned authorities and accepted traditions, free from the wilder extravagances of the *conteurs*, founded ostensibly on a very ancient manuscript, no wonder Geoffrey's *magnum opus* disarmed scepticism and was welcomed by the learned world. To be sure, at the end of the twelfth century William of Newburgh, with extraordinary perspicacity, accused Geoffrey of disguising under the honourable name of history the fables about Arthur which he took from the ancient fictions of the Britons and augmented out of his own head, and of writing to please the Bretons, of whom the majority are said to be brutishly stupid that they look still for Arthur as if he would return, and will not listen to anyone who says that he is dead. But few readers of Geoffrey's *History* had such acumen. The kings of England could be grateful to a historian who provided them with a predecessor who had conquered all of western Europe except Spain, and whose ancestors, Belinus, Constantine, and Maximian, had seized even Rome itself. For three and a half centuries the *History of the Kings of Britain* was accepted as authoritative, not only in England but on the Continent as well. Manuscripts multiplied; translations were made into Welsh, French, and Norwegian. Contrary to the claims of some scholars, it had very little influence on the French romances of the twelfth century; indeed, the gap in matter and manner between the *History* and the poems could hardly be wider. Only in the French prose romances of the thirteenth century did Geoffrey's version of Arthur's birth, the war with Rome, and the final struggle with Modred exert a strong influence on fiction.

Unit XII

A Noble Heart: The Courtly Love of Tristan and Isolde

Tristan

GOTTFRIED VON STRASSBURG

Gottfried Von Strassburg, a German clerk, wrote the Tristan *story in about the year 1210. It is said to be the most perfect form of the romance. The translation here is that of A. T. Hatto and includes the ending (missing from Gottfried's version) by Thomas of Britain. Selections include the love scene of Tristan's parents, his falling in love with Isolde, the love potion, the cave of lovers, the final battle, and the death of the two lovers. You will note that the spelling of the names changes in the last selections. Isolde becomes Ysolt. This is because Gottfried's poem is incomplete and the ending comes from Thomas of Britain who wrote in Norman-French rather than Gottfried's German.*

PROLOGUE

If I spend my time in vain, ripe for living as I am, my part in society will continue to fall short of what my experience requires of me. Thus I have undertaken a labour to please the polite world and solace noble hearts—those hearts which I hold in affection, that world which lies open to my heart. I do not mean the world of the many who (as I hear) are unable to endure sorrow and wish only to revel in bliss. (Please God to let them live in their bliss!) What I have to say does not concern that world and such a way of life; their way and mine diverge sharply. I have another world in mind which together in one heart bears its bitter-sweet, its dear sorrow, its heart's joy, its love's pain, its dear life, its sorrowful death, its dear death, its sorrowful life. To this life let my life be given, of this world let me be part, to be damned or saved with it. I have kept with it so far and with it have spent the days that were to bring me counsel and guidance through a life which has moved me profoundly. I have offered the fruits of my labour to this world as a pastime, so that with my story its denizens can bring their keen sorrow half-way to alleviation and thus abate their anguish. For if we have something before us to occupy our thoughts it frees our unquiet soul and eases our heart of its cares. All are agreed that when a man of leisure is overwhelmed by love's torment, leisure redoubles that torment and if leisure be added to languor, languor will mount and mount. And so it is a good thing that one who harbours love's pain and sorrow in his heart should seek distraction with all his mind—then his spirit will find solace and release. Yet I would never advise a man in search of pleasure to follow any pursuit that would ill become pure love. Let a lover ply a love-tale with his heart and lips and so while away the hour.

Therefore, whoever wants a story need go no further than here.—I will story him well with noble lovers who gave proof of perfect love:

A man, a woman; a woman, a man:
Tristan, Isolde; Isolde, Tristan.

RIVALIN AND BLANCHEFLOR

There was a lord in Parmenie of tender years, as I read. In birth (so his story truly tells us) he was the peer of kings, in lands the equal of princes, in person fair and charming, loyal, brave, generous, noble: and to those whom it was his duty to make happy this lord all his days was a joy-giving sun. He was a delight to all, a paragon of chivalry, the glory of his kinsmen, the firm hope of his land. Of all the qualities which a lord should have he lacked not one, except that he over-indulged himself in pleasures dear to his heart and did entirely as he pleased. For this he had to suffer in the end, since, alas, it is the way of the world, and ever was, that oncoming youth and ample fortune bear fruit in arrogance. It never occurred to him to overlook a wrong, as many do who wield great power; but returning evil for evil, matching force with force: to this he gave much thought.

Now when Mark's festival was over and the nobles had dispersed, news came to him that a king, an enemy of his, had invaded Cornwall in such force that unless he were soon repelled he would destroy all he overran. There and then Mark summoned a mighty army and met him in great strength. He fought with him and defeated him, killing and taking so many of his men that those who got away or survived on the field did so by great good fortune. There noble Rivalin was run through the side with a spear and so severely wounded that his friends bore him home to Tintagel in great grief as one half dead. They set him down, a dying man. At once it was rumoured that Rivalin of Canoel had been struck in battle and mortally wounded. This gave rise to doleful laments, at court and in the country. Those who knew of his good qualities deeply mourned his undoing. They regretted that his prowess, his handsome person, his tender youth, his admired high temper and breeding, should pass away so soon with him and have so untimely an end. His dear friend Mark lamented him with a heart-rending vehemence he had never felt before for any man. Many a noblewoman wept for Rivalin, many a lady lamented him; and all who had ever seen him were moved to pity by his plight.

But whatever the sorrow they all felt for his disaster, it was his Blancheflor alone, that faultless, noble lady, who in utter steadfastness, with eyes and heart, bewept and bewailed her dear love's pain. And indeed, when she was alone and able to vent her sorrow, she laid violent hands on herself. A thousand times she beat with them *there* and only *there, there* where she was troubled, above her heart—*there* the lovely girl struck many a blow. Thus did this charming lady torment her sweet young lovely body in such an access of grief that she would have bartered away her life for any death that did not come of love. In any event, she would have perished and died of her sorrow, had not hope refreshed her and expectancy buoyed her up, set as she was on seeing him, however, that might be: and once having seen him she would gladly suffer whatever might be in store for her. With such thoughts she held on to life, till she regained her composure and considered means of seeing him, as her sufferings required. In this way her thoughts turned to a certain nurse of hers who had sole charge of her upbringing and never let her out of her keeping. She drew this woman aside and went to a private place and there made her sad complaint to her, as those in her position have always done and still do today. Her eyes brimmed over, the hot tears fell thick and fast down her gleaming cheeks.

"Ah, woe is me," she said, clasping her hands and holding them out imploringly. "Ah," she said, "woe, woe is me! Oh dearest nurse, prove your devotion now, of which you have such a fund. And since you are so good that my whole happiness and deliverance rest on your aid alone, relying on your goodness I will tell you what troubles my heart. I shall die if you do not help me!"

"Tell me, my lady, what distresses you so and makes you complain so bitterly?"

"Darling, dare I tell you?"

"Yes, dear mistress, come tell it now."

"This dead man Rivalin of Parmenie is killing me. I would very much like to see him, if that were possible and if I knew how to go about it before he is quite dead—for alas! he is past recovery. If you can help me to this I will deny you nothing as long as I live."

"If I allow this thing," her nurse reflected, "what harm can come of it? Half dead already, this man will die tomorrow or even today, and I shall have saved my lady's life and honour and she will always love me more than other women. Sweet mistress," she said, "my pet, your misery saddens my heart, and if I can avert your sufferings by any act of mine, never doubt that I shall do so. I will go down myself and see him, and then return at once. I will spy out the whole situation, how and where he is bedded, and also take stock of his suite."

And so, pretending to go and mourn his sufferings, Blancheflor's nurse gained admittance and secretly informed him that her mistress would much like to see him, if he would allow it in accordance with honour and decorum. This done, she returned with the news. She took the girl and dressed her as a beggar-woman. Masking her lovely face in the folds of heavy veils, she took her mistress by the hand and went to Rivalin. He, for his part, had packed his people off one after another and was alone, after impressing it on them all that solitude brought him relief. For her part Blancheflor's nurse declared that she had brought a physician-woman, and she succeeded in getting her admitted to him. She then thrust the bolt across the door.

"Now, madam," she said, "go and see him!" And Blancheflor, lovely girl, went up to him, and when she looked into his eyes, "Alas," she said, "alas for this day and evermore! Oh that I was ever born! How all my hopes are dashed!"

Rivalin with great difficulty inclined his head in thanks as much as a dying man might. But Blancheflor scarcely noticed it and paid no attention, but merely sat there unseeing, and laid her cheek on Rivalin's, till for joy but also for sorrow her strength deserted her body. Her rosy lips grew wan, the hue of her flesh quite lost the glow that dwelt in it before. In her clear eyes the day turned dark and sombre as night. Thus she lay senseless in a swoon for a long time, her cheek on his cheek, as though she were dead. And when she had rallied a little from this extremity she took her darling in her arms and laying her mouth to his kissed him a hundred thousand times in a short space till her lips had fired his sense and roused his mettle for love, since love resided in them. Her kisses made him gay, they brought him such vigour that he strained the splendid woman to his half-dead body, very tenderly and close. Nor was it long before they had their way and the sweet woman received a child from him. As to Rivalin he was all but dead, both of the woman and love. But for God's helping him from this dire pass he could never have lived; yet live he did, for so it was to be.

Thus it came about that Rivalin recovered, and Blancheflor's heart was burdened and unburdened of two different kinds of pain. She left great sorrow alongside the man and bore greater sorrow away. She left the anguish of a lovelorn heart, but what she bore away was death. She left her anguish when love came; death she received with the child. Yet, however it was she recovered, in whichever way she had been burdened and unburdened by him (on the one hand to her loss, on the other to her gain), she regarded nothing else but dear love and the man who was dear to her. Of the child or tragic death within her she knew nothing: but love and the man she did know, and behaved as a living person should and as a lover does. Her heart, her mind, her desire, were centred entirely on Rivalin, and his in return on her and on the passion she inspired in him. Between them they had in their minds but one delight and one desire. Thus he was she, and she was he. He was hers and she was his. There Blancheflor, there Rivalin! There Rivalin, there Blancheflor! There both, and there true love!

Their life was now intimately shared. They were happy with each other and heartened one another with much kindness shared in common. And when they could decently arrange a rendezvous, their worldly joy was so entire, they felt so appeased and contented, that they would not have give this life of theirs for any heavenly kingdom.

MOROLD

What fresh matters shall I now set in train? When landless Tristan arrived in Cornwall he heard a rumour which greatly displeased him; that mighty Morold had come from Ireland and was demanding tribute from both Cornwall and England under threat of armed combat. In the matter of the tribute things stood thus. As I read the history and as the authentic story says, the man who was king of Ireland was called Gurmun the Gay and he was a scion of the house of Africa, where his father was king. When the latter died, his country passed into the possession of Gurman and Gurmun's brother, who was co-heir with him. But Gurmun was so ambitious and high-tempered that he disdained to share anything with anybody—his heart insisted that he should rule as monarch. And so he began to choose strong and steadfast men, the best that were known for desperate encounters, both knights and sergeants, whom he was able to attach to himself by his wealth or his charm; and he left all his land to his brother there and then.

Quickly forsaking those parts, Gurmun obtained leave and authority in writing of the mighty, illustrious Romans that he should have possession of all he could subdue, yet concede them some right and title to it. Nor did he delay any longer, but ranged over land and sea till he came to Ireland and conquered that country and compelled its people, despite themselves, to take him as lord and king; after which they came round to helping him to harry their neighbours at all times with battles and assaults. In the course of these events he subdued Cornwall and England, too. But Mark was a boy at the time and irresolute in war, as children are, and so lost his power and became tributary to Gurmun. It further helped Gurmun greatly and added to his power and prestige that he married Morold's sister. This led to his being much feared. Morold was a duke and would have liked to rule a country of his own, for he was very bold and possessed lands, much wealth, a good person, and resolute spirit. He was Gurmun's champion.

Now I will tell you truly and precisely what tribute it was that was sent from Cornwall and England to Ireland. In the first year they sent them three hundred marks of bronze and no more; in the second silver; in the third gold. In the fourth year Morold the Strong arrived from Ireland armed both for battle and single combat. Barons and their peers were summoned before him from all over Cornwall and England, and in his presence they drew lots as to which should surrender to him such of their children as were capable of service and as handsome and acceptable in their appearance as courtly usage required. There were no girls, only boys. Of the latter there were to be thirty from either land, and there would be no way of opposing such degradation other than by single combat or by a battle between peoples. Now the Cornish and English were unable to obtain justice by means of open warfare since their lands had declined in strength. On the other hand Morold was so strong, pitiless, and harsh that scarcely anyone he looked in the face dared risk his life against him more than would a woman. And when the tribute had been sent back to Ireland and the fifth year had come round again, at the solstice they had always to send to Rome such envoys as were acceptable to her to learn what instructions her mighty Senate would dispatch to each subject land. For each year there was proclaimed to them how they were to dispense the laws and statutes of the land in the manner of the Romans, and how conduct their courts of justice. And indeed, they had to live in strict accordance with the directions they received in Rome. Thus every fifth year these two lands submitted their Presentation and statutory tribute to Rome, their noble mistress. Yet they did her this honour less as a due, either in law or religion, than by command of Gurmun.

Let us now return to the story. Tristan had heard all about this infamy in Cornwall; and the pact by which this tribute was levied was well known to him before. Yet, from the talk of the people, he heard every day of the country's shame and suffering wherever he rode past castles or towns. And when he returned once more to court at Tintagel, I tell you he witnessed such woeful scenes on the roads and in the streets that he was deeply angered. The news of Tristan's arrival quickly reached Mark and his court, and they were all glad of it—I mean as glad as their sorrowful state would allow; for the greatest lords of all Cornwall were, as you have heard, now assembled at court for their dishonour. The noble peers of the realm were resorting to drawing lots for the ruin of their sons! And it was thus that Tristan found them all, kneeling and at prayer, each man on his own, but openly and without shame, with streaming eyes, and in an agony of body and soul, as he begged the good God to shield his race and offspring.

Tristan went up to them while they were all at their prayers. How was he received? You are easily answered. To tell the truth, Tristan was not welcomed by a single soul of all the household, not even by Mark, as affectionately as he would have been had this annoyance left them free. But Tristan overlooked it and boldly went to where the lots were being apportioned and where Mark and Morold were seated. "You lords, one and all," he said, "to name you all by one name, who hasten to draw lots and sell your noble blood, are you not ashamed of the disgrace you are bringing on this land? Brave as you always are in all things, every one of you, you ought by rights to make yourselves and your country honoured and respected, and advance its glory! But you have laid your freedom at the hands and feet of your enemy by means of this shameful tribute, and, as in the past, you give your noble sons into serfdom and bondage, who ought to be your joy, your delight, your very life! And yet you cannot show who is forcing you to do so, or what necessity compels it, but for a duel and one man! No other necessity is involved. Nevertheless, you cannot hit on one among you all who is willing to put his life against a single man to try whether he shall fall or prevail. Now assuming that he falls, well then, a quick death and this long-lived tribulation are valued differently in Heaven and on earth. But if he wins and the unjust cause is overthrown, he will have honour in this world and God's regard in the next for ever. After all, fathers should give their lives for their children, since their lives are one and indivisible: this is to go with God. But it is utterly against God's commandments when a man yields the liberty of his sons up to serfdom, when he hands them over as bondsmen and himself lives on in freedom! If I may advise you on a course that would be fitting both in piety and in honour, my counsel is that you choose a man, wherever you can find him among the people of this country, who is fit for single combat and willing to leave it to fortune whether he survives or no. And all of you pray for him in God's name, above all, that the Holy Ghost grant him honour and a happy outcome, and that he fear not Morold's strength and stature overmuch. Let him place his trust in God, who has never yet deserted any man that had right on his side. Take counsel quickly, deliberate with speed how to avert this disgrace and save yourselves from but a single man! Dishonour your birth and good names no more!"

"Oh, sir!" they all replied, "it is not like that with Morold—none can face him and live!"

"Enough of that!" said Tristan. "For God's sake, consider who you are! By birth you are peers of all kings, and equals of all emperors, and you mean to barter away and disown your noble sons, who are just as noble as you, and turn them into bondsmen? If it is a fact that you cannot inspirit anyone so far that he dares to fight in God's name and in a just cause against that one man for the wrong suffered by you all and the plight of this land, and if you would be good enough to leave it to God and to me, then indeed, my lords, in God's name I will give my young life as a hostage to fortune and undertake the combat for you! God grant it turn out well for you and restore you to your rights. Moreover, if somehow I should fare otherwise than well in this battle, your case will not be harmed. If I fall in this duel it will neither avert nor bring on, reduce or increase the distress of a single one of you. Matters will stand as before. But if it turns out happily, then indeed God willed it so and God alone will be thanked for it. For the man I shall face single-handed (as I am well aware) is one whose strength and spirit have long stood the test of warfare, whereas my powers are just developing and I am not so eligible for deeds of arms as our present need requires, except that I have two victory-bringing aids in battle—God and our just cause! They shall go into battle with me. I have a willing heart, too, and that is also good for duelling. So that, if these three help me, I feel sure that I shall not succumb to one man, however untried I may be in other respects."

Meanwhile a battle-ground had been appointed for the champions, an islet in the sea, yet near enough to the shore for the crowd to see with ease what happened there. It was further agreed that apart from these two men none should set foot on it till the battle was over. And indeed this was well honoured. Accordingly, two pontoons were sculled along for them, of which each could ship a man and a horse in armour. The boats were waiting. Morold boarded one and, taking the sweep, ferried himself across. Arriving at the island he quickly beached the boat and made it fast. He mounted at once, took hold of his spear, and wheeled and galloped his horse at full tilt in grand style over the length and breadth of the island. The charges he delivered with lowered lance within that battle-ground were as light and sportive as if it were a game.

When Tristan had embarked in his craft and taken his gear—that is, his horse and his lance—on board with him, he stationed himself in the bows. "Your Majesty," he said, "my lord Mark, do not be over-anxious for me or for my safety. Let us leave it in God's hands. Our fears are of no avail. What if we fared better than we are led to believe? Our victory and good fortune would be due not to skill in arms but to the power of God alone. Leave all your dreads and fears, for I may yet emerge unscathed. As to me, this affair weighs lightly on me. So let it be with you. Take heart! It will fall out only as it must. But whatever turn my affairs should take, whatever the end in store for me, commend your land this day to Him on whom I rely. May God himself go with me to the duelling-ground and restore justice to its own. God must either win or suffer defeat with me! May He watch over the issue!"

With these words he gave them the sign of the cross and, pushing off in his pontoon, left in God's name. Many lips now commended his life to God, many hands sent tender farewells after him.

When he had landed on the farther shore he set his craft adrift and quickly mounted his charger. Morold was there in an instant.

"Tell me," asked Morold, "what does this mean, what had you in mind when you set your boat adrift?"

"I did it for this reason. Here are two men and a boat. It is certain that unless both fall one of them will be killed, so that the victor's needs will be met by this one boat which brought you to the island."

"Clearly there can be no turning back from this duel," answered Morold. "If you were to stop it even at this late hour and we parted company amicably on the understanding that my dues from these lands were confirmed to me, I would think you fortunate, for indeed I should be very sorry if it fell to me to kill you. No knight that I ever set eyes on has pleased me as much as you."

"If peace is to be made between us the tribute must be abolished," was Tristan's spirited answer.

"Take my word for it," said the other, "peace will not be made on those terms! We shall not make friends that way. The tribute must go with me!"

"Then we are wasting our time parleying. Since you say you are so sure to kill me, Morold, defend yourself, if you wish to live—there is no other way about it!"

Tristan threw his horse round, flattened the curve into a dead straight line, and galloped right ahead with lowered spear with all the zest of which his heart was capable. With thighs that beat like wings, with spurs and ankles, he took his horse by the flanks. Why should the other man dally, whose life was now at stake? He did as all men do whose minds are resolved on prowess. At his heart's bidding he, too, wheeled rapidly away and more rapidly back again, raising and lowering his lance. And now he came spurring along like one drawn by the Devil. Horse and rider, they came flying at Tristan swifter than a merlin, and Tristan was as eager for Morold. They flew along with equal dash and keenness so that they broke their spears, which shivered in their shields into a thousand pieces. Then swords were snatched from sides and they went at it hand to hand from the saddle. God himself would have joyed to see it!

Morold gave him no chance to look up, such a rain of blows did he deal him. And so Morold went on hacking at him till he mastered him with blows, and Tristan, hard put to it to meet them, thrust out his shield too far and held his guard too high, so that finally Morold struck him such an ugly blow through the thigh, plunging almost to the very life of him, that his flesh and bone were laid bare through hauberk and jambs, and the blood spurted out and fell in a cloud on that island.

"What now?" asked Morold. "Will you own yourself beaten? You can see for yourself by this that you should never plead an unjust cause. The wrongness of your case is now clearly revealed. If you wish to live, think how it can be done, while you still have the time. For believe me, Tristan, your plight must irrevocably end in your death! Unless I alone avert it you will never be cured by man or woman—the sword that has wounded you is bated with deadly poison! No physician or medical skill can save you from this pass, save only my sister Isolde, Queen of Ireland. She is versed in herbs of many kinds, in the virtues of all plants, and in the art of medicine. She alone knows the secret, and no other in the world. If she does not heal you, you will be past all healing. If you will listen to me and admit your liability for tribute, I will get my sister the Queen herself to cure you. I will share all I have with you in friendship and deny you nothing you fancy."

"I shall not abandon my oath and my honour either for you or your sister. I have brought here in my free hand the freedom of two countries, and back it shall go with me, else I must suffer greater harm for them, even death itself! Know that I am not driven to such straits by a single wound, so that all must stand or fall by it. We are still very far from having decided this duel. The tribute will end in your death or mine, there is no help for it!" With this, Tristan attacked him again.

And when Morold had collected himself somewhat from his fall and was making for his horse, Tristan was already upon him and in a flash had struck his helmet a blow that sent it flying. At this, Morold took a run at him and struck off a foreleg of his horse above the knee clean through the barding, so that the beast sank on its haunches beneath him, while Morold was content to leap aside. Then, moving his shield onto his back as his seasoned instincts prompted him, the wily man groped down and retrieved his helmet. In the light of his experience he calculated that when he was back in the saddle he would don his helmet and attack Tristan again. And now that he had collected it, had run to his mount and come near enough to seize the bridle, thrust his left foot home into the stirrup and grasp the saddle with his hand—Tristan had overtaken him. He struck Morold across the pommel so that his sword and his right hand, mail and all, fell upon the sand. And as it fell, Tristan struck him another blow, high upon the coif, and the blow went home so deeply that, when Tristan jerked back his sword, the wrench left a fragment embedded in Morold's skull. (This, as it turned out later, drove Tristan to fear and desperation. It all but proved the death of him.)

THE LOVE-POTION

When this affair had been concluded the King announced to the knights and barons, the companions of the realm, throughout the Palace that the man before them was Tristan. He informed them, in the terms which he had heard, why Tristan had come to Ireland and how the latter had promised to give him guarantees in all the points which he, Gurmun, had stipulated, jointly with Mark's grandees.

The Irish court was glad to hear this news. The great lords declared that it was fitting and proper to make peace; for long-drawn enmity between them, as time went on, brought nothing but loss. The King now requested Tristan to ratify the agreement, as he had promised him, and Tristan duly did so. He and all his sovereign's vassals swore that Isolde should have Cornwall for her nuptial dower, and be mistress of all England. Hereupon Gurmun solemnly surrendered Isolde into the hands of Tristan her enemy. I say "enemy" for this reason: she hated him now as before.

Tristan laid his hand upon Isolde: "Sire," he said, "lord of Ireland, we ask you, my lady and I, for her sake and for mine, to deliver up to her any knights or pages that were surrendered to this land as tribute from Cornwall and England; for it is only right and just that they should be in my lady's charge, now that she is Queen of their country."

"With pleasure," said the King, "it shall be done. It has our royal approval that they should all depart with you!" This gave pleasure to many.

Tristan then ordered a ship to be procured in addition to his own, to be reserved for Isolde and himself and any others he might choose. And when it had been supplied, he made ready for the voyage. Wherever any of the exiles were traced, up and down the land, at court or in the country, they were sent for at once.

While Tristan and his compatriots were making ready, Isolde, the prudent Queen, was brewing in a vial a love-drink so subtly devised and prepared, and endowed with such powers, that with whomever any man drank it he had to love her above all things, whether he wished it or no, and she love him alone. They would share one death and one life, one sorrow and one joy.

The wise lady took this philtre and said softly to Brangane: "Brangane, dear niece, do not let it depress you, but you must go away with my daughter. Frame your thoughts to that, and listen to what I say. Take this flask with its draught, have it in your keeping, and guard it above all your possessions. See to it that absolutely no one gets to hear of it. Take care that nobody drinks any! When Isolde and Mark have been united in love, make it your strict concern to pour out this liquor as wine for them, and see that they drink it all between them. Beware lest anyone share with them—this stands to reason—and do not drink with them yourself. This brew is a love-philtre! Bear it well in mind! I most dearly and urgently commend Isolde to your care. The better part of my life is bound up with her. Remember that she and I are in your hands, by all your hopes of Paradise! Need I say more?"

"Dearest lady," answered Bragane, "if you both wish it, I shall gladly accompany her and watch over her honour and all her affairs, as well as ever I can."

Tristan and all his men took their leave, in one place and another. They left Wexford with jubilation. And now, out of love for Isolde, the King and Queen and the whole court followed him down to the harbour. The girl he never dreamt would be his love, his abiding anguish of heart, radiant, exquisite Isolde, was the whole time weeping beside him. Her mother and father passed the brief hour with much lamenting. Many eyes began to redden and fill with tears. Isolde brought distress to many hearts, for to many she was a source of secret pain. They wept unceasingly for their eyes' delight, Isolde. There was universal weeping. Many hearts and many eyes wept there together, both openly and in secret.

And now that Isolde and Isolde, the Sun and her Dawn, and the fair Full Moon, Brangane, had to take their leave, the One from the Two, sorrow and grief were much in evidence. That faithful alliance was severed with many a pang. Isolde kissed the pair of them many, many times.

When the Cornishmen and the ladies' Irish attendants had embarked and said good-bye, Tristan was last to go on board. The dazzling young Queen, the Flower of Ireland, walked hand in hand with him, very sad and dejected. They bowed towards the shore and invoked God's blessing on the land and on its people. Then they put to sea, and, as they got under way, began to sing the anthem "We sail in God's name" with high, clear voices, and they sang it once again as they sped onward on their course.

Now Tristan had arranged for a private cabin to be given to the ladies for their comfort during the voyage. The Queen occupied it with her ladies-in-waiting and no others were admitted, with the occasional exception of Tristan.

He sometimes went in to console the Queen as she sat weeping. She wept and she lamented amid her tears that she was leaving her homeland, whose people she knew, and all her friends in this fashion, and was sailing away with strangers, she neither knew whither nor how. And so Tristan would console her as tenderly as he could. Always when he came and found her sorrowing he took her in his arms gently and quietly and in no other way than a liege might hand his lady. The loyal man hoped to comfort the girl in her distress. But whenever he put his arm round her, fair Isolde recalled her uncle's death.

Meanwhile the ships sped on their course. They both had a favourable wind and were making good headway. But the fair company, Isolde and her train, were unused to such hard going in wind and water. Quite soon they were in rare distress. Their Captain, Tristan, gave orders to put to shore and lie idle for a while. When they had made land and anchored in a haven, most of those on board went ashore for exercise. But Tristan went without delay to see his radiant lady and pass the time of day with her. And when he had sat down beside her and they were discussing various matters of mutual interest he called for something to drink.

Now, apart from the Queen, there was nobody in the cabin but some very young ladies-in-waiting. "Look," said one of them, "here is some wine in this little bottle." No, it held no wine, much as it resembled it. It was their lasting sorrow, their never-ending anguish, of which at last they died! But the child was not to know that. She rose and went at once to where the draught had been hidden in its vial. She handed it to Tristan, their Captain, and he handed it to Isolde. She drank after long reluctance, then returned it to Tristan, and he drank, and they both of them thought it was wine. At that moment in came Brangane, recognized the flask, and saw only too well what was afoot. She was so shocked and startled that it robbed her of her strength and she turned as pale as death. With a heart that had died within her she went and seized that cursed, fatal flask, bore it off and flung it into the wild and raging sea!

"Alas, poor me," cried Brangane, "Alas that ever I was born! Wretch that I am, how I have ruined my honour and trust! May God show everlasting pity that I ever came on this journey and that death failed to snatch me, when I was sent on this ill-starred voyage with Isolde! Ah, Tristan and Isolde, this draught will be your death!"

Now when the maid and the man, Isolde and Tristan, had drunk the draught, in an instant that arch-disturber of tranquility was there, Love, waylayer of all hearts, and she had stolen in! Before they were aware of it she had planted her victorious standard in their two hearts and bowed them beneath her yoke. They who were two and divided now became one and united. No longer were they at variance: Isolde's hatred was gone. Love, the reconciler, had purged their hearts of enmity, and so joined them in affection that each was to the other as limpid as a mirror. They shared a single heart. Her anguish was his pain: his pain her anguish. The two were one both in joy and in sorrow, yet they hid their feelings from each other. This was from doubt and shame. She was ashamed, as he was. She went in doubt of him, as he of her. However blindly the craving in their hearts was centred on one desire, their anxiety was how to begin. This masked their desire from each other.

When Tristan felt the stirrings of love he at once remembered loyalty and honour, and strove to turn away. "No, leave it, Tristan," he was continually thinking to himself, "pull yourself together, do not take any notice of it." But his heart was impelled towards her. He was striving against his own wishes, desiring against his desire. He was drawn now in one direction, now in another. Captive that he was, he tried all that he knew in the snare, over and over again, and long maintained his efforts.

The loyal man was afflicted by a double pain: when he looked at her face and sweet Love began to wound his heart and soul with her, he bethought himself of Honour, and it retrieved him. But this in turn was the sign for Love, his liege lady, whom his father had served before him, to assail him anew, and once more he had to submit. Honour and Loyalty harassed him powerfully, but Love harassed him more. Love tormented him to an extreme, she made him suffer more than did Honour and Loyalty combined. His heart smiled upon Isolde, but he turned his eyes away: yet his greatest grief was when he failed to see her. As is the way of captives, he fixed his mind on escape and how he might elude her, and returned many times to this thought: "Turn one way, or another! Change this desire! Love and like elsewhere!" But the noose was always there. He took his heart and soul and searched them for some change: but there was nothing there but Love—and Isolde.

And so it fared with her. Finding this life unbearable, she, too, made ceaseless efforts. When she recognized the lime that bewitching Love had spread and saw that she was deep in it, she endeavoured to reach dry ground, she strove to be out and away. But the lime kept clinging to her and drew her back and down. The lovely woman fought back with might and main, but stuck fast at every step. She was succumbing against her will. She made desperate attempts on many sides, she twisted and turned with hands and feet and immersed them ever deeper in the blind sweetness of Love, and of the man. Her limed senses failed to discover any path, bridge, or track that would advance them half a step, half a foot, without Love being there too. Whatever Isolde thought, whatever came uppermost in her mind, there was nothing there, of one sort or another, but Love, and Tristan.

This was all below the surface, for her heart and her eyes were at variance—Modesty chased her eyes away, Love drew her heart towards him. That warring company, a Maid and a Man, Love and Modesty, brought her into great confusion; for the Maid wanted the Man, yet she turned her eyes away: Modesty wanted Love, but told no one of her wishes. But what was the good of that? A Maid and her Modesty are by common consent so fleeting a thing, so shortlived a blossoming, they do not long resist. Thus Isolde gave up her struggle and accepted her situation. Without further delay the vanquished girl resigned herself body and soul to Love and to the man.

Isolde glanced at him now and again and watched him covertly, her bright eyes and her heart were now in full accord. Secretly and lovingly her heart and eyes darted at the man rapaciously, while the man gave back her looks with tender passion. Since Love would not release him, he too began to give ground. Whenever there was a suitable occasion the man and the maid came together to feast each other's eyes. These lovers seemed to each other fairer than before—

such is Love's law, such is the way with affection. It is so this year, it was so last year and it will remain so among all lovers as long as Love endures, that while their affection is growing and bringing forth blossom and increase of all lovable things, they please each other more than ever they did when it first began to burgeon. Love that bears increase makes lovers fairer than at first. This is the seed of Love, from which it never dies.

Love seems fairer than before and so Love's rule endures. Were Love to seem the same as before, Love's rule would soon wither away.

THE CAVE OF LOVERS

Tristan had long known of a cavern in a savage mountainside, on which he had chanced when his way had led him there out hunting. The cavern had been hewn into the wild mountain in heathen times, before Corynaeus' day, when giants ruled there. They used to hide inside it when, desiring to make love, they needed privacy. Wherever such a cavern was found it was barred by a door of bronze, and bore an inscription to Love—*la fossiure a la gent amant*, which is to say "The Cave of Lovers."

The name was well suited to the thing. The story tells us that this grotto was round, broad, high, and perpendicular, snow-white, smooth, and even, throughout its whole circumference. Above, its vault was finely keyed, and on the keystone there was a crown most beautifully adorned with goldsmiths' work and encrusted with precious stones. Below, the pavement was of smooth, rich, shining marble, as green as grass. At the centre there was a bed most perfectly cut from a slab of crystal, broad, high, well raised from the ground, and engraved along its sides with letters, announcing that the bed was dedicated to the Goddess of Love. In the upper part of the grotto some small windows had been hewn out to let in the light, and these shone in several places.

Where one went in and out there was a door of bronze. Outside, above the door, there stood three limes of many branches, but beyond them not a single one. Yet everywhere downhill there were innumerable trees which cast the shade of their leafy boughs upon the mountainside. Somewhat apart, there was a level glade through which there flowed a spring—a cool, fresh brook, clear as the sun. Above that, too, there stood three limes, fair and very stately, sheltering the brook from sun and rain. The bright flowers and the green grass, with which the glade was illumined, vied with each other most delightfully, each striving to outshine the other. At their due times you could hear the sweet singing of the birds. Their music was so lovely—even lovelier here than elsewhere. Both eye and ear found their pasture and delight there: the eye its pasture, the ear its delight. There were shade and sunshine, air and breezes, both soft and gentle. Away from the mountain and its cave for fully a day's journey there were rocks unrelieved by open heath, and wilderness and wasteland. No paths or tracks had been laid towards it of which one might avail oneself. But the country was not so rough and fraught with hardship as to deter Tristan and his beloved from halting there and making their abode within that mountaincave.

Some people are smitten with curiosity and astonishment, and plague themselves with the question how these two companions, Tristan and Isolde, nourished themselves in this wasteland? I will tell them and assuage their curiosity. They looked at one another and nourished themselves with that! Their sustenance was the eye's increase. They fed in their grotto on nothing but love and desire. The two lovers who formed its court had small concern for their provender. Hidden away in their hearts they carried the best nutriment to be had anywhere in the world, which offered itself unasked ever fresh and new. I mean pure devotion, love made sweet as balm that consoles body and sense so tenderly, and sustains the heart and spirit—this was their best nourishment. Truly, they never considered any food but that from which the heart drew desire, the eyes delight, and which the body, too, found agreeable. With this they had enough. Love drove her ancient plough for them, keeping pace all the time, and gave them an abundant store of all those things that go to make heaven on earth. Their high feast was Love, who gilded all their joys; she brought them King Arthur's Round Table as homage and all its company a thousand times a day! What better food could they have for body or soul? Man was there with Woman, Woman there with Man. What else should they be needing? They had what they were meant to have, they had reached the goal of their desire.

Now some people are so tactless as to declare (though I do not accept it myself) that other food is needed for this pastime. I am not so sure that it is. There is enough here in my opinion. But if anyone has discovered better nourishment in this world let him speak in the light of his experience. There was a time when I, too, led such a life, and I thought it quite sufficient.

Now I beg you to bear with me while I reveal to you on account of what hidden significance that cave was thus constructed in the rock.

It was, as I said, round, broad, high, and perpendicular, snow-white, smooth, and even, throughout its whole circumference. Its roundness inside betokens Love's Simplicity: Simplicity is most fitting for Love, which must have no corners, that is, Cunning or Treachery. Breadth signifies Love's Power, for her Power is without end. Height is Aspiration that mounts aloft to the clouds: nothing is too great for it so long as it means to climb, up and up, to where the molten Crown of the Virtues gathers the vault to the keystone. The Virtues are invariably encrusted with precious stones, inlaid in filigree of gold and so adorned with praise, that we who are of lower aspiration—whose spirits flag and flutter over the pavement and neither settle nor fly—we gaze up intently at the masterpiece above us, which derives from the Virtues and descends to us from the glory of those who float in the clouds above us and send their refulgence down to us!—we gaze at them and marvel! From this grow the feathers by which our spirit takes wing and, flying, brings forth praise and soars in pursuit of those Virtues.

The wall was white, smooth, and even: such is Integrity's nature. Her brilliant and uniform whiteness must never be mottled with colour, nor should Suspicion find any pit or ridge in her. In its greenness and firmness the marble floor is like Constancy; this meaning is the best for it in respect of colour and smoothness. Constancy should be of the same fresh green as grass, and smooth and gleaming as glass.

At the centre, the bed of crystalline Love was dedicated to her name most fittingly. The man who had cut the crystal for her couch and her observance had divined her nature unerringly: Love *should* be of crystal—transparent and translucent!

Overhead, three little windows in all had been hewn through the solid rock into the cave, very secretly and neatly, through which the sun would shine. The first stood for Kindness, the second for Humility, the third for Breeding. Through these three, the sweet light, that blessed radiance, Honour, dearest of all luminaries, smiled in and lit up that cave of earthly bliss.

It also has its meaning that the grotto was so secluded in the midst of this wild solitude, in that one may well compare it with this—that Love and her concerns are not assigned to the streets nor yet to the open country. She is hidden away in the wilds, the country that leads to her refuge makes hard and arduous going—mountains are strewn about the way in many a massive curve. The tracks up and down are so obstructed with rocks for us poor sufferers that, unless we keep well to the path, if we make one false step we shall never get back alive. But whoever is so blessed as to reach and enter that solitude will have used his efforts to most excellent purpose, for he will find his heart's delight there. Whatever the ear yearns to hear, whatever gratifies the eye, this wilderness is full of it. He would hate to be elsewhere.

I know this well, for I have been there. I, too, have tracked and followed after wildfowl and game, after hart and hind in the wilderness over many a woodland stream and yet passed my time and not seen the end of the chase. My toils were not crowned with success. I have found the lever and seen the latch in that cave and have, on occasion, even pressed on to the bed of crystal—I have danced there and back some few times. But never have I had my repose on it. However hard the floor of marble beside it, I have so battered the floor with my steps that, had it not been saved by its greenness, in which lies its chiefest virtue, and from which it constantly renews itself, you would have traced Love's authentic tracks on it. I have also fed my eyes on the gleaming wall abundantly and have fixed my gaze on the medallion, on the vault and on the keystone, and worn out my eyes looking up at its ornament, so bespangled with Excellence! The sun-giving windows have often sent their rays into my heart. I have known that cave since I was eleven, yet I never set foot in Cornwall.

THE POISONED SPEAR

My lords, this tale is told in many ways, so I shall keep to one version in my rhymes, saying as much as is needed and passing over the remainder. But the matter diverges at this point and I do not wish to keep too much to one account. Those who narrate and tell the tale of Tristran tell it differently—I have heard various people do so. I know well enough what each says and what they have put into writing; but to judge by what I have heard, they do not follow Breri, who knew all the deeds and stories of all the kings and all the counts that had lived in Britain. Moreover, many of us are unwilling to assent to what narrators say about the dwarf, whose wife Caerdin is supposed to have loved. The dwarf in turn is said by great guile to have dealt Tristran a poisoned wound, after maiming Caerdin; and because of this wound Tristran sent Guvernal to England to fetch Ysolt. Thomas declines to accept this and is ready to prove that it could not have been the case. Guvernal was a familiar figure in all that region, and it was known throughout the kingdom that he was a partner in their intrigue and messenger to Ysolt. The King hated him for it mightily and set his men on the watch for him. Thus how could Guvernal come to offer the King, the barons, and the serjeants his service at court, as if he were a foreign merchant, without being quickly recognized, so well known was he? I do not know how he could have avoided detection or fetched Ysolt away. Such narrators have strayed from the story and departed from the truth, and if they are unwilling to admit it, I have no mind to wrangle with them—let them hold to theirs and I to mine. The tale will bear me out.

Tristran and Caerdin returned to Brittany in good spirits and amused themselves happily with their friends and followers. One day they were out hunting until it was time to return. Their companions had gone on ahead of them. There was none there but these two. They were traversing La Blanche Lande, keeping the sea to their right. They observe a knight approaching at the gallop on a piebald steed. He is armed very splendidly: he bears a shield of gold fretted with vair, and has the lance, pennant, and cognizance of the same tincture. He comes galloping down a path, covered and protected by his shield. He is tall and big and very robust; he is armed, and a fine knight. Tristran and Caerdin wait to meet him on the path together, very curious to know who he is. When he sees them he comes on towards them and then bows to them very courteously. Tristran returns his greeting and then asks him where he is going, what his business is, and why he is in such haste.

"Sir," replied the knight, "would you tell me the way to the castle of Tristran the Amorous?"

"What do you want with him?" asked Tristran. "Who are you? What is your name? We will gladly take you to his house; but if you wish to speak with Tristran you need not go any farther, for Tristran is my name. Now tell me what you want."

"This is pleasant news for me," answered the other. "I am called 'Dwarf Tristran.' I am of the Marches of Brittany and live hard by the Sea of Spain. I had a castle there and a fair mistress, whom I love as much as life; but I have lost her through great misadventure. The night before last she was stolen from me. Estult l'Orgillus of Castel Fer has had her carried off by force. He holds her in his castle and does with her just as he pleases. I suffer such grief in my heart from it that I am almost dead of sorrow, misery, and anguish. I do not know what in the world I can do. I cannot find solace without her, for I have lost my happiness, my joy, and my delight, and I value my life now but little Lord Tristran, I have heard say that whoever is bereft of what he most desires cares little for what remains. I have never been so unhappy, and this is why I have come to you. You are feared and held in great dread and are altogether the best knight, the noblest, the most just and, of all who ever lived, he that has loved most! So I beg you to have pity, sir, I call upon your magnanimity and implore you to accompany me on this task and win me back my mistress. I will do homage and allegiance to you, if you will help me to accomplish this thing!"

"Indeed I will help you all I can, my friend," answered Tristran. "But now let us go home. We shall start out towards daybreak and execute our business."

He sends for his arms and equips himself, and then rides away with Dwarf Tristran. They go to lie in wait for Estult l'Orgillus of Castel Fer, to kill him. They have travelled until they have found his strong castle. They dismount at the skirts of a wood and there await events.

Estult l'Orgillus was very haughty, and he had six brothers who were knights, bold, valiant, and excellent warriors; but in valour he surpassed them all. Two of these were returning from a tournament, and the two Tristrans lay in wait for them by the wood. They quickly shouted their challenge to them and then struck at them fiercely. The two brothers were killed there. The cry was raised through the countryside, and the lord heard the alarm. Those in the castle mounted and attacked the two Tristrans with impetuous onslaught. The latter were excellent knights, and skilled at bearing arms. They defended themselves against all as bold and valiant knights. They did not cease from fighting till they had slain the four. Tristran the Dwarf was struck down dead, the other Tristran was wounded through the loins by a lance bated with venom. Stung to anger, he took ample vengeance, for he killed the man who wounded him. And now all seven brothers are slain, one Tristrain is dead and the other in evil case, in that he is severely wounded in his body.

Tristran has turned back with great difficulty because of the anguish that grips him. By great effort he reaches home and has his wounds dressed. He sends for doctors to aid him and many are sent to him. But none can cure him of the poison, for they do not recognize it, and so they are all misled. They cannot make any plaster that will cast or draw it out. They bruise and pound roots enough, they cull herbs and make medicines, but fail to do him any good. Tristran only grows worse. The venom spreads all over his body and makes it swell up, both inside and out. He grows black and discoloured, he loses strength, his bones now show through the skin. He now knows that he will lose his life unless he is succoured at once, and he sees that none can cure him and therefore he must die. No one knows any medicine for this malady: and yet, if Queen Ysolt knew that he had this great ill and were at his side, she would heal him completely. But he is unable to go to her or to suffer the hardship of the sea. Moreover, he fears that country, since he has many enemies there. And Ysolt cannot come to him either, so he does not know how he may be healed. He suffers great sorrow at heart; for his slow torment, his malady, and the stench of his wound distress him. He laments his lot and is deeply afflicted, for the poison cruelly torments him.

CAERDIN'S MISSION

Tristran sends for Caerdin privately in the wish to reveal his sorrow to him; he had a loyal affection for Caerdin, and Caerdin loved him in return. He had the chamber in which he lay cleared of people, he would let none but themselves remain in the house while they took counsel together. But Ysolt of the White Hands wonders in her heart what it can be that he wishes to do, whether he means to forsake the world and turn monk or canon. She is greatly perturbed by this. She goes and stands beside the wall outside his chamber, opposite his bed, for she wants to listen to what they say. She gets a friend to stand guard for her for as long as she stays by the wall. With great effort Tristran has managed to lean on his elbow against the wall. Caerdin sits beside him and the two of them are weeping and lamenting piteously and that their good fellowship, their love, and their friendship will be severed after so brief a space. In their hearts they have pity and sorrow, anguish, affliction, and pain. Each is sad for the other. They weep and make much lamentation, since now their affection, which has been so loyal and true, must be parted.

Tristran addresses Caerdin. "Listen, fair friend," he says, "I am in a foreign country. Fair companion, I have no friend or relation but you. I have had no pleasure or happiness here apart from the comfort you have given me. I am convinced that, if I were in my own country, I could get help to cure me. But because I have no aid here I am dying, my fair, sweet friend. Without aid I must die; for apart from Queen Ysolt no human being can cure me. She can cure me if she wishes: she has the power and the remedy, and she has the wish, if only she knew of my wound. But, dear friend, I do not know what to do—by what strategem she may know of it. For I am sure that she could help me with this illness and heal my wound with her skill, if she but knew of it.—But how can she come here? If I knew anyone who would go and take my message to her, she would give me some good help as soon as she learned of my great need. I trust her so well that I am quite sure that she would let nothing prevent her from helping me in my distress, so firm is the love she bears me. I certainly cannot help myself. For this reason, my friend, I appeal to you: out of friendship and generosity, undertake this service for me! Bear me this message for the sake of our companionship and by the pledge you gave with your hand when Ysolt bestowed Brengvein on you, and I will give you my own pledge here that, if you will undertake this journey for me, I will become your liege man and love you above all else!"

Caerdin sees Tristran weeping, he hears him lament and despair and it makes him very sad, so that he replies with tender affection. "Dear companion," he says, "do not weep, and I will do all you wish. Believe me, my friend, to make you well I shall go very close to death and into mortal danger to win your comfort. By the loyalty which I owe you, it will not be my fault or for want of anything that I can do or for any distress or hardship if I fail to exert all my strength to carry out your wish. Tell me the message that you wish me to take to her, and I will go and make ready."

"Thank you," answered Tristran. "Now listen to what I shall tell you. Take this ring with you—it is our secret token. And when you come to that country, go to court, pretend to be a merchant, and carry fine cloths of silk with you. Be sure that she sees this ring; for as soon as she has set eyes on it and recognized you, she will seek a subtle pretext for talking with you when she can. Greet her and wish her health from me, for without her there is none in me. I send her so many wishes for her well-being that none remains with me. My heart salutes her in hopes of healing, since, without her, health will not return to me. I sent her all my health. I shall never have succour in my life, or health or healing, unless she bring them. If she does not bring me health or succour me with her own lips, then my health will remain with her and I shall die of my great pain. In fine, tell her I am dead, unless I have comfort of her! Make my distress plain to her, and the malady of which I am languishing, and tell her to come and succour me."

THE DEATH OF TRISTRAN AND YSOLT

Tristran lies on his bed languishing of his wound. He can find no succour in anything. Medicine cannot avail him; nothing that he does affords him any aid. He longs for the coming of Ysolt, desiring nothing else. Without her he can have no ease—it is because of her that he lives so long. There, in his bed, he pines and he waits for her. He has high hopes that she will come and heal his malady, and believes that he will not live without her. Each day he sends to the shore to see if the ship is returning, with no other wish in his heart. And many is the time that he commands his bed to be made beside the sea and has himself carried out to it, to await and see the ship—what way she is making, and with what sail? He has no desire for anything, except for the coming of Ysolt: his whole mind, will, and desire are set on it. Whatever the world holds he rates of no account unless the Queen is coming to him. Then he has himself carried back again from the fear which he anticipates, for he dreads that she may not come, may not keep her faith with him, and he would much rather hear it from another than see the ship come without her. He longs to look out for the ship, but does not wish to know it, should she fail to come. There is anguish in his heart, and he is full of desire to see her. He often laments to his wife but does not tell her what he longs for, apart from Caerdin, who does not come. Seeing him delay so long Tristran greatly fears that Caerdin has failed in his mission.

While they are happily sailing, there is a spell of warm weather and the wind drops so that they can make no headway. The sea is very smooth and still, the ship moves neither one way nor the other save so far as the swell draws it. They are also without their boat. And now they are in great distress. They see the land close ahead of them, but have no wind with which to reach it. And so up and down they go drifting, now back, now forward. They cannot make any progress and are very badly impeded. Ysolt is much afflicted by it. She sees the land she has longed for and yet she cannot reach it: she all but dies of her longing. Those in the ship long for land, but the wind is too light for them. Time and again, Ysolt laments her fate. Those on the shore long for the ship, but they have not seen it yet. Thus Tristran is wretched and sorrowful, he often laments and sighs for Ysolt, whom he so much desires. The tears flow from his eyes, he writhes about, he all but dies of longing.

While Tristran endures such affliction, his wife Ysolt comes and stands before him. Meditating great guile she says: "Caerdin is coming, my love! I have seen his ship on the sea. I saw it making hardly any headway but nevertheless I could see it well enough to know that it is his. God grant it brings news that will comfort you at heart!"

Tristran starts up at this news. "Do you know for sure that it is his ship, my darling?" he asks. "Tell me now, what sort of sail is it?"

"I know it for a fact!" answered Ysolt. "Let me tell you, the sail is all black! They have hoisted it and raised it up high because they have no wind!"

At this Tristran feels such pain that he has never had greater nor ever will, and he turns his face to the wall and says: "God save Ysolt and me! Since you will not come to me I must die for your love. I can hold on to life no longer. I die for you, Ysolt, dear love! You have no pity for my sufferings, but you will have sorrow of my death. It is a great solace to me that you will have pity for my death."

Three times did he say "Dearest Ysolt." At the fourth he rendered up his spirit.

Thereupon throughout the house the knights and companions weep. Their cries are loud, their lament is great. Knights and serjeants rise to their feet and bear him from his bed, then lay him upon a cloth of samite and cover him with a striped pall.

And now the wind has risen on the sea. It strikes the middle of the sailyard and brings the ship to land. Ysolt has quickly disembarked, she hears the great laments in the street and the bells from the minsters and chapels. She asks people what news? and why they toll the bells so? and the reason for their weeping? Then an old man answers: "My lady, as God help me, we have greater sorrow than people ever had before. Gallant, noble Tristran, who was a source of strength to the whole realm, is dead! He was generous to the needy, a great succour to the wretched. He has died just now in his bed of a wound that his body received. Never did so great a misfortune befall this realm!"

As soon as Ysolt heard this news she was struck dumb with grief. So afflicted is she that she goes up the street to the Palace in advance of the others, without her cloak. The Bretons have never seen a woman of her beauty; in the city they wonder whence she comes and who she may be. Ysolt goes to where she sees his body lying, and, turning towards the east, she prays for him piteously. "Tristran, my love, now that I see you dead, it is against reason for me to live longer. You died for my love, and I, love, die of grief, for I could not come in time to heal you and your wound. My love, my love, nothing shall ever console me for your death, neither joy nor pleasure nor any delight. May this storm be accursed that so delayed me on the sea, my sweetheart, so that I could not come! Had I arrived in time, I would have given you back your life and spoken gently to you of the love there was between us. I should have bewailed our fate, our joy, our rapture, and the great sorrow and pain that have been in our loving. I should have reminded you of this and kissed you and embraced you. If I had failed to cure you, then we could have died together. But since I could not come in time and did not hear what had happened and have come and found you dead, I shall console myself by drinking of the same cup. You have forfeited your life on my account, and I shall do as a true lover: I will die for you in return!"

She takes him in her arms and then, lying at full length, she kisses his face and lips and clasps him tightly to her. Then straining body to body, mouth to mouth, she at once renders up her spirit and of sorrow for her lover dies thus at his side.

Tristran died of his longing, Ysolt because she could not come in time. Tristran died for his love; fair Ysolt because of tender pity.

Here Thomas ends his book. Now he takes leave of all lovers, the sad and the amorous, the jealous and the desirous, the gay and the distraught, and all who will hear these lines. If I have not pleased all with my tale, I have told it to the best of my power and have narrated the whole truth, as I promised at the beginning. Here I have recounted the story in rhyme, and have done this to hold up an example, and to make this story more beautiful, so that it may please lovers, and that, here and there, they may find some things to take to heart. May they derive great comfort from it, in the face of fickleness and injury, in the face of hardship and grief, in the face of all the wiles of Love.

The Knight of the Cart

SIR THOMAS MALORY

This is Malory's English version of Chretien de Troyes's Knight of the Cart. *Chretien wrote many of the Arthurian legends in the court of Marie de Champagne. This story is also known as* Lancelot, *the main character of the story. While preserving all of the essentials, it is much shorter than Chretien's romance, which is 7,000 lines in length.*

So it befell in the month of May, Queen Guinevere called unto her ten knights of the Table Round, and she gave them warning that early upon the morn she would ride on-maying into woods and fields besides Westminster:

"And I warn you that there be none of you but he be well horsed, and that ye all be clothed all in green, either in silk other in cloth. And I shall bring with me ten ladies, and every knight shall have a lady by him. And every knight shall have a squire and two yeomen; and I will that all be well horsed."

So they made them ready in the freshest manner, and these were the names of the knights: Sir Kay le Seneschal, Sir Agravain, Sir Braundiles, Sir Sagramore le Desirous, Sir Dodinas le Savage, Sir Ozanna le Cure Hardy, Sir Ladinas of the Forest Savage, Sir Persaunt of Inde, Sir Ironside, that was called the Knight of the Red Launds, and Sir Pelleas the Lover. And these ten knights made them ready in the freshest manner to ride with the queen.

An so upon the morn or it were day, in a May morning, they took their horses with the queen, and rode on-maying in woods and meadows as it pleased them, in great joy and delights. For the queen had cast to have been again with King Arthur at the furthest by ten of the clock, and so was that time her purpose.

Then there was a knight which hight Sir Mellyagaunce, and he was son unto King Bagdemagus, and this knight had that time a castle of the gift of King Arthur within seven mile of Westminster. And this knight Sir Mellyagaunce loved passingly well Queen Guinevere, and so had he done long and many years. And the book saith he had lain in await for to steal away the queen, but evermore he forbare for because of Sir Lancelot; for in no wise he would meddle with the queen an Sir Lancelot were in her company other else an he were near-hand.

But this knight Sir Mellyagaunce had espied the queen well and her purpose, and how Sir Lancelot was not with her, and how she had no men of arms with her but the ten noble knights all rayed in green for maying. Then he purveyed him a twenty men of arms and an hundred archers for to distress the queen and her knights; for he thought that time was best season to take the queen.

So as the queen was out on-maying with all her knights, which were bedashed with herbs, mosses and flowers in the freshest manner, right so there came out of a wood Sir Mellyaguance with an eight score men, all harnessed as they should fight in a battle of arrest, and bade the queen and her knights abide, for maugre their heads they should abide.

"Traitor knight," said Queen Guinevere, "what cast thou to do? Wilt thou shame thyself? Bethink thee how thou art a king's son, and a knight of the Table Round, and thou thus to be about to dishonour the noble king that made thee knight! Thou shamest all knighthood and thyself and me. And I let thee wit thou shalt never shame me, for I had liefer cut mine own throat in twain rather than thou should dishonour me!"

"As for all this language," said Sir Mellyagaunce, "be as it be may. For wit you well, madam, I have loved you many a year, and never or now could I get you at such avail. And therefore I will take you as I find you."

Then spake all the ten noble knights at once and said,

"Sir Mellyagaunce, wit thou well thou art about to jeopardy thy worship to dishonour, and also ye cast to jeopardy your persons. Howbeit we be unarmed and ye have us at a great advantage, for it seemeth by you that ye have laid watch upon us, but rather than ye should put the queen to a shame and us all, we had as lief to depart from our lives, for an we otherways did we were shamed forever."

Then said Sir Mellyaguance:

"Dress you as well ye can, and keep the queen!"

Then the ten knights of the Round Table drew their swords and these other let run at them with their spears. And the ten knights manly abode them, and smote away their spears, that no spear did them no harm. Then they lashed together with swords, and anon Sir Kay, Sir Sagramore, Sir Agravain, Sir Dodinas, Sir Ladinas, and Sir Ozanna were smitted to the earth with grimly wounds. Then Sir Braundiles and Sir Persaunt, Sir Ironside, and Sir Pelleas fought long, and they were sore wounded, for these ten knights, or ever they were laid to the ground, slew forty men of the boldest and the best of them.

So when the queen saw her knights thus dolefully wounded and needs must be slain at the last, then for very pity and sorrow she cried and said,

"Sir Mellyaguance, slay not my noble knights, and I will go with thee upon this covenant: that thou save them and suffer them no more to be hurt, with this that they be led with me wheresomever thou leadest me. For I will rather slay myself than I will go with thee, unless that these noble knights may be in my presence."

"Madam," said Sir Mellyagauance, "for your sake they shall be led with you into mine own castle, with that ye will be ruled and ride with me."

Then the queen prayed the four knights to leave their fighting, and she and they would not depart.

"Madam," said Sir Pelleas, "we will do as ye do, for as for me I take no force of my life nor death."

For, as the French book saith, Sir Pelleas gave such buffets there that none armour might hold him.

Then by the queen's commandment they left battle and dressed the wounded knights on horseback, some sitting and some overthwart their horses, that it was pity to behold. And then Sir Mellyagaunce charged the queen and all her knights that none of all her fellowship should depart from her, for full sore he dread Sir Lancelot du Lake, lest he should have any knowledging. And all this espied the queen, and privily she called unto her a childe of her chamber which was swiftly horsed of a great advantage.

"Now go thou," siad she, "when thou seest thy time, and bear this ring unto Sir Lancelot du Lake, and pray him as he loveth me that he will see me and rescue me, if ever he will have joy of me. And spare not thy horse," said the queen, "nother for water nother for land."

So the child espied his time, and lightly he took his horse with spurs and departed as fast he might. And when Sir Mellyagaunce saw him so flee, he understood that it was by the queen's commandment for to warn Sir Lancelot. Then they that were best horsed chased him and shot at him, but from them all the childe went deliverly.

And then Sir Mellyaguance said unto the queen,

"Madam, ye are about to betray me, but I shall ordain for Sir Lancelot that he shall not come lightly at you."

And then he rode with her and all the fellowship in all the haste that they might. And so by the way Sir Mellyagaunce laid in bushment of the best archers that he might get in his country, to the number of a thirty, to await upon Sir Lancelot, charging them that if they saw such a manner a knight come by the way upon a white horse, "that in any wise ye slay his horse, but in no manner have ye ado with him bodily, for he is overhardy to be overcome." So this was done, and they were come to his castle; but in no wise the queen would never let none of the ten knights and her ladies out of her sight, but always they were in her presence. For the book saith Sir Mellyagaunce durst make no masteries, for dread of Sir Lancelot, insomuch he deemed that he had warning.

So when the childe was departed from the fellowship of Sir Mellyagaunce, within a while he came to Westminster, and anon he found Sir Lancelot. And when he had told his message and delivered him the queen's ring,

"Alas!" said Sir Lancelot, "now am I shamed for ever, unless that I may rescue that noble lady from dishonour."

Then eagerly he asked his arms. And ever the childe told Sir Lancelot how the ten knights fought marvellously, and how Sir Pelleas, and Sir Ironside, Sir

Braundiles, and Sir Persaunt of Inde fought strongly, but namely Sir Pelleas, there might none harness hold him; and how they all fought till they were laid to the earth, and how the queen made appointment for to save their lives and to go with Sir Mellyagaunce.

"Alas," said Sir Lancelot, "that most noble lady, that she should be so destroyed! I had liefer," said Sir Lancelot, "than all France that I had been there well armed."

So when Sir Lancelot was armed and upon his horse, he prayed the childe of the queen's chamber to warn Sir Lavain how suddenly he was departed, and for what cause. "And pray him as he loveth me, that he will hie him after me, and that he stint not until he come to the castle where Sir Mellyagaunce abideth. For there," said Sir Lancelot, "he shall hear of me an I be a man living, and then shall I rescue the queen and the ten knights the which he traitorly hath taken, and that shall I prove upon his head, and all of them that hold with him."

Then Sir Lancelot rode as fast as he might, and the book saith he took the water at Westminster Bridge, and made his horse swim over the Thames unto Lambeth. And so within a while he came to that same place thereas the ten noble knights fought with Sir Mellyagaunce. And then Sir Lancelot followed the track until that he came to a wood, and there was a strait way, and there the thirty archers bade Sir Lancelot turn again, "and follow no longer that track."

"What commandment have ye," said Sir Lancelot, "to cause me, that am a knight of the Round Table, to leave my right way?"

"These ways shalt thou leave, other else thou shalt go it on thy foot, for wit thou well thy horse shall be slain."

"That is little mastery," said Sir Lancelot, "to slay mine horse! But as for myself, when my horse is slain I give right nought of you, not an ye were five hundred more!"

So then they shot Sir Lancelot's horse and smote him with many arrows. And then Sir Lancelot avoided his horse and went on foot, but there were so many ditches and hedges betwixt them and him that he might not meddle with none of them.

"Alas, for shame!" said Sir Lancelot, "that ever one knight should betray another knight! But it is an old said saw, 'A good man is never in danger but when he is in the danger of a coward.'"

Then Sir Lancelot walked on a while, and was sore acumbered of his armour, his shield, and his spear. Wit you well he was full sore annoyed! And full loath he was for to leave anything that longed unto him, for he dread sore the treason of Sir Mellyagaunce.

Then by fortune there came by him a chariot that came thither to fetch wood.

"Say me, carter," said Sir Lancelot, "what shall I give thee to suffer me to leap into thy chariot, and that thou will bring me unto a castle within this two mile?"

"Thou shalt not enter this chariot," said the carter, "for I am sent for to fetch wood."

"Unto whom?" said Sir Lancelot.

"Unto my lord, Sir Mellyagaunce," said the carter.

"And with him would I speak," said Sir Lancelot.

"Thou shalt not go with me!" said the carter.

When Sir Lancelot leapt to him and gave him backward with his gauntlet a rearmain that he fell to the earth stark dead, then the other carter, his fellow, was afeared, and weened to have gone the same way. And then he said,

"Fair lord, save my life, and I shall bring you where ye will."

"Then I charge thee," said Sir Lancelot, "that thou drive me and this chariot unto Sir Mellyagaunce gate."

"Then leap ye up into the chariot," said the carter, "and ye shall be there anon."

So the carter drove on a great wallop, and Sir Lancelot's horse followed the chariot, with more than forty arrows in him.

And more than an hour and an half Queen Guinevere was awaiting in a bay window. Then one of her ladies espied an armed knight standing in a chariot.

"Ah, see, madam," said the lady, "where rideth in a chariot a goodly armed knight, and we suppose he rideth unto hanging."

"Where?" said the queen.

Then she espied by his shield that it was Sir Lancelot, and then she was ware where came his horse after the chariot, and ever he trod his guts and his paunch under his feet.

"Alas," said the queen, "now I may prove and see that well is that creature that hath a trusty friend. A ha!" said Queen Guinevere, "I see well that ye were hard bestead when ye ride in a chariot." And then she rebuked that lady that likened Sir Lancelot to ride in a chariot to hanging: "Forsooth, it was foul-mouthed," said the queen, "and evil-likened, so for to liken the most noble knight of the world unto such a shameful death. Ah! Jesu defend him and keep him," said the queen, "from all mischievous end!"

So by this was Sir Lancelot comen to the gates of that castle, and there he descended down and cried, that all the castle might ring:

"Where art thou, thou false traitor, Sir Mellyagaunce, and knight of the Table Round? Come forth, thou traitor knight, thou and all thy fellowship with thee, for here I am, Sir Lancelot du Lake, that shall fight with you all!"

And therewithal he bare the gate wide open upon the porter, and smote him under the ear with his gauntlet, that his neck brast in two pieces. When Sir Mellyagaunce heard that Sir Lancelot was comen he ran unto the queen and fell upon his knee and said:

"Mercy, madam, for now I put me wholly in your good grace."

So Sir Lancelot had great cheer with the queen. And then he made a promise with the queen that the same night he should come to a window outward toward a garden, and that window was barred with iron, and there Sir Lancelot promised to meet her when all folks were on sleep.

So then came Sir Lavain driving to the gates, saying, "Where is my lord, Sir Lancelot?" Anon he was sent for, and when Sir Lavain saw Sir Lancelot, he said,

"Ah, my lord, I found how ye were hard bestead, for I have found your horse that was slain with arrows."

"As for that," said Sir Lancelot, "I pray you, Sir Lavain, speak ye of other matters and let this pass, and right it another time an we may."

Then the knights that were hurt were searched, and soft salves were laid to their wounds, and so it passed on till supper time. And all the cheer that might be made them there was done unto the queen and all her knights. And when season was they went unto their chambers, but in no wise the queen would not suffer her wounded knights to be from her, but that they were laid inwith draughts by her chamber, upon beds and pallets, that she herself might see unto them that they wanted nothing.

So when Sir Lancelot was in his chamber which was assigned unto him, he called unto him Sir Lavain, and told him that night he must speak with his lady, Queen Guinevere.

"Sir," said Sir Lavain, "let me go with you an it please you, for I dread me sore of the treason of Sir Mellyagaunce."

"Nay," said Sir Lancelot, "I thank you, but I will have nobody with me."

Then Sir Lancelot took his sword in his hand and privily went to the place where he had spied a ladder toforehand, and that he took under his arm, and bare it through the garden and set it up to the window. And anon the queen was there ready to meet him.

And then they made their complaints to other of many divers things, and then Sir Lancelot wished that he might have comen in to her.

"Wit you well," said the queen. "I would as fain as ye that ye might come in to me."

"Would ye so, madam," said Sir Lancelot, "with your heart that I were with you?"

"Yea, truly," said the queen.

"Then shall I prove my might," said Sir Lancelot, "for your love."

And then he set his hands upon the bars of iron and pulled at them with such a might that he brast them clean out of the stone walls. And therewithal one of the bars of iron cut the brawn of his hands throughout to the bone. And then he leapt into the chamber to the queen.

"Make ye no noise," said the queen, "for my wounded knights lie here fast by me."

So, to pass upon this tale, Sir Lancelot went to bed with the queen, and took no force of his hurt hand, but took his pleasaunce and his liking until it was the dawning of the day; for wit you well he slept not, but watched. And when he saw his time that he might tarry no longer, he took his leave and departed at the window, and put it together as well as he might again, and so departed unto his own chamber. And there he told Sir Lavain how he was hurt. Then Sir Lavain dressed his hand and staunched it, and put upon it a glove, that it should not be espied. And so they lay long abed in the morning till it was nine of the clock.

Then Sir Mellyagaunce went to the queen's chamber and found her ladies there ready clothed.

"Ah, Jesu mercy," said Sir Mellyagaunce, "what ails you, madam, that ye sleep this long?"

And therewithal he opened the curtain for to behold her. And then was he ware where she lay, and the head-sheet, pillow and over-sheet was bebled of the blood of Sir Lancelot and of his hurt hand. When Sir Mellyagaunce espied that blood then he deemed in her that she was false to the king, and that some of the wounded knights had lain by her all that night.

"A ha, madam," said Sir Mellyagaunce, "now I have found you a false traitress unto my lord Arthur, for now I prove well it was not for nought that ye laid these wounded knights within the bounds of your chamber. Therefore I call you of treason before my lord King Arthur. And now I have proved you, madam, with a shameful deed; and that they been all false, or some of them, I will make it good, for a wounded knight this night hath lain by you."

"That is false," said the queen, "that I will report me unto them."

But when the ten knights heard of Sir Mellyagaunce's words, and then they spake all at once, and said,

"Sir Mellyagaunce, thou falsely beliest my lady the queen, and that we will make good upon thee, any of us. Now chose which thou list of us, when we are whole of the wounds thou gavest us."

"Ye shall not! Away with your proud language! For here ye may all see that a wounded knight this night hath lain by the queen."

Then they all looked and were sore ashamed when they saw that blood. And wit you well Sir Mellyagaunce was passing glad that he had the queen at such advantage, for he deemed by that to hide his own treason.

And so in this rumour came in Sir Lancelot and found them at a great affray.

"What array is this?" said Sir Lancelot.

Then Sir Mellagaunce told them what he had found, and so he showed him the queen's bed.

"Now, truly," said Sir Lancelot, "ye did not your part nor knightly, to touch a queen's bed while it was drawn and she lying therein. And I dare say," said Sir Lancelot, "my lord King Arthur himself would not have displayed her curtains, and she being within her bed, unless that it had pleased him to have lain him down by her. And therefore, Sir Mellyagaunce, ye have done unworshipfully and shamefully to yourself."

"Sir, I wot not what ye mean," said Sir Mellyagaunce, "but well I am sure there hath one of her hurt knights lain with her this night. And that will I prove with mine hands, that she is a traitress unto my lord King Arthur."

"Beware what ye do," said Sir Lancelot, "for an ye say so, and will prove it, it will be taken at your hands."

"My lord Sir Lancelot," said Sir Mellyagaunce, "I read you beware what ye do; for though ye are never so good a knight, as I wot well ye are renowned the best knight of the world, yet should ye be advised to do battle in a wrong quarrel, for God will have a stroke in every battle."

"As for that," said Sir Lancelot, "God is to be dread! But as to that, I say nay plainly, that this night there lay none of these ten knights wounded with my lady Queen Guinevere, and that will I prove with mine hands that ye say untruly in that. Now, what say ye?" said Sir Lancelot.

"Thus I say," said Sir Mellyagaunce, "here is my glove, that she is a traitress unto my lord King Arthur, and that this night one of the wounded knights lay with her."

"Well, sir, and I receive your glove," said Sir Lancelot.

And anon they were sealed with their signets, and delivered unto the ten knights.

And then Sir Lancelot and Sir Mellyagaunce dressed them together with spears as thunder, and there Sir Lancelot bare him quit over his horse's croup.

And then Sir Lancelot alight and dressed his shield on his shoulder, and took his sword in his hand, and so they dressed to each other and smote many great strokes together. And at the last Sir Lancelot smote him such a buffet upon the helmet that he fell on the one side to the earth. And then he cried upon him loud and said,

"Most noble knight, Sir Lancelot, save my life! For I yield me unto you, and I require you, as ye be a knight and fellow of the Table Round, slay me not, for I yield me as overcomen; and whether I shall live or die I put me in the king's hand and yours."

Then Sir Lancelot wist not what to do, for he had liefer than all the good in the world he might be revenged upon him. So Sir Lancelot looked upon the queen, if he might espy by any sign or countenance what she would have done. And anon the queen wagged her head upon Sir Lancelot, as who saith, "Slay him!" And full well knew Sir Lancelot by her signs that she would have him dead. Then Sir Lancelot bade him,

"Arise for shame, and perform this battle with me to the utterance."

"Nay," said Sir Mellyagaunce, "I will never arise until that ye take me as yolden and recreant."

"Well, I shall proffer you a large proffer," said Sir Lancelot, "that is for to say I shall unarm my head and my left quarter of my body, all that may be unarmed as for that quarter, and I will let bind my left hand behind me there it shall not help me, and right so I shall do battle with you."

Then Sir Mellyagaunce start up and said on hight,

"Take heed, my lord Arthur, of this proffer, for I will take it. And let him be disarmed and bounden according to his proffer."

"What say ye?" said King Arthur unto Sir Lancelot. "Will ye abide by your proffer?"

"Yea, my lord," said Sir Lancelot, "for I will never go from that I have once said."

Then the knights' partners of the field disarmed Sir Lancelot, first his head and then his left arm and his left side, and they bound his left arm to his left side fast behind his back, without shield or anything. And anon they yode together.

Wit you well there was many a lady and many a knight marvelled of Sir Lancelot that would jeopardy himself in such wise.

Then Sir Mellyagaunce came with sword all on hight, and Sir Lancelot showed him openly his bare head and the bare left side. And when he weened to have smitten him upon the bare head, then lightly he devoided to the left leg and the left side, and put his hand and his sword to that stroke, and so put it on side with great sleight. And then with great force Sir Lancelot smote him on the helmet with such a buffet that the stroke carved the head in two parties.

Then there was no more to do, but he was drawn out of the field, and at the great instance of the knights of the Table Round the king suffered him to be interred, and the mention made upon him who slew him and for what cause he was slain.

And then the king and the queen made more of Sir Lancelot, and more he was cherished than ever he was aforehand.

. . . And so I leave here of this tale and overleap great books of Sir Lancelot, what great adventures he did when he was called Le Chevalier de Chariot. For, as the French book saith, because of despite that knights and ladies called him "the knight that rode in the chariot," like as he were judged to the gibbet, therefore, in the despite of all them that named him so, he was carried in a chariot a twelvemonth; but for little after that he had slain Sir Mellyagaunce in the queen's quarrel he never of a twelvemonth come on horseback. And as the French book saith, he did that twelvemonth more than forty battles.

The Art of Courtly Love

ANDREAS CAPELLANUS

Like Chretien de Troyes, Andreas (or Andrew the Chaplain) was in service to Marie de Champagne and this book was written at her request. It gives a fascinating portrait of life at the court of Queen Eleanor in the last decades of the twelfth century. The original Latin title is De Arte Honeste Amandi *[the art of honest or honourable love]. That would certainly be a more accurate title since Andreas advocates a love that comes from the heart and not from economic convenience as was common for medieval marriage. But the term "courtly love" has been generally used since the nineteenth century. Andreas's book gained a great deal of fame throughout Europe as a serious treatment of love, until it was condemned as immoral and heretical by the Bishop of Paris in 1277.*

BOOK ONE

Introduction to the Treatise on Love

We must first consider what love is, whence it gets its name, what the effect of love is, between what persons love may exist, how it may be acquired, retained, increased, decreased, and ended, what are the signs that one's love is returned, and what one of the lovers ought to do if the other is unfaithful.

Chapter 1. What Love Is

Love is a certain inborn suffering derived from the sight of and excessive meditation upon the beauty of the opposite sex, which causes each one to wish above all things the embraces of the other and by common desire to carry out all of love's precepts in the other's embrace.

That love is suffering is easy to see, for before the love becomes equally balanced on both sides there is no torment greater, since the lover is always in fear that his love may not gain its desire and that he is wasting his efforts. He fears, too, that rumors of it may get abroad, and he fears everything that might harm it in any way, for before things are perfected a slight disturbance often spoils them. If he is a poor man, he also fears that the woman may scorn his poverty; if he is ugly, he fears that she may despise his lack of beauty or may give her love to a more handsome man; if he is rich, he fears that his parsimony in the past may stand in his way. To tell the truth, no one can number the fears of one single lover. This kind of love, then, is a suffering which is felt by only one of the persons and may be called "single love." But even after both are in love the fears that arise are just as great, for each of the lovers fears that what he has acquired with so much effort may be lost through the effort of someone else, which is certainly much worse for a man than if, having no hope, he sees that his efforts are accomplishing nothing, for it is worse to lose the things you are seeking than to be deprived of a gain you merely hope for. The lover fears, too, that he may offend his loved one in some way; indeed he fears so many things that it would be difficult to tell them.

That this suffering is inborn I shall show you clearly, because if you will look at the truth and distinguish carefully you will see that it does not arise out of any action; only from the reflection of the mind upon what it sees does this suffering come. For when a man sees some woman fit for love and shaped according to his taste, he begins at once to lust after her in his heart; then the more he thinks about her the more he burns with love, until he comes to a fuller meditation. Presently he begins to think about the fashioning of the woman and to differentiate her limbs, to think about what she does, and to pry into the secrets of her body, and he desires to put each part of it to the fullest use. Then after he has come to this complete meditation, love cannot hold the reins, but he proceeds at once to action; straightway he strives to get a helper and to find an intermediary. He begins to plan how he may find favor with her, and he begins to seek a place and a time opportune for talking; he looks upon a brief hour as a very long year, because he cannot do anything fast enough to suit his eager mind. It is well known that many things happen to him in this manner. This inborn suffering comes, therefore, from seeing and meditating. Not every kind of meditation can be the cause of love, an excessive one is required; for a restrained thought does not, as a rule, return to the mind, and so love cannot arise from it.

Chapter III. Where Love Gets Its Name

Love gets its name (*amor*) from the word for hook (*amus*), which means "to capture" or "to be captured," for he who is in love is captured in the chains of desire and wishes to capture someone else with his hook. Just as a skillful fisherman tries to attract fishes by his bait and to capture them on his crooked hook, so the man who is a captive of love tries to attract another person by his allurements and exerts all his efforts to unite two different hearts with an intangible bond, or if they are already united he tries to keep them so forever.

Chapter IV. What the Effect of Love Is

Now it is the effect of love that a true lover cannot be degraded with any avarice. Love causes a rough and uncouth man to be distinguished for his handsomeness; it can endow a man even of the humblest birth with nobility of character; it blesses the proud with humility; and the man in love becomes accustomed to performing many services gracefully for everyone. O what a wonderful thing is love, which makes a man shine with so many virtues and teaches everyone, no matter who he is, so many good traits of character! There is another thing about love that we should not praise in few words: it adorns a man, so to speak, with the virtue of chastity, because he who shines with the light of one love can hardly think of embracing another woman, even a beautiful one. For when he thinks deeply of his beloved the sight of any other woman seems to his mind rough and rude.

I wish you therefore to keep always in mind, Walter my friend, that if love were so fair as always to bring his sailors into the quiet port after they had been soaked by many tempests, I would bind myself to serve him forever. But because he is in the habit of carrying an unjust weight in his hand, I do not have full confidence in him any more than I do in a judge whom men suspect. And so for the present I refuse to submit to his judgment, because "he often leaves his sailors in the mighty waves." But why love, at times, does not use fair weights I shall show you more fully elsewhere in this treatise.

Chapter VII. Various Decisions in Love Cases

Now then, let us come to various decisions in cases of love:

I. A man who was greatly enamoured of a certain woman devoted his whole heart to the love of her. But when she saw that he was in love with her, she absolutely forbade him to love. When she discovered that he was just as much in love with her as ever, she said to him one day, "I know it is true that you have striven a very long time for my love, but you can never get it unless you are willing to make me a firm promise that you will always obey all my commands and that if you oppose them in any way you will be willing to lose my love completely." The man answered her, "My lady, God forbid that I should ever be so much in error as to oppose your commands in anything; so since what you ask is very pleasing, I gladly assent to it." After he had promised this she immediately ordered him to make no more effort to gain her love and not to dare to speak a good word of her to others. This was a heavy blow to the lover, yet he bore it patiently. But one day when this lover and some other knights were with some ladies he heard his companions speaking very shamefully about his lady and saying things about her reputation that were neither right nor proper. He endured it for a while with an ill grace, but when he saw that they kept on disparaging the lady he burst out violently against them and began to accuse them of slander and to defend his lady's reputation. When all this came to her ears she said that he ought to lose her love completely because by praising her he had violated her commands.

This point the Countess of Champagne explained as follows in her decision. She said that the lady was too severe in her command, because she was not ashamed to silence him by an unfair sentence after he had wholly submitted himself to her will and after she had given him the hope of her love by binding him to her with a promise which no honorable woman can break without a reason. Nor did the aforesaid lover sin at all when he tried to deliver a well-deserved rebuke to those who were slandering his lady. For although he did make such a promise in order the more easily to obtain her love, it seems unfair of the woman to lay upon him the command that he should trouble himself no more with love for her.

II. Again. Another man, although he was enjoying the embraces of a most excellent love, asked her for permission to obtain the embraces of a different woman. Having received this he went away and refrained longer than usual from the solaces of the first lady. But after a month had elapsed he came back to the first one and said that he had never received any solaces from the other lady, nor had he wished to receive them, but he had merely wanted to test the constancy of his loved one. This woman refused him her love on the ground that he was unworthy, saying that for him to ask and receive such permission was reason enough for her to deprive him of her love. But the opinion of Queen Eleanor, who was consulted on the matter, seems to be just the opposite of this woman's. She said, "We know that it comes from the nature of love that those who are in love often falsely pretend that they desire new embraces, that they may the better test the faith and constancy of their co-lover. Therefore a woman sins against the nature of love itself if she keeps back her embraces from her lover on this account or forbids him her love, unless she has clear evidence that he has been unfaithful to her."

III. There were two men who were equal in birth and life and morals and everything else except that one happened to have more property than the other, so that many wondered which was preferable as a lover. From this case came the dictum of the Countess of Champagne, who said, "It would not be right for one to prefer a vulgar rich man to a noble and handsome poor one. Indeed a handsome poor man may well be preferred to a rich nobleman if both are seeking the love of a rich woman, since it is more worthy for a woman who is blessed with an abundance of property to accept a needy lover than one who has great wealth. Nothing should be more grievous to all good men than to see worth overshadowed by poverty or suffering from the lack of anything. It is right, therefore, for men to praise a wealthy woman who disregards money and seeks a needy lover whom she can help with her wealth, for nothing seems so praiseworthy in a lover of either sex as to relieve the necessities of the loved one so far as may be. But if the woman herself is in need, she is more ready to accept the rich lover; for if both lovers are oppressed by poverty there is little doubt that their love will be of short duration. Poverty brings a great feeling of shame to all honourable men and gives them many an anxious thought and is even a great disturber of quiet sleep; so as a result it commonly puts love to flight.

IX. A certain man asked the same lady to make clear where there was the greater affection—between lovers or between married people. The lady gave him a logical answer. She said: "We consider that marital affection and the true love of lovers are wholly different and arise from entirely different sources, and so the ambiguous nature of the word prevents the comparison of the things and we have to place them in different classes. Comparisons of more or less are not valid when things are grouped together under an ambiguous heading and the comparison is made in regard to that ambiguous term. It is no true comparison to say that a name is simpler than a body or that the outline of a speech is better arranged than the delivery.

X. The same man asked the same lady this question. A certain woman had been married, but was now separated from her husband by a divorce, and her former husband sought eagerly for her love. In this case the lady replied: "If any two people have been married and afterwards separate in any way, we consider love between them wholly wicked."

XVII. Again. A certain knight was in love with a woman who had given her love to another man, but he got from her this much hope of her love—that if it should ever happen that she lost the love of her beloved, then without a doubt her love would go to this man. A little while after this the woman married her lover. The other knight then demanded that she give him the fruit of the hope she had granted him, but this she absolutely refused to do, saying that she had not lost the love of her lover. In this affair the Queen gave her decision as follows: "We dare not oppose the opinion of the Countess of Champagne, who ruled that love can exert no power between husband and wife. Therefore we recommend that the lady should grant the love she has promised."

XVIII. A certain knight shamefully divulged the intimacies and the secrets of his love. All those who were serving in the camp of Love demanded that this offense should be most severely punished, lest if so serious a transgression went unavenged, the example might give occasion to others to do likewise. A court of ladies was therefore assembled in Gascony, and they decided unanimously that forever after he should be deprived of all hope of love and that in every court of ladies or of knights he should be an object of contempt and abuse to all. And if any woman should dare to violate this rule of the ladies, for example by giving him her love, she should be subject to the same punishment, and should henceforth be an enemy of all honest women.

Chapter VIII. The Rules of Love

Let us come now to the rules of love, and I shall try to present to you very briefly those rules which the King of Love is said to have proclaimed with his own mouth and to have given in writing to all lovers. These are the rules.

I. Marriage is no real excuse for not loving.
II. He who is not jealous cannot love.
III. No one can be bound by a double love.
IV. It is well known that love is always increasing or decreasing.
V. That which a lover takes against the will of his beloved has no relish.
VI. Boys do not love until they arrive at the age of maturity.
VII. When one lover dies, a widowhood of two years is required of the survivor.
VIII. No one should be deprived of love without the very best of reasons.

IX. No one can love unless he is impelled by the persuasion of love.
X. Love is always a stranger in the home of avarice.
XI. It is not proper to love any woman whom one would be ashamed to seek to marry.
XII. A true lover does not desire to embrace in love anyone except his beloved.
XIII. When made public love rarely endures.
XIV. The easy attainment of love makes it of little value; difficulty of attainment makes it prized.
XV. Every lover regularly turns pale in the presence of his beloved.
XVI. When a lover suddenly catches sight of his beloved his heart palpitates.
XVII. A new love puts to flight an old one.
XVIII. Good character alone makes any man worthy of love.
XIX. If love diminishes, it quickly fails and rarely revives.
XX. A man in love is always apprehensive.
XXI. Real jealousy always increases the feeling of love.
XXII. Jealousy, and therefore love, are increased when one suspects his beloved.
XXIII. He whom the thought of love vexes eats and sleeps very little.
XXIV. Every act of a lover ends in the thought of his beloved.
XXV. A true lover considers nothing good except what he thinks will please his beloved.
XXVI. Love can deny nothing to love.
XXVII. A lover can never have enough of the solaces of his beloved.
XXVIII. A slight presumption causes a lover to suspect his beloved.
XXIX. A man who is vexed by too much passion usually does not love.
XXX. A true lover is constantly and without intermission possessed by the thought of his beloved.
XXXI. Nothing forbids one woman being loved by two men or one man by two women.

Unit XIII

In Search of the Holy Grail: The Parzival Legend

Parzival

WOLFRAM VON ESCHENBACH

Wolfram von Eschenbach, a German knight of the late twelfth century, wrote the great medieval romance Parzival. *These selections include the love of Gahmuret and Belacane, Parzival's first visit to the Grail Castle, the teachings of Tevrizent, the fight between Parzival and Feirefiz, the Healing of Amfortas and the reunion of Parzival and Condwiramurs.*

CHAPTER I

If vacillation dwell with the heart the soul will rue it. Shame and honour clash where the courage of a steadfast man is motley like the magpie. But such a man may yet make merry, for Heaven and Hell have equal part in him. Infidelity's friend is black all over and takes on a murky hue, while the man of loyal temper holds to the white.

This winged comparison is too swift for unripe wits. They lack the power to grasp it. For it will wrench past them like a startled hare! So it is with a dull mirror or a blind man's dream. These reveal faces in dim outline: but the dark image does not abide, it gives but a moment's joy. Who tweaks my palm where never a hair did grow? He would have learnt close grips indeed! Were I to cry "Oh!" in fear of that it would mark me as a fool. Shall I find loyalty where it must vanish, like fire in a well or dew in the sun?

On the other hand I have yet to meet a man so wise that he would not gladly know what guidance this story requires, what edification it brings. The tale never loses heart, but flees and pursues, turns tail and wheels to the attack and doles out blame and praise. The man who follows all these vicissitudes and neither sits too long nor goes astray and otherwise knows where he stands has been well served by mother wit.

Feigned friendship leads to the fire, it destroys a man's nobility like hail. Its loyalty is so short in the tail that if it meet in the wood with gadflies it will not quit a bite in three.

These manifold distinctions do not all relate to men. I shall set these marks as a challenge to women. Let any who would learn from me beware to whom she takes her honour and good name, beware whom she makes free of her love and precious person, lest she regret the loss of both chastity and affection. With God as my witness I bid good women observe restraint. The lock guarding all good ways is modesty—I need not wish them any better fortune. The false will gain a name for falsity.—How lasting is thin ice in August's torrid sun? Their credit will pass as soon away. The beauty of many has been praised far and wide; but if their hearts be counterfeit I rate them as I should a bead set in gold. But I do not reckon it a tawdry thing when the noble ruby with all its virtues is fashioned into base brass, for this I would liken to the spirit of true womanhood. When a woman acts to the best of her nature you will not find me surveying her complexion or probing what shields her heart: if she be well proofed *within* her breast her good name is safe from harm.

CHAPTER 5

Whoever cares to hear where the knight is arriving whom Dame Adventure sent on his travels, can now take note of marvels unparalleled. Let the son of Gahmuret ride on. True-hearted people everywhere will wish him luck, since he is destined now to suffer great anguish, but at times also honour and joy. One thing was distressing him—that he was far from the woman than whom none was ever said by book or tale to have been more virtuous or fairer. Thoughts of the Queen began to unsettle his wits, and had he not been a stout-hearted man he must have lost them quite. His charger trailed his reins impetuously through bog and over fallen trees, for no man's hand was guiding it. The tale informs us that a bird would have been hard put to it to fly the distance he rode that day. Unless my source has deceived me, the journey he made on the day when he killed Ither with his javelin, and later when he reached the land of Brobarz from Graharz, was shorter by far.

Would you now like to hear how he is faring?

In the evening he came to a lake. Some sportsmen whose lake it was had anchored there. When they saw him ride up they were near enough to the shore to hear anything he said. One of those he saw in the boat was wearing clothes of such quality that had he been lord of the whole earth they could not have been finer. His hat was of peacock's feathers and lined inside. Parzival asked the Angler in God's name and of his courtesy to tell him where he could seek shelter for the night, and thus did that man of sorrows answer: "Sir," he said, "I know of no habitation beside the lake or inland for thirty miles. Nearby stands a lone mansion. I urge you to go there. What other place could you reach before nightfall? After passing the rock-face, turn right. When you come up to the moat, where I suspect you will have to halt, ask them to lower the drawbridge and open up the road to you."

He accepted the Angler's advice and took his leave. "If you do find the right way there," added the Angler, "I shall take care of you myself this evening: then suit your thanks to the entertainment you will have received. Take care—some tracks lead to unknown country, you could miss your path on the mountainside, and I would not wish *that* to happen to you."

Parzival set off and moved into a brisk trot along the right path as far as the moat. There he found the drawbridge raised. Nothing had been spared to make an impregnable stronghold: it stood smooth and rounded as though from a lathe. Unless attackers were to come on wings or be blown there by the wind, no assault could harm it. Clusters of towers and numerous palaces stood there marvellously embattled. Had all the armies in the world assailed it, the hurt they would have inflicted would not have ruffled the defenders once in thirty years.

A page attended to asking him what he wanted and where he had journeyed from.

"The Angler sent me here," he answered. "I thanked him in the hope of finding bare shelter for the night. He asked for the drawbridge to be lowered and told me to ride in to you."

"Seeing that it was the Angler who said so, you are wlecome, sir!" said the page. "You shall be honoured and have comfortable quarters for the sake of the man who sent you here." And he let down the drawbridge.

The bold knight rode into the fortress and on to a spacious courtyard. Its green lawn had not been trampled down in chivalric sport, for there was no vying at the bohort there, jousters never rode over it with pennants flying as they do over the meadow at Abenberg. Zestful deeds had not been done there for many a day, for they had come to know heartfelt grief.

Parzival was not made to feel this in any way. He was welcomed by knights young and old. A crowd of very young gentlemen ran forward to take his bridle, each trying to seize it first. They held his stirrup, so down he had to come from his horse. Knights invited him to step forward and they then conducted him to his room. They unarmed him swiftly but decorously, and when they set eyes on the beardless young man and saw how charming he was, they declared him rich in Fortune's favours.

The young man asked for water and had soon washed the rusty grime from hands and face, with the result that it seemed to them as though another day were dawning from him, with such refulgence did he sit there, the perfect image of a handsome young consort.

They brought him a cloak of cloth-of-gold of Araby, which the good-looking fellow put on—not lacing it to—which earned him many compliments.

"My lady the Princess Repanse de Schoye was wearing it," said the discreet Master of the Wardrobe. "It is lent to you from off her person, for as yet no clothes have been cut for you. I could decently ask it of her since, if I judge you correctly, you are a man of worth."

"May God reward you, sir, for saying so. If you have judged me rightly I am indeed fortunate. The power of God bestows such reward."

They filled his cup and entertained him in such a way that despite their grief they shared his pleasure. They treated him with honour and esteem. And indeed there was greater store of meat and drink there than he found at Belrepeire before delivering it from its plight.

They took his equipment away, a thing he was soon to regret when a prank took him unawares. For a man deft in speech summoned our mettlesome guest to court to join his host overfreely, with a show of anger, and for this almost lost his life at the hands of young Parzival. Not finding his splendid sword there, Parzival clenched his fist so hard that the blood shot through his nails and splashed his sleeve.

"Hold, sir!" cried the knights. "This man is licensed to jest, however dismal we may be. Bear with him, as a gentleman.—You were merely meant to understand that the Angler is here. Go and join him—he esteems you a noble guest—and shake off your load of anger."

They mounted the stairs to a hall where a hundred chandeliers were hanging with many candles set upon them high over the heads of the company, and with candle-dips round its walls. On the floor he saw a hundred couches with as many quilts laid over them, furnished by those who had that duty, each seating four companions and with spaces in between and a round carpet before it. King Frimutel's son could well afford it. One thing was not omitted there: they had not thought it too extravagant to have three square andirons in marble masonry on which was the element of fire burning wood of aloes. Here at Wildenberg none ever saw such great fires at any time. Those were magnificent pieces of workmanship!

The lord of the castle had himself seated on a sling-bed over against the middle of the fireplace. He and happiness had settled accounts with each other, he was more dead than alive. Parzival with his radiant looks now entered the hall and was well received by him who had sent him there—his host did not keep him standing but bade him approach and be seated ". . . close beside me. Were I to seat you a way off it would be treating you too much as a stranger!" Such were this sorrowful lord's words.

Because of his ailment his lordship maintained great fires and wore warm clothes of ample cut, with sable both outside and in the lining both of his pellice and the cloak above it. Its meanest fur would have been highly prized, being of the black-with-grey variety. On his head he wore a covering of that same fur, sable bought at great price, doubled upon itself. Around its top it had an Arabian orphrey, and at the centre a button of translucent ruby.

A great company of grave knights were sitting where they were presented with a sad spectacle. A page ran in at the door, bearing—this rite was to evoke grief—a Lance from whose keen steel blood issued and then ran down the shaft to his hand and all but reached his sleeve.

At this there was weeping and wailing throughout that spacious hall, the inhabitants of thirty lands could not have wrung such a flood from their eyes. The page carried the Lance round the four walls back to the door and then ran out again, whereupon the pain was assuaged that had been prompted by the sorrow those people had been reminded of.

If it will not weary you I will set about taking you to where service was rendered with all due ceremony.

At the far end of the Palace a steel door was thrown open. Through it came a pair of noble maidens such—now let me run over their appearances for you—that to any who had deserved it of them they would have made Love's payment in full, such dazzling young ladies were they. For head-dress each wore a garland of flowers over her hair and no other covering, and each bore a golden candelabra. Their long flaxen hair fell in locks, and the lights they were carrying were dazzling-bright. But let us not pass over the gowns those young ladies made their entry in! The gown of the Countess of Tenabroc was of fine brown scarlet, as was that of her companion, and they were gathered together above the hips and firmly clasped by girdles round their waists.

Then there came a duchess and her companion, carrying two trestles of ivory. Their kips glowed red as fire. All four inclined their heads, and then the two set up their trestles before their lord. They stood there together in a group, one as lovely as the other, and gave him unstinting service. All four were dressed alike.

But see, four more pairs of ladies have not missed their cue! Their function was that four were to carry large candles whilst the other four were to apply themselves to bringing in a precious stone through which the sun could shine by day. Here is the name it was known by—it was a garnet-hyacinth! Very long and broad it was, and the man who had measured it for a tabletop had cut it thin to make it light. The lord of this castle dined at it as a mark of opulence. With an inclination of their heads all eight maidens advanced in due order into their lord's presence, then four placed the table on the trestles of snow-white ivory that had preceded it, and decorously returned to stand beside the first four. These eight ladies were wearing robes of samite of Azagouc greener than grass, of ample cut for length and breadth, and held together at their middles by long narrow girdles of price. Each of these modest young ladies wore a dainty garland of flowers above her hair.

The daughters of Counts Iwan of Nonel and Jernis of Ryl had been taken many a mile to serve there, and now these two princely ladies were seen advancing in ravishing gowns! They carried a pair of knives keen as fish-spines, on napkins, one apiece, most remarkable objects.—They were of hard white silver, ingeniously fashioned, and whetted to an edge that would have cut through steel! Noble ladies who had been summoned to serve there went before the silver, four faultless maidens, and bore a light for it. And so all six came on.

Now hear what each of them does. They inclined their heads. Then two carried the silver to the handsome table and set it down, and at once returned most decorously to the first twelve, so that if I have totted it up correctly we should now have eighteen ladies standing there.

Just look! You can now see another six advancing in sumptuous gowns, half cloth-of-gold, half brocade of Niniveh. These and the former six already mentioned were wearing their twelve gowns cut parti-wise, of stuffs that had cost a fortune.

After these came the Princess. Her face shed such refulgence that all imagined it was sunrise. This maiden was seen wearing brocade of Araby. Upon a green achmardi she bore the consummation of heart's desire, its root and its blossoming—a thing called "The Gral," paradisal, transcending all earthly perfection! She whom the Gral suffered to carry itself had the name of Repanse de Schoye. Such was the nature of the Gral that she who had the care of it was required to be of perfect chastity and to have renounced all things false.

Lights moved in before the Gral—no mean lights they, but six fine slender vials of purest glass in which balsam was burning brightly. When the young ladies who were carrying the vials with balsam had come forward to the right distance, the Princess courteously inclined her head and they theirs. The faithful Princess then set the Gral before his lordship. (This tale declares that Parzival gazed and pondered on that lady intently who had brought in the Gral, and well he might, since it was her cloak that he was wearing.) Thereupon the seven went and rejoined the first eighteen with all decorum. They then opened their ranks to admit her who was noblest, making twelve on either side of her, as I am told. Standing there, the maiden with the crown made a most elegant picture.

Chamberlains with bowls of gold were appointed at the rate of one to every four of the knights sitting there throughout the hall, with one handsome page carrying a white towel. What luxury was displayed there!

A hundred tables were fetched in at the door, each of which was set up before four noble knights and carefully spread with white table-linen.

The host—the man crippled in his pride—washed hands, as did Parzival too, which done, a count's son hastened to offer them a fine silk towel on bended knee.

At every table there were pages with orders to wait assiduously on those who were seated at them. While one pair knelt and carved, the other carried meat and drink to table and saw to the diners' other needs.

Let me tell you more of their high living. Four trolleys had been appointed to carry numerous precious drinking-cups of gold, one for each knight sitting there. They drew the trolleys along the four walls. Knights were seen by fours to stretch out their hands and stand the cups on their tables. Each trolley was dogged by a clerk whose job it was to check the cups on to the trolley again after supper. Yet there is more to come.

A hundred pages were bidden to receive loaves into white napkins held with due respect before the Gral. They came on all together, then fanned out on arriving at the tables. Now I have been told and I am telling you on the oath of each single one of you—so that if I am deceiving anyone you must all be lying with me—that whatever one stretched out one's hand for in the presence of the Gral, it was waiting, one found it all ready and to hand—dishes warm, dishes cold, new-fangled dishes and old favourites, the meat of beasts both tame and wild . . .

"There never was any such thing!" many will be tempted to say. But they would be misled by their ill temper, for the Gral was the very fruit of bliss, a cornucopia of the sweets of this world and such that it scarcely fell short of what they tell us of the Heavenly Kingdom.

In tiny vessels of gold they received sauces, peppers, or pickles to suit each dish. The frugal man and the glutton equally had their fill there served to them with great ceremony.

For whichever liquor a man held out his cup, whatever drink a man could name, be it mulberry wine, wine or ruby, by virtue of the Gral he could see it there in his cup. The noble company partook of the Gral's hospitality.

Parzival well observed the magnificence and wonder of it all, yet, true to the dictates of good breeding, he refrained from asking any question.

"Gurnemanz advised me with perfect sincerity against asking many questions," he thought. "What if I stay here for as long as I stayed with him? I shall then learn unasked how matters stand with this household." While he was musing thus a page approached carrying a sword whose sheath was worth a thousand marks and whose hilt was a ruby, whilst its blade could have been a source of marvels. His lordship bestowed it on his guest.

"Sir," he said, "I took this into the thick of battle on many a field before God crippled my body. Let it make amends for any lack of hospitality you have suffered here. You will wear it to good effect always. Whenever you put it to the test in battle it will stand you in good stead."

Alas that he asked no Question then! Even now I am cast down on his account! For when he was given the sword it was to prompt him to ask a Question! I mourn too for his gentle host, who is dogged by misfortune from on high of which he could be rid by a Question.

Enough had been dispensed there. Those whose function it was, set to work and removed the tables after the four trolleys had received their loads again. And now in reverse order of entry, each lady performs her service. They again assigned their noblest, the Princess, to the Gral, and she and all the young ladies bowed gracefully to both their lord and Parzival, and so carried back through the door what they had brought in with such ceremony.

Parzival glanced after them. On a sling-bed in a chamber, before the doors had been shut behind them, Parzival glimpsed the most handsome old man he had ever seen or heard of, whose hair I can assert without exaggeration was more silvery even than hoar-frost.

As to who that old man was, you shall learn the story later. And I shall also tell you the names of the lord, his castle and lands hereafter, when the time has come to do so, in all detail, authoritatively, and without playing on your curiosity.

CHAPTER 9

"Open!"

"To whom? Who is there?"

"I wish to enter your heart."

"Then you want too narrow a space."

"How is that? Can't I just squeeze in? I promise not to jostle you. I want to tell you marvels."

"Can it be you, Lady Adventure? How do matters stand with that fine fellow?—I mean with noble Parzival, whom with harsh words Cundrie drove out to seek the Gral, a quest from which there was no deterring him, despite the weeping of many ladies. He left Arthur the Briton then: but how is he faring now? Take up the tale and tell us whether he has renounced all thought of happiness or has covered himself with glory, whether his fame has spread far and wide or has shrivelled and shrunk. Recount his achievements in detail. Has he seen Munsalvæsche again and gentle Anfortas, whose heart was so fraught with sighs? Please tell us—how it would console us!—whether *he* has been released from suffering? Let us hear whether Parzival has been there, he who is your lord as much as mine. Enlighten me as to the life he has been leading. How has sweet Herzeloyde's child, Gahmuret's son, been faring? Tell us whether he has won joy or bitter sorrow in his battles. Does he hold to the pursuit of distant goals? Or has he been lolling in sloth and idleness? Tell me his whole style of living."

Now the adventure tells us that Parzival has ranged through many lands on horseback and over the waves in ships. None who measured his charge against him kept his seat, unless he were compatriot or kinsman—in such fashion does he down the scales for his opponents and, whilst making others fall, raise his own renown! He has defended himself from discomfiture in many fierce wars and so far spent himself in battle that any man who wished to lease fame from him had to do so in fear and trembling.

Herzeloyde's child rides on. His manly discipline enjoined modesty and compassion in him. Since young Herzeloyde had left him a loyal heart, remorse now began to stir in it. Only now did he ponder Who had brought the world into being, only now think of his Creator and how mighty He must be. "What if God has such power to succour as would overcome my sorrow?" he asked himself. "If He ever favoured a knight and if any knight ever earned His reward or if shield and sword and true manly ardour can ever be so worthy of His help that this could save me from my cares and if this is His Helpful Day, then let Him help, if help He can!"

He turned back in the direction whence he had ridden. They were still standing there, saddened by his departure, for they were loyal-hearted people. The young ladies followed him with their eyes, while he in turn confessed in his heart that they pleased his eyes—for their bright looks declared them beautiful.

"If God's power is so great that it can guide horses and other beasts and people, too, then I will praise His power. If the wisdom of God disposes of that help, let it guide my castilian to the best success of my journey—then in His goodness He will show power to help! Now go where God chooses!" He laid the reins over his horse's ears and urged him on hard with his spurs.

The beast made for Fontane la Salvæsche, where Orilus had received the oath. This was the abode of the austere Trevrizent, who ate miserably many a Monday and no better all through the week. He had forsworn wine, mulberry and bread. His austerity imposed further abstinence: he had no mind for such food as fish or meat or anything with blood. Such was the holy life he led. God had inspired this gentleman to prepare to join the heavenly host. He endured much hardship from fasting. Self-denial was his arm against the Devil.

From Trevrizent, Parzival is about to learn matters concerning the Gral that have been hidden. Those who questioned me earlier and wrangled with me for not telling them earned nothing but shame. Kyot asked me to conceal it because his source forbade him to mention it till the story itself reached that point expressly where it *has* to be spoken of.

Despite the snow on the ground Parzival recognized a spot where once upon a time dazzling flowers had stood. It was at the foot of an escarpment where, with his manly right hand, he had made Orilus relent towards Lady Jeschute, and Orilus's anger had evaporated. But the tracks did not let him stop there: Fontane la Salvæsche was the locality towards which his journey tended. Parzival found its lord at home, and he received him.

"Alas, sir," said the hermit, "that you should be in this condition at Holytide. Was it some desperate encounter that forced you into this armour? Or had you no fighting to do?—In which case other garb would have been seemlier if your pride permitted it. Pray dismount, sir—I fancy you will have no objection—and warm yourself beside the fire. If thirst for adventure has brought you out with an eye to winning Love's reward and it is True Love you favour, then love as Love is now in season, and in keeping with the Love of this Day! After that, serve women for their favour. But please do dismount, if I may invite you."

The warrior Parzival alighted at once and stood before him with great courtesy. He told him of the people who had pointed out the way and how they had praised his guidance.

"Sir," he said, "guide me now: I am a sinner."

In answer to these words the good man said: "I shall give you guidance. Now tell me who directed you here to me."

"Walking towards me in the forest, sir, there came a grey-haired man. He saluted me kindly, as did his retinue. That same honest person sent me here to you, and I rode along his tracks till I found you."

"That was Gabenis," said his host. "He is versed to perfection in noble ways. The Prince is a Punturteis, the mighty King of Kareis married his sister. No fruit of human body was ever purer than his daughters who came walking towards you in the forest! The Prince is of royal line. He visits me here each year."

"When I saw you standing in my path, were you at all afraid as I rode up to you?" Parzival asked his host. "Did my coming irk you?"

"Believe me, sir," replied the hermit, "bears and stags have startled me more often than man. I can tell you truly: I fear nothing of human kind, since I, too, possess human ability. If you will not think me boastful, I declare I never fled the field; nor am I innocent of love. My heart never knew the villainy of turning tail in battle. While I bore arms I was a knight like you and strove to win the love of noble ladies. From time to time I paired chaste with sinful thoughts. I lived in dazzling style to win a lady's favour. But I have forgotten these things. Give me your bridle. Your horse shall rest at the foot of that cliff. Then, soon, we shall go and gather some young fir-tips and bracken for him—I have no other fodder. Nevertheless we shall keep him in good fettle."

Parzival made as though to prevent him from taking the bridle.

"Your good manners do not permit you to struggle with your host short of lowering themselves," said the good man. And so Parzival yielded the bridle to his host, who then led the horse beneath the overhanging rock where the rays of the sun never came—a wild stable indeed! A waterfall gushed down through it. A weak man would have been hard put to it, wearing armour where the bitter cold could strike him in this fashion. His host led Parzival into a grotto, well protected from the wind and with a fire of glowing charcoal which the stranger could well put up with! The master of the house lit a candle, and the warrior removed his armour and reclined on a bed of straw and ferns, while all his limbs grew warm and his skin shone clear. No wonder he was weary from the forest, since he had ridden along few roads and passed the night with no roof over his head till day-break, and many another, too. But now he had found a kind host.

There was a coat lying there. The hermit lent it him to put on and then took him to another grotto, where the austere man kept the books he read. An altar-stood there bare of its cloth, in keeping with the Good Friday rite. On it a reliquary could be seen which was instantly recognized—Parzival had laid his hand on it to swear an unsullied oath on the occasion when Lady Jeschute's suffering was changed to joy, and her happiness took an upward turn.

"I know this casket, sir," said Parzival to his host, "for I once swore an oath on it when passing by. I found a painted lance beside it. Sir, I took that lance and was told later that I advanced my reputation with it. I was so absorbed in thoughts of my wife that I lost my self-awareness. I rode two mighty jousts with it—I fought them both in utter obliviousness! Honour had not yet deserted me. But now I have more cares than were seen in any man. Kindly tell me how long is it since the time I took the lance from here?"

"My friend Taurian left it behind," replied the good man. "He told me he missed it, later. It is now four-and-a-half years and three days since you took it. If you care to listen I will reckon it out for you." And from his psalter he read him the full count of the years and weeks that had elapsed in the meantime.

"Only now," said Parzival, "do I realize how long I have been wandering with no sense of direction and unsustained by any happy feelings. Happiness for me is but a dream: I bear a heavy pack of grief. And I will tell you more. All this time I was never seen to enter any church or minster where God's praise was sung. All I sought was battle. I am deeply resentful of God, since He stands godfather to my troubles: He has lifted them up too high, while my happiness is buried alive. If only God's power would succour me, what an anchor my happiness would be, which now sinks into sorrow's silt! If my manly heart is wounded—can it be *whole* when Sorrow sets her thorny crown on glory won by deeds of arms from formidable foes?—then I set it down to the shame of Him who has all succour in His power, since if He is truly prompt to help He does not help me—for all the help they tell of Him!"

His host sighed and looked at him. "Sir," he said, "if you have any sense you will trust in God. He will help you, since help He must. May God help both of us! You must give me a full account, sir—but *do* sit down first! Tell me soberly how your anger began so that God became the object of your hatred. Yet kindly bear with me while I tell you He is innocent, before you accuse Him in my hearing. His help is always forthcoming.

"Although I was a layman I could read and indite the message of the Scriptures: that to gain His abundant help mankind should persevere in God's service, Who never wearied of giving His steadfast aid against the soul's being plunged into Hell. Be unswervingly constant towards Him, since God Himself is perfect constancy, condemning all falsity. We should allow Him to reap the benefit of having done so much for us, for His sublime nature took on human shape for our sakes. God is *named* and *is* Truth, He was Falsity's foe from the Beginning. You should ponder this deeply. It is not in Him to play false. Now school your thoughts and guard against playing Him false.

"You can gain nothing from Him by anger. Anyone who sees you hating Him would think you weak of understanding. Consider what Lucifer and his comrades achieved! As angels they had no gall: so where in God's name did they find the malice that makes them wage ceaseless war, whose reward in Hell is so bitter? Astiroth and Belcimon, Belet and Radamant and others I could name—this bright heavenly company took on a hellish hue as the result of their malice and envy.

"When Lucifer made the descent to Hell with his following, a Man succeeded him. For God made noble Adam from earth. From Adam's body He then broke Eve, who consigned us to tribulation by not listening to her Maker and thus shattered our bliss. Through birth these two had progeny. One son was driven by his discontent and by vainglorious greed to deflower his grandmother. Now it might please many to ask, before they understood this account, how this is possible? It nevertheless came to pass, and sinfully."

"I doubt that it ever happened," interposed Parzival. "From whom was the man descended by whom, according to you, his grandmother lost her maidenhead? You ought never to have said such a thing."

"I will remove your doubts," his host replied. "If I do not tell the unvarnished truth you must object to my deceiving you! The earth was Adam's mother, by her fruits Adam was nourished. The earth was still a virgin then. It remains for me to tell you who took her maidenhead. Adam was father to Cain, who slew Abel for a trifle. When blood fell upon the pure earth her virginity was gone, taken by Adam's son. This was the beginning of hatred among men, and thus it has endured ever since.

"There is nothing in the whole world so pure as an honest maiden. Consider the purity of maidens: God Himself was the Virgin's child. Two men were born of virgins: God Himself took on a countenance like that of the first virgin's son, a condescension from His sublimity. With Adam's race there began both sorrow and joy, for he whom all angels see above them does not deny our consanguinity, and his lineage is a vehicle of sin; so that we, too, have to bear our load of it. May the power of Him Who is compassionate show mercy here! Since His faithful Humanity fought faithfully against unfaithfulness you should put your quarrel with Him by. Unless you wish to forfeit your heavenly bliss admit penance for your sins. Do not be so free of word or deed—let me tell you the reward of one who slakes his anger in loose speech. He is damned by his own mouth! Take old sayings for new, if they teach you constancy. In ancient times the vates Plato and Sibyl the Prophetess truly foretold beyond all error that a surety would come to us for greatest debts. In His divine love He that is highest of all released us in Hell and left the wicked inside.

"These glad tidings tell of the True Lover. He is a light that shines through all things, unwavering in His love. Those to whom He shows His love find contentment in it. His wares are of two sorts: He offers the world love and anger. Now ask yourself which helps more. The unrepentant sinner flees God's love: but he that atones for his sins serves Him for His noble favour.

"He that passes through men's thoughts bears such Grace. Thoughts keep out the rays of the sun, thoughts are shut away without a lock, are secure from all creatures. Thoughts are darkness unlit by any beam. But of its nature, the Godhead is translucent, it shines through the wall of darkness and rides with an unseen leap unaccompanied by thud or jingle. And when a thought springs from one's heart, none is so swift but that it is scanned ere it pass the skin—and only if it be pure does God accept it. Since God scans thoughts so well, alas, how our frail deeds must pain him!

"When a man forfeits God's benevolence so that God turns away in shame, to whose care can human schooling leave him? Where shall the poor soul find refuge? If you are going to wrong God, Who is ready with both Love and Wrath, you are the one who will suffer. Now so direct your thoughts that He will requite your goodness."

"Sir, I shall always be glad that you have taught me about Him Who leaves nothing unrewarded, whether virtue or misdeed," said Parzival. "I have spent my youth in care and anxiety until this day and endured sorrow for the sake of loyalty."

"Unless you do not wish to divulge them, I should like to hear your sins and sorrows," replied his host. "If you will let me judge of them I might well be able to give advice you could not give yourself."

"My deepest distress is for the Gral," replied Parzival. "After that it is for my wife, than whom no fairer creature was ever given suck by mother. I languish and pine for them both."

"You are right, sir," said his host. "The distress you suffer is as it should be, since the anguish you give yourself comes from longing for the wife that is yours. If you are found in holy wedlock, however you may suffer in Purgatory, your torment shall soon end, and you will be loosed from your bonds immediately through God's help. You say you long for the Gral? You foolish man—this I must deplore! For no man can win the Gral other than one who is acknowledged in Heaven as destined for it. This much I have to say about the Gral, for I know it and have seen it with my own eyes."

"Were you there?" asked Parzival.

"Indeed, sir," was his host's reply.

Parzival did not reveal to him that he, too, had been there, but asked to be told about the Gral.

"It is well known to me," said his host, "that many formidable fighting-men dwell at Munsalvæsche with the Gral. They are continually riding out on sorties in quest of adventure. Whether these same Templars reap trouble or renown, they bear it for their sins. A warlike company lives there. I will tell you how they are nourished. They live from a Stone whose essence is most pure. If you have never heard of it I shall name it for you here. It is called 'Lapsit exillis'. By virtue of this Stone the Phoenix is burned to ashes, in which he is reborn.—Thus does the Phoenix moult its feathers! Which done, it shines dazzling bright and lovely as before! Further: however ill a mortal may be, from the day on which he sees the Stone he cannot die for that week, nor does he lose his colour. For if anyone, maid or man, were to look at the Gral for two hundred years, you would have to admit that his colour was as fresh as in his early prime, except that his hair would grey!—Such powers does the Stone confer on mortal men that their flesh and bones are soon made young again. This Stone is also called 'The Gral.'

"Today a Message alights upon the Gral governing its highest virtue, for today is Good Friday, when one can infallibly see a Dove wing its way down from Heaven. It brings a small white Wafer to the Stone and leaves it there. The Dove, all dazzling white, then flies up to Heaven again. Every Good Friday, as I say, the Dove brings it to the Stone, from which the Stone receives all that is good on earth of food and drink, a paradisal excellence—I mean whatever the earth yields. The Stone, furthermore, has to give them the flesh of all the wild things that live below the aether, whether they fly, run or swim—such prebend does the Gral, thanks to its indwelling powers, bestow on the chivalric Brotherhood.

"As to those who are appointed to the Gral, hear how they are made known. Under the top edge of the Stone an Inscription announces the name and lineage of the one summoned to make the glad journey. Whether it concern girls or boys, there is no need to erase their names, for as soon as a name has been read it vanishes from sight! Those who are now full-grown all came here as children. Happy the mother of any child destined to serve there! Rich and poor alike rejoice if a child of theirs is summoned and they are bidden to send it to that Company! Such children are fetched from many countries and forever after are immune from the shame of sin and have a rich reward in Heaven. When they die here in this world, Paradise is theirs in the next.

"When Lucifer and the Trinity began to war with each other, those who did not take sides, worthy, noble angels, had to descend to earth to that Stone which is forever incorruptible. I do not know whether God forgave them or damned them in the end: if it was His due He took them back. Since that time the Stone has been in the care of those whom God appointed to it and to whom He sent his angel. This, sir, is how matters stand regarding the Gral."

"If knightly deeds with shield and lance can win fame for one's earthly self, yet also Paradise for one's soul, then the chivalric life has been my one desire!," said Parzival. "I fought wherever fighting was to be had, so that my warlike hand has glory within its grasp. If God is any judge of fighting He will appoint me to that place so that the Company there know me as a knight who will never shun battle."

"There of all places you would have to guard against arrogance by cultivating meekness of spirit," replied his austere host. "You could be misled by youthfulness into breaches of self-control.—Pride goes before a fall!" Thus his host, whose eyes filled with tears as he recalled the story he was now to tell in full.

"Sir, there was a king who went by the name of Anfortas, as he does today," he said. "The agony with which he was punished for his pride should move you and wretched me to never-ending pity! His youth and wealth and pursuit of love beyond the restraints of wedlock brought harm to the world through him. Such ways do not suit the Gral. In its service knights and squires must guard against licentiousness: humility has always mastered pride. A noble Brotherhood lives there, who by force of arms have warded off men from every land, with the result that the Gral has been revealed only to those who have been summoned to Munsalvæsche to join the Gral Company. Only one man ever came there without first having been assigned. *He had not reached years of discretion!* He went away saddled with sin in that he said no word to his host on the sad plight in which he saw him. It is not for me to blame anyone: but he will be bound to pay for his sin of failing to inquire about his host's hurt. For Anfortas bore a load of suffering, the like of which had never been seen. Before this man's visit, King Lähelin had ridden to Brumbane. Here the noble knight Lybbeals of Prienlascors had waited to joust with him, and by joust had met his death. Lähelin led the warrior's charger away, thus plainly despoiling the dead.

"Sir, are you Lähelin? In my stable there is a horse of the same coat as those belonging to the Gral Company. The horse comes from Munsalvæsche because its saddle shows the Turtledove, the device which Anfortas gave for horses when happiness was still his, though their shields have always borne it. Titurel handed it down to his son King Frimutel, who, brave knight, was displaying it when he lost his life in a joust. Frimutel loved his wife so dearly that no wife was ever loved more by husband, I mean with such devotion. You should renew his ways and love your spouse with all your heart. Follow his example—you bear him a close resemblance! He was also Lord of the Gral. Ah, sir, from where have you journeyed? Kindly tell me from whom you are descended."

Each looked the other in the eyes.

"I am the son of a man who, impelled by knightly ardour, lost his life in a joust," Parzival told his host. "I beg you to include him in your prayers, sir. My father's name was Gahmuret, and by birth he was an Angevin. Sir, I am not Lähelin. If I ever stripped a corpse it was because I was dull of understanding. However, I did this thing, I confess myself guilty of the crime. I slew Ither of Cucumerlant with my sinful hand, I stretched him out dead on the grass and took what there was to take."

"Alas, wicked World, why do you so?" cried his host, saddened by this news. "You give us more pain and bitter sorrow than ever joy! So this is the reward you offer, such is the end of your song? Dear nephew," he went on, "what counsel can I give you now? You have slain your own flesh and blood. If you take this misdeed unatoned to the Judgment into the presence of God and He judges you with strict justice, it will cost you your own life, since you and Ither were of one blood. What payment will you make Him for Ither of Gaheviez? God had made manifest in him the fruits of true nobility which enhanced life's quality. All wrong-doing saddened him who was the very balm of constancy! All obloquy of this world fought shy of him, all that is noble made its way into his heart! Worthy ladies ought to hate you for the loss of his lovable person. His service of them was so entire that when they saw the charming man their eyes shone. May God have pity on it that you were ever the cause of such distress! Add to that, your mother, my sister Herzeloyde, died of anguish for you!"

"Oh no, good sir!" cried Parzival. "What are you saying now? Were that so and I were Lord of the Gral it could not console me for what you have just told me! If I am your nephew, do as all sincere people do and tell me straight: are these two things true?"

"It is not in me to deceive," answered the good man. "No sooner had you left your mother than she died—that was what she had for her love. You were the Beast she suckled, the Dragon that flew away from her. It had come upon her as she slept, sweet lady, before giving birth to you. I have a brother and a sister living. My sister Schoysiane bore a child and died bearing that fruit. Her husband was Duke Kyot of Katelangen, who henceforth renounced all happiness. His little daughter Sigune was entrusted to your mother's care. Schoysiane's death afflicts me utterly—how could it fail to? Her womanly heart was so virtuous, it might have been an ark afloat on the flood of wantonness! A sister of mine is as yet unwed and keeps her chastity. She is Repanse de Schoye and has charge of the Gral, which is so heavy that sinful mortals could not lift it from its place. Her brother and mine is Anfortas, who was Lord of the Gral by heredity and so remains. Alas, happiness lies far beyond his reach, apart from his firm hope that his sufferings will earn him bliss eternal! Things came to this sad pass in a way scarce short of marvellous, as I shall tell you, nephew. If you have a good heart you will be moved to pity by his sorrows.

"When my father Frimutel lost his life, his eldest son was summoned to the Gral as King and Lord Protector both of the Gral and its Company. This was my brother Anfortas, who was worthy of the Crown and its dominion. At that time we were still quite small. But when my brother approached the age at which the first bristles begin to show, Love assailed him, as is her way with striplings—she presses her friends so hard that one may call it dishonourable of her. But any Lord of the Gral who seeks love other than that allowed him by the Writing will inevitably have to pay for it with pain and suffering fraught with sighs.

"As the object of his attentions my lord and brother chose a lady whom he judged of excellent conduct—as to who she was, let it rest. He served her with unflinching courage, and many shields were riddled by his fair hand. As knight-errant the charming, comely youth won fame so exalted that he ran no risk of its being surpassed by any in all the lands of chivalry. His battle-cry was 'Amor,' yet that shout is not quite right for humility.

"One day—his nearest and dearest did not at all approve—the King rode out alone to seek adventure under Love's compulsion and joying in her encouragement. Jousting, he was wounded by a poisoned lance so seriously that he never recovered, your dear uncle—through the scrotum. The man who was fighting there and rode that joust was a heathen born of Ethnise, where the Tigris flows out from Paradise. This pagan was convinced that his valour would earn him the Gral. His name was engraved on his lance. He sought chivalric encounters in distant countries, crossing seas and lands with no other thought than to win the Gral. As a result of his prowess, our happiness vanished. Yet your uncle's prowess must be commended too. He carried the lance-head away with him in his body, and when the noble youth returned to his familiars his tragic plight was clear to see. He had slain that heathen on the field—let us not waste our tears on *him*.

"When the King returned to us so pale, and drained of all his strength, a physician probed his wound till he found the lance-head and a length of bamboo shaft which was also buried there. The physician recovered them both. I fell on my knees in prayer and vowed to Almighty God that I would practise chivalry no more, in the hope that to His own glory He would help my brother in his need. I also foreswore meat, bread and wine, and indeed promised that I would never again relish anything else that had blood. I tell you, dear nephew, parting with my sword was another source of sorrow to my people. 'Who is to be Protector of the Gral's secrets?' they asked, while bright eyes wept.

"They lost no time in carrying the King into the presence of the Gral for any aid God would give him. But when the King set eyes on it, it came as a second affliction to him that he might not die. Nor was it fitting he should after I had dedicated myself to a life of such wretchedness, and the dominion of our noble lineage had been reduced to such frailty.

"The King's wound had festered. None of the various books of medicine we consulted furnished a remedy to reward our trouble. All that was known by way of antidotes to asp, ecidemon, ehcontius, lisis, jecis and meatris—these vicious serpents carry their venom hot—and other poisonous snakes, all that the learned doctors extract from herbs by the art of physic—let me be brief—were of no avail: it was God Himself who was frustrating us. We called in the aid of Gehon, Phison, Tigris and Euphrates, and so near to Paradise from which the four rivers flow that their fragrance was still unspent, in the hope that some herb might float down in it that would end our sorrow. But this was all lost effort, and our sufferings were renewed. Yet we made many other attempts. We obtained that same twig to which the Sibyl referred Aeneas, to ward off the hazards of Hell and Phlegethon's fumes, not to name other rivers flowing there. We devoted time to possessing ourselves of that twig as a remedy, in case the sinister lance that slays our happiness had been envenomed or tempered in Hellfire: but it was not so with that lance.

"There is a bird called Pelican. When it has young it loves them to excess. Instinctive love impels it to pick through its own breast and let the blood flow into its chicks' mouths. This done, it dies. We obtained some blood of this bird to see if its love would be efficacious, and anointed the wound to the best of our ability: but it helped us not at all.

"There is a beast called Monicirus, which esteems virginal purity so highly that it falls asleep in maiden's laps. We acquired this animal's heart to assuage the King's pain. We took the carbuncle-stone on this beast's brow where it grows at the base of its horn. We stroked the wound with it at the front, then completely immersed the stone in it: but the wound kept its gangrened look. This mortified the King and us.

"We then took a herb called trachonte—it is said to grow from any dragon that is slain and to partake of the nature of air—in order to discover whether the revolution of the Dragon would avail against the planets' return and the change of the moon, which caused the pain of the wound: but the sublime virtue of this herb did not serve our purpose.

"We fell on our knees before the Gral, where suddenly we saw it written that a knight would come to us and were he heard to ask a Question there, our sorrows would be at an end; but that if any child, maiden or man were to forewarn him of the Question it would fail in its effect, and the injury would be as it was and give rise to deeper pain. 'Have you understood?' asked the Writing. 'If you alert him it could prove harmful. If he omits the Question on the first evening, its power will pass away. But if he asks his Question in season he shall have the Kingdom, and by God's will the sorrow shall cease. Thereby Anfortas will be healed, but he shall be King no more.'

"In this way we read on the Gral that Anfortas's agony would end when the Question came to him. We anointed his wound with whatever might soothe it—the good salve nard, whatever is decocted with theriac, and the smoke of lignaloes: yet he was always in pain. I then withdrew to this place. Scant happiness is all my passing years afford me. Since then a knight rode that way, and it would have been better had he not done so—the knight I told you of before. All that he achieved there was shame, for he saw all the marks of suffering yet failed to ask his host 'Sire, what ails you?' Since youthful inexperience saw to it that he asked no Question, he let slip a golden opportunity."

CHAPTER 16

Anfortas and his people were still suffering an agony of grief. From loyal love they left him in his plight. For he often asked them to let him die and indeed would soon have done so had they not, as often, shown him the potent Gral.

The King often kept his eyes shut tight for as many as four days on end. Then they carried him to the Gral whether he liked it or not, and with the malady racking him to the point where he had to open his eyes, he was made to live against his will and not die. This was how they proceeded with him until the day when Parzival and particoloured Feirefiz rode joyfully to Munsalvæsche.

The newcomers found a great multitude of people there: fine old knights in number, noble pages, many men-at-arms. The mournful Household had good cause to rejoice at their coming! Feirefiz Angevin and Parzival were well received on the flight of steps leading up to the Palace, into which they then all went.

Here, according to custom, lay a hundred large round carpets, each with a cushion of down on it and a long quilt of samite. If the pair went about it tactfully they could find seats somewhere or other till their armour was taken from them . . .

A chamberlain now went up to them bringing them robes of equal splendour. All the knights present sat down, and many precious cups of gold—*not* glass—were set before them. After drinking, Feirefiz and Parzival went to the sorrowful Anfortas.

You have already heard all about his reclining instead of sitting, and how richly his bed was adorned. Anfortas now received the pair joyfully, yet with signs of anguish, too.

"I have suffered torments of expectation, wondering if you were ever going to restore me to happiness. Now, the last time, you left me in such a way that if yours is a kind and helpful nature you will show remorse for it. If you are a man of reputation and honour, ask the knights and maidens here to let me die, and so end my agony. If you are Parzival, keep me from seeing the Gral for seven nights and eight days—then all my sorrows will be over! I dare not prompt you otherwise. Happy you, if people were to say you succoured me! Your companion here is a stranger: I am not content that he should stand in my presence. Why do you not let him go to take his ease?"

Parzival wept. "Tell me where the Gral is," he said. "If the goodness of God triumphs in me, this Company here shall witness it!" Thrice did he genuflect in its direction to the glory of the Trinity, praying that the affliction of this man of sorrows be taken from him. Then, rising to his full height, he added: "Dear Uncle, what ails you?"

He Who for St Sylvester's sake bade a bull return from death to life and go, and Lazarus stand up, now helped Anfortas to become whole and well again. The lustre which the French call "fleur" entered his complexion—Parzival's beauty was as nothing beside it, and that of Absalom son of David, and Vergulaht of Ascalun, and of all who were of handsome race, and the good looks conceded to Gahmuret when they saw the delightful sight of him marching into Kanvoleiz—the beauty of none of these was equal to that which Anfortas carried out from his illness. God's power to apply his artistry is undiminished today.

No other Election was made than of the man the Gral Inscription had named to be their lord. Parzival was recognized forthwith as King and Sovereign. If I am any judge of wealth, I imagine no one would find a pair of men as rich as Parzival and Feirefiz in any other place. The Lord and Master and his guest were served assiduously.

I do not know how many leagues Condwiramurs had ridden by then towards Munsalvæsche in happy mood.—She had learnt the truth earlier on, a message had come to her that her sad state of deprivation was over. Duke Kyot and many other worthy men had thereupon conducted her thence into the forest at Terre salvæsche, where Segramors had been felled by a lance-thrust and the snow and blood had so resembled her. There Parzival was to fetch her, an excursion he could well endure!

A Templar reported to him as follows. "A group of courtly knights have brought the Queen with all ceremony." Parzival decided to take some of the Gral Company and ride out to Trevrizent's, whose heart rejoiced at the news that Anfortas's fortunes now stood at the pont where he was not to die of his lance-wound, and the Question had won him peace.

"God has many mysteries," Trevrizent told Parzival. "Whoever sat at His councils or who has fathomed His power? Not all the Host of Angels will ever get to the bottom of it. God is Man and His Father's Word, God is both Father and Son, His Spirit has power to bring great succour. A greater marvel never occurred, in that, after all, with your defiance you have wrung the concession from God that His everlasting Trinity has given you your wish. I lied as a means of distracting you from the Gral and how things stood concerning it. Let me atone for my error—I now owe you obedience, Nephew and my lord. You heard from me that the banished angels were at the Gral with God's full support till they should be received back into His Grace. But God is constant in such matters: He never ceases to war against those whom I named to you here as forgiven. Whoever desires to have reward from God must be in feud with those angels. For they are eternally damned and chose their own perdition. But I am very sorry you had such a hard time. It was never the custom that any should battle his way to the Gral: I wished to divert you from it. Yet your affairs have now taken another turn, and your prize is all the loftier! Now guide your thoughts towards humility."

"I wish to see the woman I have not seen once in five years," said Parzival to his uncle. "When we were together she was dear to me, as she indeed still is.—Of course I wish to have your advice as long as we are both alive: you advised me well in the past, when I was in great need. Now I wish to ride and meet my wife who, as I have heard, has reached a place on the Plimizœl on her way to me."

Parzival asked Trevrizent for leave to go, and the good man commended him to God.

Parzival rode through the night, for the Forest was well-known to his companions. When it dawned, he was approaching a place where many tents had been pitched, a find that pleased him greatly. Many pennants of the land of Brobarz had been planted there, with many shields that had marched behind them. They were the Princes of his own country who were encamped there. Parzival inquired where the Queen herself was quartered, and if she had her own separate ring, and they showed him where she lay surrounded by tents in a sumptuous ring.

Now Duke Kyot of Katelangen had risen early. Parzival and his men were riding up. The ray of dawn was still silver-grey, yet Kyot at once recognized the Gral escutcheon worn by the company, for they were displaying nothing but Turtle-doves. The old man fetched a sigh when he saw it, since his chaste Schoysiane had won him great happiness at Munsalvæsche and then died giving birth to Sigune.

Kyot went up to Parzival and received him and his people kindly. He sent a page to the Queen's Marshal to ask him to provide good lodgment for whatever knights he saw had reined in there. Parzival himself he led by the hand to where the Queen's wardrobe stood, a small tent of buckram. There they unarmed him completely.

Of this the Queen as yet knew nothing. In a tall and spacious pavilion in which numerous fair ladies were lying, here, there, and everywhere, Parzival found Loherangrin and Kardeiz beside her, and—joy perforce overwhelmed him!—Kyot rapped on the coverlet and told the Queen to wake up and laugh for sheer happiness. She opened her eyes and saw her husband. She had nothing on her but her shift, so she swung the coverlet round her and sprang from the bed on to the carpet, radiant Condwiramurs! As to Parzival, he took her into his arms, and I am told they kissed.

"Welcome! Fortune has sent you to me, my heart's joy," she said. "Now I ought to scold you, but I cannot. All honour to this day and hour that have brought me this embrace, banishing all my sadness! I have my heart's desire. Care will get nothing from me!"

The boys Kardeiz and Loherangrin, who lay there naked in the bed, now woke up. Parzival, nothing loth, kissed them affectionately. Tactful Kyot then had the boys carried out. He also hinted to those ladies that they should leave the pavilion, and this they did after welcoming their lord back from his long journey. Kyot then courteously commended the Queen's husband to her and led the young ladies away. It was still very early. The chamberlains closed the flaps.

If ever on a past occasion the company of his wits had been snatched away from him by blood and snow (he had in fact seen them on this very meadow!), Condwiramurs now made amends for such torment: she had it there. He had never received Love's aid for Love's distress elsewhere, though many fine women had offered him their love. As far as I know, he disported himself there till towards mid-morning.

Balin or the Knight with the Two Swords

SIR THOMAS MALORY

This story is largely Malory's invention in order to explain the origin of the Wasteland. *It differs sharply from the origin in Chretien, Wolfram, and the* Queste del Saint Graal. *Malory is more analytic; he presents Balin, a super-socialized knight whose over-zealous desire to accomplish socially determined goals, as opposed to searching for his own path, destroys every land he visits.*

After the death of Uther reigned Arthur, his son, which had great war in his days for to get all England into his hand; for there were many kings within the realm of England and of Scotland, Wales and Cornwall.

So it befell on a time when King Arthur was at London, there came a knight and told the king tidings how the King Rions of North Wales had reared a great number of people, and were entered in the land and brent and slew the king's true liege people.

"If this be true," said Arthur, "it were great shame unto mine estate but that he were mightily withstood."

"It is truth," said the knight, "for I saw the host myself."

"Well," said the king, "I shall ordain to withstand his malice."

Then the king let make a cry that all the lords, knights and gentlemen of arms should draw unto the castle called Camelot in those days, and there the king would let make a council-general and a great jousts. So when the king was come thither with all his baronage and lodged as they seemed best, also there was come a damsel the which was sent from the great Lady Lyle of Avalon. And when she came before King Arthur she told from whence she came, and how she was sent on message unto him for these causes. Then she let her mantle fall that was richly furred, and then was she girt with a noble sword, whereof the king had marvel and said,

"Damsel, for what cause are ye girt with that sword? It beseemeth you nought."

"Now shall I tell you," said the damsel. "This sword that I am girt withal doth me great sorrow and cumbrance, for I may not be delivered of this sword but by a knight, and he must be a passing good man of his hands and of his deeds, and without villainy other treachery, and without treason. And if I may find such a knight that hath all these virtues, he may draw out this sword out of the sheath."

Then it befell so that time there was a poor knight with King Arthur that had been prisoner with him half a year for slaying of a knight which was cousin unto King Arthur. And the name of this knight was called Balin, and by good means of the barons he was delivered out of prison, for he was a good man named of his body, and he was born in Northumberland. And so he went privily into the court and saw this adventure whereof it raised his heart, and would essay as other knights did. But for he was poor and poorly arrayed, he put himself not far in press. But in his heart he was fully assured to do as well, if his grace happed him, as any knight that there was. And as the damsel took her leave of Arthur and of all the barons, so departing, this knight Balin called unto her and said,

"Damsel, I pray you of your courtesy suffer me as well to essay as these other lords. Though that I be poorly arrayed yet in my heart meseemeth I am fully assured as some of these other, and meseemeth in mine heart to speed right well."

This damsel then beheld this poor knight and saw he was a likely man; but for his poor arrayment she thought he should not be of no worship without villainy or treachery. And then she said unto that knight,

"Sir, it needeth not you to put me to no more pain, for it seemeth not you to speed thereas all these other knights have failed."

"Ah, fair damsel," said Balin, "worthiness and good tatches and also good deeds is not only in arrayment, but manhood and worship is hid within a man's person; and many a worshipful knight is not known unto all people. And therefore worship and hardiness is not in arrayment."

"By God," said the damsel, "ye say sooth, therefore ye shall essay to do what ye may."

Then Balin took the sword by the girdle and sheath and drew it out easily; and when he looked on the sword it pleased him much. Then had the king and all the barons great marvel that Balin had done that adventure; many knights had great despite at him.

"Certes," said the damsel, "this is a passing good knight and the best that ever I found, and most of worship without treason, treachery, or felony. And many marvels shall he do. Now, gentle and courteous knight, give me the sword again."

"Nay," said Balin, "for this sword will I keep but it be taken from me with force."

"Well," said the damsel, "ye are not wise to keep the sword from me, for ye shall slay with that sword the best friend that ye have and the man that ye most love in the world, and that sword shall be your destruction."

"I shall take the adventure," said Balin, "that God will ordain for me. But the sword ye shall not have at this time, by the faith of my body!"

"Ye shall repent it within short time," said the damsel, "for I would have the sword more for your advantage than for mine; for I am passing heavy for your sake, for an ye will not leave that sword it shall be your destruction, and that is great pity."

So at that time there was a knight the which was the king's son of Ireland, and his name was Lanceor, the which was an orgulous knight and accounted himself one of the best of the court. And he had great despite at Balin for the achieving of the sword, that any should be accounted more hardy or more of prowess, and he asked King Arthur license to ride after Balin and to revenge the despite that he had done.

"Do your best," said Arthur. "I am right wroth with Balin. I would he were quit of the despite that he hath done unto me and my court."

Then this Lanceor went to his ostry to make him ready. So in the meanwhile came Merlin unto the court of King Arthur, and anon was told him the adventure of the sword and the death of the Lady of the Lake.

"Now shall I say you," said Merlin; "this same damsel that here standeth, that brought the sword unto your court, I shall tell you the cause of her coming. She is the falsest damsel that liveth—she shall not say nay! For she hath a brother, a passing good knight of prowess and a full true man, and this damsel loved another knight that held her as paramour. And this good knight, her brother, met with the knight that held her to paramour, and slew him by force of his hands. And when this false damsel understood this she went to the Lady Lyle of Avalon and took her his sword and besought her of help to be revenged on her own brother.

"And so this Lady Lyle of Avalon took her this sword that she brought with her, and told there should no man pull it out of the sheath but if he be one of the best knights of this realm, and he should be hardy and full of prowess; and with that sword he should slay his brother. This was the cause, damsel, that ye came into this court. I know it as well as ye. God would ye had not come here; but ye came never in fellowship of worshipful folk for to do good, but always great harm. And that knight that hath achieved the sword shall be destroyed through the sword; for which will be great damage, for there liveth not a knight of more prowess than he is. And he shall do unto you, my lord Arthur, great honour and kindness; and it is great pity he shall not endure but a while, for of his strength and hardiness I know him not living his match."

So this knight of Ireland armed him at all points and dressed his shield on his shoulder and mounted upon horseback and took his glaive in his hand, and rode after a great pace as much as his horse might drive. And within a little space, on a mountain, he had a sight of Balin, and with a loud voice he cried,

"Abide, knight! for else ye shall abide whether ye will either no! And the shield that is tofore you shall not help you," said this Irish knight, "therefore come I after you."

"Peradventure," said Balin, "ye had been better to have hold you at home. For many a man weeneth to put his enemy to a rebuke, and oft it falleth on himself. Out of what court be ye come from?" said Balin.

"I am come from the court of King Arthur," said the knight of Ireland, "that am come hither to revenge the despite ye did this day unto King Arthur and to his court."

"Well," said Balin, "I see well I must have ado with you; that me forthinketh that I have grieved King Arthur or any of his court. And your quarrel is full simple," said Balin, "unto me; for the lady that is dead did to me great damage, and else I would have been loath as any knight that liveth for to slay a lady."

"Make you ready," said the knight Lanceor, "and dress you unto me, for that one shall abide in the field."

Then they feautred their spears in their rests and came together as much as their horses might drive. And the Irish knight smote Balin on the shield that all went to shivers off his spear. And Balin smote him again through the shield, and the hauberk perished, and so bore him through the body and over the horse crupper; and anon turned his horse fiercely and drew out his sword, and wist not that he had slain him.

Then he saw him lie as a dead corpse, he looked about him and was ware of a damsel that came riding full fast as the horse might drive, on a fair palfrey. And when she espied that Lanceor was slain she made sorrow out of measure and said,

"Ah! Balin, two bodies thou hast slain in one heart, and two hearts in one body, and two souls thou hast lost."

And therewith she took the sword from her love that lay dead, and fell to the ground in a swough. And when she arose she made great dole out of measure, which sorrow grieved Balin passingly sore. And he went unto her for to have taken the sword out of her hand; but she held it so fast he might not take it out of her hand but if he should have hurt her. And suddenly she set the pommel to the ground, and rove herself throughout the body.

When Balin espied her deeds he was passing heavy in his heart and ashamed that so fair a damsel had destroyed herself for the love of his death. "Alas!" said Balin, "me repenteth sore the death of this knight for the love of this damsel, for there was much true love betwixt them." And so for sorrow he might no longer behold them, but turned his horse and looked toward a fair forest.

And then was he ware by his arms that there came riding his brother Balan. And when they were met they put off their helms and kissed together and wept for joy and pity. Then Balan said,

"Brother, I little weened to have met you at this sudden adventure, but I am right glad of your deliverance of your dolorous prisonment: for a man told me in the Castle of Four Stones that ye were delivered, and that man had seen you in the court of King Arthur. And therefore I came hither into this country, for here I supposed to find you."

And anon Balin told his brother of his adventure of the sword and the death of the Lady of the Lake, and how King Arthur was displeased with him.

"Wherefore he sent this knight after me that lieth here dead. And the death of this damsel grieveth me sore."

"So doth it me," said Balan, "but ye must take the adventure that God will ordain you."

"Truly," said Balan, "I am right heavy that my lord Arthur is displeased with me, for he is the most worshipfullest king that reigneth now in earth; and his love I will get other else I will put my life in adventure. For King Rions lieth at the siege of the Castle Terrabil, and thither will we draw in all goodly haste to prove our worship and prowess upon him."

"I will well," said Balan, "that ye do so; and I will ride with you and put my body in adventure with you, as a brother ought to do."

"Now go we hence," said Balin, "and well we be met."

The meanwhile as they talked there came a dwarf from the city of Camelot on horseback as much as he might, and found the dead bodies; wherefore he made great dole, and pulled his hair for sorrow, and said,

"Which of two knights have done this deed?"

"Whereby asketh thou?" said Balan.

"For I would wit," said the dwarf.

"It was I," said Balin, "that slew this knight in my defendant; for hither he came to chase me, and either I must slay him either he me. And this damsel slew herself for his love, which repenteth me. And for her sake I shall owe all women the better will and service all the days of my life."

"Alas!" said the dwarf, "thou hast done great damage unto thyself. For this knight that is here dead was one of the most valiant men that lived. And trust well, Balin, the kin of this knight will chase you through the world till they have slain you."

"As for that," said Balin, "them I fear not greatly; but I am right heavy that I should displease my lord, King Arthur, for the death of this knight."

So as they talked together there came a king of Cornwall riding, which hight King Mark. And when he saw these two bodies dead, and understood how they were dead by the two knights above-said, then made the king great sorrow for the true love that was betwixt them, and said, "I will not depart till I have on this earth made a tomb." And there he pight his pavilions and sought all the country to find a tomb, and in a church they found one was fair and rich. And then the king let put them both in the earth, and laid the tomb upon them, and wrote the names of them both on the tomb, how "here lieth Lanceor, the king's son of Ireland, that at his own request was slain by the hands of Balin," and how "this lady Columbe and paramour to him slew herself with his sword for dole and sorrow."

And so departed King Mark unto Camelot to King Arthur.

And Balin took the way to King Rions, and as they rode together they met with Merlin disguised so that they knew him nought.

"But whitherward ride ye?" said Merlin.

"We had little ado to tell you," said these two knights.

"But what is thy name?" said Balin.

"At this time," said Merlin, "I will not tell."

"It is an evil sign," said the knights, "that thou art a true man, that thou wilt not tell thy name."

"As for that," said Merlin, "be as it be may. But I can tell you wherefore ye ride this way: for to meet with King Rions. But it will not avail you without ye have my counsel."

"Ah," said Balin, "ye are Merlin! We will be ruled by your counsel!"

"Come on," said Merlin, "and ye shall have great worship. And look that ye do knightly, for ye shall have need."

"As for that," said Balin, "dread you not, for we will do what we may."

Then there lodged Merlin and these two knights in a wood among the leaves beside the highway, and took off the bridles of their horses and put them to grass, and laid them down to rest till it was nigh midnight. Then Merlin bade them rise and make them ready: "for here cometh the king nighhand, that was stolen away from his host with a three score horses of his best knights, and twenty of them rode tofore the lord to warn the Lady de Vaunce that the king was coming." For that night King Rions should have lain with her.

"Which is the king?" said Balin.

"Abide," said Merlin, "for here in a strait way ye shall meet with him." And therewith he showed Balin and his brother the king.

And anon they met with him, and smote him down and wounded him freshly, and laid him to the ground. And there they slew on the right hand and on the left hand more than forty of his men; and the remnant fled. Then went they again unto King Rions and would have slain him, had he not yielded him unto their grace. Then said he thus:

"Knights full of prowess, slay me not! For by my life ye may win, and by my death little."

"Ye say sooth," said the knights, and so laid him on a horse-litter.

So with that Merlin vanished, and came to King Arthur aforehand and told him how his most enemy was taken and discomfit.

"By whom?" said King Arthur.

"By two knights," said Merlin, "that would fain have your lordship. And to-morrow ye shall know what knights they are."

So anon after came the Knight with the Two Swords and his brother, and brought with them King Rions of North Wales, and there delivered him to the porters, and charged them with him.

And so they two returned again in the dawning of the day.

Then King Arthur came to King Rions and said,

"Sir king, ye are welcome. By what adventure came ye hither?"

"Sir," said King Rions, "I came hither by a hard adventure."

"Who won you?" said King Arthur.

"Sir," said he, "the Knight with the Two Swords and his brother, which are two marvellous knights of prowess."

"I know them not," said Arthur, "but much am I beholding unto them."

"Ah, sir," said Merlin, "I shall tell you. It is Balin that achieved the sword and his brother Balan, a good knight: there liveth not a better of prowess, nother of worthiness. And it shall be the greatest dole of him that ever I knew of knight; for he shall not long endure."

"Alas," said King Arthur, "that is great pity; for I am much beholding unto him, and I have evil deserved it again for his kindness."

"Nay, nay," said Merlin, "he shall do much more for you, and that shall ye know in haste."

Then they rode three or four days and never met with adventure. And so by fortune they were lodged with a gentleman that was a rich man and well at ease. And as they sat at supper Balin heard one complain grievously by him in a chamber.

"What is this noise?" said Balin.

"Forsooth," said his host, "I will tell you. I was but late at a jousting and there I jousted with a knight that is brother unto King Pellam, and twice I smote him down. And then he promised to quit me on my best friend. And so he wounded thus my son that cannot be whole till I have of that knight's blood. And he rideth all invisible, but I know not his name."

"Ah," said Balin, "I know that knight's name, which is Garlon, and he hath slain two knights of mine in the same manner. Therefore I had liefer meet with that knight than all the gold in this realm, for the despite he hath done me."

"Well," said his host, "I shall tell you how. King Pellam of Listenoise hath made do cry in all the country a great feast that shall be within these twenty days, and no knight may come there but he bring his wife with him other his paramour. And that your enemy and mine ye shall see that day."

"Then I promise you," said Balin, "part of his blood to heal your son withal."

"Then we will be forward to-morn," said he.

So on the morn they rode all three toward King Pellam, and they had fifteen days' journey or they came thither. And that same day began the great feast. And so they alight and stabled their horses and went into the castle, but Balin's host might not be let in because he had no lady. But Balin was well received and brought unto a chamber and unarmed him. And there was brought him robes to his pleasure, and would have had Balin leave his sword behind him.

"Nay," said Balin, "that will I not, for it is the custom of my country a knight always to keep his weapon with him. Other else," said he, "I will depart as I came."

Then they gave him leave with his sword, and so he went into the castle and was among knights of worship, and his lady afore him. So after this Balin asked a knight and said,

"Is there not a knight in this court which his name is Garlon?"

"Yes, sir, yonder he goeth, the knight with the black face, for he is the marvellest knight that is now living. And he destroyeth many good knights, for he goeth invisible."

"Well," said Balin, "is that he?" Then Balin advised him long, and thought: "If I slay him here, I shall not escape. And if I leave him now, peradventure I shall never meet with him again at such a steven, and much harm he will do an he live."

And therewith this Garlon espied that Balin visaged him, so he came and slapped him on the face with the back of his hand and said,

"Knight, why beholdest thou me so? For shame, eat thy meat and do that thou come for."

"Thou sayest sooth," said Balin, "this is not the first spite that thou hast done me, and therefore I will do that I come for." And rose him up fiercely and clave his head to the shoulders.

"Now give me the truncheon, said Balin to his lady, "that he slew your knight with."

And anon she gave it him, for always she bore the truncheon with her. And therewith Balin smote him through the body and said openly,

"With that truncheon thou slewest a good knight, and now it sticketh in thy body." Then Balin called unto his host and said, "Now may ye fetch blood enough to heal your son withal."

So anon all the knights rose from the table for to set on Balin. And King Pellam himself arose up fiercely and said,

"Knight, why hast thou slain my brother? Thou shalt die therefore or thou depart."

"Well," said Balin, "do it yourself."

"Yes," said King Pellam, "there shall be no man have ado with thee but I myself, for the love of my brother."

Then King Pellam caught in his hand a grim weapon and smote eagerly at Balin, but he put his sword betwixt his head and the stroke, and therewith his sword brast in sunder. And when Balin was weaponless he ran into a chamber for to seek a weapon, and from chamber to chamber, and no weapon could he find. And always King Pellam followed after him. And at last he entered into a chamber which was marvellously dight and rich, and a bed arrayed with cloth of gold, the richest that might be, and one lying therein. And thereby stood a table of clean gold with four pillars of silver that bore up the table, and upon the table stood a marvellous spear strangely wrought.

So when Balin saw the spear he got it in his hand and turned to King Pellam and felled him and smote him passingly sore with that spear, that King Pellam fell down in a swough. And therewith the castle broke, roof and walls, and fell down to the earth. And Balin fell down and might not stir hand nor foot, and for the most party of that castle was dead through the Dolorous Stroke.

Right so lay King Pellam and Balin three days.

Then Merlin came thither, and took up Balin and gat him a good horse, for his was dead, and bade him void out of that country.

"Sir, I would have my damsel," said Balin.

"Lo," said Merlin, "where she lieth dead."

And King Pellam lay so many years sore wounded, and might never be whole till that Galahad the Haut Prince healed him in the quest of the Sankgreall. For in that place was part of the blood of Our Lord Jesu Christ, which Joseph of Arimathea brought into this land. And there himself lay in that rich bed. And that was the spear which Longius smote Our Lord with to the heart. And King Pellam was nigh of Joseph his kin, and that was the most worshipfullest man on life in those days, and great pity it was of his hurt, for through that stroke it turned to great dole, tray and teen.

Then departed Balin from Merlin, "for," he said, "never in this world we part neither meet no more." So he rode forth through the fair countries and cities and found the people dead and slain on every side, and all that ever were on live cried and said,

"Ah, Balin! Thou has done and caused great damage in these countries! For the Dolorous Stroke thou gave unto King Pellam these three countries are destroyed. And doubt not but the vengeance will fall on thee at the last!"

"O knight Balin, why have ye left your own shield? Alas! ye have put yourself in great danger, for by your shield ye should have been known. It is great pity of you as ever was of knight, for of thy prowess and hardiness thou hast no fellow living."

"Me repenteth," said Balin, "that ever I came within this country, but I may not turn now again for shame, and what adventure shall fall to me, be it life or death, I will take the adventure that shall come to me."

And then he looked on his armour and understood he was well armed, and therewith blessed him and mounted upon his horse. Then afore him he saw come riding out of a castle a knight, and his horse trapped all red, and himself in the same colour. When this knight in the red beheld Balin, him thought it should be his brother Balin because of his two swords, but because he knew not his shield he deemed it was not he.

And so they aventred their spears and came marvellously fast together, and they smote other in the shields, but their spears and their course were so big that it bare down horse and man, that they lay both in a swoon; but Balin was bruised sore with the fall of the horse, for he was weary of travel. And Balan was the first that rose on foot and drew his sword, and went toward Balin, and he arose and went against him; but Balan smote Balin first, and he put up his shield and smote him through the shield and tamed his helm. Then Balin smote him again with that unhappy sword, and well-nigh had felled his brother Balan, and so they fought there together till their breaths failed.

Then Balin looked up to the castle and saw the towers stand full of ladies. So they went unto battle again, and wounded everych other dolefully, and then they breathed ofttimes, and so went unto battle that all the place thereas they fought was blood red. And at that time there was none of them both but they had either smitten other seven great wounds, so that the least of them might have been the death of the mightiest giant in this world.

Then they went to battle again so marvellously that doubt it was to hear of that battle for the great blood-shedding, and their hauberks unnailed, that naked they were on every side. At the last Balan, the younger brother, withdrew him a little and laid him down. Then said Balin le Savage,

"What knight art thou? For or now I found never no knight that matched me."

"My name is," said he, "Balan, brother unto the good knight Balin."

"Alas," said Balin, "that ever I should see this day!" and therewith he fell backward in a swoon.

Then Balan yode on all four, feet and hands, and put off the helm of his brother, and might not know him by the visage, it was so full hewn and bled; but when he awoke he said,

"O Balan, my brother! Thou hast slain me and I thee, wherefore all the wide world shall speak of us both."

"Alas," said Balan, "that ever I saw this day, that through mishap I might not know you! For I espied well your two swords, but because ye had another shield I deemed ye had been another knight."

"Alas," said Balin, "all that made an unhappy knight in the castle, for he caused me to leave mine own shield to our both's destruction. And if I might live I would destroy that castle for ill customs."

"That were well done," said Balan, "for I had never grace to depart from them syn that I came hither, for here it happed me to slay a knight that kept this island, and syn might I never depart. And no more should ye, brother, and ye might have slain me as ye have and escaped yourself with the life."

Right so came the lady of the tower with four knights and six ladies and six yeomen unto them, and there she heard how they made their moan either to other and said, "We came both out of one womb, that is to say one mother's belly, and so shall we lie both in one pit." So Balan prayed the lady of her gentleness for his true service that she would bury them both in that same place there the battle was done, and she granted them, with weeping, it should be done richly in the best manner.

"Now, will ye send for a priest, that we may receive our sacrament, and receive the blessed body of Our Lord Jesu Christ?"

"Yea," said the lady, "it shall be done;" and so she sent for a priest and gave them their rites.

"Now," said Balin, "when we are buried in one tomb, and the mention made over us how two brethren slew each other, there will never good knight nor good man see our tomb but they will pray for our souls." And so all the ladies and gentlewomen wept for pity.

Then anon Balan died, but Balin died not till the midnight after. And so were they buried both, and the lady let make a mention of Balan how he was there slain by his brother's hands, but she knew not Balin's name.

In the morn came Merlin and let write Balin's name on the tomb with letters of gold, that HERE LIETH BALIN LE SAVAGE THAT WAS THE KNIGHT WITH THE TWO SWORDS, AND HE THAT SMOTE THE DOLOROUS STROKE. Also Merlin let make there a bed, that there should never man lie therein but he went out of his wit. Yet Lancelot du Lake forbid that bed through his noblesse.

And anon after Balin was dead Merlin took his sword and took off the pommel and set on another pommel. So Merlin bade a knight that stood before him to handle that sword, and he essayed it and might not handle it. Then Merlin laughed.

"Why laugh ye?" said the knight.

"This is the cause," said Merlin: "there shall never man handle this sword but the best knight of the world, and that shall be Sir Lancelot other else Galahad, his son. And Lancelot with this sword shall slay the man that in the world he loveth best, that shall be Sir Gawain."

And all this he let write in the pommel of the sword.

Then Merlin let make a bridge of iron and steel into that island, and it was but half a foot broad, "and there shall never man pass that bridge, nother have hardiness to go over it but if he were a passing good man and a good knight without treachery or villainy." Also the scabbard of Balin's sword Merlin left it on this side the island, that Galahad should find it. Also Merlin let make by his subtlety that Balin's sword was put in a marble stone standing upright as great as a millstone, and hoved always above the water, and did many years. And so by adventure it swam down the stream to the City of Camelot, that is in English called Winchester. And the same day Galahad the Haute Prince came with King Arthur, and so Galahad brought with him the scabbard and achieved the sword that was in the marble stone hoving upon the water. And on Whitsunday he achieved the sword, as it is rehearsed in *The Book of the Sankgreall.*

Soon after this was done Merlin came to King Arthur and told him of the Dolorous Stroke that Balin gave King Pellam, and how Balin and Balan fought together the most marvellous battle that ever was heard of, and how they were buried both in one tomb.

"Alas!" said King Arthur, "this is the greatest pity that ever I heard tell of two knights, for in this world I knew never such two knights."

THUS ENDETH THE TALE OF BALIN AND OF BALAN, TWO BRETHREN THAT WERE BORN IN NORTHUMBERLAND, THAT WERE TWO PASSING GOOD KNIGHTS AS EVER WERE IN THOSE DAYS.

Confrontation with the Unconscious

CARL G. JUNG

In this selection from his autobiography, Carl G. Jung, the famous psychiatrist, recounts his discovery of his own path in psychoanalysis after leaving the discipleship of Sigmund Freud. Earlier in the autobiography, Jung compares his life at this stage to the Grail Quest: "In the period following these dreams I did a great deal of thinking about the mysterious figure of the knight. But it was only much later, after I had been meditating on the dream for a long time, that I was able to get some idea of its meaning. . . . I had an inkling that a great secret lay hidden behind those stories. Therefore it seemed quite natural to me that the dream should conjure up the world of the Knights of the Grail and their quest—for that was, in the deepest sense, my own world, which had scarcely anything to do with Freud's. My whole being was seeking for something still unknown which might confer meaning upon the banality of my life."

After the parting of the ways with Freud, a period of inner uncertainty began for me. It would be no exaggeration to call it a state of disorientation. I felt totally suspended in mid-air, for I had not yet found my own footing. Above all, I felt it necessary to develop a new attitude toward my patients. I resolved for the present not to bring any theoretical premises to bear upon them, but to wait and see what they would tell of their own accord. My aim became to leave things to chance. The result was that the patients would spontaneously report their dreams and fantasies to me, and I would merely ask, "What occurs to you in connection with that?" or, "How do you mean that, where does that come from, what do you think about it?" The interpretations seemed to follow of their own accord from the patients' replies and associations. I avoided all theoretical points of view and simply helped the patients to understand the dream-images by themselves, without application of rules and theories.

Soon I realized that it was right to take the dreams in this way as the basis of interpretation, for that is how dreams are intended. They are the facts from which we must proceed. Naturally, the aspects resulting from this method were so multitudinous that the need for a criterion grew more and more pressing—the need, I might also put it, for some initial orientation.

About this time I experienced a moment of unusual clarity in which I looked back over the way I had traveled so far. I thought, "Now you possess a key to mythology and are free to unlock all the gates of the unconscious psyche." But then something whispered within me, "Why open all gates?" And promptly the question arose of what, after all, I had accomplished. I had explained the myths of peoples of the past; I had written a book about the hero, the myth in which man has always lived. But in what myth does man live nowadays? In the Christian myth, the answer might be, "Do *you* live in it?" I asked myself. To be honest, the answer was no. For me, it is not what I live by. "Then do we no longer have any myth?" "No, evidently we no longer have any myth." "But then what is your myth—the myth in which you do live?" At this point the dialogue with myself became uncomfortable, and I stopped thinking. I had reached a dead end.

Then, around Christmas of 1912, I had a dream. In the dream I found myself in a magnificent Italian loggia with pillars, a marble floor, and a marble balustrade. I was sitting on a gold Renaissance chair; in front of me was a table of rare beauty. It was made of green stone, like emerald. There I sat, looking out into the distance, for the loggia was set high up on the tower of a castle. My children were sitting at the table too.

Suddenly a white bird descended, a small sea gull or a dove. Gracefully, it came to rest on the table, and I signed to the children to be still so that they would not frighten away the pretty white bird. Immediately, the dove was transformed into a little girl, about eight years of age, with golden blond hair. She ran off with the children and played with them among the colonnades of the castle.

I remained lost in thought, musing about what I had just experienced. The little girl returned and tenderly placed her arms around my neck. Then she suddenly vanished; the dove was back and spoke slowly in a human voice. "Only in the first hours of the night can I transform myself into a human being, while the male dove is busy with the twelve dead." Then she flew off into the blue air, and I awoke.

I was greatly stirred. What business would a male dove be having with twelve dead people? In connection with the emerald table the story of the Tabula Smaragdina occurred to me, the emerald table in the alchemical legend of Hermes Trismegistos. He was said to have left behind him a table upon which the basic tenets of alchemical wisdom were engraved in Greek.

I also thought of the twelve apostles, the twelve months of the year, the signs of the zodiac, etc. But I could find no solution to the enigma. Finally I had to give it up. All I knew with any certainty was that the dream indicated an unusual activation of the unconscious. But I knew no technique whereby I might get to the bottom of my inner processes, and so there remained nothing for me to do but wait, go on with my life, and pay close attention to my fantasies.

One fantasy kept returning: there was something dead present, but it was also still alive. For example, corpses were placed in crematory ovens, but were then discovered to be still living. These fantasies came to a head and were simultaneously resolved in a dream.

I was in a region like the Alyscamps near Arles. There they have a lane of sarcophagi which go back to Merovingian times. In the dream I was coming from the city, and saw before me a similar lane with a long row of tombs. They were pedestals with stone slabs on which the dead lay. They reminded me of old church burial vaults, where knights in armor lie outstretched. Thus the dead lay in my dream, in their antique clothes, with hands clasped, the difference being that they were not hewn out of stone, but in a curious fashion mummified. I stood still in front of the first grave and looked at the dead man, who was a person of the eighteen-thirties. I looked at his clothes with interest, whereupon he suddenly moved and came to life. He unclasped his hands; but that was only because I was looking at him. I had an extremely unpleasant feeling, but walked on and came to another body. He belonged to the eighteenth century. There exactly the same thing happened: when I looked at him, he came to life and moved his hands. So I went down the whole row, until I came to the twelfth century—that is, to a crusader in chain mail who lay there with clasped hands. His figure seemed carved out of wood. For a long time I looked at him and thought he was really dead. But suddenly I saw that a finger of his left hand was beginning to stir gently.

Of course I had originally held to Freud's view that vestiges of old experiences exist in the unconscious. But dreams like this, and my actual experiences of the unconscious, taught me that such contents are not dead, outmoded forms, but belong to our living being. My work had confirmed this assumption, and in the course of years there developed from it the theory of archetypes.

The dreams, however, could not help me over my feeling of disorientation. On the contrary, I lived as if under constant inner pressure. At times this became so strong that I suspected there was some psychic disturbance in myself. Therefore I twice went over all the details of my entire life, with particular attention to childhood memories; for I thought there might be something in my past which I could not see and which might possibly be the cause of the disturbance. But this retrospection led to nothing but a fresh acknowledgment of my own ignorance. Thereupon I said to myself, "Since I know nothing at all, I shall simply do whatever occurs to me." Thus I consciously submitted myself to the impulses of the unconscious.

The first thing that came to the surface was a childhood memory from perhaps my tenth or eleventh year. At that time I had had a spell of playing passionately with building blocks. I distinctly recalled how I had built little houses and castles, using bottles to form the sides of gates and vaults. Somewhat later I had used ordinary stones, with mud for mortar. These structures had fascinated me for a long time. To my astonishment, this memory was accompanied by a good deal of emotion. "Aha," I said to myself, "there is still life in these things. The small boy is still around, and possesses a creative life which I lack. But how can I make my way to it?" For as a grown man it seemed impossible to me that I should be able to bridge the distance from the present back to my eleventh year. Yet if I wanted to re-establish contact with that period, I had no choice but to return to it and take up once more that child's life with his childish games. This moment was a turning point in my fate, but I gave in only after endless resistances and with a sense of resignation. For it was a painfully humiliating experience to realize that there was nothing to be done except play childish games.

Nevertheless, I began accumulating suitable stones, gathering them partly from the lake shore and partly from the water. And I started building: cottages, a castle, a whole village. The church was still missing, so I made a square building with a hexagonal drum on top of it, and a dome. A church also requires an altar, but I hesitated to build that.

Preoccupied with the question of how I could approach this task, I was walking along the lake as usual one day, picking stones out of the gravel on the shore. Suddenly I caught sight of a red stone, a four-sided pyramid about an inch and a half high. It was a fragment of stone which had been polished into this shape by the action of the water—a pure product of chance. I knew at once: this was the altar! I placed it in the middle under the dome, and as I did so, I recalled the underground phallus of my childhood dream. This connection gave me a feeling of satisfaction.

I went on with my building game after the noon meal every day, whenever the weather permitted. As soon as I was through eating, I began playing, and continued to do so until the patients arrived; and if I was finished with my work early enough in the evening, I went back to building. In the course of this activity my thoughts clarified, and I was able to grasp the fantasies whose presence in myself I dimly felt.

Naturally, I thought about the significance of what I was doing, and asked myself, "Now, really, what are you about? You are building a small town, and doing it as if it were a rite!" I had no answer to my question, only the inner certainty that I was on the way to discovering my own myth. For the building game was only a beginning. It released a stream of fantasies which I later carefully wrote down.

In order to seize hold of the fantasies, I frequently imagined a steep descent. I even made several attempts to get to very bottom. The first time I reached, as it were, a depth of about a thousand feet; the next time I found myself at the edge of a cosmic abyss. It was like a voyage to the moon, or a descent into empty space. First came the image of a crater, and I had the feeling that I was in the land of the dead. The atmosphere was that of the other world. Near the steep slope of a rock I caught sight of two figures, an old man with a white beard and a beautiful young girl. I summoned up my courage and approached them as though they were real people, and listened attentively to what they told me. The old man explained that he was Elijah, and that gave me a shock. But the girl staggered me even more, for she called herself Salome! She was blind. What a strange couple: Salome and Elijah. But Elijah assured me that he and Salome had belonged together from all eternity, which completely astounded me. . . . They had a black serpent living with them which displayed an unmistakable fondness for me. I stuck close to Elijah because he seemed to be the most reasonable of the three, and to have a clear intelligence. Of Salome I was distinctly suspicious. Elijah and I had a long conversation which, however, I did not understand.

Naturally I tried to find a plausible explanation for the appearance of Biblical figures in my fantasy by reminding myself that my father had been a clergyman. But that really explained nothing at all. For what did the old man signify? What did Salome signify? Why were they together? Only many years later, when I knew a great deal more than I knew then, did the connection between the old man and the young girl appear perfectly natural to me.

In such dream wanderings one frequently encounters an old man who is accompanied by a young girl, and examples of such couples are to be found in many mythic tales. Thus, according to Gnostic tradition, Simon Magus went about with a young girl whom he had picked up in a brothel. Her name was Helen, and she was regarded as the reincarnation of the Trojan Helen. Klingsor and Kundry, Lao-tzu and the dancing girl, likewise belong to this category.

I have mentioned that there was a third figure in my fantasy besides Elijah and Salome: the large black snake. In myths the snake is a frequent counterpart of the hero. There are numerous accounts of their affinity. For example, the hero has eyes like a snake, or after his death he is changed into a snake and revered as such, or the snake is his mother, etc. In my fantasy, therefore, the presence of the snake was an indication of a hero-myth.

Salome is an anima figure. She is blind because she does not see the meaning of things. Elijah is the figure of the wise old prophet and represents the factor of intelligence and knowledge; Salome, the erotic element. One might say that the two figures are personifications of Logos and Eros. But such a definition would be excessively intellectual. It is more meaningful to let the figures be what they were for me at the time—namely, events and experiences.

Soon after this fantasy another figure rose out of the unconscious. He developed out of the Elijah figure. I called him Philemon. Philemon was a pagan and brought with him an Egypto-Hellenistic atmosphere with a Gnostic coloration. His figure first appeared to me in the following dream.

There was a blue sky, like the sea, covered not by clouds but by flat brown clods of earth. It looked as if the clods were breaking apart and the blue water of the sea were becoming visible between them. But the water was the blue sky. Suddenly there appeared from the right a winged being sailing across the sky. I saw that it was an old man with the horns of a bull. He held a bunch of four keys, one of which he clutched as if he were about to open a lock. He had the wings of the kingfisher with its characteristic colors.

Since I did not understand this dream-image, I painted it in order to impress it upon my memory. During the days when I was occupied with the painting, I found in my garden, by the lake shore, a dead kingfisher! I was thunderstruck, for kingfishers are quite rare in the vicinity of Zürich and I have never since found a dead one. The body was recently dead—at the most, two or three days—and showed no external injuries.

Philemon and other figures of my fantasies brought home to me the crucial insight that there are things in the psyche which I do not produce, but which produce themselves and have their own life. Philemon represented a force which was not myself. In my fantasies I held conversations with him, and he said things which I had not consciously thought. For I observed clearly that it was he who spoke, not I. He said I treated thoughts as if I generated them myself, but in his view thoughts were like animals in the forest, or people in a room, or birds in the air, and added, "If you should see people in a room, you would not think that you had made those people, or that you were responsible for them." It was he who taught me psychic objectivity, the reality of the psyche. Through him the distinction was clarified between myself and the object of my thought. He confronted me in an objective manner, and I understood that there is something in me which can say things that I do not know and do not intend, things which may even be directed against me.

Psychologically, Philemon represented superior insight. He was a mysterious figure to me. At times he seemed to me quite real, as if he were a living personality. I went walking up and down the garden with him, and to me he was what the Indians call a guru.

Whenever the outlines of a new personification appeared, I felt it almost as a personal defeat. It meant: "Here is something else you didn't know until now!" Fear crept over me that the succession of such figures might be endless, that I might lose myself in bottomless abysses of ignorance. My ego felt devalued—although the successes I had been having in worldly affairs might have reassured me. In my darknesses (*horridas nostrae mentis purga tenebras*—"cleanse the horrible darknesses of our mind"—the *Aurora Consurgens* says) I could have wished for nothing better than a real, live guru, someone possessing superior knowledge and ability, who would have disentangled for me the involuntary creations of my imagination. This task was undertaken by the figure of Philemon, whom in this respect I had willy-nilly to recognize as my psychagogue. And the fact was that he conveyed to me many an illuminating idea.

More than fifteen years later a highly cultivated elderly Indian visited me, a friend of Gandhi's, and we talked about Indian education—in particular, about the relationship between guru and chela. I hesitantly asked him whether he could tell me anything about the person and character of his own guru, whereupon he replied in a matter-of-fact tone, "Oh yes, he was Shankaracharya."

"You don't mean the commentator on the Vedas who died centuries ago?" I asked.

"Yes, I mean him," he said, to my amazement.

"Then you are referring to a spirit?" I asked.

"Of course it was his spirit," he agreed.

At that moment I thought of Philemon.

"There are ghostly gurus too," he added. "Most people have living gurus. But there are always some who have a spirit for teacher."

This information was both illuminating and reassuring to me. Evidently, then, I had not plummeted right out of the human world, but had only experienced the sort of thing that could happen to others who made similar efforts.

Later, Philemon became relativized by the emergence of yet another figure, whom I called Ka. In ancient Egypt the "king's ka" was his earthly form, the embodied soul. In my fantasy the ka-soul came from below, out of the earth as if out of a deep shaft. I did a painting of him, showing him in his earth-bound form, as a herm with base of stone and upper part of bronze. High up in the painting appears a kingfisher's wing, and between it and the head of Ka floats a round, glowing nebula of stars. Ka's expression has something demonic about it—one might also say, Mephistophelian. In one hand he holds something like a colored pagoda, or a reliquary, and in the other a stylus with which he is working on the reliquary. He is saying, "I am he who buries the gods in gold and gems."

Philemon had a lame foot, but was a winged spirit, whereas Ka represented a kind of earth demon or metal demon. Philemon was the spiritual aspect, or "meaning." Ka, on the other hand, was a spirit of nature like the Anthroparion of Greek alchemy—with which at the time I was still unfamiliar. Ka was he who made everything real, but who also obscured the halcyon spirit, Meaning, or replaced it by beauty, the "eternal reflection."

In time I was able to integrate both figures through the study of alchemy.

Shortly before this experience I had written down a fantasy of my soul having flown away from me. This was a significant event: the soul, the anima, establishes the relationship to the unconscious. In a certain sense this is also a relationship to the collectivity of the dead; for the unconscious corresponds to the mythic land of the dead, the land of the ancestors. If, therefore, one has a fantasy of the soul vanishing, this means that it has withdrawn into the unconscious or into the land of the dead. There it produces a mysterious animation and gives visible form to the ancestral traces, the collective contents. Like a medium, it gives the dead a chance to manifest themselves. Therefore, soon after the disappearance of my soul the "dead" appeared to me, and the result was the *Septem Sermones*. This is an example of what is called "loss of soul"—a phenomenon encountered quite frequently among primitives.

From that time on, the dead have become ever more distinct for me as the voices of the Unanswered, Unresolved, and Unredeemed; for since the questions and demands which my destiny required me to answer did not come to me from outside, they must have come from the inner world. These conversations with the dead formed a kind of prelude to what I had to communicate to the world about the unconscious: a kind of pattern of order and interpretation of its general contents.

When I look back upon it all today and consider what happened to me during the period of my work on the fantasies, it seems as though a message had come to me with overwhelming force. There were things in the images which concerned not only myself but many others also. It was then that I ceased to belong to myself alone, ceased to have the right to do so. From then on, my life belonged to the generality. The knowledge I was concerned with, or was seeking, still could not be found in the science of those days. I myself had to undergo the original experience, and, moreover, try to plant the results of my experience in the soil of reality; otherwise they would have remained subjective assumptions without validity. It was then that I dedicated myself to service of the psyche. I loved it and hated it, but it was my greatest wealth. My delivering myself over to it, as it were, was the only way by which I could endure my existence and live it as fully as possible.

Today I can say that I have never lost touch with my initial experiences. All my works, all my creative activity, has come from those initial fantasies and dreams which began in 1912, almost fifty years ago. Everything that I accomplished in later life was already contained in them, although at first only in the form of emotions and images.

My science was the only way I had of extricating myself from that chaos. Otherwise the material would have trapped me in its thicket, strangled me like jungle creepers. I took great care to try to understand every single image, every item of my psychic inventory, and to classify them scientifically—so far as this was possible—and, above all, to realize them in actual life. That is what we usually neglect to do. We allow the images to rise up, and maybe we wonder about them, but that is all. We do not take the trouble to understand them, let alone draw ethical conclusions from them. This stopping-short conjures up the negative effects of the unconscious.

It is equally a grave mistake to think that it is enough to gain some understanding of the images and that knowledge can here make a halt. Insight into them must be converted into an ethical obligation. Not to do so is to fall prey to the power principle, and this produces dangerous effects which are destructive not only to others but even to the knower. The images of the unconscious place a great responsibility upon a man. Failure to understand them, or a shirking of ethical responsibility, deprives him of his wholeness and imposes a painful fragmentariness on his life.

In the midst of this period when I was so preoccupied with the images of the unconscious, I came to the decision to withdraw from the university, where I had lectured for eight years as *Privatdozent* (since 1905). My experience and experiments with the unconscious had brought my intellectual activity to a standstill. After the completion of *The Psychology of the Unconscious* I found myself utterly incapable of reading a scientific book. This went on for three years. I felt I could no longer keep up with the world of the intellect, nor would I have been able to talk about what really preoccupied me. The material brought to light from the unconscious had, almost literally, struck me dumb. I could neither understand it nor give it form. At the university I was in an exposed position, and felt that in order to go on giving courses there I would first have to find an entirely new and different orientation. It would be unfair to continue teaching young students when my own intellectual situation was nothing but a mass of doubts.

I therefore felt that I was confronted with the choice of either continuing my academic career, whose road lay smooth before me, or following the laws of my inner personality, of a higher reason, and forging ahead with this curious task of mine, this experiment in confrontation with the unconscious. But until it was completed I could not appear before the public.

Consciously, deliberately, then, I abandoned my academic career. For I felt that something great was happening to me, and I put my trust in the thing which I felt to be more important *sub specie aeternitatis.* I knew that it would fill my life, and for the sake of that goal I was ready to take any kind of risk.

What, after all, did it matter whether or not I became a professor? Of course it bothered me to have to give this up; in many respects I regretted that I could not confine myself to generally understandable material. I even had moments when I stormed against destiny. But emotions of this kind are transitory, and do not count. The other thing, on the contrary, is important, and if we pay heed to what the inner personality desires and says, the sting vanishes. That is something I have experienced again and again, not only when I gave up my academic career. Indeed, I had my first experiences of this sort as a child. In my youth I was hot-tempered; but whenever the emotion had reached its climax, suddenly it swung around and there followed a cosmic stillness. At such times I was remote from everything, and what had only a moment before excited me seemed to belong to a distant past.

The consequence of my resolve, and my involvement with things which neither I nor anyone else could understand, was an extreme loneliness. I was going about laden with thoughts of which I could speak to no one: they would only have been misunderstood. I felt the gulf between the external world and the interior world of images in its most painful form. I could not yet see that interaction of both worlds which I now understand. I saw only an irreconcilable contradiction between "inner" and "outer."

However, it was clear to me from the start that I could find contact with the outer world and with people only if I succeeded in showing—and this would demand the most intensive effort—that the contents of psychic experience are real, and real not only as my own personal experiences, but as collective experiences which others also have. Later I tried to demonstrate this in my scientific work, and I did all in my power to convey to my intimates a new way of seeing things. I knew that if I did not succeed, I would be condemned to absolute isolation.

It was only toward the end of the First World War that I gradually began to emerge from the darkness. Two events contributed to this. The first was that I broke with the woman who was determined to convince me that my fantasies had artistic value; the second and principal event was that I began to understand mandala drawings. This happened in 1918–19. I had painted the first mandala in 1916 after writing the *Septem Sermones;* naturally I had not, then, understood it.

In 1918–19 I was in Château d'Oex as Commandant de la Région Anglaise des Internés de Guerre. While I was there I sketched every morning in a notebook a small circular drawing, a mandala, which seemed to correspond to my inner situation at the time. With the help of these drawings I could observe my psychic transformations from day to day. One day, for example, I received a letter from that esthetic lady in which she again stubbornly maintained that the fantasies arising from my unconscious had artistic value and should be considered art. The letter got on my nerves. It was far from stupid, and therefore dangerously persuasive. The modern artist, after all, seeks to create art out of the unconscious. The utilitarianism and self-importance concealed behind this thesis touched a doubt in myself, namely, my uncertainty as to whether the fantasies I was producing were really spontaneous and natural, and not ultimately my own arbitrary inventions. I was by no means free from the bigotry and hubris of consciousness which wants to believe that any halfway decent inspiration is due to one's own merit, whereas inferior reactions come merely by chance, or even derive from alien sources. Out of this irritation and disharmony within myself there proceeded, the following day, a changed mandala: part of the periphery had burst open and the symmetry was destroyed.

Only gradually did I discover what the mandala really is: "Formation, Transformation, Eternal Mind's eternal recreation." And that is the self, the wholeness of the personality, which if all goes well is harmonious, but which cannot tolerate self-deceptions.

My mandalas were cryptograms concerning the state of the self which were presented to me anew each day. In them I saw the self—that is, my whole being—actively at work. To be sure, at first I could only dimly understand them; but they seemed to me highly significant, and I guarded them like precious pearls. I had the distinct feeling that they were something central, and in time I acquired through them a living conception of the self. The self, I thought, was like the monad which I am, and which is my world. The mandala represents this monad, and corresponds to the microcosmic nature of the psyche.

I no longer know how many mandalas I drew at this time. There were a great many. While I was working on them, the question arose repeatedly: What is this process leading to? Where is its goal? From my own experience, I knew by now that I could not presume to choose a goal which would seem trustworthy to me. It had been proved to me that I had to abandon the idea of the superordinate position of the ego. After all, I had been brought up short when I had attempted to maintain it. I had wanted to go on with the scientific analysis of myths which I had begun in *Wandlungen und Symbole*. That was still my goal—but I must not think of that! I was being compelled to go through this process of the unconscious. I had to let myself be carried along by the current, without a notion of where it would lead me. When I began drawing the mandalas, however, I saw that everything, all the paths I had been following, all the steps I had taken, were leading back to a single point—namely, to the mid-point. It became increasingly plain to me that the mandala is the center. It is the exponent of all paths. It is the path to the center, to individuation.

During those years, between 1918 and 1920, I began to understand that the goal of psychic development is the self. There is no linear evolution; there is only a circumambulation of the self. Uniform development exists, at most, only at the beginning; later, everything points toward the center. This insight gave me stability, and gradually my inner peace returned. I knew that in finding the mandala as an expression of the self I had attained what was for me the ultimate. Perhaps someone else knows more, but not I.

Some years later (in 1927) I obtained confirmation of my ideas about the center and the self by way of a dream. I represented its essence in a mandala which I called "Window on Eternity." The picture is reproduced in *The Secret of the Golden Flower.* A year later I painted a second picture, likewise a mandala, with a golden castle in the center. When it was finished, I asked myself, "Why is this so Chinese?" I was impressed by the form and choice of colors, which seemed to me Chinese, although there was nothing outwardly Chinese about it. Yet that was how it affected me. It was a strange coincidence that shortly afterward I received a letter from Richard Wilhelm enclosing the manuscript of a Taoist-alchemical treatise entitled *The Secret of the Golden Flower,* with a request that I write a commentary on it. I devoured the manuscript at once, for the text gave me undreamed-of confirmation of my ideas about the mandala and the circumambulation of the center. That was the first event which broke through my isolation. I became aware of an affinity; I could establish ties with something and someone.

In remembrance of this coincidence, this "synchronicity," I wrote underneath the picture which had made so Chinese an impression upon me: "In 1928, when I was painting this picture, showing the golden, well-fortified castle, Richard Wilhelm in Frankfurt sent me the thousand-year-old Chinese text on the yellow castle, the germ of the immortal body."

This is the dream I mentioned earlier: I found myself in a dirty, sooty city. It was night, and winter, and dark, and raining. I was in Liverpool. With a number of Swiss—say, half a dozen—I walked through the dark streets. I had the feeling that there we were coming from the harbor, and that the real city was actually up above, on the cliffs. We climbed up there. It reminded me of Basel, where the market is down below and then you go up through the Totengässchen ("Alley of the Dead"), which leads to a plateau above and so to the Petersplatz and the Peterskirche. When we reached the plateau, we found a broad square dimly illuminated by street lights, into which many streets converged. The various quarters of the city were arranged radially around the square. In the center was a round pool, and in the middle of it a small island. While everything round about was obscured by rain, fog, smoke, and dimly lit darkness, the little island blazed with sunlight. On it stood a single tree, a magnolia, in a shower of reddish blossoms. It was as though the tree stood in the sunlight and were at the same time the source of light. My companions commented on the abominable weather, and obviously did not see the tree. They spoke of another Swiss who was living in Liverpool, and expressed surprise that he should have settled here. I was carried away by the beauty of the flowering tree and the sunlit island, and thought, "I know very well why he has settled here." Then I awoke.

On one detail of the dream I must add a supplementary comment: the individual quarters of the city were themselves arranged radially around a central point. This point formed a small open square illuminated by a larger street lamp, and constituted a small replica of the island. I knew that the "other Swiss" lived in the vicinity of one of these secondary centers.

This dream represented my situation at the time. I can still see the grayish-yellow raincoats, glistening with the wetness of the rain. Everything was extremely unpleasant, black and opaque—just as I felt then. But I had had a vision

Copyrights and Acknowledgments

INTRODUCTORY UNIT The Hero's Journey: The World of Joseph Campbell

2 *The Inner Reaches of Outer Space: Myth as Metaphor and as Religion* by Joseph Campbell. Copyright © 1986 Harper & Row, Publishers, Inc. Reprinted by permission. 8 Excerpt by Bill Moyers from *The Power of Myth by Joseph Campbell with Bill Moyers*, edited by Betty Sue Flowers. Copyright © 1988 by Alfred van der Marck Editions and Apostrophe S. Productions, Inc. Reprinted by permission of Doubleday, a division of Bantam Doubleday Dell Publishing Group, Inc.

UNIT I In the Beginning: Origins of Man and Myth

47 "The Emergence of Mankind" from *Myths to Live By*, by Joseph Campbell. Copyright © 1972 by Joseph Campbell. All rights reserved. Reprinted by permission of Viking Press, a division of Penguin Books USA, Inc. 55 "Art As A Revelation," Forward to *Mythologies of the Great Hunt, Part II of The Way of the Animal Powers*, Volume I of The Historical Atlas, by Joseph Campbell. Copyright © 1983 by Harper & Row. Reprinted by permission of the publisher.

UNIT II Where People Lived Legends: American Indian Myths

68 "Mythogenesis" from *The Flight of the Wild Gander: Explorations in the Mythological Dimension*, by Joseph Campbell. Copyright © 1969 by Joseph Campbell. Reprinted by permission of Russell & Volkening as agents for the author. 76 "The Metaphorical Journey," from *The Inner Reaches of Outer Space: Myth as Metaphor and as Religion*, by Joseph Campbell. Copyright © 1986 by Alfred van der Marck Editions. Reprinted by permission of Harper & Row. 84 "How Many Sheep Will It Hold?" from *Seeing with a Native Eye* by Walter Holden Capps. Copyright © 1976 by Walter Holden Capps. Reprinted by permission of Harper & Row, Publishers, Inc. 94 Sam D. Gill, "The Shadow of a Vision Yonder" from *Seeing with a Native Eye* by Walter Holden Capps. Copyright © 1976 by Walter Holden Capps. Reprinted by permission of Harper & Row, Publishers, Inc. 103 "In the Beginning" from *The Tewa World: Space, Time, Being, and Becoming in a Pueblo Society*, by Alfonso Ortiz. Copyright © 1964 by The University of Chicago Press. Reprinted with permission of The University of Chicago Press.

UNIT III And We Washed Our Weapons in the Sea: Gods and Goddesses of the Neolithic Period

108 From Anthony F. Aveni, "Introduction: Archaeoastronomy and Its Components" from *Skywatchers of Ancient Mexico.* Copyright 1980. Reprinted by permission of the publisher, University of Texas Press. **112** "The Prehistoric Goddess" by Anne L. Barstow from *The Book of the Goddess· Past and Present.* Copyright © 1983 by Carl Olson. Reprinted by permission of the Crossroad Publishing Company. **119** From Marija Gimbutas, *The Goddesses and Gods of Old Europe: 6500–3500 B.C. Myths and Cult Images.* New and Updated Edition. Published by The University of California Press. Copyright © 1981 Thames and Hudson Ltd. Used by permission.

UNIT IV Pharaoh's Rule: Egypt, the Exodus, and the Myth of Osiris

132 From Henri Frankfort, "The Egyptian Gods" from *Ancient Egyptian Religion: An Interpretation.* Copyright © 1948, 1975 Columbia University Press. Used by permission. **138** From E. A. Wallis Budge, "The Gods of Egypt" from *The Gods of the Egyptians.* Copyright 1969 by Dover Publications, Inc. Reprinted by permission. **147** From James Henry Breasted, "The Osirianization of the Hereafter" from *Development of Religion and Thought in Ancient Egypt.* Copyright 1912 by Charles Scribners' Sons. Copyright 1940 by Charles Breasted. Copyright 1959 by Harper & Row. Reprinted by arrangement with Charles Scribners' Sons which published the original edition in 1912.

UNIT V The Sacred Source: The Perennial Philosophy of the East

154 From Surendranath Dasgupta, "The Samkhya and the Yoga Doctrine of Soul or Purusha" from *A History of Indian Philosophy.* Copyright 1957 by Cambridge University Press. Reprinted with permission of Cambridge University Press. **159** From F. R. Allchin, "The Legacy of the Indus Civilization" from the collection of *Harappan Civilization: A Contemporary Perspective* edited by Gregory L. Possehl. Copyright 1982 by Oxford and IBH Publishing Co., New Delhi, India. Reprinted with permission of Oxford and IBH Publishing Co. **170** "Revelation" from *A Source Book of Advaita Vedanta* by Eliot Deutsch and J. A. B. van Buitenen. Copyright 1971 by University of Hawaii Press. Reprinted with permission. **176** Copyright © Wendy Doniger O'Flaherty, 1973. "The Interpretation of Hindu Mythology" from *Asceticism and Eroticism in the Mythology of Shiva* (1973), used by permission of Oxford University Press. **193** *Hindu Myths: A Sourcebook Translated from the Sanskrit,* by Wendy Doniger O'Flaherty (Penguin Classics, 1975). Copyright © 1975, Wendy Doniger O'Flaherty. Used by permission of Penguin Books.

UNIT VI The Way to Enlightenment: Buddhism

196 Heinrich Zimmer, "Hinayana and Mahayana" from *Philosophies of India,* edited by Joseph Campbell, Bollingen Series 26. Copyright © 1951, 1979 renewed by Princeton University Press. Excerpts reprinted with permission of Princeton University Press.

203 Walpola Sri Rahula, "The Doctrine of No-Soul: *Anatta," What the Buddha Taught*. Copyright © 1959, 1974 by W. Rahula. Reprinted by permission. **208** "Some Buddhist Concepts of Kuan Yin" from *Bodhisattva of Compassion: The Mystical Traditions of Kuan Yin*, by John Blofeld. Copyright © 1978 by Shambala Publications. Reprinted by permission. **219** From E. A. Burtt: "Famous Mahayana Parables: Burning House, Prodigal Son" from *The Teachings of the Compassionate Buddha*. Copyright 1955 by New American Library. Reprinted with permission by Edwin A. Burtt.

UNIT VII From Id to the Ego in the Orient: Kundalini Yoga Part I

238 From Herbert V. Guenther, "The Way and the Apparent Eroticism of Tantrism" from *The Tantric View of Life*. Copyright 1972 by Shambala Publications. Reprinted with permission by Shambala Publications. **246** From Charles S. J. White, "Swami Muktananda and the Enlightenment Through Sakti-Pat" from *The History of Religion, Vol. 13, No. 4*. Copyright 1974 by University of Chicago. All rights reserved. **257** "Tantrism" by Agehananda Bharati from *Abingdon Dictionary of Living Religions*, edited by Keith Crim. Copyright © 1981 by Abingdon Press. Used by permission.

UNIT VIII From Psychology to Spirituality: Kundalini Yoga Part II

264 "The Synthesis of Yogas: Hatha, Raja, and Tantra" reproduced by kind permission of the publisher, Unwin Hyman Ltd. Extract taken from *Integral Yoga: The Concepts of Harmonious and Creative Living*, by Haridas Chaudhuri, published December 12, 1975. **272** From Heinrich Zimmer, "Tantra" from *Philosophies of India* edited by Joseph Campbell. Copyright © 1969 by Princeton University Press. Reprinted by permission of Princeton University Press. **278** *Yoga: Immortality and Freedom*, by Mircea Eliade, translated by Willard R. Trask. Copyright © 1958, 1969 by Princeton University Press. Reprinted by permission of Princeton University Press. **284** *Buddhist Texts Through the Ages*, by Edward Conze. Copyright 1954 by Philosophical Library. Reprinted by permission.

UNIT IX The Descent to Heaven: The Tibetan Book of the Dead

296 "Birth, Marriage, Sickness, and Death" by Giuseppe Tucci from *Tibet: Land of Snows*, translated by J. E. Stapleton Driver. Copyright © 1967 Stein and Day. Reprinted by permission of Stein and Day. **306** "The *Bardo* of the Moments of Death" from *The Tibetan Book of the Dead or The After-Death Experiences on the Bardo Plane*, according to Lama Kazi Dawa-Samdup's English rendering, edited by W. Y. Evans-Wentz. Copyright © 1960 by W. Y. Evans-Wentz. **314** "I am Recognized" from *Return to Tibet* by Heinrich Harrer, translated by Ewald Osers. Translation copyright © 1984 by Ewald Osers. Reprinted by permission of Schocken Books, published by Pantheon Books, a Division of Random House, Inc. **318** From F. Sierksma, "Buddhism: Mantrayana and Vajrayana" from *Tibet's Terrifying Deities: Sex and Aggression in Religious Acculturation*. Copyright 1966 by Fokke Sierksma. Reprinted by permission.

UNIT X From Darkness to Light: The Mystery Religion of Ancient Greece

332 From Euripides, "The Bacchae" from *An Anthology of Greek Drama* edited with an introduction by Charles Alexander Robinson, Jr. Copyright 1954 by Holt, Rinehart, and Winston. Reprinted with permission. **344** From Walter Burkert, "Mysteries and Asceticism" from *Greek Religions*. Copyright 1985 by Harvard University Press. Reprinted by permission of Harvard University Press. **357** From George deForest Lord, *Hymn to Demeter* from the text in *Hesiod, The Homeric Hymns, and Homerica*. Edited by H. G. Evelyn-White. Copyright 1987 by Harvard University Press. Reprinted by permission. **365** From Ovid, "The Teachings of Pythagoras," from *Metamorphoses*, translated by Rolfe Humphries. Copyright 1955 Indiana University Press. Reprinted with permission of Indiana University Press.

UNIT XI Where There Was No Path: Arthurian Legends and the Western Way

376, 377, 378 "De Excidio et Conquestu Britanniae" by Gildas, "Historia Brittonum" by Nennius, "Historia Regum Britanniae" by Geoffrey of Monmouth. Selections from Richard L. Brengle, *Arthur, King of Britain: History, Chronicle, Romance, and Criticism*. Copyright by Simon & Schuster. Reprinted by permission. **389** From Sir Thomas Malory, *King Arthur and His Knights*. Copyright 1956 by Oxford University Press. Reprinted with permission of Oxford University Press. **404** Roger Sherman Loomis, "The Mabinogion" and "The Intermediaries" from *The Development of Arthurian Romance*. Published by W. W. Norton & Company. Copyright © 1963 by Roger Sherman Loomis. Reprinted by permission.

UNIT XII A Noble Heart: The Courtly Love of Tristan and Isolde

416 *Tristan* by Gottfried Von Strassburg, translated by A. T. Hatto (Penguin Classics, 1960). Copyright 1960, A. T. Hatto. Used by permission. **436** Sir Thomas Malory, "The Knight of the Cart" from *King Arthur and His Knights*. Copyright 1956 by Oxford University Press. Reprinted by permission of Oxford University Press. **444** Andreas Capellanus, *The Art of Courtly Love* with introduction, translation, and notes by John Jay Parry. Copyright © 1941 Columbia University Press. Used by permission.

UNIT XIII In Search of the Holy Grail: The Parzival Legend

452 From Wolfram von Eschenbach, *Parzival*, translated by A. T. Hatto. Copyright 1980 by A. T. Hatto. Reprinted by permission of Penguin Books. **471** From Sir Thomas Malory, "Balin or the Knight with the Two Swords" *King Arthur and His Knights*. Copyright 1956 by Oxford University Press. Reprinted by permission of Oxford University Press. **481** Carl G. Jung, "Confrontation with the Unconscious" from *Memories, Dreams, and Reflection*. Copyright 1962 Vintage Book Edition, Random House. Reprinted with permission of Random House.

A 9
B 0
C 1
D 2
E 3
F 4
G 5
H 6
I 7
J 8